Welcome

Welcome to the technical communications lab at the University of North Texas.

The following pages give you information about
- The Technical Communication Lab and its Policies
- Guidelines for Using the Lab
- Plagiarism and Other Forms of Academic Dishonesty
- Essential Competencies for Technical Writing Classes

The lab staff and your instructor want to ensure that you have a successful semester.

If you have questions related to the lab, please contact
Mr. Robert Congrove
congrove@unt.edu

If you have questions about the Technical Communication program, please contact
Dr. Brenda Sims
sims@unt.edu

THE TECHNICAL COMMUNICATION LAB AND ITS POLICIES

The Technical Communication Lab is a teaching lab for technical writing students only and is available exclusively for students currently enrolled in technical writing.

Because this lab is a teaching lab, classes have priority over individuals. To insure that a computer will be available when you come to the lab, you can view a weekly schedule at the desk. The schedule will show open hours.

General Lab Policies

To serve you better, the lab staff has set up the following policies:
- Place removable drives in the computer first so that the computer can scan them for viruses.
- Leave food, drinks, and tobacco products outside the lab.
- If you need to pick up a print-out or ask the staff a question, stay in front of the desk. Only lab staff may go behind the desk.
- Follow the Student Code of Conduct.

UNT owns anything, private or public, you store on any UNT computer, including:
- Mail
- Documents
- Graphics
- Web pages

You are legally responsible for the content of the items you store.

Virus Disclaimer

The Technical Communication lab staff does its best to guard computers from viruses, but we cannot guarantee a complete virus-free environment because of the number of students who use this lab. If you detect a virus, please inform the lab staff so that we can clean the computers.

Using the Lab Computers

Approximately 1,000 technical writing students will use the lab this semester. To protect your documents and your computer:

- Leave the computers on.
- Save your document often.
- Do not alter, change or modify any computer settings.

PLAGIARISM AND OTHER FORMS OF ACADEMIC DISHONESTY

According to the University of North Texas Undergraduate Catalog, the term 'plagiarism' includes, but is not limited to:

(a) the knowing or negligent use by paraphrase or direct quotation of the published or unpublished work of another person without full and clear acknowledgment; and/or

(b) the knowing or negligent unacknowledged use of materials prepared by another person or agency engaged in the selling of term papers or other academic materials.

If you turn in a document—either electronic or hard copy—that is plagiarized:

- You will receive an F in the course.
- Your actions will be reported to the Dean of Students Office.

If you use unauthorized assistance in writing or preparing documents or graphics, or taking quizzes, tests, or examinations:

- You will be dismissed with an F in the course.
- Your actions will be reported to the Dean of Students Office.

ESSENTIAL COMPETENCIES FOR TECHNICAL WRITING CLASSES

In accordance with the terms and spirit of the Americans with Disabilities Act and Section 504, Rehabilitation Act, your instructor will cooperate with the Office of Disability Accommodation to make reasonable accommodations for you if you qualify as a student with disabilities. Please register with the Office of Disability Accommodation and present your written request for accommodation to your instructor by the fourth class day.

Essential competencies for technical writing classes include the ability to:
- Read and analyze technical documents.
- Write clear, concise, and visually appropriate technical prose for the intended readers in response to various types of assignments.
- Discuss, in class and in small groups, technical documents.
- Use software for word processing, electronic mail, presentations, and graphic applications.
- Work effectively as part of a team.
- Speak in front of the class and use multimedia to support the speech.
- Follow the rules of standard grammar, usage, spelling, and punctuation.
- Complete assignments in a timely manner.

Please sign the form below and return it to the lab staff at the end of your opening orientation.

I have received and read the "Technical Communication Lab and Its Policies," "Plagiarism and Other Forms of Academic Dishonesty," and "Essential Competencies for Technical Writing Classes."

- I understand the policies regarding plagiarism and disability accommodation.
- I understand the policies related to computing resources.
- I understand the policies of the Technical Communication Lab.
- I understand that it is my responsibility to follow the "Code of Student Conduct and Discipline" as outlined in the University of North Texas Undergraduate Catalog.

Print name_____

Student signature_____

Date_____

Student ID # _____

Course Section_____

COURSE ASSESSMENT—FOR THE STUDENT OF 2700

Each semester, we assess your progress to determine if our ENGL 2700 classes are adequately meeting your needs. To determine if our courses are as effective as possible, each student must:

1. Navigate to P:\SCRATCH\Assessment\Course Assessment Form.
2. Double click on the file.
3. Fill out the form.
4. Save to your desktop.
5. Email the completed form as an attachment to sims@unt.edu.

You will complete this assessment twice:
- Before the fourth class period.
- During the final two weeks of the semester.

SOFTWARE TECHNIQUES

In this section, you will learn about the following software techniques:
- Using Basic Microsoft Word Functions
- Using Microsoft Word Graphics Functions
- Creating Brochures
- Creating Manuals
- Creating a Gantt Chart

Using Basic Microsoft Word Functions

The following sets of instructions outline the basic procedures for working with Microsoft Word.

Opening a Document

Use the Open function to open a previously saved document from a flash drive/USB, or P: drives. To open a document, follow these steps:
1. Click on MS Office7 Button/Open.
2. Select the Public (P:) drive from the dialog box labeled "Look In:" and double click on the Scratch folder.
3. Select the file Maslow.docx and click Open.

Correcting Documents

If you make a mistake in your document, follow these steps:
1. Click the arrow next to Undo, on the Quick Access Toolbar.
2. Click the action you want to undo. When you undo an action, you also undo all actions above it in the list.

Note: If you decide you didn't want to undo an action, click Redo on the Quick Access Toolbar.

Highlighting Text

To highlight text follow these steps:
1. Place the mouse pointer at the beginning of the text you want to highlight.
2. Click the left mouse button and hold it down.
3. Drag the mouse pointer until the area you want to highlight is selected.
4. Release the mouse button.

Cutting, Copying and Pasting Text

To cut, copy, and paste text, begins by highlighting the text. Then follow the instructions below.

- To cut the blocked text and place it in the clip board, click on Home Ribbon, Cut.
- To copy the blocked text, click on Home Ribbon, Copy.
- To paste the text copied or cut from the clip board, click on Home Ribbon, Paste.

Spell Checking

Word will scan your document and underline misspelled words in red and suggest alternate spellings. To check the spelling in a document:

1. Click on Review Ribbon.
2. Click on Spelling and Grammar.

Once Word finds a misspelled word, you have the following options:

- Ignore: ignores the selected word.
- Ignore All: ignores all occurrences of the selected word.
- Add: adds the selected word to the spell check memory bank.
- Change: replaces the selected word with your choice.
- Change All: replaces all occurrences of the selected text.
- Autocorrect: automatically replaces all instances of the selected word with your choice.

Modifying Fonts

To modify basic font features, use the Home Ribbon. Once inside the Font menu, you can modify any font feature.

To format the font face, follow these steps:

1. Click on the arrow found at the bottom right of font block.
2. Scroll through the list of fonts.
3. Click on the new font name you want.

To change the font size, follow these steps:

1. Click inside the font size box and type in the new point size.
2. Press Enter.

Changing Font Appearance

You can use the menu to:

- Bold text.
- Italicize text.
- Underline text.
- Change font color.

To modify advanced font features:

1. Highlight the text that you want to change.
2. Click on Home Ribbon.
3. Click on the arrow at the bottom right corner of the font box.
4. Make your changes.
5. Click OK.

Embedding Fonts (to keep your choice of fonts with the document):

1. Click on the MS Button.
2. Click on Word Options.
3. Click on Save.
4. Click on Embed Fonts in the File.
5. Click OK.

Using the Disk Drives Available in the Lab

The following drives are available for student use in the lab:

Media drives

Media drives store information to a physical drive you can then take with you.

- CD-R or DVD-R drives (allow up to 20 minutes to copy information)
  Use these drives to save multiple files or presentations to a CD/DVD.
- USB ports
  Use these ports to connect your portable USB storage device.

Network drives

Use network drives to back-up your data from a physical drive or to save files for others.

- Scratch drive (P:/Scratch/)
  Save files that you wish to share here. All technical writing students can access it from inside the technical writing lab.

Using the Save Function

Use the Save function to save and name your document.
To save and name your document, follow these steps:
1. Click on MS Office7 Button.
2. Click on Save.
3. Type in a short file name in the box labeled File Name.
4. Click OK.

Using the Save As Function

Use the Save As function to:
• Save a document under another name.
• Save a document to another drive.

To use the Save As function:
1. Click on MS Office7 Button.
2. Click on Save As.
3. Select the drive.
4. Type in the new name of the document in the File Name box.
5. Click on Save.

Printing a Document

Use the Print function to print an entire document, a range of pages in a document, or a single page of the document.
To print a document, follow these steps:
1. Click on MS Office7 Button.
2. Click on Print.
3. Make your print selection: Full document, Current Page, or Multiple Pages.
4. Choose the appropriate printer for either black and white or color, depending on the type of document.
5. Click on Print.

Closing a Document

To close a document, follow these steps:
1. Click on MS Office7 Button.
2. Click on Close.

Exiting Word

The Exit function allows you to leave Word and go into the Windows environment. To exit Word, follow these steps:
1. Click on MS Office7 Button.
2. Click on Exit.

Using Microsoft Word Graphics Functions

The following instructions outline how to insert and modify clip art and pictures using Microsoft Word.

Inserting a Drawing Canvas

To insert a drawing canvas: (drawing canvas must be larger than the graphic selected)
1. Choose Insert Ribbon Shapes/New Drawing Canvas.
2. Choose Text Wrapping in the Drawing Tools Ribbon.
3. Choose Tight.

Note: All pictures and clip art must be placed in a drawing canvas and the drawing canvas must be selected before placing an image into the document.

Inserting and Deleting Clip Art and Pictures

To insert Clip Art:
•   Click Insert Ribbon/ Clip Art.

To insert a picture from a graphic file:
1. Choose Insert Ribbon Shapes/New Drawing Canvas.
2. Choose the Insert Ribbon/Picture/From File.
3. Choose a file folder to view.
4. Select a picture to insert.
5. Click Insert.

To delete a picture:
1. Select it.
2. Press delete.

Inserting Internet Pictures

From the website:
1. Right-click on the desired graphic.
2. Select Save Picture as...in the menu.
3. Save the picture to the desired drive.
4. Use the instructions under "Inserting a Picture" to place the picture in your document.

Moving a Picture
1. Click on the drawing canvas to get the Image Handles.
2. Position the mouse over the edge of the drawing canvas to get a 4-sided arrow.
3. Click on the edge of the drawing canvas and hold the left mouse button.
4. Drag the mouse to move the picture and the drawing canvas.
5. Release the left mouse button to finish moving the picture.

Resizing a Picture
1. To resize a picture, position your mouse over an image handle to get a 2-sided arrow.
2. Click on the picture and hold the left mouse button.
3. Drag it to change the size of the picture.
4. Release the left mouse button to finish the resizing.

Inserting a Picture as a Watermark
1. Click on the Page Layout Ribbon.
2. Using the Page Background box/ Watermark.
3. Select Custom Watermark.
4. Click on Picture Watermark.
5. Select Picture.
6. Click OK.

Adding a Border
1. Right-click on the picture.
2. Select Format Picture.
3. Change to the Colors and Lines tab.
4. Under Line, choose a color, a style and weight for your border.
5. Click OK.

Adding a Caption

1. Click on the Reference Ribbon/Caption/ Insert Table of Figures.
2. Type your own caption after 'Figure 1'.
3. Click OK.
4. To delete the border, right click on the edge of the text box, and select Format Text Box.
5. Select No Line under Line Color options.
6. Click OK.

Copying and Pasting a Picture

1. Select a picture.
2. Click Home Ribbon/Copy.
3. Click where you want to insert the picture.
4. Click Home Ribbon/Paste.

Grouping

1. Left-click on the picture.
2. Hold down control and/or shift button and click on the object you wish to group with the picture.
   Note: This function is useful for grouping the picture with the caption.
3. Right-click one of the items selected.
4. Select Grouping/Group.

Modifying Advanced Picture Properties

Rotating and Flipping Pictures

1. Change the picture wrapping to tight.
2. Left-click to select the object.
3. Click Format Ribbon.
4. Inside the Arrange box select a rotating or flip.

Note: If using Free Rotate, click and move the green image handle that appears at the top of the picture to rotate the picture to an exact degree.

Aligning the Picture

1. Holding down the shift key, click the pictures to align.
2. From the Format Ribbon/Align.
3. Select an alignment option.
4. Click outside of the pictures to deselect.

Creating Brochures

Follow these instructions to create an 8.5"x11", double-sided, tri-fold brochure. For best results, follow these instructions in order.

Setting Margins

1. Go to the Page Layout Ribbon/Page Setup box.
2. Choose the Margins tab.
3. Set all margins to .25" (Top, Bottom, Left, Right).
4. Click OK.

Setting Paper Orientation

1. Go to the Page Layout Ribbon.
2. Under Page Setup box click Orientation.
3. Under Orientation, Click on Landscape.

Creating Columns

1. Go to the Page Layout Ribbon/Columns.
2. Change the number of columns to 3.

Your document should now have three columns, landscape orientation, and corrected margins.

Understanding Layout

A typical brochure has six panels—three on each page. To create the second page:

1. Click in the first column.
2. Click on the Page Layout Ribbon/ Breaks/ Column.

View the following diagram below before you start typing. The layout of your document should match the diagram.

This is your **Inside Flap**	The second column will be your **Back Page**	**Start typing here.** This is your **Cover.**

	On the second page your will layout the **Inside-columns** of your brochure.	

Switching Between Columns

To move from column to column, you need to insert column breaks. To insert column breaks:

1. Click on Page Layout Ribbon/Break/Column Break.
2. Repeat step 1 for the next five columns.

Hint: Start typing the body text on the inside columns. Once you finish typing the third inside column, go back to the first page.

Creating Manuals

To create a manual using Microsoft Word, follow these steps:

1. Set up even/odd page layouts.
2. Set up the table of contents.
3. Create the fly pages.
4. Create the Chapter 1 body pages.
5. Create the Chapter 2 body pages.
6. Create the table of contents.
7. Create the index.

Setting Up Even/Odd Page Layout

1. Select the Page Layout Ribbon.
2. In the Page Setup Box click in the bottom right hand corner on the arrow.
3. Click the Layout Tab.
4. In the Section Start area, choose Odd Page.
5. In the Headers and Footers area, choose Different Odd and Even.
6. Select Apply to Whole Document.
7. Click OK.

Turning on Formatting Marks

1. Click on the Home Ribbon. Under the Paragraph Box select the Paragraph Icon.

Creating a Cover Page

1. On the first page type the word "Cover Page".

Inserting a Break

1. Click the Page Layout Ribbon/Page Setup Box/Breaks/Odd Page.
2. Repeat this step with the TOC/Introduction/Fly Ch 1/Ch 1/Fly Ch 2/Ch 2.
3. You should end 13 of 13 pages.
4. Scroll back to the top and click on the Cover Page.

Creating Headers and Footers

1. Click on the Insert Ribbon/header/ Blank Header.
2. Make sure the Header and Footer Tools Ribbon pops up.
3. Under the Navigation Box click Next Section.
4. Go through the whole document and unlink each page by clicking the Link to Previous button (under the Navigation Box).
5. Click the Next Section button to go to the next page.
6. Repeat until done with document.

Front Matter: Adding Page Numbers and Correct Titles

1. Scroll to the top of the document.

Remember: That the Cover Page does not get anything so go to the next page.

2. Under the Headers and Footers Tool tab, select Link to Previous to make sure the pages are unlinked to each other. Note: The pages are linked if Linked to Previous is orange.
3. Type the title of that page (Table of Contents) in the header.
4. Hit Tab twice.
5. Under the Header and Footer Tools/Header and Footer Box click on Page Number.
6. Click on Format Page Numbers under the number style choose Roman Numerals i, ii, iii… etc.
7. Under Start at tell it to start at iii.
8. Click OK.
9. Make sure your cursor is at the right hand corner.
10. Click on Page Number/Current Position/Plain Number.
11. Repeat for Introduction.

Body Pages: Adding Page Numbers and Correct Titles

1. No titles on Fly Pages.
2. Click Next Section.
3. Under the Headers and Footers Tool tab, select Link to Previous to make sure the pages are unlinked to each other. Note: The pages are linked if Linked to Previous is orange.
4. Type Chapter 1, hit tab twice.
5. Under the Header and Footer Tools/Header and Footer Box click on Page Number.
6. Click on Format Page Numbers under number style choose Arabic numbers 1,2,3…etc.
7. Under Start type 3.
8. Click OK.
9. Make sure your cursor is at the right hand corner. Click on Page Number/Current Position/Plain Number.
10. Repeat for Ch 2.

Adding Even Pages

1. Go back to Ch 1 page.
2. Enlarge your font to 72 and hold a key down until you get to page 12 of 15.
3. Double click in the header.
4. Under the Header and Footer Tools/Header and Footer Box click on Page Number.

5. Choose Current Position/Plain Number. Hit tab twice and type the Title of Manual in the right hand corner.
6. Double click in the body and you are done.
   The following examples illustrate two different formats for headers and footers. Ask your instructor which format you should use.

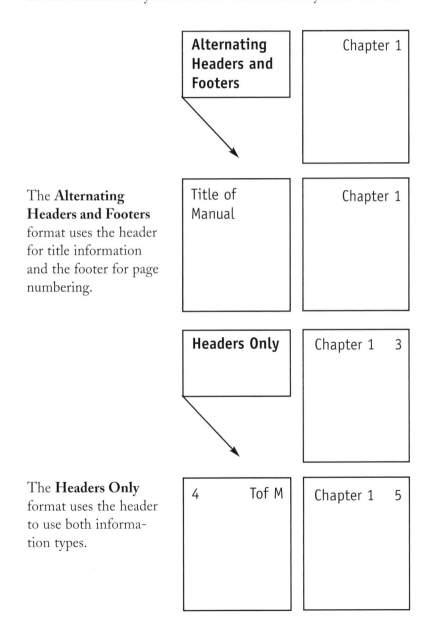

The **Alternating Headers and Footers** format uses the header for title information and the footer for page numbering.

The **Headers Only** format uses the header to use both information types.

The shaded pages show the blank back pages created by Word when you insert an Odd Page Section Break [SB(OP)].

You can only see these pages in Print Preview.

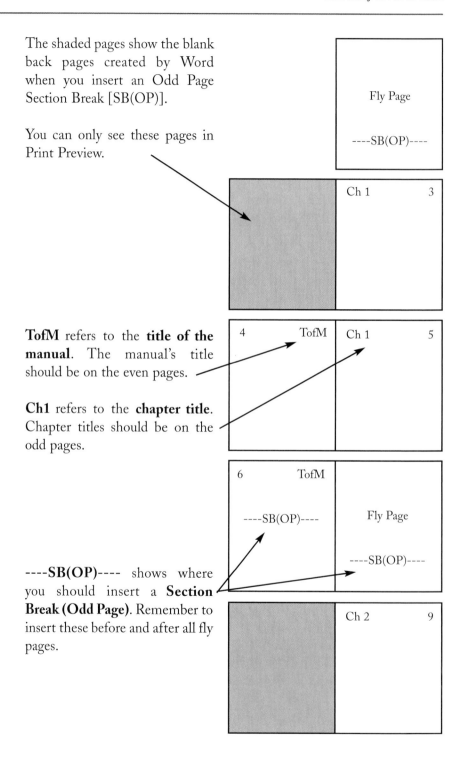

TofM refers to the **title of the manual**. The manual's title should be on the even pages.

Ch1 refers to the **chapter title**. Chapter titles should be on the odd pages.

----**SB(OP)**---- shows where you should insert a **Section Break (Odd Page)**. Remember to insert these before and after all fly pages.

Creating the Table of Contents

To create the table of contents, follow these steps:

1. Remove all tabs and all spaces between the section name and the page number.
2. Highlight everything in the Table of Contents.
3. Click on Page Layout Ribbon/Click on Right-hand Arrow in Paragraph Dialogue Box/Click on Tabs (at bottom left corner).
4. Click Clear All.
   - Change Tab Stop Position: to 5.88.
   - Under Alignments choose Align Right.
   - Under Leader choose µ2.........
   - Click Set.
5. Click OK.
6. Place the cursor in front of the number.
7. Hit the Tab key on the keyboard to get the dot leader.
   You will see this ....#

Creating the Index

Please note you must have the correct page numbers before starting the index, or the index will be inaccurate.

1. Click on References Ribbon/Under Index Dialogue Box, Click on Insert Index.
2. Select Mark Entry from the new window. To mark an entry:
   - Highlight the word on the page.
   - Click on the Mark Entry window.
   - Select your marking option.
   - Click either Mark (which will mark that single word) or Mark All (which will mark that word throughout the document).
3. Close the Mark Entry window after completing the Index entries.
4. Place the cursor at the exact point where you want the Index to start.
5. Go to Reference Ribbon/Index/ Insert Index.
6. Use the Formats menu to choose your Index style.
7. Click OK.

Creating a Gantt Chart

A Gantt chart allows you to follow the time management of a project allowing readers to track the length of time devoted to individual tasks. To create a Gantt Chart, follow these steps:

Selecting the Chart Type

1. Click on Insert Ribbon/Under Illustrations Dialogue Box, Click on Chart.
2. Click to Chart/Bar/Stacked Bar.
3. Click OK.
   (Note: Excel should open right next to the chart in word with all the data)
4. On Bar Screen right click on Left Color.
5. Click on Format Data Series.
6. Click on Fill.
7. Click on No Fill.
8. Click on Close.
9. Right click on (far) Right Color.
10. Click on Format Data Series.
11. Click on Fill.
12. Click on No Fill.
13. Click on Close.
14. The data for Series 3 should be deleted.
15. Data Series 2 is the floating bar that is visible.
16. Altering Data Series 1 & 2 will create the Gant Chart.

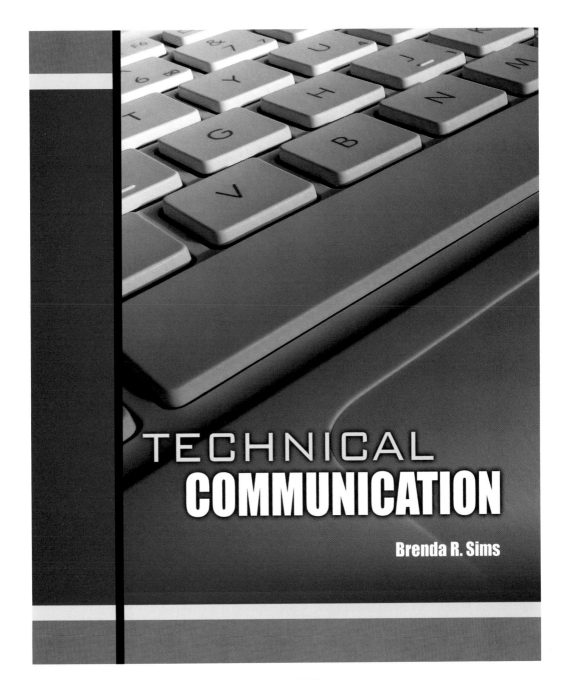

TECHNICAL
COMMUNICATION

Brenda R. Sims

Kendall Hunt
publishing company

Kendall Hunt
publishing company

www.kendallhunt.com
Send all inquiries to:
4050 Westmark Drive
Dubuque, IA 52004-1840

ISBN 978-0-7575-6498-7

Printed in the United States of America
10 9 8 7 6 5 4 3 2 1

Dedication

To my parents for teaching me the value of hard work

TABLE OF CONTENTS

Preface

Technical Communication has three goals:
* to prepare students for communicating in the workplace
* to prepare students for communicating information in a variety of mediums
* to provide technical communication instructors with a flexible, comprehensive teaching tool

Beneath these two goals lies the foundation of this book: students most effectively learn to communicate by understanding their readers, their purpose, and the communication context. This book, therefore, contains samples of student and workplace writing and exercises by which students can gain this foundation and then apply the principles of technical communication.

Technical communication goes far beyond reporting facts. Technical communication is a series of deliberate problem-solving activities that requires critical thinking. Before students can communicate effectively, they must understand why they are communicating, who is reading (or listening), and what the reader expects to learn. Without this information, communication will most likely fail to achieve its desired purpose. When students understand their purpose for writing and their reader's, purpose for reading, they will communicate more effectively.

Technical Communication presents guidelines designed to give students the tools and practice they need to communicate effectively in the workplace. Using these guidelines, students can determine the organization, layout, and content that will best meet the needs of their readers.

Overview of this Book

In addition to a brief introduction to technical communication in Chapter 1, this text contains five major sections.

Understanding Your Role as a Writer

In this section, students learn to understand their roles as writers in the workplace. In Chapter 2, "Facing Ethical and Legal Challenges," students learn to consider the ethical and legal dimensions of their communications and the workplace. In Chapter 3, "Collaborating in the Workplace," students see how to adapt to the interpersonal challenges and opportunities of collaborative writing. This chapter also suggests ways for students to use electronic media to facilitate collaborative writing.

Planning the Document

In this section, students learn the major steps to planning a document: understanding the readers, researching the topic, and organizing the information. In Chapter 4, "Writing for Your Readers," students learn principles for examining workplace writing from several vantage points: that of the writer, the readers, and the workplace. In Chapter 5, "Researching Information," students learn strategies for formulating research questions and exploring and evaluating primary and secondary sources. Chapter 6, "Organizing Information for Your Readers," presents techniques for structuring documents that readers can understand and use.

Writing the Document

In this section, students learn how to write reader-focused documents and how to communicate persuasively. In Chapter 7, "Writing Easy-to-Read Documents" and in Chapter 8, "Using Reader-Focused Language," students learn and practice style principles at the sentence, paragraph, and document levels. These principles provide students with strategies they can use to write clear, concise, reader-focused documents. Chapter 9, "Building Persuasive Arguments," gives students guidelines for understanding how to construct persuasive arguments and how to present those arguments effectively.

Creating the Visual Elements of the Document

In this section, students learn how to present visual information. Chapter 10, "Designing Reader-Focused Documents," and Chapter 11, "Creating Visual Information," demonstrate the rhetorical implications of document design and graphics. These chapters give students "how-to" information that they can easily apply to their own documents and information.

Preparing Documents and Presentations

This section applies the guidelines presented in the earlier chapters to specific types of technical documents and presentations. Students learn to apply these principles and tools to the following commonly written workplace documents:

- letters, memos, and e-mail (Chapter 12)
- job correspondence (Chapter 13)
- definitions and descriptions (Chapter 14)
- instructions and manuals (Chapter 15)
- proposals (Chapter 16)
- informal reports (Chapter 17)
- formal reports (Chapter 18)
- Web sites (Chapter 19)

The section ends with a chapter on delivering memorable oral presentations (Chapter 20). In Chapters 12-19, students examine sample documents written by other students and workplace professionals. These documents include annotations that point out how the documents demonstrate the principles presented in the chapters.

Appendixes

This book has two appendixes. Appendix A, "Documenting Your Sources," presents information on citing sources using APA, MLA, and CSE style. Appendix B, "Review of Common Sentence Errors, Punctuation, and Mechanics," provides a convenient, brief handbook.

Features to Enhance Learning

Technical Communication offers the following features that enhance student learning:

- *Tips boxes* in every chapter summarize key information that students need to create effective documents or to think about critical issues. Visually distinguished from the rest of the text and indexed in the front matter, the tips are easy for students to find and reference.
- *Taking It into the Workplace* focuses on technical communication from the vantage point of the workplace professional. These boxes include

assignments that require students to learn about communicating and writing in the workplace.

- **Case studies** at the end of each chapter give students the opportunity to apply the principles of a particular chapter in extended workplace scenarios. These case studies focus on communication situations that may have gone wrong and ask students to respond.
- **Ethics Notes** highlight an ethical issue related to the topic of that chapter. These notes appear in blue boxes throughout the book.
- **Worksheets** provide a checklist that students can use as they apply the guidelines presented in each chapter. These worksheets appear online so that students can use them and hand them in with their documents.
- **Companion Web site** provides *PowerPoint*® presentations for each chapter, quizzes, sample documents, and interactive document analyses.

Acknowledgments

Technical Communication has benefited greatly from the insights and suggestions of many colleagues and instructors. I have also worked with an excellent team at Kendall Hunt: Jen Wreisner and Kendra Miller. They have encouraged me and shown a refreshing "can-do" spirit.

At the University of North Texas, I have had the privilege to work with talented graduate students and faculty members who have provided a creative, innovative workplace. Under the supervision of Robert "Bob" Congrove, my technical communication lab staff have patiently and cheerfully answered my questions. I want to thank all of you.

Most of all, I want to thank my husband, Bill, and my son, Patrick, for their patience, their listening, and their ideas.

I'd Enjoy Hearing from You

If you have any comments or suggestions for improving this book, I'd enjoy hearing them. Please contact me at the Department of Linguisticis and Technical Communication at the University of North Texas, Denton, TX 76203. My e-mail is Brenda.Sims@unt.edu. I look forward to hearing from you.

Brenda R. Sims

iStockphoto 2008.

Technical Communication and Your Career

As you begin reading this book, you may wonder: *"Why do I need a technical communication course? I came to college to learn about my major, not writing."* Indeed, many of you reading this book probably came to college to study engineering, computer science, biology, chemistry, or other fields. You probably didn't come to college solely to learn to write. However, in any field, you will demonstrate your competence, in part, through your writing (Couture and Rymer). You will write in two contexts:

- **Routine tasks:** Routine tasks might involve setting up a meeting by way of an e-mail, drafting a memo to your coworkers about a project, or taking minutes at a weekly meeting.
- **Non-routine tasks** (Couture and Rymer): Non-routine tasks might include writing a feasibility report for the CEO of your company or drafting a proposal to persuade your managers to develop new software.

Whether the task is routine or non-routine, your managers and coworkers will use your writing to develop a perception of your competence as a professional.

In this chapter, you will learn what technical communication is, why technical communication is important to your career, and how your workplace may affect your writing. You will also learn the characteristics of effective technical communication.

WHAT IS TECHNICAL COMMUNICATION?

You probably use or read technical communication regularly. For example, you may use the online help of a software program or read the instructions for downloading tunes to your MP3 player. You might read the literature accompanying a medication. All of these documents are technical communication.

You communicate technically when you write, design, and produce technical information. Technical communication clearly and accurately conveys technical information so that readers can understand it quickly and easily and use it safely and efficiently.

Technical communication may take many forms: Web sites, owner's manuals, online help, meeting minutes, instructions accompanying medications, catalogs, etc. In each of these documents, the text, design, and graphics help readers gather information, answer questions, or complete a task.

Two types of professionals create technical communication:
- **Technical communicators:** Sometimes called technical writers, technical communicators can create any form of technical communication: Web sites, manuals, proposals, sales and marketing materials, oral presentations, online help, and white papers. A technical communicator's primary job function is to create and, often, to oversee the production of technical documents.
- **Technical professionals:** Technical professionals write to get their jobs done. They might write letters, manuals, proposals, or reports. Writing is part of their job, but it is not their primary job function.

HOW WILL TECHNICAL COMMUNICATION IMPACT YOUR CAREER?

Whether you are a technical communicator or a technical professional, you will need to communicate your ideas effectively to perform your job and to succeed in your career. Throughout your career, your job will require that you tell others about your work and its value to your company and, perhaps, to your profession.

You can't assume that your managers and your coworkers will value and approve of your work simply because you did it. Instead, you must effectively communicate your work, ideas, and progress to those with the power to implement your ideas or to those who supervise you (Barabas). You may also need to persuade others that your work has value. When you can effectively communicate ideas in writing, you will have a greater opportunity to ensure that your coworkers and managers understand the value of your work. Your managers may even evaluate you indirectly or directly on how well you communicate in writing.

> *When you can effectively communicate ideas in writing, you have a greater opportunity to ensure that your coworkers and managers understand the value of your work.*

In many workplaces, "it is not the performance of an employee that counts, but rather managers' perceptions of that performance" (Couture and Rymer). Managers often develop this perception through an employee's written and oral communication skills; "supervisory evaluations of communication skills correlate well with employees' overall competence within the organization" and "effective writing" is a typical measure of a professional's performance (Scudder and Guinan; Couture and Rymer). Your managers will develop a perception of your skills, your knowledge, and your value to the organization, in part, through your writing.

HOW DOES THE WORKPLACE AFFECT WHAT AND HOW YOU COMMUNICATE?

If you are like the typical college graduate, technical communication—specifically, writing—will fill about 20–60 percent of your time as a professional (University of Maryland). For professionals in technical fields, writing will fill at least 40 percent of your time at work (University of Maryland).

As a professional, you not only will spend much of your time writing, but you also will need to write effectively to succeed in your career. Surveys tell us that 94 percent of college graduates believe that writing well is important for success in their workplace; 42 percent believe it is of great importance (Anderson). Although communicating well does not solely lead to success in the workplace, it is an important factor. You may find that you can enhance your reputation with your managers, your peers, and your organization through your written and oral communication. In a labor force filled with mediocre writers, a professional who communicates effectively will stand out and succeed (Hansen, R. and Hansen, K.).

Whether you are proposing a new idea to your manager or recording a project's history for the permanent files, clear communication gives you visibility and credibility with your managers, your peers, and, ultimately, your organization. Poor communication gives you visibility, too—but without credibility. If you communicate poorly, others may have difficulty understanding your ideas; and your ideas and your work may ultimately fail to receive the recognition they deserve.

WHAT FACTORS AFFECT WHAT AND HOW YOU COMMUNICATE?

Several factors affect what and how you communicate in your workplace. These factors include:
- organizational and managerial expectations
- readers' needs and expectations
- a collaborative environment
- time and budget
- ethical issues

What Do My Organization and Manager Expect?

When you become a professional, your organization and your manager may have certain expectations about your documents. These expectations may be communicated explicitly, or they may be implied.

Your organization or manager may expect the format, organization, or style of a document to meet certain criteria or established guidelines, or a manager may have certain preferences about format and style. For example, many organizations have a standard format or established template for progress reports or for meeting minutes. Some organizations have a standard cover page and a formatted layout for letters and memos. Many organizations also have a style sheet that dictates style and design (type, color, and graphics guidelines) that the organization expects you to follow when creating documents for internal employees or external readers. Find out whether your organization or manager has specific expectations about style and design and meet those expectations to the best of your ability.

TAKING IT INTO THE WORKPLACE
Visiting with a Professional in Your Field

You can best learn how communication will impact your career by talking with professionals. To help you learn more about communication in your field, locate a professional working in your major field. For example, if your major is computer science, find a computer programmer or systems analyst. If your field is construction management or building construction, find a project manager for a construction project. You might contact these professionals in person, by telephone, or by e-mail.

Assignment

Once you have located a professional, set up an interview to discuss how communication, specifically writing, impacts his or her career. If you cannot interview the professional in person, you might suggest a telephone or e-mail interview. At the interview, ask the professional the following questions:

- What types of writing do you do at work?
- What steps do you take when creating a document?
- How do time and budget affect your writing?
- What percentage of your week do you spend on communication tasks?
- Do you collaborate with others when you write? If yes, describe the collaboration.
- How has communication impacted your career?

You may also develop some questions of your own. You may want to ask the professional to give you some samples of documents that he or she has written. After the interview, do the following:

- Write a memo to your classmates about what you learned.
- E-mail your memo to your instructor and to your classmates.

iStockphoto 2008.

Look for this icon throughout the text to signal online sources for worksheets, sample documents, and style guidelines.

This book suggests style and design guidelines for many workplace documents. If your organization or manager does not have explicit written guidelines, you can use the sample documents and the style guidelines presented in this book and online at www.kendallhunt.com/technicalcommunication. Look for the icon shown at left. These documents and guidelines will help you become familiar with conventions that you may encounter in the workplace. To supplement these sample documents and guidelines, you can gather examples of effective documents written by your coworkers to use as models. Your organization and your manager will appreciate that you are creating documents that fit with other documents the company produces and that you are working to be part of the organization's corporate culture.

What Do My Readers Need and Expect?

As a student, you generally know your reader—your instructor—and what he or she expects. However, in the workplace, you may or may not know your readers. They may be your managers or your coworkers. They may be company executives whom you've never met. They may be clients, users, or even potential customers. You may never meet your readers.

The readers for many of your documents may be more than one person or group, and each individual or group may have different expectations of your documents. Your readers may include men and women who live and work in countries and cultures other than your own and whose expectations of you and your documents differ from readers in your own country. As a professional, you will want to account for these differences to create effective documents.

This book will help you learn how to determine what your readers expect and how to meet those expectations. It also will help you to consider what international readers will expect from your documents and how to write for those readers.

How Will a Collaborative Environment Affect My Communication?

When communicating in the workplace, you frequently will work with others to produce a document: 87 percent of the college graduates surveyed by Lisa Ede and Andrea Lunsford, authors of *Singular Texts/Plural Authors: Perspectives on Collaborative Writing,* said that they sometimes collaborated with others to produce documents. Technical professionals and technical communicators collaborate on most documents except correspondence, progress reports, and meeting minutes, which they write individually. They collaborate in these ways:

- planning a document with others, either within their organization or outside their organization
- coauthoring or writing as part of a team
- reviewing and revising documents

Planning Documents with Others

Before a large document project, many organizations create a team to determine the purpose, readers, schedule, and organization. The team may contain only writers; however, in most cases, the team consists of members from different areas of the organization. An organization might select team members according to the function they will perform in the production of the document; the team might have a subject-matter expert, a writer, a graphics expert, and an editor.

When team members plan documents together, they can identify and answer important questions about global issues early in the document cycle, before writing begins—issues such as budget limitations, deadlines, document design, and readers' expectations and needs. In early planning sessions, team members can establish a schedule for producing the document and agree about areas of responsibility.

Once the team has planned the document, set the schedule, and assigned areas of responsibility, one or more persons may actually produce the document. In some collaborative situations, the team may plan the document, but only one person may actually do the writing (Raign and Sims).

Coauthoring or Writing as Part of a Team

In some organizations and writing situations, several people write a document. These people collaborate in one of two ways:

- Each person is responsible for writing a particular section of the document while one team member serves as a final editor (Raign and Sims).

- Each person is responsible for writing a particular section, and the team edits the document. The team members send the draft of their sections to the entire team for comments and edits.
- Team members write the document together.

Most teams find the first and second methods of collaborating more efficient. Regardless of the method, the more-successful teams decide on the style, design, and schedule of the document early in the writing process.

Reviewing and Revising Documents Collaboratively

Even if you don't work as part of a team, you probably will collaborate with others when reviewing and revising most documents. You may even collaborate with others by reviewing their documents—documents that you didn't help plan or write.

The review can be a formal process by which the writers and other interested parties meet to review the document and suggest revisions. These people may meet more than once before formally approving the document. Even if you are not writing as part of a team, you may take part in a formal review of your documents. William Sims, a licensed professional engineer, reports that newly hired and unlicensed engineers write under the signature of a licensed engineer; so a senior or licensed engineer must review many documents. Collaboration of this type does not involve the teamwork mentioned earlier; instead, collaboration takes place at the reviewing or revising stage, rather than at the planning or drafting stage.

The review process also may be informal. When it is, coauthors and interested parties receive a copy of a document to review. Often, these reviewers make comments and suggestions through e-mail or in a shared file accessible to all reviewers. The authors then use this file to make revisions.

This book will help you develop techniques for successfully collaborating with others to create effective documents.

What Do My Time and Budget Allow?

The workplace will limit the amount of time you can spend on creating documents. Your manager and your organization will expect you to write quickly and efficiently. You will be expected to finish documents on time and within budget. Every professional must contend with time and budget limitations; but, like other successful professionals in your field, you can learn to write effective documents despite these constraints.

Your budget and schedule may force you to spend extra hours at work or to submit a document before you are ready. For example, your manager may ask you to write the documentation for a new software application that your company is marketing. He or she may require you to have the document ready for user testing within a month, even though you normally would need two months. You may have to adapt an idea or a document to meet budget requirements.

This book suggests ways to streamline your writing process, to prioritize layout and design decisions, and to use online resources to help you submit documents on time and within budget.

What Ethical Issues Should I Consider?

As a professional, you may face ethical considerations about the language, the graphics, or the information that you or your coworkers use in workplace documents. For example, how will you report the results of tests on a new airbag design when the testing shows some serious design flaws and redesigning the airbag would delay the production of a new car model?

The language you use could affect how your readers perceive the design problem and, ultimately, how they decide to act. The language you choose could force the company to spend thousands of dollars correcting the design flaws. Your decision could also cause the company to lose sales to a competitor or to install the flawed airbag in automobiles—possibly endangering consumers.

As a professional, you may face similar ethical issues. This book will help you to analyze the ethical implications of the language, graphics, and information you select for your writing and to understand how language can affect readers' perceptions or endanger users. It will also give you four moral standards that you can use when facing ethical challenges in the workplace.

WHAT MAKES TECHNICAL COMMUNICATION EXCELLENT?

You read and use technical documents every day; yet not all of these documents are excellent. As a professional, you will want your technical documents to be measured as excellent. Technical communication is excellent when it successfully conveys your intended message and meets the needs and expectations of your readers. You can best convey your message and meet the needs of your readers when your technical communication has the following characteristics:

* includes honest, ethical information
* addresses specific readers
* uses clear, concise language
* uses a professional, accessible design
* includes complete, accurate information
* follows the conventions of grammar, punctuation, spelling, and usage

Figure 1.1 presents an example of excellence in technical communication.

> *Technical communication is excellent when it successfully conveys your intended message and meets the needs and expectations of your readers.*

Includes Honest, Ethical Information

Excellent technical communication is honest, ethical, and complete. Honesty is at the heart of ethical information. When you communicate ethically, you have "done the right thing." You have communicated out of the "intrinsic rightness of the behavior," not just to keep your job or to receive personal or monetary gain (Dombrowski). You have communicated ethically if you have given the reader honest, complete information and if you have not misled the reader.

Technical communication is dishonest when you misinform readers or intentionally leave out important information—perhaps information that could kill or injure someone. If you are dishonest, you and your organization may face legal charges.

FIGURE 1.1

User's Guide Demonstrates Excellence in Technical Communication

This excerpt from the TurboChef *Use and Care Guide* provides users with an overview of the oven's control panel. The text, graphics, and design illustrate the characteristics of excellence in technical communication:

- **Includes honest, ethical information:** Tells users how to use the controls. Specifically identifies the information for only the top oven, not the bottom oven.
- **Addresses specific readers:** Addresses readers who have just purchased an oven and who are learning to use the features available on the oven.
- **Uses clear, concise language:** Uses a minimum of words and combines clear, accurate graphics to help readers identify features of the oven.
- **Uses a professional, accessible design:** Includes appropriately sized graphics to help users identify features of the oven. Incorporates clear labels to identify the controls and their functions.
- **Includes complete, accurate information:** Identifies each control and tells users where to gather additional information on the "Speedcook" modes and the preprogrammed recipes.
- **Follows the conventions of grammar, punctuation, spelling, and usage:** Uses correct grammar, punctuation, and spelling.

4   The Top Oven

Control Identification & Operation – The Top Oven

Cook up to 15 times faster in the top Speedcook oven. The oven includes a state-of-the-art interface that provides built-in, on-demand information and over 450 recipes with pre-calculated cook times and temperatures. For an overview of the seven Speedcook modes, see pages 5-6. For a list of all preprogrammed recipes, see pages 30-33.

Classic Style Analog Clock and Timer
Runs with advanced electronic control knobs. To operate see Set Clock Knob and Set Timer Knob.

Set Clock Knob
Used to set time.
1. Press knob to extend.
2. Turn knob to set time.
3. Press knob back into original position.

Cook Navigator™
Consists of the Cook Navigator Display and the Info, Back, Cancel, and Start keys.

CookWheel™ (outer knob)
Turn to select a Speedcook mode, Self-Clean, or to turn the oven off. Modes include: Air-Crisp, Bake, Broil, Favorites, Dehydrate, Roast, and Toast.

CookWheel™ (inner knob)
Turn to navigate through the options on the Cook Navigator Screen. Press knob to select an option, begin a preheat cycle, or initiate a cook cycle.

Set Timer Knob
Use to set timer.
1. Press knob to extend.
2. Turn knob to set timer – up to 60 minutes.
3. Press knob back into original position.
Timer does not begin counting down until knob is pressed back into position.

Info Key
Press to display information on the Oven Setup (volume, screen brightness and contrast, language, and standard or metric measurement), Help Topics, or Oven Tips.

Back Key
Press to return to previous screen.

Cancel Key
Press to terminate a cook cycle or to return to the main screen.

Oven Light Switch
Press to illuminate top cook cavity.

Start Key
Press to select an option, begin a preheat cycle, or initiate a cook cycle.

NOTES:
1. Pressing the "Start" key does the same thing as pressing in the inner CookWheel knob.
2. To turn the oven on, you must select a Speedcook mode. See CookWheel (outer knob).

Source: Downloaded from the World Wide Web. www.turbochef.com/residential/shared/pdf/TC30DWO_QuickReference.pdf. TurboChef *Use and Care Guide*, page 1. Used with permission of TurboChef.

Addresses Specific Readers

Your technical documents can accomplish their purpose only when they
- meet the needs and expectations of your intended readers
- convey your intended message in terms that the readers will understand

Before you can create documents that will succeed, you must identify your readers. This task is easy when you know your readers. For example, if you are writing instructions to help your coworkers create a Web site, you will know (or can easily find out) what they know about the task. You can find out if they are familiar with the software you will be using or if they have created Web sites using other software or with HTML. You can then determine how much detail to include and how to best structure the instructions. However, if you are writing the same instructions for a group of consumers, you will not know the readers. You may not know if they have previously created a Web site using different software; you may not even know how familiar they are with using the computer. In this situation, you can create a profile of your readers. With this profile, you can determine the appropriate level of detail to include in your instructions.

Uses Clear, Concise Language

To convey your intended message, your technical documents must be clear. For readers to use your technical documents, the writing must also be concise. Let's look at an example from some instructions to contractors working with electrical transformers:

> **The transformers are configured such that operating personnel are exposed to live 12.47kv when any of the enclosure doors are opened.**

This instruction is not clear or concise, possibly endangering the users. The instruction would be clearer if written as follows:

> **DANGER: To avoid being exposed to live 12.47kv, keep all enclosure doors closed.**

When technical communication isn't clear and concise,

- **it can be dangerous.** The original instruction to the contractors does not tell them to keep the doors closed. If one of the operating personnel opened the doors, he or she could be severely burned or, worse, killed.
- **it can be unethical.** When technical communication is unethical, readers can get hurt; and you and your organization may face serious legal charges.
- **it can be expensive.** When technical communication isn't clear and concise, either the writer or the reader wastes time; and in the workplace, time is money. For example, Melissa Brown, a manager of documentation for a marketing company, reports that by including a concise tips supplement in software documentation, her company was able to reduce the number of calls to technical support (Blain and Lincoln). Her company saved substantial money simply by including these tips sections (Redish).

Uses a Professional, Accessible Design

You can use design to make your documents more effective and to achieve your intended purpose. An effective design has three objectives:

- **It helps readers locate information and understand how you have organized your document.** Most readers of technical documents do not read the whole document; instead, they look for specific information within the document. When a document is effectively designed, readers can efficiently locate information and navigate through a document. The design features—such as headings, lists, and graphics—help readers see how the document is organized, so they can easily find the information they need.
- **It creates a positive, professional impression of your document and your organization.** When a technical document has an attractive, professional design, readers are more likely to read it, and you are more likely to achieve the intended purpose: an attractive, professional design can create a positive impression of you, your information, and your organization. A sloppy, unprofessional design, likewise, can create a negative impression of you to your readers.
- **It gives your documents an attractive, inviting appearance.** When faced with a page or screen filled with only words, you (and your readers) may not read a document. Readers are more likely to read and use your document when it uses design features that create an attractive, inviting appearance.

Includes Complete, Accurate Information

Even when the design is effective and the language is clear and concise, a technical document can only succeed if the information included is complete and accurate. A successful technical document gives readers all the information they need to understand the problem, to perform the required task, to understand an unfamiliar topic, or to make a decision. You will best know what information to include and not to include when you identify and create a profile of your readers. Successful technical communicators don't *assume* what the readers know; they *find out* what the readers know. Then, they can include complete information to help the readers accomplish their goals.

Effective technical documents also give readers accurate information. If your technical document gives readers inaccurate information, you may merely confuse or annoy them. Documents with inaccurate information can be expensive or dangerous. For example, an executive with a U.S. construction company didn't proofread a contract before it was signed. In the contract, the company agreed to complete a project for $200,000 instead of $2,000,000. The contract writer simply left out a zero. Although the company was able to amend the contract, the company unnecessarily spent thousands of dollars in legal fees and lost much goodwill with its client.

Follows the Conventions of Grammar, Punctuation, Spelling, and Usage

Effective technical communication follows the conventions of grammar, punctuation, spelling, and usage. When your technical documents and your correspondence don't follow these conventions, readers may misread your communications. When you don't follow these conventions, you send negative, unprofessional signals to your readers. For example, if you send an e-mail filled with spelling and punctuation errors to a potential client, he or she may assume that you and your organization do sloppy work and may question whether your technical information is accurate. These errors may also cause readers to focus on your writing rather than on the information you are trying to convey. These same errors may cost you promotions, as your managers may evaluate you on your ability to communicate. In a survey of 402 companies, executives identified writing as the most valued skill in an employee (Hansen, R. and Hansen, K.). Although following the conventions of correctness isn't all that makes up good writing, many managers will evaluate your writing solely on its correctness.

CASE STUDY ANALYSIS
Embarrassing Typo Costs County $40,000*

Background

In 2006, county officials in Ottawa County, Michigan, were preparing for November 7 elections. In the election, citizens would be deciding on a proposed state constitutional amendment to ban affirmative action programs that give preferential treatment to individuals or groups based on race, gender, and other items.

The county printed 180,000 ballots, at a cost of approximately 30 cents each. The county mailed about 10,000 of these ballots to absentee voters. On October 3, Ottawa County Clerk, Daniel Krueger, noticed a typo—a very embarrassing typo. The "l" was missing in the word "public."

"My first thought was, 'Oh, crap,'" Krueger said, as reported in the *Holland Sentinel*, Holland, MI, newspaper. "We had about five or six people proofread it. It's just one of those words. Even after we told people it was there, they still read over it. It happens occasionally."

Because the error occurred on a statewide proposed amendment, Krueger decided to reprint the ballots. The cost to the county general fund was $40,000. Krueger said, "It needed to be reprinted," citing that the proposal was statewide and controversial. In other cases of misprints, the county had decided to use the ballots with errors, but those typos typically consisted of misspelled names, omissions, or incorrect numbers.

The Michigan Secretary of State Bureau of Elections representative, Kelly Chesney, told the *Holland Sentinel*, "The county made the right decision. It happens every election. There are 1,500 local election officers running our elections. They check and double check, but mistakes happen. Unfortunately, sometimes there is human error involved."

Assignment

1. Pretend that you are Ottawa County Clerk, Daniel Krueger. Write a letter to the county commissioners, explaining what went wrong in the ballot printing process and how you plan to ensure it doesn't happen again. Hand in your letter to your instructor. (For information on writing letters, see Chapter 12).

2. Develop a plan for proofreading an important document on campus housing safety measures that you must submit to the university for approval. Include at least three other individuals as editors/proofreaders in addition to yourself. Create a flow chart of the plan, and turn it in to your instructor. (For information on flow charts, see Chapter 11).

*Compiled from information downloaded from the World Wide Web: http://hollandsentinel.com/stories/101006/local_20061010013.shtml. Nims, Tereasa. "Ballot error costs county $40,000; Word 'public' was misspelled on 170,000 ballots that must be reprinted." Tuesday, October 10, 2006.

EXERCISES

1. **Web exercise:** Locate a Web page that demonstrates some or all of the characteristics of excellent technical communication. In a memo to your instructor, discuss the following:
    *   Who will read or use the Web page?
    *   How is the page an example of technical communication?
    *   Does the page demonstrate any or all of the characteristics of excellent technical communication?

    Attach a printout of the Web page to your memo.

2. **Collaborative exercise:** Form a team with two or three members of your class. Locate a manual for a consumer product. You might select a manual for a microwave, a bicycle, an MP3 player, or a cell phone. In a memo to your instructor, answer the following questions:
    *   Who will use the manual?
    *   Does the manual demonstrate the characteristics of excellent technical communication? If so, which characteristics does it demonstrate? If not, why?
    *   How would you improve the manual to make it excellent?

    Include a copy of the manual with your memo.

iStockphoto 2008.

Facing Ethical and Legal Challenges

In the workplace, you may face decisions where you will have to consider your values—what you believe is right and wrong. At the heart of these decisions is "doing the right thing." However, "doing the right thing" may not always be clear-cut. For example, you may feel pressured to compete at any cost, to sacrifice safety to get a product out on time, or to sidestep environmental regulations to cut costs. You might be tempted to distort statistical data to make the results of a study appear more favorable, or you might be intimidated into manipulating the language of an annual report to make your company look more profitable.

When faced with these situations, your decision to "do the right thing" may be difficult. Your organization expects you to follow standard business ethics and to know and to do the right thing.

UNDERSTANDING ETHICS

Ethics is the study of values, often called principles of conduct, that apply to a person or group. *Ethics*, according to Aristotle, is the study of what is involved in doing good. When we say that someone has acted ethically, we usually mean that the person has done the right thing. When you behave ethically, you act out of the "intrinsic rightness of the behavior"; you don't do something just to keep your job or to receive personal or monetary gain (Dombrowski).

When you face ethical decisions that are not clear-cut, the following moral standards can help you weigh the consequences of your actions:[1]

- the morality of an action
- the consequences of an action
- the rights of the people involved
- the care for relationships

Along with the moral standards, you will also want to consider the laws related to your action. The law cannot always tell you how to act ethically; certainly, the law cannot give you ethical constraints that come only from the fine-tuning of your own judgment and conscience (Shimberg). The law can give you clearly defined legal restraints. Paul Dombrowski, author of *Ethics in Technical Communication*, explains that "ethics cannot be reduced to politics or the law because it must guide us when the law or political rules are silent." Ethics must then fill these gaps to help us know how to act when laws are silent. The law often sets up only minimal legal restraints for behavior, whereas ethics implies "high standards of honest and honorable dealing and of methods used" in professions and businesses (Golen, Powers, and Titkemeyer).

Morality of an Action

When you look at the morality of an action, you consider whether the action itself is morally wrong.

If you look at the **morality of an action**, you consider whether it violates your moral duty. Instead of looking at the consequences of the action, you consider the action itself to decide whether it is morally wrong. Some

actions are wrong "just for what they are and not because of their bad consequences" (Wicclair and Farkas). For example, we consider lying and stealing to be morally wrong even when no one is harmed or the lying and stealing produce positive consequences.

Consequences of an Action

If you look solely at the **consequences of an action**, you use what ethical theorists call the **standard of**

iStockphoto 2008.

---

[1] I base these moral standards on the scholarship of Paul Dombrowski; Mark Wicclair and David Farkas; Steven Golen, Celeste Powers, and M. Agnes Titkemeyere; and Tom Beauchamp and Norman Bowie.

utility. According to this standard, you select the course of action that produces the greatest good for the greatest number of people or the least amount of harm for the fewest number of people (Wicclair and Farkas). Regardless of whether an action is morally right or wrong, the standard of utility "prohibits actions that produce more bad than good" (Wicclair and Farkas).

Let's consider how the government of Great Britain in 2001 dealt with an outbreak of foot-and-mouth disease, a highly contagious viral disease that infects cattle, sheep, and other animals. Because foot-and-mouth disease is difficult to control, it can devastate herds of livestock, damaging agricultural industries and severely limiting the production of meat and dairy products. To stop the outbreak of this disease, the government of Great Britain called for the slaughter of thousands of livestock in contaminated areas. By slaughtering livestock, government officials hoped to save thousands of other livestock and family farms. The government was using the standard of utility: it chose the least amount of harm for the fewest livestock and farmers. The government didn't focus solely on whether the action was right or wrong; it considered how it could produce the greatest good for the greatest number of people.

> *The standard of utility helps you select the course of action that produces the greatest good for the greatest number of people or the least amount of harm for the fewest number of people.*

Rights of the People Involved

If you look at the impact of an action on others, you consider whether you are violating the *rights of the people involved* (Wicclair and Farkas). This standard means that people have the right to be treated fairly and to expect similar cases to be treated alike.

Employees in any company have a right to a reasonably safe workplace and to be informed of any potential dangers in that workplace, and consumers expect to be informed of risks they may encounter as a result of a product. For example, when you visit an amusement park, you may see a sign that reads: "People with back or heart problems should not ride this roller coaster. Pregnant women should not ride this roller coaster." By posting this sign, the park is considering the welfare of visitors who may ride the roller coaster. These visitors have a right to know the inherent dangers of the roller coaster. The park cannot know the health of its visitors, but it can fully disclose the dangers so that visitors can make informed decisions. Similarly, when you buy a product, you have the right to know that the information accompanying that product is complete and accurate.

Care for Relationships

If you look at the *care for relationships*, you consider how the action affects relationships with others, especially those closest to us—our families, our coworkers, and our communities. For example, after the hijacked planes crashed into the World Trade Centers in New York in September 2001, some motion picture companies stopped the release of movies that portrayed hijacking and terrorism. Movie executives based their decisions not on whether the movies were good or bad, but instead on their care for those who had lost loved ones in the tragedy.

APPLYING THE MORAL STANDARDS

When facing an ethical decision, you may find that the moral standards discussed above conflict with one another. This dilemma can make an ethical decision difficult to assess and even more difficult to make.

For example, let's consider Susan, a quality-control engineer for an automobile manufacturer. She is responsible for testing a newly designed side impact airbag, which company executives are eager to put into next year's models. Susan's tests of the new design have not been completely successful. Test results showed that all the airbags inflated on impact, but 10 airbags out of every 100 tested inflated to 60 percent capacity. The partially inflated airbags would protect passengers from most of a collision's impact, but those passengers might receive more injuries than passengers whose airbags inflated fully. If the passengers with partially inflated airbags were small children, the injuries could be especially serious.

Susan asks for more time to further test the airbags; however, company executives reply that they must get the airbags in the new models. Susan feels pressured to certify that the airbags are safe. Her ethical decision is not clear-cut.

- If she looks at the situation using the *moral standard*, her choice is simple. She cannot report that the airbags are safe and ready to be installed.

- If she looks at the situation through the *standard of utility* (the greatest good for the greatest number), her choice is more complex. If she says the airbags are safe, more lives will be saved with the airbags than without the airbags. However, if an installed airbag only partially inflates, someone—especially a child—could be injured or even killed.
- If she considers *the rights of those involved*, the consumers and her company have the right to know the results of the tests, regardless of the impact on sales.

What should Susan do? What would you do? Because moral standards often conflict, this chapter will help you ask the right questions so you can assess a situation and make a difficult ethical decision like the one facing Susan.

UNDERSTANDING YOUR LEGAL RESPONSIBILITIES

As a professional, you are responsible for knowing and following four bodies of law:
- copyright law
- trademark law
- liability law
- contract law

Copyright Laws

In school, you have a responsibility to research and write papers ethically, without plagiarizing. *Plagiarism* is the intentional or unintentional use of another person's words, photographs, music, or graphics without acknowledging that you have used that work. If you plagiarize, you could receive a failing grade or be expelled from school.

Plagiarism is an ethical issue, not a legal issue. Unlike plagiarism, copyright is a legal matter. *Copyright laws* give the "owner" of intellectual property—such as words, graphics, music, or photographs—the sole right to "copy" the work that he or she has created and to profit from or prohibit the sale or distribution of that work. For example, if your roommate writes a report, he/she has the right to copy that report; but if you want to use that report, you must get permission from your roommate.

TIPS FOR FOLLOWING COPYRIGHT LAWS

- **Follow fair use guidelines**. Do not rely on excessive amounts of material borrowed from other sources unless that information is repurposed from your company. Fair use guidelines do not apply to graphics, so you must always seek written permission to use a graphic.
- **When in doubt about fair use, obtain written permission**. The fair use guidelines can be confusing because the law does not provide a specific number of words, lines, or notes that you may legally use without permission. If you are not sure whether you are following fair use guidelines, ask for permission to use the material. If you cannot gain permission, do not use the copyrighted material.
- **Cite the source of material that does not belong to you or your organization**. The Copyright Office notes that simply acknowledging the source of the copyrighted material does not substitute for obtaining permission. By citing your sources, you also fulfill your ethical responsibility to be honest.
- **Ask for guidance from legal counsel.** If you need help understanding copyright laws, ask your organization's legal counsel for help. Don't simply guess or assume that you are following the law correctly.

Some companies encourage employees to reuse information. The reused information is sometimes called *boilerplate*. Companies call the practice of reuse *repurposing* or *single sourcing* of information. They allow it because the repurposing saves time and eliminates errors.

For example, if you work for AT&T, you can legally copy information from the AT&T Web site and use that information in AT&T marketing documents. However, if you work for AT&T and find information or a graphic on the Verizon Web site, you cannot copy that information or graphic to the AT&T Web site without Verizon's written permission. Without written permission, you would be violating Verizon's copyright.

You may wonder why you have to get permission from Verizon instead of from the person who created the information or graphic. Anything that an employee creates while on the job belongs to the company, not to the employee. This concept is called "work made for hire." Because the creator of the information or graphic worked for Verizon, the information or graphic copyright belongs to Verizon, not to the individual.

Copyright law does not give the owner all rights. In some limited instances, copyrighted material may be used without permission. The law provides for the *fair use* of small parts of copyrighted material without the owner's written permission. For example, if you include a statement from a Web site in your report and you cite the source, you are following the fair use guidelines. These guidelines also allow you to use copyrighted material without written permission for purposes such as education or news reporting. The guidelines in Figure 2.1 will help you determine fair use. Be wary of using other people's written information and making only cosmetic changes to it. You could be guilty of violating copyright laws, and you certainly would be guilty of violating the author's rights and your moral duty to be honest. If you need more information on copyright laws, go to the Web site for the U.S. Copyright Office, www.copyright.gov.

Trademark Laws

Trademark laws protect the owner(s) of a name or logo used for a product. Companies use trademarks and registered trademarks to protect the identity of their products:

- *Trademarks* protect words, names, symbols, sounds, or colors that distinguish a company's goods and services from those manufactured or sold by others and indicate the source of the goods (U.S. Patent and Trademark Office). If the product is trademarked, the company includes the [tm] sym-

---

FIGURE 2.1

What is Fair Use?

The U.S. Copyright Web site explains that the law sets out four factors to consider when determining fair use:[2]

- **The purpose and character of the use, including whether such use is of commercial nature or is for nonprofit educational purposes.** For example, fair use guidelines are applied more liberally to schools and more conservatively to for-profit organizations.
- **The nature of the copyrighted work.** If the work is essential to the public, fair use guidelines are applied more liberally than if the work is non-essential to the public.
- **The amount and substantiality of the portion used in relation to the copyrighted work as a whole.** For example, two pages of a 100-page document is a small portion, while two pages of a four-page document is a large portion. The law does not specify a specific amount or percentage that you may use without written permission.
- **The effect of the use upon the potential market for or value of the copyrighted work.** If the use of the copyrighted material may hurt the owner's potential for profit, then you have probably violated fair use guidelines.

[2] Source: Downloaded from the World Wide Web, October 25, 2007: www.copyright.gov/fls/fl102.html. U.S. Copyright Office, "Fair Use," p. FL-102.

TIPS FOR PROTECTING A TRADEMARK

- **Use the trademark or registered trademark symbol.** Each time you include the name of a trademarked product, use the appropriate symbol. If you are unsure whether a product is trademarked, go to the Web site for the U.S. Office of Patent and Trademarks, www.uspto.gov.
- **Use a footnote the first time you use a trademark.** At least once in a document, preferably near the beginning, follow the trademark or registered trademark symbol with an asterisk or footnote number. At the bottom of the page in a footnote, state that the product is a trademark or registered trademark. For example, a footnote might read: ¹Kleenex is a registered trademark of Kimberly-Clark Corporation.
- **Use the trademark as an adjective, not as a noun.** For example, you would write: Doritos® tortilla chips, not simply Doritos®. Likewise, you would write: Dr. Pepper™ soft drink, not Dr. Pepper™.
- **Don't do anything to hurt the spirit of the trademark or to alter the trademark.** Do not change the trademark in any way. For example, if the trademark uses a particular color or font, do not change that color or font.

bol after the name of the product. This symbol indicates that only that company has the right to use that name.
- *Registered Trademarks* are a word, name, symbol, sound, or color that a company has registered with the U.S. Patent and Trademark Office. The company can then use the ® after the name of the product. For example, Frito-Lay has a registered trademark for Doritos® brand tortilla chips.

You are responsible for protecting your company's or your client's trademark and for using the trademark and registered trademark symbols accurately.

Liability Laws

Liability laws protect the public from inaccurate information from authors, editors, or publishers. These groups are responsible for injury that occurs from defective information, even if they give out the information unknowingly. If a company violates a liability law, it could face a liability lawsuit for "personal injury, death, property damage, or financial loss caused by a defective product" (Helyar).

Contract Laws

A contract formalizes an agreement between two parties by making it a legal promise. You will want to consider two areas of contract law: [3]
- **Express Warranties**: an *express warranty* is an oral or written statement (such as an advertisement) for a product. For example,

[3]Copyright by and used with permission of FreeAdvice.com

TIPS FOR FOLLOWING LIABILITY LAWS

You can help protect your organization and yourself from possible liability suits by following these tips adapted from Pamela S. Heylar, author of *Product Liability: Meeting Legal Standards for Adequate Instructions*.

- **Use language and graphics that the users will understand.** For example, if you are writing a manual that children will use, include simple, clear graphics that children can easily follow. If you are writing a manual for non-native speakers of English, use simple language free of *idioms* (expressions whose meanings are different from the standard or literal meanings of the words they contain: e.g., "going cold turkey").
- **Tell the user how the product functions and what it can and cannot do.** Make sure that your users understand what the product does. You can be liable if you do not also explain what the product cannot do.
- **Warn the users of potential risks when using the product.** State the specific dangers in using the product. Use direct, clear language. Don't assume that readers will know the danger. You are responsible for directly stating that danger.
- **Make sure that users can easily see the warnings.** For example, if you are warning users of the risk of cutting their fingers or toes with a lawn mower, put the warning both in the instruction manual and, more importantly, on the mower.
- **Tell the users what the product can do and what it can't or shouldn't do.** Tell users what the product is designed to do and what it isn't designed to do. Put this information not only in the instruction manual that will accompany the product, but also in materials that a potential buyer will see. For example, a manual for a gas barbecue grill states: "For outdoor use only. Never operate grill in enclosed areas, as an explosion or a carbon monoxide buildup might occur, which could result in injury or death" (Coleman). While most users would know that they should operate a barbecue grill only outside, the manufacturer could be liable if it didn't warn users of the risk of using the grill indoors.
- **Inform users of all aspects of owning the product, from maintaining to disposing of the product.** When purchasing some products or services, the user may have ongoing responsibilities; you must inform the user of these responsibilities. For example, most car manuals provide owners with a maintenance schedule. Along with the schedule, the manual usually includes a statement such as, "If your vehicle is damaged because you failed to follow the recommended Maintenance Schedule and/or to use or maintain fluids, fuel, lubricants, or refrigerants recommended in this Owner's manual, your warranty may be invalid and may not cover the damage."
- **Test the product and the accompanying product information.** Perform usability testing to make sure that the product is safe and that the instructions and product information are accurate. For information on usability testing, see Chaper 13.

TIPS FOR MAKING ETHICAL DECISIONS

- **Gather all the related information.** Make sure that you have all the facts and that your facts are accurate. You don't want to risk losing your job or your reputation by basing a decision on inaccurate or incomplete information.

- **Think first; then act.** Once you are satisfied that you have all the facts and that they are accurate, think about all the possible choices and use the questions in the following section, "Ask the Right Questions." Once you are satisfied that you have made the ethical choice, then take the action or communicate your decision.

- **Find out all you can about the people affected by your decision and those who will read your communication.** You can then determine the best way to approach these people if you want to argue for change or you need to suggest a course of action that they may not want to consider.

- **Talk to people whom you trust.** They may help you consider alternative, yet ethical, choices. If you feel that you cannot trust anyone in your organization, talk to someone whom you trust outside the organization. Don't try to face the situation alone.

- **Aim to establish a reputation as a hard-working, loyal coworker with integrity.** Then, when you do take a stand on an ethical issue, your coworkers and managers won't take your stand lightly.

if a producer of washing machines claims that a washer/dryer combination doesn't require an exterior vent, that claim is an expressed warranty. If a salesperson says the combination washer/dryer works great, he or she is not making an express warranty; such a statement would be considered a personal opinion.

- **Implied Warranties**: an *implied warranty* is an implied promise that a product is fit for the particular purpose of the buyer. This warranty is not written or spoken, but is implied. For example, by its very nature, a cell phone is fit for making phone calls, and a car is fit to carry passengers. You might see an implied warranty in the graphics included in product advertisements or manufacturing information.

MAKING ETHICAL DECISIONS

Now that we have defined ethics and considered the moral and legal standards, let's examine the following guidelines to help you work through ethical dilemmas:

- **Use a decision-making model** to help you work through ethical dilemmas.
- **Ask the right questions** to help you decide whether an action is ethical.

Use a Decision-Making Model

Before making a decision, you need to gather all the related information about the situation and the people involved. Many companies suggest that their employees follow a model to help them work through an ethical dilemma. By following a model, you can make more informed ethical decisions. In Figure 2.2 on page 30, Raytheon gives its employees a quick test to use when facing an ethical problem. If your workplace does not have a stated model for making ethical decisions, follow the tips in the sidebar on the previous page.

Ask the Right Questions

When you face an ethical decision, you can help determine the appropriate action by asking the right questions. Your company executives may have a series of questions that they expect you to consider when making decisions. For example, Kraft Foods gives employees four simple questions to ask themselves when confronted with an ethical workplace dilemma (see Figure 2.3, page 31). Proctor & Gamble has a *Worldwide Business Conduct Manual*. In this manual, the company states the core of its business ethic is "doing the right thing." To help employees do the right thing, Proctor and Gamble instructs employees to ask the questions in Figure 2.4, page 31.

If your company does not have a manual or guidelines for making ethical decisions, you can use the following questions to help you recognize your fundamental ethical responsibilities:

- Is it legal?
- Is it consistent with company policy and my professional code of conduct?
- Am I doing the right thing?
- Am I acting in the best interests of all involved?
- How will it appear to others? Am I willing to take responsibility publicly and privately?
- Will it violate anyone's rights?

Is It Legal?

You must follow all laws that apply to you, your product, and your organization. You must follow all laws that relate to ideas and products. For example, if you use a graphic from a Web site, you should determine if the information is copyrighted. If it is, you are legally obligated to obtain permission to use that graphic. If

When making an ethical decision, ask yourself these questions:

- *Is it legal?*

- *Is it consistent with company policy and my professional code of conduct?*

- *Am I doing the right thing?*

- *Am I acting in the best interests of all involved?*

- *How will it appear to others?*

- *Am I willing to take responsibility publicly and privately?*

- *Will it violate anyone's rights?*

FIGURE 2.2

Raytheon Quick Test for Ethical Decisions

When facing an ethical problem, ask yourself these questions:
- Is the action legal?
- Is it right?
- Who will be affected?
- Does it fit Raytheon's values?
- How will I feel afterwards?
- How would it look in the newspaper?
- Will it reflect poorly on the company?

When in doubt, ask—and keep asking until you get an answer.

Source: Downloaded from the World Wide Web, October 25, 2007:
www.raytheon.com/stewardship/ethics/ethics_answers/test/index.html. Raytheon Company, Raytheon Business Ethics and
Compliance Information, Ethics Quick Test and Decision-Making Model.

---

you use copyrighted ideas or graphics without permission, you are, in essence, stealing from the owners.

Is It Consistent with Company Policy and My Professional Code of Conduct?

Your company or organization expects you to follow its policies and guidelines. You are responsible for knowing those policies. If you do not know if an action is consistent with company policy, ask a manager or a mentor. Likewise, your profession may have a code of conduct. If so, you should also abide by that code. To learn more about professional codes of conduct, see "Taking It into the Workplace" on page 32.

Am I Doing the Right Thing?

You have a duty to be honest and truthful to your employer and the public and to report problems and information that could either negatively or positively influence your company, its employees, and its products or services. Likewise, the public expects you to use money, equipment, and supplies honestly and to give accurate, complete information.

FIGURE 2.3

Kraft Foods Guide to Ethical Dilemmas

Integrity: Doing What Is Right

Ask Before Acting

- Is it legal?

- Does it follow company policy?

- Is it right?

- How would it look to those outside the company? For example, how would it look to our customers, the people in the communities where we work, and the general public?

Remember These Rules

- Know the legal and company standards that apply to your job.

- Follow these standards—always.

- Ask if you are ever unsure what's the right thing to do.

- Keep asking until you get the answer.

Kraft Foods is the world's largest manufacturer and marketer of consumer packaged goods.

Provided courtesy of Kraft Foods and downloaded from the World Wide Web, Jan. 1, 2008: http://www.kraft.com/assets/pdf/KraftFoods_CodeofConduct.pdf.

FIGURE 2.4

Proctor and Gamble's Business Code of Conduct

Summary of Company policy statement

The core of the Company's business ethic is "doing the right thing." In addition to complying with any applicable legal requirements and the requirements described elsewhere in this Manual, you should ask the following questions in making decisions:

- Is my action the "right thing to do?"
- Would I feel comfortable if my action were reported broadly in the news, or were reported to a person whose principles I respect?
- Will my action protect the Company's reputation as an ethical company?
- Am I being truthful and honest?

If the answer to any of these questions about the action you are considering is not an unqualified "Yes," then simply do not take the action.

Source: Downloaded from the World Wide Web, Jan. 1, 2008: www.pg.com/company/our_commitment/corp_gov/WBCM-REDUCED_Single_Page.pdf. Proctor & Gamble Worldwide Business Conduct Manual, p.10.

TAKING IT INTO THE WORKPLACE
Your Profession and Its Code of Ethics

Many professional organizations have developed a code of conduct or a code of ethics for their members. These codes give members standards for ethical behavior in their profession. While these codes are hard to enforce, they do specifically guide the organization's members in ethical professional behavior. Many businesses have also adopted codes of conduct or ethics. These codes give members guidelines for ethical behavior and inform the public about how its members will conduct business.

iStockphoto 2008.

Assignment

Assume that you are just graduating from college or have just started a new job and that you plan to join a professional organization in your field. You want to find out if the organization has a code of ethics. Your assignment is as follows:

- Find a professional organization in your field.
- Find out if the organization has a code of ethics. (If not, find another organization related to your field.) Most professional organizations have a Web site and will post their code on the site. You might begin with these online resources:
    - National Institute for Engineering Ethics (www.murdough.ttu.edu/pd.cfm?pt=NIEE)
    - Center for the Study of Ethics in Professions (http://ethics.iit.edu)
- Bring a copy of the code of ethics to class.
- Answer these questions about the code:
    - Does the code provide a model for making ethical decisions?
    - How effectively does the code protect the interests of the public? Of the organization? What specific words or phrases demonstrate this effectiveness?
    - How can the organization enforce the code?
    - Does the code help the member make ethical decisions? If so, how? If not, what would you include in the code?

You will be dishonest and untruthful if you manipulate the language to hide facts, to make data say what you or your managers want, or to leave out unfavorable information. For example, you would be acting unethically if you only reported the positive results of patients being treated with an experimental drug while hiding or otherwise not reporting negative results.

When facing an ethical dilemma, ask yourself whether the proposed action is the right thing to do. If it is right, is it right only for this particular situation or for all similar situations? If the action is right now, is it right for everyone else in the same situation (Golen, Powers, and Titkemeyer)? For many situations, this answer will be quite clear. For example, if you manufacture a baby stroller and you know that it can collapse and injure or kill a child, you can clearly see that your company cannot sell the stroller.

Other situations may be more complex. For example, suppose you see a coworker use her company credit card to charge personal items. The company has a policy that employees may use the card only for company business. The company policy specifically states that employees may not charge personal items on the card. The coworker conscientiously pays the balance on the credit card each month, so the company doesn't incur any interest charges. Should you tell her supervisor? If you were her supervisor, would you want to know? After all, she is paying the balance each month; the company isn't being hurt. Is it right for her to use the credit card for personal charges?

You can also ask yourself this question: If it is right in this situation, is it right for everyone else in the same situation? If you can answer "yes," then your action is probably ethical. If you answer "no," then the action is probably unethical. If you are still uncertain, consider your options, ask more questions, or even talk to someone you trust and respect. This person may be able to help you look at the situation objectively.

Am I Acting in the Best Interests of All Involved?
Ideally, in any given situation, the best interests of all involved will coincide, making it easy to identify the right action. However, you may face a situation where these interests conflict. If so, you may have an ethical dilemma. When you face ethical problems, think about all the consequences and how these consequences will affect those involved both now and in the future.

Consider how the Red Cross handled donations to the victims of the September 11, 2001 terrorist acts. In radio and television ads across the United States, the Red Cross asked for donations to help the victims. The Red Cross received millions of dollars—more money than it had ever received for disaster relief. Executives with the Red Cross decided that because they had received record amounts of money, they would put some of the money aside for future disasters. They reasoned that the money would still go to disaster victims, just not to victims of the September 11 tragedies. When donors discovered that their money wasn't going to the victims, they felt that the Red Cross had misled them and was mishandling the money.

Did the Red Cross act in the best interests of the victims or the organization? By earmarking some of the September 11 donations, the Red Cross intended to protect the interests of future victims. However, did it act in the best interests of the donors and the September 11 victims? By not disclosing to the public what it was doing and by not using the money as the donors expected, the Red Cross lost the trust of many current and potential donors and, in turn, possibly hurt the interests of future victims.

How Will It Appear to Others? Am I Willing to Take Responsibility Publicly and Privately?

When you face an ethical decision, ask yourself this question: How will this decision or action appear to others? Even what you may consider to be an innocent action can result in the appearance of wrongdoing (Dow). If you don't want others to know what you have decided or done, then you should reconsider your decision and perhaps seek advice. With any decision that you make, you should be willing to take responsibility both publicly and privately. You might consider these additional questions when analyzing a possible action or communication:

- Will it reflect poorly on me or my organization?
- How would it look in the local newspaper?
- How will I feel afterward?

Will It Violate Anyone's Rights?

When faced with ethical dilemmas at work, consider whether the action violates the rights of your organization, other employees, the public, or others involved with the situation or communication (Wicclair and Farkas). These people have a right to be treated fairly and to receive "what is due or owed" and what they deserve or "can legitimately claim." Your organization, your coworkers, and the public also have the right to expect similar situations to be treated similarly, so consider whether you would act or communicate in the same way in similar situations. As you consider others' rights, look at any expressed or implied warranties that your decision may affect.

FOLLOW THE PRINCIPLES FOR ETHICAL COMMUNICATION

As an employee, you are expected to follow company policies and to avoid any actions that conflict with those policies. You are also expected to act with integrity. When you act with integrity, you put honesty above any financial objective, marketing target, or effort to compete (Kraft Foods). To communicate with integrity, follow the principles for ethical communication listed in the ethics note below.

Follow All Relevant Laws

Follow all laws related to intellectual property and to your product or service. Consider these guidelines to ensure that you are following laws related to intellectual property:

- **Get permission for any copyrighted information or graphics.** Do not plagiarize. If you don't know if the information is copyrighted, find out and get written permission to use it. If you can't find out, don't use it.
- **Protect yourself and your organization by honoring all trademarks.** Use trademark and registered trademark symbols. Do not alter the trademark. Use the tips listed on page 26.
- **Follow all guidelines related to liability laws.** Use the tips listed on page 27.
- **Honor the expressed and implied warranties related to your products, goods, and services.**

> ## ETHICS NOTE
>
> ### Principles for Ethical Communication
>
> - Follow all relevant laws.
> - Follow company policies and/or your professional code of conduct.
> - Be honest.
> - Do not mislead your readers.
> - Use clear, precise language.
> - Include all the information that readers need or have a right to know.
> - Take ownership of your writing.
> - Acknowledge the work of others.
> - Avoid discriminatory language.

Follow Company Policies and/or Your Professional Code of Conduct

Determine if your company (or organization) has a code of conduct; ask your manager, ask a human resources representative, or search your company's Web site. If your company does not have a stated code of conduct, read company policies to determine if your action will violate company policy. You can also turn to your professional code of conduct for guidance. If you cannot find a stated company or professional policy or code of conduct, ask your manager or another trusted company official for help.

Be Honest

You have an ethical responsibility to be honest. You may feel pressure to lie about your company's or another company's products, information, or services. If you lie, you are not doing the right thing. You may have to resist a manager's pressure to lie about an action, a product, a service, or a test result. Even if you have to go over the manager's head, you have a responsibility to tell the truth.

Do Not Mislead Your Readers

A misleading statement or graphic allows readers to make false conclusions or to think that conditions exist when they don't. These misleading statements or graphics are the same as lying: you, in essence, give readers untruthful information.

You can mislead readers in the following ways:
- **Creating false impressions**. You can create false impressions with language and graphics. For example, an owner's manual for a glass cooktop states, "Clean only with the OvenBrite." This statement leads owners to assume that only the OvenBrite brand will work. This statement creates a false impression; it is not true. You may be tempted to use terms such as "highly energy efficient," "best on the market," "innovative," or "state of the art" to make a product or service sound better than it is. Rather than using these terms, include specific information. For example, instead of saying "highly energy efficient" for an appliance, give the appliance's Energy Star rating.

You can also create false impressions with graphics. For example, look at the graphic comparing cereals in Figure 2.5. The graphic begins at

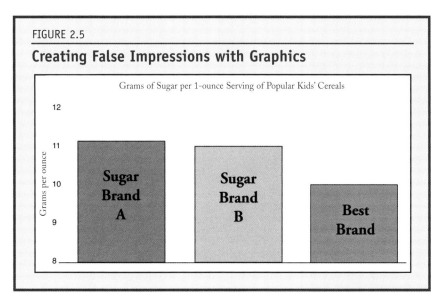

FIGURE 2.5

Creating False Impressions with Graphics

Grams of Sugar per 1-ounce Serving of Popular Kids' Cereals

eight, rather than zero. The chart implies that Best Brand has one-third less sugar than Brands A and B because the readers don't see the entire chart. Actually, Best Brand has only about 5 percent less sugar than the competing brands.

- **Exaggerating**. If you write, "This new microwave offers features to meet all your cooking needs," you are exaggerating because no microwave will meet users' every cooking need. Indeed, the microwave may have some advanced features not common to all microwaves, but it can't possibly meet every need. This statement creates false impressions, especially for novice cooks.
- **Using euphemisms**. You use a euphemism when you substitute a mild or less-negative word for a harsh or blunt word. For example, a company might use the phrase "involuntarily separated" instead of "fired" or "dismissed." Euphemisms can mislead readers.
- **Deemphasizing important information or emphasizing misleading information**. You mislead readers when you deemphasize information, especially negative information, or emphasize misleading or incorrect information through page layout, type size, or color. For example, an advertisement for a drug that controls heart rate emphasizes that the drug is safe. The word "safe" appears in a bright-red band and in excessively large, 36-point type. Yet, at the bottom of the page, in excessively small, 6-point black type, the advertisement warns that a defibrillator and emergency equipment

should be available for the user. Clearly, the drug is not completely safe; however, the reader may never see the warning because the advertisement deemphasizes the important, life-saving information.

Use Clear, Precise Language

You have a responsibility to write as clearly and precisely as possible and to make your information easy to find. Clear, precise language helps your readers understand your message.

You can mislead or deceive readers if your language is not clear. For example, if you are writing an instruction manual for a cordless vacuum cleaner, you might tell readers that the vacuum is "great for unexpected spills and messes and convenient for small, routine cleanup jobs around the house." This information is not clear unless you explain what you mean by "spills and messes." Many readers may consider spills of milk or juice a job for the vacuum when, in fact, the vacuum is designed to clean up only dry materials, not liquids. Instead, the instruction manual should say, "Use your vacuum to pick up dry materials. Do not use the vacuum to clean up spills of liquids and other wet materials. If you use the vacuum to pick up wet materials, you could be electrocuted."

Include All the Information that Readers Need or Have a Right to Know

You have an ethical responsibility to give your readers all the information they need or have a right to know. If you intentionally or unintentionally omit information, you are not acting ethically. For example, you may be tempted to tamper with research data to make it look precise by smoothing out or omitting irregularities so that the data appears statistically significant. You have a responsibility to give readers the untampered data with its irregularities, regardless of its statistical significance.

Throughout history, professionals have been faced with the pressure to omit information that readers had a right to know. For example, in a memo about the O-rings that led to the Challenger space shuttle disaster in 1986, the writers used the following sentence: "The conclusion is that secondary sealing capability in the SRM field joint cannot be guaranteed." This sentence misleads readers because it doesn't give them enough information to correctly assess the potential risks. A more accurate sentence might say: "The conclusion is that the secondary seal is not effective at temperatures below 50 degrees Fahrenheit, so the joint is highly vulnerable to catastrophic failure at such temperatures" (Winsor). The first sentence is technically accurate, but it doesn't give readers enough information to arrive at a correct conclusion.

Take Ownership of Your Writing

When appropriate, include references to the writer. Your writing will be more effective when you include personal pronouns or references, such as "I," "we," or the organization's name. Identifying the speaker or actor is a "necessary ingredient for ethical communication" (Rubens).

When you omit references to a speaker or actor, you are misleading the readers. When a writer says, "Each of the participants was interviewed for 30 minutes," he or she is not taking responsibility for the data; and readers may not know who did the interviewing. Instead, the writer should say, "I interviewed each participant for 30 minutes." The revised sentence does not destroy the factual nature of the information; it identifies the writer as the interviewer and places responsibility for the interview data on the writer.

Acknowledge the Work of Others

If others helped you do the work, acknowledge their contribution. If you use the work of others, cite your sources accurately. When you don't acknowledge the work of others, your readers assume that you have done all the work. When you take credit for work you have not done, you are acting unethically. If you use someone else's work in your research or writing without proper acknowledgement, you are plagiarizing. Consider how you would feel if one of your coworkers didn't acknowledge your work to your manager or didn't cite your part in a big project.

Avoid Discriminatory Language

You have an ethical responsibility to use language that does not discriminate against people because of their gender, age, physical appearance, physical or mental ability, religion, sexual orientation, or ethnicity. Discriminatory language reflects negatively on you and your organization.

CASE STUDY ANALYSIS
The Human Radiation Experiments

Background

From the 1940s to the 1970s, patients, some terminally ill, were injected with plutonium—mostly without their knowledge or consent—at Oak Ridge Hospital, the University of Rochester, the University of Chicago, and the University of California.[4] In the years following the atomic bombings of the Japanese cities Hiroshima and Nagasaki, the U.S. military and nuclear weapons industry wanted data on the biological effects of plutonium and radioisotopes. To determine these effects, scientists of the Manhattan Project injected 18 unsuspecting patients with plutonium. These patients had all been diagnosed with terminal diseases and weren't expected to live more than 10 years. Some of the patients, according to investigators at the Atomic Energy Commission, had not granted informed consent. The patients who had granted consent did so under false pretenses because the word "plutonium" was classified during World War II. Those patients who survived did not even know that they had been injected with plutonium until 1974.

These experiments continued during the Cold War, when the U.S. military wanted to know how much radiation a soldier could endure before becoming disabled even though researchers were aware of the hazards of working with plutonium as early as January 5, 1944. From 1960 to 1971, scientists at the University of Cincinnati performed experiments on 88 cancer patients ranging in age from 9 to 84. These patients were repeatedly exposed to massive doses of radiation, yet medical researchers from the 1930s through the 1950s had determined that whole body radiation was not effective in treating most cancers. These patients had tumors that would resist radiation—a fact that the doctors already knew (Braffman-Miller, Department of Defense). Most of these patients were uneducated, had low IQs, and were poor. The researchers in charge of these experiments wrote in 1969 that "directional radiation will be attempted since this type of exposure is of military interest"

---

[4]I gathered information on the human radiation experiments from Judith Braffman-Miller, "When Medicine Went Wrong: How Americans were Used Illegally as Guinea Pigs," *USA Today*, and the Department of Energy, "DOE Openness: Human Radiation Experiment."

(quoted in Braffman-Miller). The doctors did not use this procedure to treat tumors, but instead they used it to study how radiation exposure might affect soldiers. These researchers also denied patients treatment for the side effects of nausea and vomiting that resulted from the radiation. These researchers instructed the hospital staff to ignore these symptoms: "DO NOT ASK THE PATIENT WHETHER HE HAS THESE SYMPTOMS" (see Figure 2.6, page 43). Instead, the staff was to record the time, duration, and severity of these symptoms without treating the patient. Publicly, the researchers at the University of Cincinnati claimed that the purpose of their research was to study ways to treat cancer. However, in a report to the Department of Defense, dated September 1, 1966, they wrote that the purpose of their study was "to understand better the influence of radiation on combat effectiveness of troops and to develop more suitable methods of diagnosis, prognosis, pro-phylaxis and treatment of radiation injuries" (quoted in Braffman-Miller).

In 1966, Dr. George Shields and Dr. Thomas E. Gaffney at the University of Cincinnati formally questioned the purpose of the experiments. These doctors pointed out that the researchers deceived patients and were hiding the real purpose of their experiments. Their memos appear in Figures 2.7 and 2.8 on pages 44-45. Ultimately, the experiments were continued with some revisions.

Assignment

1. Read the memo written by Dr. George Shields (Figure 2.7, page 44). Be prepared to answer the following questions in class:
   a. What is the purpose of this memo?
   b. What was Dr. Shields's ethical dilemma?
   c. Dr. Shields recommends disapproving the study; however, he states that if the study is continued, the researchers should change the language used to inform patients of the risk. What language does he propose? Would this language ensure that all involved are treated ethically? Explain your answer.
2. Read the memo written by Dr. Thomas Gaffney (Figure 2.8, page 45). Be prepared to answer the following questions in class.
   a. What is the purpose of this memo?
   b. What was Dr. Gaffney's ethical dilemma?

3. Consider the ethical dilemma the scientists who continued the experiments faced. Be prepared to answer the following questions in class:

a. What was their ethical dilemma?

b. Were these scientists acting in the best interests of the patients or of the staff caring for these patients? Explain your answer.

c. Were these scientists and those who funded the research acting for the greater good—that of protecting soldiers who might be exposed to radiation? Were they acting ethically because they might save the lives of thousands of soldiers by sacrificing a few terminally ill cancer patients?

d. Did the patients need to know the purpose of the experiments? Explain your answer.

e. Should the scientists have disclosed the adverse effects of whole body radiation to the patients? To the public? Explain your answer.

FIGURE 2.6

Instructions to the Hospital Staff at the University of Cincinnati

INSTRUCTIONS FOR RECORDING OF SYMPTOMS
FOLLOWING IRRADIATION

DEPARTMENT OF RADIOLOGY
GENERAL HOSPITAL

Date: ______________________________

PATIENT ______________________________ NO. __________ WARD ________

TIME OF THERAPY __________

This patient has just received total body radiation for therapeutic purposes. It is possible that the symptoms listed below may develop within the next several days. Please note carefully the time at which these symptoms develop and note their duration and severity.

DO NOT ASK THE PATIENT WHETHER HE HAS THESE SYMPTOMS

	Time of Onset		Duration	Severity
	Date	Hr.		
Anorexia				
Nausea				
Vomiting				
Abdominal Pain				
Diarrhea				
Weakness				
Prostration				
Mental Confusion				

E.L. Saenger, M.D., line 207
H. Perry, M.D., line 200

Daily—Card No. 1

Downloaded from the World Wide Web, September 30, 2001: http://www.gwu.edu/~nsarchiv/radiation/dir/. Saenger, E.L., "Research proposal for sub-task in nuclear weapons effect research." Instructions for Recording of Symptoms Following Irradiation.

FIGURE 2.7

Memo from Dr. George Shields

TO: Dr. Edward A. Gall

FROM: Dr. George Shields

DATE: March 13, 1967

SUBJECT: Protection of Humans with Stored Autologous Marrow

I regret that I must withdraw myself from the subcommittee studying this proposal, for reasons of close professional and personal contact with the investigators and with some of the laboratory phases of this project. The following comments are sent to you in confidence, at your request.

This protocol is difficult to evaluate. The purpose of the study is obscure, as is the relationship of the experimental groups to the purposes. The significance of the study in relation to the health of the patients under study may be considerable if the investigators succeed in prolonging the lives of these patients with malignant disease, but the risk of treatment may be very high if the authors' hypothesis (that bone marrow transfusions will ameliorate bone marrow depression due to radiation) is incorrect. The radiation proposed has been documented in the authors' own series to cause a 25% mortality.

I recommend that this study be disapproved, because of the high risk of this level of radiation. Admittedly it is very difficult, in fact impossible, to balance potential hazard against potential benefit in experiments of this sort. The stakes are high. Our current mandate is that we evaluate the risks on some arbitrary scale. I believe a 25% mortality is too high (25% of 36 patients is 9 deaths), but this is of course merely an opinion.

If it is the consensus of the investigators and the review committee that a 25% mortality risk is not prohibitive, then the experiment could be reconsidered from the standpoint of informed consent - provided the patient is appraised of this risk in a quantative fashion. I believe that the conditions of informed consent will have been observed if the authors change "all patients are informed that a risk exists, but that all precautions to prevent untoward results will be taken" to the equivalent of "all patients are informed that a 1 in 4 chance of death within a few weeks due to treatment exists, etc."

Finally, although it is not our concern directly, a comment as to experimental design is indicated in this particular protocol. The authors' stated purposes are vague in the first page of the application, but on the last page three purposes are listed since it would require an untreated group and no reference has been made by the authors to such an untreated group of patients.

The second purpose can be fulfilled by this protocol only with the retrospective group (Group 1). The evaluation of bone marrow transfusion in the treatment of bone marrow depression would require a concomitant control group of patients treated only with radiation. It is apparent that the authors feel the radiation risk is too high to re-expose another group to this level of radiation without some effort at radio-protection, and therefore the authors have chosen to use the retrospective group as a control. There is considerable question whether this retrospective group will be entirely similar and therefore whether it will serve the second purpose.

The third purpose, "to determine whether autologous bone marrow therapy may play a role in treatment of bone marrow depression following acute radiation exposure in warfare or occupationally induced accidents," is not the subject of this experiment because normal individuals are not being tested. It is problematic whether the information gained in this study will apply to normal individuals following acute radiation exposure. Therefore it is my definite opinion that the third purpose of this experiment would not justify the risk entailed.

For these several reasons I feel that the experimental design is inadequate, and because of the high risk inherent in this level of radiation, I think experimental design should be a proper subject for our consideration in this instance.

Source: Downloaded from the World Wide Web, January 4, 2008: http://www.gwu.edu/~nsarchiv/radiation/dir/mstreet/commeet/me. Memo from Dr. George Shields to Edward A. Gall, SUBJECT: Protection of Humans with Stored Autologous Marrow, reprint of letter.

FIGURE 2.8
Memo from Dr. Thomas E. Gaffney

To: Dr. Edward Gall, Chairman
           Clinical Research Committee

From: Dr. Thomas E. Gaffney,

Date: 4/17/67

I cannot recommend approval of the proposed study entitled "The Therapeutic Effect of Total Body Irradiation Followed by Infusion of Stored Autologous Marrow in Humans" for several reasons.

The stated goal of the study is to test the hypothesis that total body irradiation at a dose of 200 rad followed by infusion of stored autologous marrow is effective, palliative therapy for metastatic malignancy in human beings. I don't understand the rationale for this study. The applicants have apparently already administered 150-200 rad to some 18 patients with a variety of malignancies and to their satisfaction have not found a beneficial effect. In fact, as I understand it, they found considerable morbidity associated with this high dose radiation. Why is it now logical to expand this study?

Even if the study is expanded, its current design will not yield meaningful data. For instance, the applicants indicate their intention to evaluate the influence of 200 rad total body radiation on survival in patients with a variety of neoplasms. This "variety," or heterogeneity, will be present in a sample size of only 16 individuals. It will be difficult if not impossible to observe a beneficial effect in such a small sample containing a variety of diseases all of which share only CANCER in common.

This gross deficiency in design will almost certainly prevent making meaningful observations. When this deficiency in experimental method is placed next to their previously observed poor result and high morbidity with this type of treatment in a "variety of neoplasms," I think it is clear that the study as proposed should not be done.

I have the uneasy suspicion, shared up by the revised statement of objective, that this revised protocol is a subterfuge to allow the investigators to achieve the purpose described in their original application; mainly, to test the ability of autologous marrow to "take" in patients who have received high doses of total body radiation. This latter question may be an important one to answer but I can't justify 200 rad total body radiation simply for this purpose, "even in terminal case material" (italics are mine).

I think there is sufficient question as to the propriety of these studies to warrant consideration by the entire Research Committee. I recommend therefore that this protocol and the previous one be circulated to all members of the Committee and that a meeting of the entire Committee be held to review this protocol prior to submitting a recommendation to the Dean.

Sincerely,

Thomas E. Gaffney, M.D.

EXERCISES

1. **Collaborative exercise:** You and your team will decide on the language to use in a report on the testing of a new battery-powered smoke detector. The situation is as follows:

   Testing revealed that the battery-powered smoke detectors did not always sound when the battery was low. Specifically, 75 percent of the smoke detectors emitted a sound to indicate a low battery, and 20 percent of the detectors emitted a sound so weak that a homeowner could not hear it beyond 20 feet. Your supervisor, Donna Dimaggio, and her manager want to start production of the smoke detectors within two weeks. They are waiting for your test results. Because earlier reports from other employees did not indicate problems with the smoke detectors, your supervisor is assuming that your test results will be insignificant. She and her manager will not be pleased if their division of the company can't begin manufacturing these smoke detectors because the division has not shown a profit in the last three quarters. If the division doesn't begin showing a profit, the company may downsize or eliminate the division.

   Your team should complete the following:
   - Determine what language choices you have for reporting the test results.
   - Use the principles for ethical communication to analyze these language choices. Write a memo to your instructor listing each choice and explaining the consequences of each choice. (For more information on writing memos, see Chapter 12.)
   - Write a memo to the supervisor reporting the test results and recommending a course of action.

2. **Collaborative exercise:** Working with a team, research an issue that involves ethics. For example, you might study an issue related to bioethics, such as stem cell research. If you need some ideas for information on bioethics, go to the National Bioethics Advisory Commission's Publications page (http://www.bioethics.gov) or the Center for Applied Ethics (http://www.ethics.ubc.ca). If you are interested in a topic related to ethics and engineering, go to the

online Ethics Center for Engineering and Science (http://onli-neethics.org). For other issues in ethics, visit with a reference librarian at your college or university library.

a. In a memo to your classmates,
   - describe the issue
   - explain the ethical dilemma
   - analyze the possible actions and consequences of those actions
   - suggest the ethical course of action and the consequences
b. In a memo to your instructor,
   - present your analysis and suggest the ethical action you would take. Your instructor may ask you to present your analysis in class.

3. **Web exercise:** Find a page from a Web site, advertisement, or product information that you believe misleads the reader.
   - Print the page and bring it to class.
   - Write a memo to your instructor explaining why the page misleads the reader.
   - Attach a copy of the page to your memo.

4. Interview a professional in your field who has faced ethical dilemmas. Ask how he or she dealt with these dilemmas. After the interview, write a memo to your classmates. In the memo,
   - introduce the professional and describe his or her job
   - describe how the professional deals with ethical dilemmas. Does he or she have guidelines to follow? Are they his or her guidelines or the employer's guidelines? Are the guidelines from a professional organization?
   - if possible, give examples of dilemmas that the professional has faced

5. **Collaborative exercise:** Form a team with two or three of your classmates. Find a code of conduct for a company and analyze how effectively (or ineffectively) the code states the goals of the company. (Many companies post their codes on their Web sites.)
   - Does the code describe appropriate behavior expected of employees? Does the code describe inappropriate behavior?
   - Does the code provide instruction and/or guidance to help employees decide if behavior is appropriate or inappropriate?

- Does the code specifically state the penalties for not following the code? If so, analyze the penalties. Are they fair? Do they protect all involved?
- How does the code address employees who report misconduct?

As a team, write a memo to your instructor summarizing what you have learned. Include a copy of the code with your memo.

6. During the summer, you begin work as an intern for Centurian, Inc., a software company. You got the job on the recommendation of your best friend's dad, Mr. Roger Thomas. You really appreciate the opportunity to work for this company and hope that they will ask you to return during Christmas break. As part of your responsibilities, you help maintain the company's Web site. As you begin your work on the site, you find that the site contains several copyrighted graphics and images. The image on the homepage is a copyrighted image that the company has apparently not asked for or received permission to use. You also find that the company has used passages from product literature without citing sources.
   - Is it unethical for you to ignore your findings? Why or why not?
   - If you report the findings, what are the possible consequences?
   - Be prepared to discuss your ideas with the class.

7. Read the information in Figure 2.9, pages 49-50. This information accompanies a drug used to treat arthritis. Consider whether the information is presented ethically. Be prepared to discuss your answers in class.
   - Does this information leave the readers with a clear impression of the side effects of the drug?
   - Does it directly answer the question about whether the drug damages the heart valves?
   - Has the manufacturer presented the information in an ethical manner? Explain your answer.

FIGURE 2.9

Document for Exercise 7

<u>Medication Guide</u>
for
<u>Non-Steroidal Anti-Inflammatory Drugs (NSAIDs)</u>
<u>(See the end of this Medication Guide for a list of prescription NSAID medicines.)</u>

What is the most important information I should know about medicines called Non-Steroidal Anti-Inflammatory Drugs (NSAIDs)?

NSAID medicines may increase the chance of a heart attack or stroke that can lead to death.
This chance increases:
- with longer use of NSAID medicines
  - in people who have heart disease

NSAID medicines should never be used right before or after a heart surgery called a "coronary artery bypass graft (CABG)."

NSAID medicines can cause ulcers and bleeding in the stomach and intestines at any time during treatment. Ulcers and bleeding:
- can happen without warning symptoms
- may cause death

The chance of a person getting an ulcer or bleeding increases with:
- taking medicines called "corticosteroids" and "anticoagulants"
- longer use
- smoking
- drinking alcohol
- older age
- having poor health

NSAID medicines should only be used:
- exactly as prescribed
- at the lowest dose possible for your treatment
- for the shortest time needed

What are Non-Steroidal Anti-Inflammatory Drugs (NSAIDs)?
NSAID medicines are used to treat pain and redness, swelling, and heat (inflammation) from medical conditions such as:
- different types of arthritis
- menstrual cramps and other types of short-term pain

Who should not take a Non-Steroidal Anti-Inflammatory Drug (NSAID)?
Do not take an NSAID medicine:
- if you had an asthma attack, hives, or other allergic reaction with aspirin or any other NSAID medicine
- for pain right before or after heart bypass surgery

Tell your healthcare provider:
- about all of your medical conditions.
- about all of the medicines you take. NSAIDs and some other medicines can interact with each other and cause serious side effects. **Keep a list of your medicines to show to your healthcare provider and pharmacist.**
- if you are pregnant. **NSAID medicines should not be used by pregnant women late in their**

29

FIGURE 2.9 CONTINUED

Document for Exercise 7

pregnancy.
- if you are breastfeeding. **Talk to your doctor.**

What are the possible side effects of Non-Steroidal Anti-Inflammatory Drugs (NSAIDs)?

Serious side effects include:	Other side effects include:
• heart attack • stroke • high blood pressure • heart failure from body swelling (fluid retention) • kidney problems including kidney failure • bleeding and ulcers in the stomach and intestine • low red blood cells (anemia) • life-threatening skin reactions • life-threatening allergic reactions • liver problems including liver failure • asthma attacks in people who have asthma	• stomach pain • constipation • diarrhea • gas • heartburn • nausea • vomiting • dizziness

Get emergency help right away if you have any of the following symptoms:
- shortness of breath or trouble breathing
- chest pain
- weakness in one part or side of your body
- slurred speech
- swelling of the face or throat

Stop your NSAID medicine and call your healthcare provider right away if you have any of the following symptoms:
- nausea
- more tired or weaker than usual
- itching
- your skin or eyes look yellow
- stomach pain
- flu-like symptoms
- vomit blood
- there is blood in your bowel movement or it is black and sticky like tar
- skin rash or blisters with fever
- unusual weight gain
- swelling of the arms and legs, hands and feet

These are not all the side effects with NSAID medicines. Talk to your healthcare provider or pharmacist for more information about NSAID medicines.

Other information about Non-Steroidal Anti-Inflammatory Drugs (NSAIDs)
- Aspirin is an NSAID medicine but it does not increase the chance of a heart attack. Aspirin can cause bleeding in the brain, stomach, and intestines. Aspirin can also cause ulcers in the stomach and intestines.

- Some of these NSAID medicines are sold in lower doses without a prescription (over – the –counter). Talk to your healthcare provider before using over –the –counter NSAIDs for more than 10 days.

NSAID medicines that need a prescription

Generic Name	Tradename
Celecoxib	Celebrex
Diclofenac	Cataflam, Voltaren, Arthrotec (combined with misoprostol)
Diflunisal	Dolobid

30

iStockphoto 2008.

CHAPTER THREE

3

Collaborating in the Workplace

As a professional in the workplace, you will work as part of a team. The team might collaborate to design a new product, to propose a project, to solve a problem, or to write a procedure. The type and level of the collaboration will depend on the project, its purpose, and the team's work style. When professionals collaborate, they can better prepare solutions, products, designs, and documents; they can also improve how an organization functions. However, for many professionals, collaboration is frustrating and time-consuming. As a student, you may have experienced this frustration when working with a team. This chapter will help you to collaborate effectively and efficiently and to understand collaborative writing in the workplace.

COLLABORATIVE WRITING IN THE WORKPLACE

In the workplace, professionals may collaborate in two settings:
- when writing for others
- when working as part of a formal team

Collaborating when Writing for Others

When you collaborate by writing for others, you write all or most of a document while other professionals supply the information you need to write, design, and, perhaps, publish. A good example of this type of collaboration is Paul's work on a proposal for a new telecommunications system for the Osteopathic Medical Center.

Paul collaborates with technical specialists and salespeople to gather the information for the proposal and to plan the document. He may even talk with the medical center's administrators or work with graphic designers for the layout and graphics. After Paul has gathered the information and discussed his plans with his associates, he will draft, revise, and publish the document himself. The technical specialists and the salespeople will not write any of the proposal. Once Paul has completed the proposal, the salespeople will present it to the Osteopathic Medical Center; Paul's name won't appear on the document.

Workplace teams may use their expertise, the stages of the writing process, or the sections of the document to set up team members' responsibilities.

You can also write collaboratively by preparing a document for someone else's signature. You might collaborate with that person to determine what he or she wants in the document; then, as you write, you and that person may collaborate to be sure the document is meeting his or her expectations. In these situations, you have the ultimate responsibility for gathering and analyzing the information needed for the document. You might collaborate with other professionals to obtain information, but you are responsible for using your expertise to write a document that will carry someone else's signature. A good example of this type of collaboration occurs in the scenario below.

As a new engineer for an architectural engineering company, Rhonda is responsible for preparing construction specifications. She recently wrote specifications for protecting the underground pipes of a water-pumping station that the company is designing. She researched the protection required for the pipes and wrote the specifications, and the project manager ultimately signed the specifications.

Collaborating as a Team Member

When you collaborate as a team member, you work with one or more people to produce a document. Workplace teams may use any of the following criteria to set up team members' responsibilities:

- team members' expertise
- stages of the writing process
- sections of the document

Collaborating Based on Team Members' Expertise

Some teams will divide the work based on team members' expertise or job functions. For example, if you work in marketing, you would be responsible for information related to marketing. Consider the team in Figure 3.1, page 54: the biologist is the subject-matter expert and is responsible for gathering the information needed to draft the document while the other experts are responsible for their job functions—marketing, legal, and writing. For this team, the biologist gathers the information, the legal specialist answers the legal questions, and the marketing specialist provides the marketing information. The technical communication specialist writes, revises, and produces the document. In this setup, the team divides the work solely on each team member's expertise or job function.

When collaborating based on job function, team members may not share the writing responsibility equally. For example, Neil collaborated with coworkers in engineering and management to prepare a proposal. They analyzed the writing situation together, brainstormed about what to include in the proposal, and outlined the proposal; yet Neil wrote all sections of the proposal, except the plan of action and the budget. The team members did not share the writing responsibility equally, but each one contributed. In some team writing projects, you may have more or less responsibility because of your job function or technical expertise.

Collaborating Based on the Stages of the Writing Process

Instead of dividing the work based on members' job functions, teams may collaborate based on the stages of the writing process. As Figure 3.2, page 54, shows, team members work together to plan the document: together they analyze the writing situation and organize the information. After the team has planned the document, some team members will draft the document. Other team members will prepare the graphics. Once those team members responsible for drafting the document and preparing the graphics have completed their work, some (or, perhaps, all) of the team members will revise the document. Finally, the entire team will proofread the document. Teams often collaborate in this fashion to work more efficiently. When teams draft as a

FIGURE 3.1

Collaborating Based on Team Members' Expertise

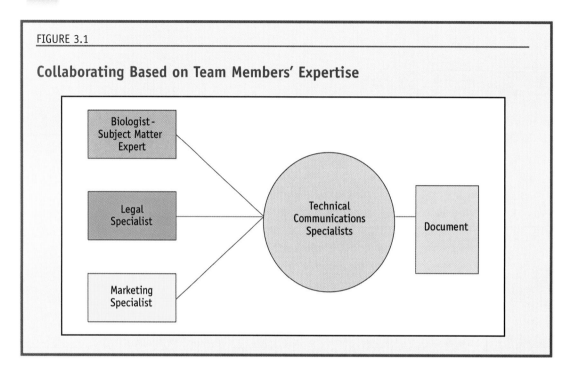

FIGURE 3.2

Collaborating Based on the Stages of the Writing Process

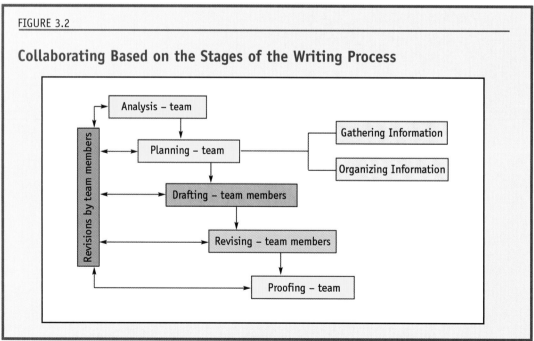

group, the work is inefficient and time-consuming. By dividing the work based on the stages of the writing process, teams can work more efficiently.

Collaborating Based on the Sections of the Document

Teams may collaborate based on the sections of the document. Each team member is responsible for preparing one or more sections of the document. For example, if a team of four is writing a manual composed of four chapters, each team member might write one chapter. This method of collaboration is common with large projects, such as proposals. However, this method can create uneven style, inconsistencies, or unnecessary repetition. To avoid these problems, teams should plan and edit the document carefully and use style sheets.

MANAGING THE PROJECT AND CONDUCTING EFFECTIVE TEAM MEETINGS

At the beginning of a team project, you will need to decide how you will manage the project and conduct the meetings. You will first want to determine how your team will operate by holding an initial team meeting. At this meeting, team members can define the team's task, take care of basic housekeeping, and set the team's agenda. These duties appear in the tips on page 56. By taking care of these tasks and the agenda at the first meeting, you foster effective and efficient collaboration. After the first meeting, you will want to make sure that your team conducts efficient face-to-face meetings.

Conducting Effective and Efficient Face-to-Face Meetings

When possible, your team should meet face-to-face. As Michael Schrage, author of *No More Teams! Mastering the Dynamics of Creative Collaboration*, explains, his favorite collaboration tool is the face-to-face meeting. For example, a Silicon Valley engineering firm decided it would only hold face-to-face meetings when problems arose; otherwise, employees would handle all business over the Internet. After "60 days, they went back to holding weekly meetings" because they were only getting together to argue (Quain quoting Schrage). The more we use technology to collaborate, the more important face-to-face meetings become.

TIPS FOR THE FIRST TEAM MEETING

- **Define your team's task.** What is your team assigned to do? What document, or deliverable, will your team complete? All team members have to understand and agree on the team's task. The team also needs to agree on the readers, the purpose, and the scope of the document. For example, if you are writing the documentation for a vehicle, your team must agree that the documentation will be completed by a set date and that the final product must fit into a vehicle's glove box. Be sure to allow ample time to define your team's task as this step lays the foundation for the project.
- **Trade phone numbers, e-mail addresses, and, if appropriate, fax numbers.** All team members need to know how to contact other team members. This contact information is especially important if team members work in different geographic locations—for example, if you have team members in different buildings, cities, states, or even countries.
- **Select a team leader, sometimes called a project manager.** The team leader serves as a liaison (link) between the team and the clients and/or management. (In school, the team leader represents the team when contacting the instructor.) The team leader keeps the team on task, sets up and conducts team meetings, coordinates the work of the team, and often coordinates communication with management and clients.
- **Decide on each team member's task.** Each team member will have a primary responsibility for a task. Your team should first determine how tasks will be divided: by team member's expertise, by the stages of the writing process, or by the sections of the document. As you determine how you will divide the tasks, consider how the tasks best fit each team members' technical expertise and abilities.
- **Set up procedures for how you will complete the project.** Before you begin working on the project, your team needs to determine the following:
    - When and where will you meet? Will you meet face-to-face? Will you meet through videoconferencing? Will you meet using an electronic tool?
    - How often will team members check in with the team to monitor progress? All team members should clearly understand the team's expectations about how often they are to check in.
    - How will your team communicate? Will you communicate by e-mail, Listserv®, Web site? All team members should clearly understand how the team expects them to communicate.

Team members need to know how each member's work is progressing so the team and the team leader can track the project. Team meetings and communication can also help the team identify and resolve any problems.

- **Decide what word-processing and/or graphics software your team will use to prepare the document.** All team members should have easy access to the software the team has decided to use.
- **Create a style sheet.** Style sheets save you time from the outset because all team members use the same format when writing their individual sections, so revisions and proofreading are simplified. As your team creates its style sheet, consider the design and style questions listed in Figure 3.3, page 58. Consider using the styles function of your software.
- **Set a schedule.** A project schedule should list specific and reasonable dates for completing individual tasks and allow time for meetings or other communication to discuss individual work—especially drafts. The schedule also should encourage team members to communicate frequently. Your team might establish a due date for the project and work backward to set intermediate deadlines. For example, if the project is due on April 12, April 8 may be the deadline for a completed draft, April 1 for a revised draft, and March 20 for a first draft.

  A schedule will clarify
    - each team member's responsibility
    - how each member's task relates to the work of the whole team
    - when each member should complete his or her work

  If possible, the schedule should allow for the unexpected, such as team members being transferred to other projects or not being available because of illness or personal emergencies. At any stage of the writing process, the team may face unexpected problems. For example, the information team members have gathered may turn out to be insufficient, or a draft may fail to meet the team's or management's expectations. Unexpected problems will frustrate even the most organized team; however, a schedule with room for flexibility will alleviate some of the frustration.
- **Decide how you will evaluate each team member's contribution.** In most workplaces, team members will not evaluate each other's contributions. Instead, the team will be evaluated on the result of their work: the deliverable. However, in some work environments and in some classrooms, you may be asked to evaluate each team member's contribution. In those cases, your team should decide on the criteria that will be used to evaluate members' contributions. Sometimes, you will be given a form to use; your team will not create the form or establish the criteria. You also might consider keeping a log of your work and your contributions. This log will help you document your work on the project. Figure 3.4, page 59, shows a sample log. You can find a blank project log online.

Download a sample team evaluation form and a blank project log template online at www.kendallhunt.com/technicalcommunication.

FIGURE 3.3

Questions to Consider When Creating a Style Sheet

Design Questions
- What page layout and page size will be most effective for the readers and the purpose?
- What type and type size will we use for the text and the headings?
- Where will the headings appear on the page? Will they be indented? If so, how far will they be indented?
- What page margins will we use?
- What bullet style will we use? Will we indent the list? If so, how far?
- What file format will each team member use to submit drafts? Will we use a word-processing or desktop-publishing program?
- Will we use color? If so, where? What will our budget allow?

Style Questions
- Will we use abbreviations, technical terminology, or language?
- Will we use words or numerals for chapter numbers, section numbers, and other enumerations? Will we use Arabic or Roman numerals?
- What terminology will we use when referring to the readers, to the organization, or to the team? Will we use second-person pronouns when referring to the readers?
- What conventions, if any, will we use when writing? (For example, if your team is writing software documentation, how will you refer to specific function keys or to the arrow keys on the keyboard? Will you use icons?)
- How will we refer to figures and tables?

In the workplace, many people consider meetings their biggest waste of time (Quain quoting Michael Schrage). You may have been on a team where meetings were, indeed, a waste of time. Perhaps because some team members did not arrive on time, other team members were not prepared, or a few team members talked on their cell phones or searched the Web during meetings. However, meetings can be valuable tools for effective collaboration if teams establish and stick to some simple guidelines. Review the tips for conducting effective face-to-face meetings on page 60.

Collaborating Effectively

When you work on a team, you may be concerned about collaborating, about the project, or about some or all of the team members. You may prefer to work alone, or you may enjoy collaborating. You may have had excellent experience collaborating, or you may have found that collaborating was frustrating, even stressful. Regardless of your experiences with teams, you will, in

FIGURE 3.4

Sample Log of a Team Member's Work

Project Log

Your Name: _____

Project Title: _____

Log of Your Independent Work

Activity	Date	Number of Hours
Conducted research for results section	March 10	3
Wrote results section	March 12	2.5
Reviewed draft of document	March 15	1.5

Log of Your Work with the Team

Activity	Date	Number of Hours
Met with Debra to discuss the research for the results	March 11	1.5
Met with team to discuss the results	March 11	1
Met with team to go over the draft of the document	March 16	2

Comments about your contributions:

TIPS FOR CONDUCTING EFFECTIVE FACE-TO-FACE MEETINGS

- **Be on time.** When you arrive on time, you show respect for the other team members, and the meeting can begin on time. When team members arrive late, the team leader must take time to summarize what the team has accomplished during the late members' absence.
- **Tell your team leader if you will miss the meeting.** If you know you will miss a meeting, let your team leader know as soon as possible.
- **Be prepared.** If you have a task to complete before the meeting, complete it. You show disrespect for your teammates when you are not prepared.
- **Stick to an agenda.** The person in charge of the meeting, usually the team leader, should do the following:
    - Establish an agenda before the meeting.
    - Send the agenda to team members before the meeting.
    - Keep the team focused on that agenda during the meeting. If the team needs to discuss an issue not on the agenda, set up another meeting or wait until you have completed all of the items on the agenda. Then, if you have time, discuss the new item.
- **Take minutes.** One team member should take minutes during the meeting and send copies of the minutes to each team member after the meeting. *Minutes* are a record of decisions made at the meeting. Minutes give you a written record of what the team decided; they are especially valuable on long projects and for those who miss the meeting. (See Chapter 17 for information on meeting minutes.)
- **Make sure every team member understands his or her task at the end of the meeting.** At the end of the meeting, the team leader should do the following:
    - Summarize the decisions made during the meeting.
    - Make sure that each team member knows his or her assigned task.

all likelihood, have to collaborate in your workplace. You can improve your collaboration experiences by following these guidelines:

- Encourage team members to share their ideas.
- Listen intently and respectfully to all team members.
- Share information willingly.
- Consider cultural differences.
- Disagree respectfully and politely.
- Critique team members' work respectfully.

TAKING IT INTO THE WORKPLACE
Collaboration in Your Profession

Nearly every profession involves collaborative work. To learn about collaboration in your field, talk with a professional who works in your major. This professional can help you gather information on the kinds of collaboration that you might encounter in the workplace. For example, if you are majoring in biology, interview a biologist about collaborative projects in his or her job. Before the interview, call or e-mail the professional for an appointment; offer to conduct the interview in person, by phone, or by e-mail.

Assignment

Gather some background information on the professional's job. Create a list of questions regarding collaborative work to ask during the interview. The following list can help get you started. Come up with additional questions of your own. (For information on interviews, see Chapter 5.)

- How often do you collaborate with coworkers?
- What are the types of projects on which you often collaborate?
- What are some advantages and disadvantages you've encountered in collaborative work?
- Do you think the final product produced through collaboration is exceptional work? Why or why not?
- What do you think is the most important consideration when working as a team?

After the interview, write a memo to your instructor summarizing the information you gathered about collaboration in your field of study. (For information on memos, see Chapter 12.)

Encourage Team Members to Share Their Ideas

You and the other team members bring unique creativity and expertise to the project. Together, you and the other team members represent a wealth of knowledge and ideas. To best use this knowledge, all members must share their knowledge and ideas. Even when ideas clash, team members can learn from one another. From this conflict, teams create more innovative products, solutions, and ideas. All team members should feel free to express their ideas—even when those ideas differ from the prevailing views expressed by the majority.

Some team members may be shy or quiet and may not feel comfortable expressing their ideas during team meetings. If you see that some team

Evaluate a collaborative document in the Interactive Student Analysis online at www.kendallhunt.com/ technicalcommunication.

members aren't participating in team discussions, ask shy or quiet members to participate and help the team include them in the discussions. You might say something like, "Rob, what do you think of that approach?" or "What are your ideas about the project, Susie?" You might direct the discussion toward the shy or quiet members by saying, "Let's hear what John has to say about this topic," or "We've heard some good ideas from Linda and Patrick. Let's hear what Juan has to say."

Listen Intently and Respectfully to All Team Members

Have you ever tried to express your ideas about something in which you were truly interested while some of your listeners talked to others as you were speaking? Have you ever spoken to a group of people who didn't give you their undivided attention? Have you ever talked to listeners who appeared to hear what you were saying, but when you asked for comments, you discovered that they hadn't really heard you? Their comments revealed that their minds had been elsewhere.

Such situations are always frustrating; in a collaborative setting, they can discourage members from sharing their ideas. Each team member needs to know that the team will listen to and consider his or her ideas. Team members shouldn't expect the team to accept any or every idea that they present, but they should expect the team to listen to and consider each idea.

To encourage other team members to share their ideas, listen intently and respectfully to everyone's ideas. Use both nonverbal and verbal signals to let speakers know that you are paying attention. Such signals indicate that you are an active listener who is paying attention.

Share Information Willingly

You can encourage others to share their ideas if you are willing to share information. If you discover information that may help another team member or that may affect what another team member is doing, share that information in a timely manner. When you share such information, your team members will be more likely to share information that may affect your work, and you will develop good working relationships.

Consider Cultural Differences

In the classroom and in the workplace, you will probably collaborate with people from other cultures. A person's culture can affect how he or she collaborates. Some cultures, for example, follow strict conventions related to gender, age, and seniority. The key is to remain open to collaboration with people from other cultures and to learn about those cultures without

 TIPS FOR BEING AN ACTIVE LISTENER

- **Maintain eye contact with the speaker.** You let speakers know that you are actively listening when you maintain eye contact with them. Speakers who receive little eye contact feel that their listeners aren't paying attention.
- **Avoid body language that may distract the speaker or other team members.** To signal that you are listening intently, look at the speaker and avoid distracting gestures.
- **Let the speaker finish his or her statements before you ask questions.** You encourage all team members to participate by allowing speakers to finish their statements before you ask questions or make comments. If you continually interrupt, speakers may become distracted or assume that you aren't listening; or they may think that you believe what you have to say is more important than what they have to say. Such interruptions may also discourage team members from sharing their ideas, especially team members who may be shy or timid.
- **Use phrases that indicate that you agree with the speaker.** If you agree with the speaker, use expressions such as "I agree," "Good idea," or "I like that" to let the speaker know that you agree. These phrases also encourage those who may be shy or uncomfortable sharing their ideas.
- **Ask questions when you want something clarified or when you want more information.** By asking questions, you get more information to determine whether you agree or disagree with the speaker. Your questions also encourage other team members to share their ideas and ask questions.
- **Ask relevant questions.** Effective team members ask questions to understand what a speaker has said and to connect ideas. These questions relate to the speaker's topic.
- **Occasionally paraphrase or summarize what the speaker has said.** By paraphrasing, you make sure that you understand what the speaker has said and give the speaker the opportunity to correct any misunderstanding. If you disagree, try to paraphrase what the speaker has said. You may discover that you actually agree.

> *Team members from other cultures may*
> - *have difficulty speaking up*
> - *be unwilling to disagree*
> - *hesitate to ask for clarification.*

jumping to incorrect conclusions. You might try to learn something about your team members' cultures before your first meeting.

Team members from other cultures may
- have difficulty speaking up or asserting themselves. Their culture may value silence more than speech.
- be unwilling to disagree, to question others, or to be questioned. Some people of other cultures will not respond with a definite "no."
- have difficulty asking the team to clarify information or expressing that they don't understand.

Disagree Respectfully and Politely

Robert I. Sutton, Stanford professor and author on innovation and business practices, writes, "Treating people with respect rather than contempt makes good business sense." One of the benefits of collaboration is the diversity of ideas and the conflicts that naturally occur. Conflicts can be a healthy part of working with others. Conflict can help a team discover the best way to handle a project or the best way to organize a report. To encourage healthy conflict, review the tips for disagreeing respectfully on the next page.

Critique Team Members' Work Respectfully

When you critique your team members' work, think about how you like to have your work critiqued. You will gain the respect of your team members when you critique their work respectfully. You want to comment on their work without offending or attacking them. When you offend a team member, you may prevent progress on the project. Your goal is to create an effective document that meets the expectations of your team and of management, not to evaluate the writer. As you critique the work of your team members, follow the tips on page 66 for being respectful.

USING ELECTRONIC TOOLS TO COLLABORATE MORE EFFICIENTLY

Electronic tools help teams communicate efficiently. With the help of these tools, team members can work at different times and locations and still col-

TIPS FOR DISAGREEING RESPECTFULLY AND POLITELY

- **Focus on the issue, not on the individual.** When disagreeing with or criticizing the work of a team member, avoid personal remarks and insults. Remember that your team members have feelings. You can show respect for those feelings by criticizing the ideas, not the person. Consider the difference in the two approaches below:
    - **Focuses on the individual:** *If you paid attention to details, you would notice that I sent the e-mail at 3:32 p.m. today.*
    - **Focuses on the issue:** *I sent the e-mail today at around 3:30.*
- **Don't take positions you cannot support.** Be careful not to overstate or exaggerate your ideas or your position.
    - **Overstated:** *This process will solve all of our problems.*
    - **Not overstated:** *With this process, we can begin to solve some of our problems.*
- **Look for points with which you can agree when disagreements arise.** You may only disagree with part of a team member's idea or comment; if so, look for and focus on points with which you can agree. In this way, you can avoid conflict or at least minimize it.
- **Be open to criticism of your ideas.** Just as you will at times criticize and disagree with your team members, they, too, will at times criticize and disagree with your ideas. Be prepared to listen, and be open to this criticism. Think of your writing and your ideas as belonging to the team, not to you. By distancing yourself emotionally from your work, you won't automatically take it personally when team members criticize your writing or ideas. This objectivity frees you to discuss the best way to write the team's document or present the best ideas for the team's project, rather than trying to protect and defend your document and your ideas. When the team begins to evaluate your ideas, participate in the discussion and be willing to look at your draft or your ideas objectively—even offer suggestions for improving your own work. However, be prepared to try to persuade team members to use your ideas if you feel they are overlooking important points. It's natural to resist change; perhaps the team can use elements of your ideas, if not the entire proposal.
- **Support the team decision.** Once you have had your say and the team has made a decision, support that decision fully. If you feel that the decision is unethical, then, talk with a mentor.

TIPS FOR RESPECTFULLY CRITIQUING THE WORK OF YOUR TEAM MEMBERS

- **Include positive comments.** When possible, begin with a positive comment. When you critique the work of a team member, comment on the strengths of his or her work. For example, you might explain that you are suggesting ways to improve a basically good draft. Focus on the positive qualities of a draft or an idea, and tell the team member about those qualities when you suggest changes.
- **Focus on the big picture before commenting on the local issues.** For example, critique the content, the organization, or the layout before critiquing the specifics of style or grammar.
- **Recommend specific ways to improve the work.** When critiquing team members' work, give them specific information that they can use to revise their work. Include a brief reason why you recommend the change.
- **Comment on the work, not on the team member/writer.** Avoid subjective comments about the writer or the writer's work. Instead, give objective comments. For example, you might criticize a specific part of an idea instead of making a broad, subjective comment that you can't support. Consider the difference in tone of the two comments below:
  - **Subjective and offensive:** *I don't like the writing in your draft. It is not well done and too formal.*
  - **Objective and polite:** *The draft might be stronger and friendlier if we included more second-person pronouns.*

  In the subjective and offensive comment, the team member uses subjective, unnecessarily negative language. In the objective, polite comment, the team member uses more diplomatic criticism and offers specific information for improving the document.
- **Be honest, yet polite.** Your goal as an individual and as a team member is to produce an effective deliverable that meets the expectations of your readers and your managers. Even when you know a recommendation or critique may be upsetting to a team member, you should be honest. However, always present your comment politely.

laborate; they can continue work on a project when a face-to-face meeting isn't convenient or possible. For instance, you might be working on your draft at 7:30 a.m. in New York while your team members are still sleeping in California. You e-mail your draft to your team members by 8:00 a.m.,

Eastern time, so they can work on it when they arrive at the office at 8:00 a.m., Pacific time. Electronic tools also allow team members to track and store revisions and changes to documents as a project progresses.

A team can use any of the following electronic tools to make collaborating easier and more efficient:
• review features of word-processing software
• e-mail
• groupware
• teleconferences

Using the Review Features of Word-Processing Software

Word-processing software has three features that teams can use to collaborate effectively:
• **Comment feature**
• **Track changes feature**
• **Highlighting feature**

Figure 3.5, page 68, gives you ideas for using these features to review a document electronically.

Using E-Mail to Share Files

You can collaborate more efficiently by electronically sharing files, especially document drafts. You can use e-mail to share files with team members or other collaborators. You simply use the attachment feature of your e-mail software. You can attach many file types, such as graphics, digital photographs, word processing, or spreadsheets. If you are sending a file to someone who does not have the software used to create the file, you can send the file as a PDF. *PDF* stands for "portable document format," a file format created in 1993 by Adobe Systems for exchanging documents. If you or your readers do not have the software *Adobe Reader* to create or open a PDF file, you can download it free from Adobe at www.adobe.com.

You might select the PDF format in the following circumstances:
• You want to send a file to a reader, and you don't want the reader to alter the file. You can "lock" a PDF file so the reader cannot alter it.

FIGURE 3.5

Using Document Review Features

Using the Comment Feature

The comment feature allows you to electronically bracket questioned text and include comments or suggestions for change. Comment bars can be flagged with several colors, so each team member can be easily identified. The comment bar includes the team members' names, as well as the dates and times of their comments. After the document has been reviewed and changes made, you can turn off the comment feature to hide the bars.

"Every good boy does well" is a phrase learned by every new piano student.

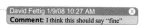

David Fettig 1/9/08 10:27 AM
Comment: I think this should say "fine"

Using Track Changes

The track changes feature allows you to electronically track who made what changes to the document and when changes were made. Track changes bars can be flagged with several colors, so team members can be easily identified. After the document has been reviewed and changes made, the track changes feature allows you to accept all changes, examine changes one by one, or hide all change bars.

"Every good boy does fine" is a phrase learned by every new piano student.

David Fettig 1/9/08 9:40 AM
Deleted: well

Using Highlighting

Highlighting allows you to electronically call out text in color. Since many colors are available, each team member can use a different color to highlight his or her comments. After the copy has been reviewed, you can easily turn off the highlighting.

"Every good boy does well" is a phrase learned by every new piano student.

- You don't know what software or version of the software the reader will use to open a file.
- The reader does not have the software used to create the file.

Using Groupware to Collaborate

Groupware is a class of software that helps users attached to a local-area network organize their collaborative activities. Groupware supports several types of collaborative tools:

- **File sharing and distribution.** Team members can post files to a shared space where all team members can access the files.

- **Password-protection.** Teams can protect files by requiring a password to open them, so only team members can access the files.
- **Synchronous conferences.** Team members can conduct conferences in real time by using instant messaging.
- **Asynchronous conferences.** Team members can conduct conferences that do not occur in real time. Instead, team members can post their comments any time; and the team members can read the comments when it is convenient for them.
- **Project management and knowledge management.** Teams can schedule and track the stages in a project with some groupware project management tools. Members can also collect, organize, and manage various types of information with groupware knowledge management tools.
- **Electronic calendars.** Team members can schedule events and automatically notify and remind team members.
- **Videoconferencing.** Teams can meet face-to-face without being in the same room. In a videoconference, video cameras can be attached to computers or placed in a room. Team members might go to a videoconferencing facility (Figure 3.6), or they might sit at their workstations (Figure 3.7).

FIGURE 3.6

Videoconferencing via Facility

IndexOpen 2008.

FIGURE 3.7

Videoconferencing via Computer Workstation

iStockphoto 2008.

- **Digital whiteboards.** Team members can share data in real time using a digital whiteboard. With the whiteboard tool, members in different locations can "mark" on the same document or screen as if they were in the same room. The whiteboard becomes each person's computer screen. For example, a team member in Seattle can write on the board, and a team member in Kansas City can see what is being written. This tool also allows teams to show the same file to team members in different locations. For example, a team member might show a *PowerPoint*® slideshow to office members at another locale during a teleconference.

ETHICS NOTE

Ethical Collaboration

On collaborative projects, make sure that you act and interact ethically. Once you move into the workplace, you may find that the competitive environment tempts you to collaborate in an unethical manner. To help you collaborate effectively, follow these guidelines:

- **Do your work and do it well.** If you do not do your work, or if you do poor or "half-hearted" work, you hurt your team and its deliverable. The result is a less-than-excellent team product, an incomplete deliverable, or work that has to be redone by your team members.
- **Be honest about what you can do.** All of us have times when unexpected problems prevent us from attending a meeting or completing the work we were assigned. If you experience such a problem during a collaborative project, contact your team and your team leader as soon as possible. For example, call your team members and your team leader if you cannot get your document to them on time. Be honest about why you cannot get the work done on time, and suggest a solution. Even if this honesty is uncomfortable, you are acting ethically by explaining the problem and suggesting a solution.
- **Treat your team members with respect; don't intimidate or belittle them.** When you intimidate or belittle your team members, you are acting unethically. Team members who are intimidated may not feel free to share their ideas or to comment on your work.
- **Claim credit only for the work you have done.** In the workplace, some people will claim that a team's project is their own. Make sure that you give proper credit to your team members' contributions.

Using Teleconferences to Collaborate

A *teleconference* is a telephone call that allows several people to talk in real time; they just cannot see each other. Teleconferences—sometimes called conference calls—allow teams to conduct real-time meetings. Teleconferences are valuable tools for economically and efficiently conducting a meeting when team members are in different locations.

CASE STUDY ANALYSIS
Collaborating on a Public Relations Dilemma

Background[1]

Big Lake Steam Electric Station, a power plant near a small Texas town, uses steam to produce electricity. The main steam pipe at the plant carries steam to a turbine, which drives a generator that produces electricity. The manager at Big Lake observed that areas of this main steam piping system were sagging.

The main steam system uses constant support hangers (a spring-type hanger) to carry the weight of its pipes and reduce the effect of the piping system on the plant equipment. Since the late 1960s when the plant was built, engineers have learned that these spring-type hangers alone cannot support the weight of the pipes over time. To compensate, engineers recommend using a combination of spring-type hangers and rigid supports (hangers without springs; see Figure 3.8). Without rigid supports, the sagging worsens, and the weight of the pipes transfers to the

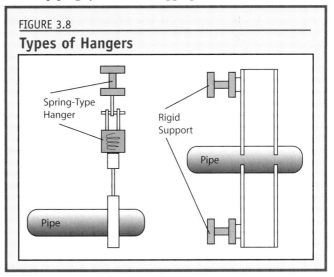

FIGURE 3.8

Types of Hangers

Spring-Type Hanger

Rigid Support

Pipe

Pipe

[1] This case study is based on an actual situation that occurred at a power plant owned by a utility company in Dallas, Texas. This case, written by Brenda R. Sims, is adapted from Richard Louth and Ann Martin Scott, eds., *Collaborative Technical Writing: Theory and Practice* (St. Paul, Minn.: ATTW, 1989). Courtesy of the Association of Teachers of Technical Writing.

plant equipment. This equipment is not designed to carry the weight and will be permanently damaged if the sagging continues. Knowing the damage that continued sagging could cause, the Big Lake manager asked the engineers in the corporate office to study the problem and recommend solutions. The corporate office's survey of the Big Lake plant revealed that the main steam piping system was sagging over 6.5 inches. Their survey also indicated that the sagging would get worse and would eventually damage the plant equipment.

Corporate's Solution for Repairing the Plant

In early July, the corporate office suggested two solutions: the first solution would only stabilize the pipes while the second would stabilize the pipes and correct the sagging.

- **Solution 1**: Replace two of the spring-type hangers with rigid-supports in the area where the main steam system was sagging. The cost: $132,100.
- **Solution 2**: Shorten the piping system by cutting out and removing a short section of the piping and then adding rigid supports, thus pulling up the lowest portion and eliminating the sag. The cost: $403,600.

Solution 1: Replacing Two Spring-Type Hangers

This solution would stabilize the pipes, but would not eliminate the existing sagging. To implement this solution, engineers would run a computer stress analysis on the pipes to verify that the existing sagging did not result in stresses that exceeded allowable limits. This analysis would require about 40 hours, but the plant would not have to shut down during the analysis. Once the engineers verified that the stress levels were acceptable, subcontractors could replace the spring-type hangers during the regularly scheduled fall maintenance shutdown. The plant personnel would not lose any work time with this solution.

Although this solution was relatively inexpensive and could occur without an unscheduled plant shutdown, corporate did not recommend this solution for the following reasons:

- It would not correct the existing sag, resulting in a permanent low point in the piping system. Low points can trap water and cause water hammer.
- The plant would have to install a manual drain at the low point. Periodically, plant personnel would have to drain the low point manually.
- Corporate feared that during busy periods the plant personnel would forget to drain the low point creating additional problems.

Solution 2: Shortening the Main Piping System and Adding Rigid Supports

This solution would stabilize the pipes and eliminate the sagging and the possibility of a low point. Therefore, even though the cost was higher, corporate recommended this solution. In the long term, the company would save money and time by shortening the pipes and by adding the rigid supports. This solution would take two weeks beyond the regularly scheduled fall maintenance shutdown. Therefore, some of the plant personnel would have to take vacation time or time off without pay.

Discussions about the Solutions

The Big Lake manager met with representatives from corporate on July 23 to discuss the feasibility of the solutions and to express his concerns over the cost in down time and money. The manager believed that shortening the piping system would cost the plant and the electric consumer too much money. Because the manager was concerned about the time involved with both solutions, corporate sent a memo to him after the meeting outlining the work needed to shorten and reweld a section of the piping system and to replace the spring-type hangers with rigid supports.

An Emergency Plant Shutdown and the Chosen Solution

On August 30, before the plant manager had decided how to repair the sagging pipes, the Big Lake plant experienced a boiler tube leak unrelated to the sagging pipes. The plant personnel had to shut down the plant to repair the leak. While waiting for the plant to cool down to begin repairing the leak, one of the spring-type hangers broke, causing the steam piping system to sag an additional two feet. The damaged hanger was temporarily repaired so the plant could resume operation and the plant manager could have more time to decide how to handle the sagging pipes.

Ten days after the unscheduled shutdown, the Big Lake manager rejected the second solution (shortening the piping system and adding rigid supports), primarily because of its high cost. He selected the first solution (replacing the spring-type hangers with rigid supports) because
• the repairs could occur during the regularly scheduled fall maintenance shutdown
• the plant would avoid additional downtime and the expense of the second solution.

Citizens' Concern about the Plant Shutdowns and the Sagging Pipes

The August 30 shutdown at the plant concerned the citizens of the surrounding town because the plant employs a large portion of the town's population. When the plant manager first discovered the pipe sagging problem, the citizens feared that

- the sagging pipes could close the plant permanently
- the steam carried by the sagging pipes could injure plant personnel if the sagging caused the pipes to rupture.

Because these fears were unwarranted, the Big Lake manager placed an announcement on the local radio and television stations to explain that Big Lake would remain open and that the plant personnel were safe. However, the shutdown on August 30 unnecessarily renewed the citizens' fears. The citizens again feared that the plant would possibly close permanently, leaving the plant personnel without jobs. They also feared that if the plant did not close, the personnel would be in danger from the sagging pipes and would be laid off without pay when the pipes were eventually repaired. Now, the Big Lake manager had to convince the citizens that the plant needed only minor repairs that could be handled without an unscheduled shutdown and that the plant personnel were safe.

Assignment

Your team will write a letter from the plant manager to the citizens of the small town near Big Lake. This letter will appear in the local newspaper. The purpose of the letter is to

- convince the citizens that the plant will remain open
- convince the citizens that the sagging of the pipes could not cause the pipes to rupture and would not endanger plant personnel
- explain the type of repairs the plant manager has planned for the plant

You will collaborate with your team to

- decide what to include in the letter
- discuss the ethical dilemma, if any, the plant manager faces
- draft a letter from your team
- comment on the individual letters that each team member will write

Deciding What to Include in the Letter and Considering the Ethics of the Situation

Your team should discuss the Big Lake situation and complete Steps 1 and 2 on the next page.

1. List the characteristics of the readers of the letter, the citizens of the small Texas town near Big Lake. The team can answer the following questions to determine the reader's characteristics:
   - Why are the citizens interested in the Big Lake situation?
   - What is their current attitude toward the plant?
   - How would closing the plant affect the town's citizens?
   - What will the citizens want to know about how the sagging pipes will be repaired?
   - What rumors have many of the citizens heard about the sagging pipes and how these pipes affect the plant and its personnel? Are these rumors accurate?
2. Decide what information your team will include in the letter.
3. Using the Chapter 2 online worksheet, Ethical Communication, determine if the information that you plan to include is ethical. If not, determine how your team will revise that information. Each team member should write an individual version of the letter and e-mail a copy of the letter to the other team member and to the instructor.

Download the Chapter 2 worksheet, Ethical Communication, at www.kendallhunt.com/ technicalcommunication.

Critiquing the Drafts

Your team will respond to each other's letters using the review features of your word-processing software.

1. Read each team member's letter.
2. Using the comment and track changes features, critique each team member's draft.
3. E-mail the marked draft to the writer.
4. Be prepared to discuss your comments with your teammates.

Writing the Final Draft

Your team will give one letter to your instructor. To prepare this final draft, your team should follow the guidelines for effective collaboration presented in this chapter. Your team has several options for preparing the draft:

- Write the draft using sections from the individual letters written by the team members.
- Revise sections of the individual letters to create the final draft.
- Write the final draft using only ideas from the individual letters.

Once your team has completed the draft, give your instructor a copy.

EXERCISES

Download the Worksheet for Effective Collaboration found online at www.kendallhunt.com/technicalcommunication.

1. **Web exercise**: Visit your college or university's Web site to find out what types of groupware are available for students and faculty. If your college or university does not have such software, use a search engine to locate some examples of free groupware that students can use when collaborating. Once you have located the software, write a memo to your classmates and include the following information:
   - the name of the groupware
   - tools in the groupware that will help student teams collaborate
   - where students can find the software

2. You have probably worked in a team in other courses, in a volunteer organization, or on a job. Think of some of your best and worst experiences when collaborating.
   - Be prepared to discuss these experiences with your class.
   - Based on these experiences, prepare a list of guidelines for an effective team experience; assume these guidelines will be used either by people in your class, in your volunteer organization, or at your job. As you prepare your list, remember to include what to do and what not to do.

3. **Collaborative exercise:** Work with three of your classmates to write a brief document for prospective students to your college or university. The document should explain why they should consider attending your college or university. Once you have written the document, use the review features in your word-processing software to comment on and review the draft.

4. You will work on a team during this course, or you will work on a team in a volunteer organization. Prepare an evaluation form that your team or the organization might use to evaluate the contributions of each team member. Once you have completed the form,
   - save it as a PDF
   - write an e-mail to your instructor
   - attach the PDF to the e-mail

5. **Web exercise**: Select a culture different from your own. Find Web sites with information about the culture and customs of that country. Look for customs that might affect how people from that coun-

try would collaborate. Be prepared to share your findings with your classmates.

REAL WORLD EXPERIENCE
Working with a Team

Assignment

The city council is trying to get more people to visit your city. The council has hired you and your classmates to write a brochure promoting the city. The council wants information about the city's history, attractions, entertainment, shopping, restaurants, and hotels. Once you have your team assignment, your team should meet. At the first meeting, your team should
- select a team leader
- trade phone numbers and e-mail addresses
- decide on each team member's task
- set a schedule for the research and future meetings
- create a style sheet
- set up a procedure for completing the project

Logging Your Work

Keep a record of your work on the brochure, using the sample log found online. Each time you work on the brochure, do the following:
- Note the date and the amount of time that you spend working.
- Note whether you worked alone or with another person.
- Describe the work that you expected to complete.
- Describe the work that you accomplished and the problems, if any, that you encountered.

Record not only the facts, but also your experiences collaborating.

Access a sample work log online at www.kendallhunt.com/technicalcommunication. Also reference the Worksheet for Effective Collaboration (same Web address as above).

Reporting Your Work to Your Instructor

After you and your team have completed the brochure,
- write a memo to your instructor describing how collaboration affected your work and your team's brochure
- use specific information from your log to support and illustrate the information that you include in the memo
- attach a copy of the brochure to your memo

C H A P T E R   F O U R

4

iStockphoto 2008.

Writing for Your Readers

A U.S. company that manufactures deodorant created a marketing campaign for its product in Japan. The campaign featured a smiling octopus dabbing deodorant under its eight arms. A typical U.S. consumer might think the octopus was cute putting deodorant under each of its arms; however, the Japanese didn't see the octopus as having eight arms. They saw the octopus as having eight legs—not somewhere you would typically apply deodorant. To the Japanese consumer, this marketing campaign looked odd.

Similarly, a company introduced a baby food product in Africa. The baby food jar included a picture of a baby—appropriate in many Western countries. In Africa, however, consumers expect the label on a food product to have a picture of exactly what is in the jar: "WYSIWYG" (what you see is what you get). To them, the product looked like food made from babies— a horrific idea. To clearly market their product to African consumers, the food company should have included a picture of the food on the jar, not a picture of a baby (Barthon).

How can you avoid similar problems in your technical communication? You need to understand the purpose and the readers to meet their needs and to achieve your communication purpose.

DETERMINE YOUR PURPOSE FOR WRITING

Before you begin writing, think about why you are writing and what you want the document to accomplish. For some documents, the purpose will be clear from the beginning; but for other documents, the purpose will be apparent only after you have asked questions about your readers and about the writing situation. Sometimes, the purpose that you identified at the beginning of the writing process will change as you gather information about readers' needs. In this situation, you may discover that the document has a long-term goal such as maintaining the goodwill of the reader or establishing a positive working relationship.

To determine your purpose for writing, try asking yourself these questions:
- What do I want this document to accomplish?
- Why am I writing this document?
- What do I want readers to know or to do after reading this document?

After answering these questions, you should be able to write the purpose of your document in one sentence. For example, if you are writing a memo to your managers asking them to buy you a laptop, your purpose statement might be: *The purpose of this memo is to persuade my managers to buy me a laptop.*

Just as you have a purpose for writing, your readers have a purpose for reading. Readers have one or more of these purposes for reading a technical document:
- to gather information
- to answer a specific question
- to make a decision
- to perform a task or specific action

Let's consider your memo to persuade your managers to buy you a laptop. Your managers will read your memo to decide whether or not to buy you a laptop. You want to make sure that you understand your managers' purpose for reading and what questions they may ask. In the following section, you will learn how to identify your readers so you can understand what questions they will ask.

IDENTIFY YOUR READERS

As you determine your purpose for writing, you can begin to identify your readers—one of your most important tasks. Your document will accomplish your purpose only when it meets the needs and expectations of your readers. To understand how to meet these needs and expectations, you must answer the following three questions:

- Are your readers internal or external?
- What do your readers know about the subject?
- What individual characteristics of your readers could affect how they will read and use your document?

Are Your Readers Internal or External?

Internal readers work for your organization; *external readers* work outside your organization. Your readers' internal or external location affects the information, formality, and language that you use. For example, if you are writing to internal readers, you may first want to determine how those readers relate to you in the organization's hierarchy horizontally and vertically. When you know these relationships, you have information about the organizational roles of readers who may have the same level of authority, but perform different duties.

Writing for Internal Readers

The organizational hierarchy may give you clues about the readers' technical knowledge and educational background; this hierarchy may tell you what the readers need to know about the subject matter.

Let's consider the organizational hierarchy at a power plant (see Figure 4.1, page 82). Below the plant manager, the organization has four areas: environmental staff, production superintendent, support superintendent, and administrative staff. The technical expertise of the people in the production and support groups varies, based on their roles, experience, and education. Individuals at the same horizontal level in production have the same level of expertise and similar educational backgrounds. However, the plant operators in this group have a different level of expertise and education. They are technicians who have high-school diplomas and extensive on-the-job experience. They don't have the educational background of the engineers and production supervisors, who have engineering degrees.

> *Know Your Readers*
>
> •
>
> *Are they internal or external?*
>
> •
>
> *What do they know about the subject?*
>
> •
>
> *What could affect how they will read and use your document?*

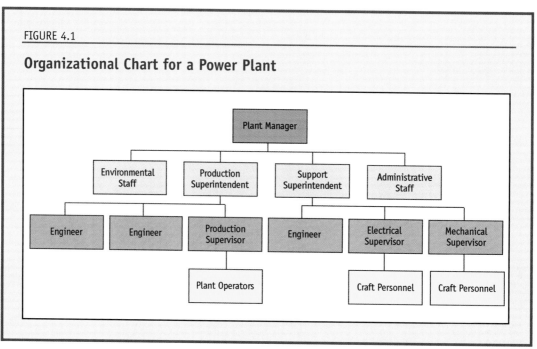

FIGURE 4.1

Organizational Chart for a Power Plant

Your readers' vertical level in the organizational hierarchy will tell you whether the readers have more authority in the organization than you. If you know where they are in the hierarchy, you can better determine the appropriate tone for a document. If you and your readers are at the same level, you probably will use an informal tone; but if your readers are several levels above you, you will use a more formal tone. Because these readers may know little about the specific project that you are writing about, you might also have to provide background information.

Writing for External Readers

External readers may be customers, clients, or other professionals. In large organizations, external readers may also work for your organization, but not in your specific department or division. When writing for most external readers, use a formal tone and format. For example, organizations generally use informal memos, both electronic and paper, when corresponding with internal readers; but when corresponding with external readers, they may use a more formal tone and format. For example, if writing an e-mail to an external reader, you might include an electronic letterhead or company logo; but you might not include this letterhead or company logo in an e-mail to an internal reader. If you are corresponding with an external reader whom you have worked with for an extended time, you may use a less formal, friendlier tone.

When using e-mail to correspond with external readers, remember to use a professional tone. If your working relationship with these readers is new or if e-mail will serve as a matter of record, use a formal tone. Some organizations will expect you to use a formal tone in any e-mail to external readers. In traditional paper documents, organizations generally use different formats for internal and external readers. For example, if you are writing a proposal for someone within the company, you might use a memo format with the pages stapled together. However, if you are sending a proposal to a customer outside your company, you might use a color layout, a table of contents, tabs between the sections, and an attractive, well-designed cover.

What Do Your Readers Know about the Subject?

Will your readers understand the technical terms that you use? Will they understand the concepts that you present? Readers of technical documents have varying levels of technical expertise. Some of them may have the same level of technical expertise as you; others may know little about your field. Your job as a writer is to determine your readers' levels of technical expertise and then to use terms and concepts that they will understand or to define the terms and concepts that they are likely to find unfamiliar. When you know your readers' level of technical expertise, you can choose words, concepts, and information that your readers will understand. Once you have determined what your readers know about your subject, you can give them only the information that they need.

You can look for the following general categories of readers:
- **Experts**–readers with a high level of technical knowledge
- **Technicians**–readers with practical technical knowledge
- **Managers**–readers who make decisions
- **General Readers**–readers with no technical expertise

Experts

Experts are readers with a high level of technical knowledge. Experts have a broad and deep knowledge of their field, based on many years of practical experience or study. Most experts have a postgraduate degree or an undergraduate degree in a technical field. An expert might be a surgeon reading an article on laparoscopy in a medical journal or an engineer working on designing a more efficient fuel system for a car. Experts share the following characteristics:

- They understand technical terms and information in their fields.
- They understand abbreviations and technical vocabulary commonly used by experts in their field.
- They expect few, if any, explanations of technical terms or information.
- They expect a direct presentation of the information.
- They want to know and understand the theory and research behind conclusions and recommendations.

Figure 4.2 presents a document for expert readers. The document uses technical terms and abbreviations familiar to medical experts. It directly presents the information and includes information on the research.

Technicians

Technicians are readers with practical technical knowledge gained from hands-on experience and training. These readers may apply or implement the expert's ideas. Technicians may perform procedures in labs; they may maintain mechanisms and systems. They may even construct what an expert has designed. For example, a technician might build the fuel system designed by the engineer or work in a lab to identify disease-causing organisms. Technicians share the following characteristics:

- They understand technical vocabulary and concepts in the field because of their practical experience.
- They expect practical, procedural information, not theory.
- They may need explanations of some technical vocabulary and concepts.

Figures 4.3 and 4.4, pages 86 and 87, illustrate documents that technicians will read and use. Figure 4.3 explains how to identify microfilaria; the technicians reading this document understand the technical vocabulary and graphics. Figure 4.4 explains how to prepare blood smears in a laboratory. Technicians reading these procedures have practical experience in taking blood samples and in analyzing those samples. These procedures include step-by-step instructions for carrying out the procedure, but they do not explain the theory behind the procedures.

FIGURE 4.2

Excerpt from a Document for Experts

The writer uses technical terms and does not define or explain them.

Zoonotic filarial infestations occur worldwide, and in most reported cases the involved species are members of the genus *Dirofilaria*. However, zoonotic *Onchocerca* infections are rare and to date only 13 cases (originating from Europe, Russia, the United States, Canada, and Japan) have been described. In all of these cases only 1 immature worm was found, and the causative species was identified as *O. gutturosa*, *O. cervicalis*, *O. reticulata*, or *O. dewittei japonica* on the basis of morphologic and in some cases serologic parameters (*1–4*). *O. cervicalis* and *O. reticulata* are found in the ligaments of the neck and extremities of horses, *O. gutturosa* is typically found in the nuchal ligaments of cattle, and *O. dewittei japonica* is found in the distal parts of the limbs and adipose tissue of footpads of wild boars.

The writer presents information directly.

The writer uses abbreviations.

We identified the causative agent of a zoonotic *Onchocerca* infection with multiple nodules in a patient with systemic lupus erythematosus (SLE) who had been receiving hemodialysis. The parasite was identified in paraffin-embedded tissue samples by PCR and DNA sequence analysis.

The Study

The patient was a 59-year-old woman with SLE who had developed multiple nodules on the neck and face over several years. Because of major renal insufficiency, she also had been receiving hemodialysis 3 times per week (3.5 hours) for >10 years. The first clinical differential diagnoses were cutaneous SLE, nephrogenous dermatopathy, calciphylaxis, and calcinosis. The clinical picture was obscured by secondary inflammations and ulcerations caused by self-inflicted trauma. Multiple sampling attempts by cutaneous core biopsies resulted in histologic diagnosis of unspecific, secondary inflammatory changes. Deep surgical excision of 1 subcutaneous nodule on the scalp indicated subcutaneous helminthosis (<u>Figure</u>). The patient was treated with ivermectin and subjected to 2 plastic surgeries for facial reconstruction, after which she recovered.

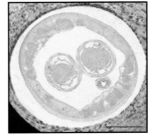

Figure. Transverse section of a female worm and surrounding tissue isolated from the patient (hematoxylin and eosin stained). Scale bar = 100 μm.

Source: Koehsler, Martina; Soleiman, Afschin; Aspock, Horst, Auer, Herbert, and Walochnik, Julia. "*Onchocerca jakutensis* Filariasis in Humans." *Emerging Infectious Diseases Journal*, CDC, volume 13.11 (November 2007).

Managers

Managers and executives (upper-level managers) run the organization or a division or department of that organization; their jobs are to ensure that

FIGURE 4.3

Document for Technicians

Bench Aids for the diagnosis of filarial infections   Introduction

World Health Organization 1997

Introduction

Several species of filarial worms infect humans in the tropical and subtropical regions of the world. The adult worms inhabit various tissues and organs of the body and are inaccessible for identification. Consequently, diagnosis of filarial infections depends primarily on the identification of the larval stage of the parasite (microfilaria). Most species of microfilaria circulate in peripheral blood; however, some are found in the skin.

The microfilaria

At the light-microscopic level and with the aid of a variety of stains, a microfilaria appears as a primitive organism, serpentine in shape and filled with the nuclei of many cells. In many, but not all, species, the body may be enveloped in a membrane called a sheath (**sh**). Where a sheath is present it may extend a short or long distance beyond either extremity of the microfilaria. In some species, depending on the stain used, the sheath displays a characteristic staining quality which aids in species identification. The nuclei of the cells that fill the body are usually darkly stained and may be crowded together or dispersed. The anterior extremity is typically devoid of nuclei and is called the cephalic or head space (**hs**); it may be short or long. Along the body of the microfilaria there are additional spaces and cells that serve as anatomical landmarks. These include the nerve ring (**nr**), excretory pore (**ep**), excretory cell (**ec**), and anal port (**ap**). In some species, an amorphous mass called the innerbody (**ib**) and four small cells called the rectal cells (**R-1, R-2, R-3, R-4**) can be seen, usually with the aid of special stains. These structures and their positions are sometimes useful for species identification. The shape of the tail and the presence or absence and distribution of nuclei within it are also important in species identification.

Periodicity

Some species of microfilariae circulate in peripheral blood at all hours of the day and night, while others are present only during certain periods. The fluctuation in numbers of microfilariae present in peripheral blood during a 24-hour period is referred to as periodicity. Species that are found in the blood during night-time hours but are absent at other times are designated *nocturnally periodic* (e.g. *Wuchereria bancrofti, Brugia malayi)*; those that are present only during certain daytime hours are designated *diurnally periodic* (e.g. *Loa loa)*. Microfilariae that are normally present in the blood at all hours but whose density increases significantly during either the night or the day are referred to as *subperiodic*. Microfilariae that circulate in the blood throughout a 24-hour period without significant changes in their numbers are referred to as *nonperiodic* or *aperiodic* (e.g. *Mansonella* spp.).

The periodicity of a given species or geographical variant is especially useful in determining the best time of day to collect blood samples for examination. To determine microfilarial periodicity in an individual, it is necessary to examine measured quantities of peripheral blood collected at consecutive intervals of 2 or 4 hours over a period of 24–30 hours.

Source: Downloaded from the World Wide Web, November 2008: www.cdc.gov/eid/content/13/ 11/1749.htm. *CDC Bench Aids.*

The writers include practical information to help technicians identify microfilaria.

The writers use technical terminology, yet define some terms and concepts technicians will not understand.

the organization runs effectively and efficiently. Managers and executives want to know the bottom line: how the information will affect their organizations now and in the future. They need information to make decisions; they do not have time to wade through theory and research in the same way experts do. Instead, managers want to know how the informa-

FIGURE 4.4

Excerpt from Procedures for Technicians

Specimen Processing

Preparing Blood Smears
If you are using venous blood, blood smears should be prepared as soon as possible after collection (delay can result in changes in parasite morphology and staining characteristics).

Thick smears
Thick smears consist of a thick layer of dehemoglobinized (lysed) red blood cells (RBCs). The blood elements (including parasites, if any) are more concentrated (app. 30?) than in an equal area of a thin smear. Thus, thick smears allow a more efficient detection of parasites (increased sensitivity). However, they do not permit an optimal review of parasite morphology. For example, they are often not adequate for species identification of malaria parasites: if the thick smear is positive for malaria parasites, the thin smear should be used for species identification.

> The writers include procedural information.

Prepare at least 2 smears per patient!

1. Place a small drop of blood in the center of the pre-cleaned, labeled slide.
2. Using the corner of another slide or an applicator stick, spread the drop in a circular pattern until it is the size of a dime (1.5 cm^2).
3. A thick smear of proper density is one which, if placed (wet) over newsprint, allows you to barely read the words.
4. Lay the slides flat and allow the smears to dry thoroughly (protect from dust and insects!). Insufficiently dried smears (and/or smears that are too thick) can detach from the slides during staining. The risk is increased in smears made with anticoagulated blood. At room temperature, drying can take several hours; 30 minutes is the minimum; in the latter case, handle the smear very delicately during staining. You can accelerate the drying by using a fan or hair dryer (use cool setting). Protect thick smears from hot environments to prevent heat-fixing the smear.
5. Do not fix thick smears with methanol or heat. If there will be a delay in staining smears, dip the thick smear briefly in water to hemolyse the RBCs.

Source: Downloaded from the World Wide Web, November 2008:
www.dpd.cdc.gov/DPDx/HTML/DiagnosticProcedures.htm. *DPDx Blood Diagnostic Procedures.*

tion relates to the organization, so they can quickly make a decision. Some managers, especially mid-level managers, will decide to take the information to upper-level management or executives. If you know that a manager will use your information in a document to executives, include an executive summary at the beginning of the document and use headings and lists to highlight information that will help the manager reuse the information or make a decision.

Managers may have backgrounds related to your field, or they may have earned a degree in your field. Before writing for managers and executives, find out if they will understand technical vocabulary. Managers and executives share the following characteristics:

• They read to gather information for making decisions.

- They expect conclusions, recommendations, and implications to appear near the beginning of the document or in an executive summary. See Chapter 18 for information on executive summaries.
- They scan documents for the information they need to make decisions.
- They may need definitions and explanations of most technical terms and information.
- They prefer simple graphics that quickly summarize information.

Figure 4.5 instructs managers on how to plan for emergencies. This excerpt from FEMA's *Emergency Management Guide for Business and Industry* gives managers information to use when establishing an emergency planning team.

General Readers

General readers have little, if any, technical knowledge. The general reader, sometimes called a layperson, reads technical information outside his or her area of expertise. All of us become general readers at some time; for example, if engineers read a brochure about blood clots, they are general readers because they are not medical experts or technicians. Likewise, if a medical expert reads a document about testing the structural integrity of pipes, he or she is a general reader in that situation. General readers are also consumers who may read and use instructions for products. For example, the instructions for changing the default settings in a software program would be aimed at general readers, not at experts or technicians. General readers share the following characteristics:

- They do not understand basic technical vocabulary and concepts.
- They need basic technical vocabulary and concepts defined.
- They expect examples and analogies.
- They expect a simple, direct presentation.
- They learn from simple graphics.

When you write for general readers, you have to determine what they already know and understand. You want to gather as much information as possible about these readers because the more you know about their background, education, reading ability, and attitudes, the better you can anticipate what they will understand and what type of vocabulary you can use.

Figure 4.6, page 90, is an excerpt from a document prepared by Homeland Security and FEMA to instruct families in planning for disasters. This

FIGURE 4.5

Excerpt from a Document for Managers

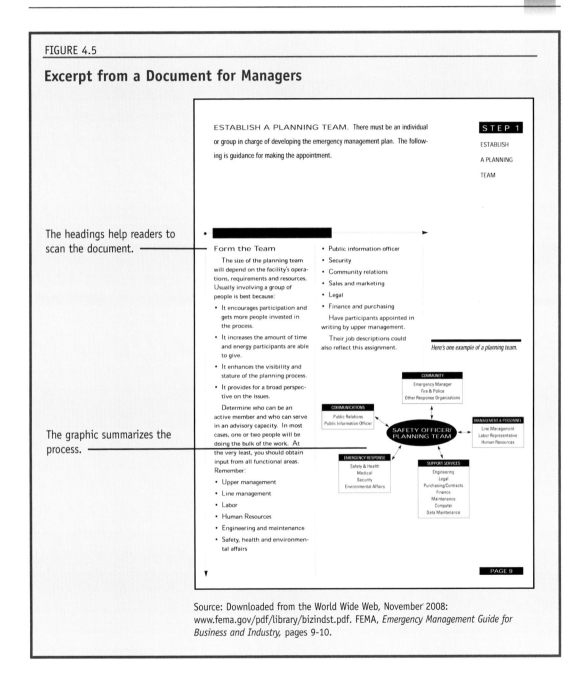

The headings help readers to scan the document.

The graphic summarizes the process.

Source: Downloaded from the World Wide Web, November 2008: www.fema.gov/pdf/library/bizindst.pdf. FEMA, *Emergency Management Guide for Business and Industry,* pages 9-10.

document uses simple, non-technical language. Compare the language in this document with the language in Figure 4.5 for managers.

FIGURE 4.6

A Document for General Readers

The table helps readers quickly locate information.

Evacuation Guidelines

Always:	If time permits:
Keep a full tank of gas in your car if an evacuation seems likely. Gas stations may be closed during emergencies and unable to pump gas during power outages. Plan to take one car per family to reduce congestion and delay.	Gather your disaster supplies kit.
Make transportation arrangements with friends or your local government if you do not own a car.	Wear sturdy shoes and clothing that provides some protection, such as long pants, long-sleeved shirts, and a cap.
Listen to a battery-powered radio and follow local evacuation instructions.	Secure your home: • Close and lock doors and windows. • Unplug electrical equipment, such as radios and televisions, and small appliances, such as toasters and microwaves. Leave freezers and refrigerators plugged in unless there is a risk of flooding.
Gather your family and go if you are instructed to evacuate immediately.	Let others know where you are going.
Leave early enough to avoid being trapped by severe weather.	
Follow recommended evacuation routes. Do not take shortcuts; they may be blocked.	
Be alert for washed-out roads and bridges. Do not drive into flooded areas.	
Stay away from downed power lines.	

Simple language helps readers understand the guidelines.

The simple graphic helps readers identify evacuation routes.

Source: Downloaded from the World Wide Web, November 2008: www.fema.gov/pdf/library/bizindst.pdf. FEMA, *Are You Ready?*, page 11.

What are Readers' Individual Characteristics?

Just as you have a purpose for writing your document, your readers have a purpose for reading it. They approach your document with specific questions and expectations about the information and perhaps the style of the document. They have specific needs that you should consider when writing and designing your document.

You may wonder how you can determine your readers' expectations and needs. If possible, talk to your readers. This method, however, may not always be possible because of the number of readers, their geographic location, or your relationship with them. You may not know your readers, or they may not be expecting the document; so gathering information directly from them might not be feasible or appropriate. If you are writing to a large group, you can find out about your readers by talking with a representative sample. You also can contact people who know your readers or have previously written for or worked with them. Use the following questions to determine your readers' characteristics:

- Who are your readers?
- What are your readers' attitudes and expectations?
- How will readers use the document?
- Will multiple categories of readers read or use the document?

Who are Your Readers?

To answer this question, consider the following reader characteristics:

- **Education**: Consider whether your reader has a degree and when he or she earned that degree. For example, a computer programmer who earned a degree in 1990 has a much different educational background than a computer programmer who received a degree in 2008. You should also consider any formal and informal training completed by your reader. When you know the reader's educational background, you can better determine what technical vocabulary and supporting materials are appropriate. When seeking information about a reader's educational background, be professional and considerate: don't simply ask for a resumé. Try to bring the topic up in conversation, or ask the reader's coworkers.
- **Job position**: Your reader's primary job responsibility will help you determine the types of information that you should include and how you should present that information. For example, if you are writing a proposal for setting up a videoconference for a manager and you know that the manager must convince upper-management of the value of this technology, you might include information on how this technology can save time and money.

- **Professional experience**: Consider your reader's professional experience—both on the job and beyond his or her job responsibilities. For example, an engineer with 15 years of experience may bring different needs and expectations to a feasibility report; a manager who has lived through Hurricane Katrina will bring a high level of expertise to a manual on preparing for emergencies. You should consider that expertise as you write.

- **Personal preferences**: Your readers will have preferences about style, format, design, and media (paper, e-mail, and other electronic forms). For example, some people like to receive documents by way of e-mail, and others prefer paper copy. Organizations also have "preferences." Some want their logo to appear in particular colors on every cover page. Some expect all documents to be printed in a specific typeface. How can you find out about these preferences?
  - Talk to your readers when possible.
  - Talk with others who have written for or worked with your readers.
  - Look at documents that your readers or their organizations have written.

As you talk with readers and their coworkers and examine their documents, look for their preferences in style, format, design, and media. You may discover other preferences, but these four areas will give you a place to start. Once you have identified preferences, decide whether you can accommodate them. When possible, try to accommodate all of them. If you are writing to more than one person or group, these preferences may be incompatible. They may contradict your organization's policies for technical documents, or they may be inconsistent in style.

- **Culture**: Your readers' cultural characteristics can affect the organization, vocabulary, design, and tone of your document. You should consider the readers' cultures to ensure that you don't unknowingly offend or mislead them. Ask these questions:
  - Are the readers from a non-U.S. culture? If so, what characteristics of this culture may affect your document?
  - How fluent are the readers in written English?
  - What choices do you need to make about organization? Deductive or inductive? Degree of specificity? Type of informa-

tion to include about you and your organization? Appropriate tone, formality, and graphics?[1]

For more information on considering readers' cultures, see the tips on understanding the differences between U.S. and non-U.S. written communication found on the following page.

What are Your Readers' Attitudes and Expectations?

When you understand your readers' attitudes and expectations, you can better select the appropriate information, tone, design, and organization to make your document more effective. You can easily determine the attitudes and expectations of readers whom you or your coworkers know; but when you or your coworkers don't know your readers, analyzing their attitudes is more difficult. You probably won't be able to analyze their attitudes toward you or your organization; so try to imagine yourself in their place to determine how interested they may be in the subject or how they might react to your document or to your organization. As you consider your readers' attitudes and expectations, ask yourself these questions:

- **What are the readers' attitudes toward the subject?** Your readers' attitudes about your subject may be negative, positive, neutral, skeptical, or enthusiastic. If the attitude is positive or enthusiastic, you won't have to entice the readers into your document or figure out how to convince them to read or use it. If the attitude is negative, skeptical, or neutral, you must decide how to motivate readers to read your document or to persuade them to accept your recommendations. For example, if you know readers will resist your recommendations, you might present the benefits or reasons for the recommendations before actually presenting the recommendations. When readers have positive attitudes about your subject, you want to reinforce those attitudes. When attitudes are negative, however, you want to change them. To anticipate readers' attitudes toward a document, think about how they will feel about the topic. For example, a software user who is troubleshooting while installing software is motivated to read a troubleshooting guide. In contrast, a company executive who receives an unsolicited letter from

---

[1] Elizabeth Tebeaux and Linda Driskill "Culture and the Shape of Rhetoric: Protocols of International Document Design," *Exploring the Rhetoric of International Professional Communication: An Agenda for Teachers and Researchers.* Ed. Carl R. Loritt with Dixie Goswami. New York: Baywood. 1999. Used by permission of Baywood Publishing Company, Inc.

TIPS FOR UNDERSTANDING THE DIFFERENCES IN U.S. AND NON-U.S. WRITTEN COMMUNICATION[2]

Characteristic	How U.S. Readers and Writers Approach this Characteristic	When Writing for U.S. Readers...	How Non-U.S. Readers and Writers Approach this Characteristic	When Writing for Non-U.S. Readers...
Importance of individual or groups	• Organizations expect written communication to show accuracy and to document activities and decisions. • Employees see themselves as individuals who are not defined by their jobs or their organizations.	• Use "I" rather than "we." • Address letters to a primary reader. • Sign your letters.	• Employees prefer face-to-face communication. • Employees see themselves as members of an organization, not as individuals.	• Use "we" rather than "I." • Sign your letters with your company name, not your name.
Value of business life and personal relationships	• Business life and goals have a higher priority than personal life and goals. • Work and personal life are separate. • Documents are precise and have an objective, direct tone.	• Focus on the technical information with little, if any, reference to your personal life or goals. • If you mention personal information related to you or your reader, place it after the technical information.	• Personal relationships are more important than business relationships. • Work and personal life are intertwined. • Readers and writers emphasize the responsibility shared by the group. • Documents have a subjective and oblique tone.	• Work to build a personal relationship with the readers and their organization. • Expect to conduct business during social activities.
Effect of job rank or status	• The distance between managers and subordinates is less formal. • Managers and subordinates usually develop close working relationships.	• Address your managers with their first names. • Generally, use less formal communications, such as e-mail and memos, to correspond with your manager.	• The relationship between manager and subordinate is formal. Job status is important. • Managers and subordinates have a formal working relationship.	• Be careful about being too informal. If you have never met your reader, use his or her title, such as Ms., Mr., or Dr. • Generally, use more formal documents, such as letters, not e-mail, to communicate.
Expectations related to context	• Readers expect documents to contain comprehensive, complete information–all the reader needs to understand the document and make a decision. These readers are from a low-context culture.	• Spell out all the details. • Include only business-related information.	• Readers value documents where the details are implicit. The implicit information is conveyed through other forms of communication developed through a personal relationship. These readers are from a high-context culture.	• Omit obvious information to refrain from insulting the reader. For example, in a high-context culture, a cup of hot coffee would not include the warning, "Beware of hot contents" because everyone knows that a cup of coffee is hot.
Expections about uncertainty	• Readers value and expect clear guidelines, goals, and information. They are uncomfortable with uncertainty. • Readers expect specific timelines and deadlines.	• Begin with the most important ideas. • State your points clearly, directly, and objectively. • Set firm deadlines. • Highlight important information in bulleted lists. • Confirm details and verbal decisions in writing.	• Readers are comfortable with uncertainty. • Readers do not expect firm deadlines and timelines. Instead, deadlines and timelines are relative and flexible. • Readers rely less on written guidelines and goals, and more on a mission statement or code of conduct.	• Place the main point at the end of the document. • De-emphasize the main points. • Suggest a deadline or a timeline, rather than state a required, firm deadline or timeline.

[2] Adapted from Elizabeth Tebeaux and Linda Driskill "Culture and the Shape of Rhetoric: Protocols of International Document Design," *Exploring the Rhetoric of International Professional Communication: An Agenda for Teachers and Researchers.* Ed. Carl R. Loritt with Dixie Goswami. New York: Baywood. 1999. Used by permission of Baywood Publishing Company, Inc.

an unknown inventor requesting funding for a new way of measuring ozone in the atmosphere may be skeptical or unenthusiastic. Again, imagine you are the reader to analyze attitudes toward your document. You can also do the following:

- Ask for help from people who have worked with your readers or who know them.
- Talk with your readers about what information to include.
- Read background documents and information related to the topic and find out what your readers thought of those documents and information.

- **What are your readers' attitudes toward you and your organization?** Your readers' attitudes toward you and your organization influence their reactions to your document. These attitudes may reflect the readers' previous experiences with you or your organization. If readers have had a good experience working with you or your organization, they will probably look favorably on your document, even before reading it. If readers have had a negative experience with you or your organization, you face a greater challenge: you will have to use part of your document to gain back the readers' confidence. If you don't know your readers, find out if your coworkers have worked with them. If your coworkers have had positive experiences with your readers, you probably can safely assume that the readers will have a favorable attitude toward you and your organization. If the experience was negative, you may have to design and organize the document to gain the readers' confidence. Even if your readers don't know you, they may have a preconceived attitude about your organization. Think about your attitude toward the Internal Revenue Service (IRS). Would you be enthusiastic about receiving a letter from the IRS? Most of us don't look forward to receiving a letter from the IRS; correspondence from the IRS often is unwelcome information about filing tax forms or paying back taxes and penalties. Even though most of us don't know anyone at the IRS, we have a preconceived attitude about the organization and what it does. Likewise, your readers may have a preconceived attitude about your organization because of past experiences with it or because of its reputation.

- **What do your readers expect from your document?** Your readers may expect your document to contain specific information, to be organized

> *Learn about Your Readers*
>
> •
>
> *Ask for help from people who have worked with them or know them.*
>
> •
>
> *Talk with the readers themselves.*
>
> •
>
> *Examine other documents prepared for your readers and ask about their effectiveness.*

in a particular manner, to present information in a particular level of detail, or to be sent in a particular media. For example, if your readers expect progress reports to be sent via e-mail, send your reports via e-mail. If you are responding to a request for a proposal, respond to every question and requirement in that request.

How Will Readers Use the Document?

Think about how and where readers will use the document and any time constraints that may affect them. Readers' physical surroundings can influence your choice of cover, binding, headings, page size, line length, paragraph divisions, and type size. For example, some readers work in noisy, distracting areas; others work in environments where documents are likely to get dirty and pages can get bent and torn. By considering readers' surroundings, you can design and organize the document to meet their needs.

For example, if you know your readers will use your instructions in a warehouse, you might put the instructions in a sturdy binding or notebook and laminate the pages so readers can easily clean the pages. If you know your readers will use your reference manual at a small computer workstation, use a relatively small page size that will fit next to the computer and put the manual in a three-ring binder that will lie flat.

Along with considering where readers will use the document, think about the time readers have to consider your document. For example, your readers may receive many documents each week and have little time to read them. To help readers locate information quickly, organize and design your document to be easily accessible. If your readers are executives who receive many proposals, include tabs for each of the major sections, so they can quickly flip to the sections they need. You could also include an executive summary that briefly states your proposal. If you are presenting your information on the Web, include concise summaries of long passages, so readers can decide whether to read "in more detail" (Fugate). You can use several design devices to help busy readers find information in your document:
- headings and subheadings
- tabs that separate the major sections
- table of contents and index for longer, formal documents
- page designs with ample white space
- graphics that summarize information quickly

- summaries at the beginning of documents and major sections
- overviews at the beginning of documents and major sections

Will Multiple Groups of Readers Read or Use the Document?

Often, more than one person or group will read your documents. *Initial readers*, such as an administrative assistant or even the head of a department, may skim the document to determine who in their organization should receive it. Judging from the summary, cover letter, title, or introduction, the initial reader sends the document to the primary reader.

The *primary reader* is the intended reader: the technician who will use the instructions, the executive who will decide to implement the proposal, and so on. The primary reader will use the document to complete a task, to gather information, or to make a decision. Some primary readers may use the document to make a decision, but may read only certain sections and then send the document to a secondary reader, who has only a minor interest in the document.

The *secondary reader* might be a technical adviser whose opinion the decision-maker seeks. Such a reader assumes the role of primary reader when reading the sections that draw on

TIPS FOR WRITING FOR MULTIPLE READERS

- **Divide the document into distinct sections so readers can go directly to the sections that apply to them.** For example, you might have a "Getting Started" section for general readers and an "Advanced Techniques" section for technicians.
- **Use devices that help readers locate information in the document.** These devices include indexes, tables of contents, executive summaries, headings, and tabs (Holland, Charrow, and Wright).
- **Put definitions of technical vocabulary, explanations of technical concepts, and other technical details in footnotes, appendices, or other special sections that readers can easily find** (Holland, Charrow, and Wright). Clearly label these sections with headings, icons, or color.
- **Direct the language and presentation of a single document to readers with the lowest level of technical knowledge.** This technique works especially well for instruction and policy manuals.
- **Write separate documents for each group of readers if you have the time and the budget.** This option will benefit your readers by allowing you to organize and direct the document to one group of readers. However, most organizations rarely have the budget or time for this option.
- **Put the document online so you can compartmentalize it for readers with various levels of knowledge.** Include good navigation tools so readers can easily locate the sections they need.

FIGURE 4.7

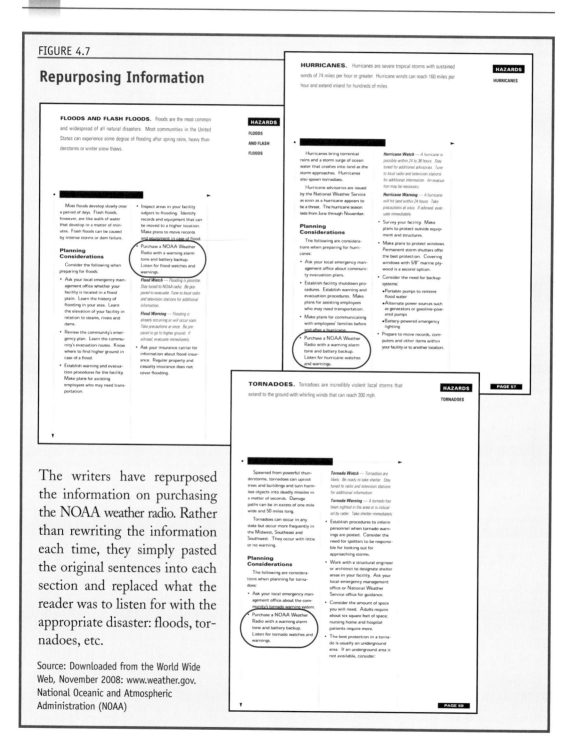

Repurposing Information

FLOODS AND FLASH FLOODS. Floods are the most common and widespread of all natural disasters. Most communities in the United States can experience some degree of flooding after spring rains, heavy thunderstorms or winter snow thaws.

HAZARDS
FLOODS
AND FLASH
FLOODS

Most floods develop slowly over a period of days. Flash floods, however, are like walls of water that develop in a matter of minutes. Flash floods can be caused by intense storms or dam failure.

Planning Considerations

Consider the following when preparing for floods:

• Ask your local emergency management office whether your facility is located in a flood plain. Learn the history of flooding in your area. Learn the elevation of your facility in relation to steams, rivers and dams.

• Review the community's emergency plan. Learn the community's evacuation routes. Know where to find higher ground in case of a flood.

• Establish warning and evacuation procedures for the facility. Make plans for assisting employees who may need transportation.

• Inspect areas in your facility subject to flooding. Identify records and equipment that can be moved to a higher location. Make plans to move records and equipment in case of flood.

• Purchase a NOAA Weather Radio with a warning alarm tone and battery backup. Listen for flood watches and warnings.

Flood Watch — Flooding is possible. Stay tuned to NOAA radio. Be prepared to evacuate. Tune to local radio and television stations for additional information.

Flood Warning — Flooding is already occurring or will occur soon. Take precautions at once. Be prepared to go to higher ground. If advised, evacuate immediately.

• Ask your insurance carrier for information about flood insurance. Regular property and casualty insurance does not cover flooding.

HURRICANES. Hurricanes are severe tropical storms with sustained winds of 74 miles per hour or greater. Hurricane winds can reach 160 miles per hour and extend inland for hundreds of miles.

HAZARDS
HURRICANES

Hurricanes bring torrential rains and a storm surge of ocean water that crashes into land as the storm approaches. Hurricanes also spawn tornadoes.

Hurricane advisories are issued by the National Weather Service as soon as a hurricane appears to be a threat. The hurricane season lasts from June through November.

Planning Considerations

The following are considerations when preparing for hurricanes:

• Ask your local emergency management office about community evacuation plans.

• Establish facility shutdown procedures. Establish warning and evacuation procedures. Make plans for assisting employees who may need transportation.

• Make plans for communicating with employees' families before and after a hurricane.

• Purchase a NOAA Weather Radio with a warning alarm tone and battery backup. Listen for hurricane watches and warnings.

Hurricane Watch — A hurricane is possible within 24 to 36 hours. Stay tuned for additional advisories. Tune to local radio and television stations for additional information. An evacuation may be necessary.

Hurricane Warning — A hurricane will hit land within 24 hours. Take precautions at once. If advised, evacuate immediately.

• Survey your facility. Make plans to protect outside equipment and structures.

• Make plans to protect windows. Permanent storm shutters offer the best protection. Covering windows with 5/8" marine plywood is a second option.

• Consider the need for backup systems:
  ◆ Portable pumps to remove flood water
  ◆ Alternate power sources such as generators or gasoline-powered pumps
  ◆ Battery-powered emergency lighting

• Prepare to move records, computers and other items within your facility or to another location.

PAGE 57

TORNADOES. Tornadoes are incredibly violent local storms that extend to the ground with whirling winds that can reach 300 mph.

HAZARDS
TORNADOES

Spawned from powerful thunderstorms, tornadoes can uproot trees and buildings and turn harmless objects into deadly missiles in a matter of seconds. Damage paths can be in excess of one mile wide and 50 miles long.

Tornadoes can occur in any state but occur more frequently in the Midwest, Southeast and Southwest. They occur with little or no warning.

Planning Considerations

The following are considerations when planning for tornadoes:

• Ask your local emergency management office about the community's tornado warning system.

• Purchase a NOAA Weather Radio with a warning alarm tone and battery backup. Listen for tornado watches and warnings.

Tornado Watch — Tornadoes are likely. Be ready to take shelter. Stay tuned to radio and television stations for additional information.

Tornado Warning — A tornado has been sighted in the area or is indicated by radar. Take shelter immediately.

• Establish procedures to inform personnel when tornado warnings are posted. Consider the need for spotters to be responsible for looking out for approaching storms.

• Work with a structural engineer or architect to designate shelter areas in your facility. Ask your local emergency management office or National Weather Service for guidance.

• Consider the amount of space you will need. Adults require about six square feet of space; nursing home and hospital patients require more.

• The best protection in a tornado is usually an underground area. If an underground area is not available, consider:

PAGE 59

The writers have repurposed the information on purchasing the NOAA weather radio. Rather than rewriting the information each time, they simply pasted the original sentences into each section and replaced what the reader was to listen for with the appropriate disaster: floods, tornadoes, etc.

Source: Downloaded from the World Wide Web, November 2008: www.weather.gov. National Oceanic and Atmospheric Administration (NOAA)

his or her area of expertise. For example, the primary reader of a proposal might be a manager, and the secondary reader might be a technical expert who assumes the role of primary reader for the "Problems" section of a document. The Dow Chemical Web site, in the interactive document analysis found online illustrates a document for multiple readers.

View the Interactive Student Analysis on Writing for Multiple Readers online at www.kendallhunt.com/ technicalcommunication.

REPURPOSE INFORMATION

Professionals often take information that they write and repurpose (reuse) it for a different reader. For example, a telecommunications company may write a process description and, later, adapt it for a technician, a manager, a consumer, or a general reader. Review Figure 4.7 to see how information can be repurposed.

CREATE A READER PROFILE

Think about your readers throughout the writing process. Spend time identifying them and anticipating as well as you can how they will react and respond to your document. After drafting a document, ask one or more of the primary readers to read your draft and tell you how they would respond. If their response is not what you want, ask them for advice on how to change your document to meet your goals.

To help you anticipate your readers' reactions, create a reader profile sheet like the one you will find online at this book's Web site.

Download a Reader Profile Sheet online at www.kendallhunt.com/ technicalcommunication.

CASE STUDY ANALYSIS
Microsoft's Clippy

Background

In Microsoft *Office* 1997, Microsoft introduced Clippy. If you use Microsoft *Word*, you may have seen Clippy, an animated cartoon character with large, winking eyes. Clippy acts as the software user's link to the online help system. Clippy is operated by a type of software called a wizard. Wizards help readers complete various tasks by asking questions.

Originally, Clippy offered online help to the user without being asked. Clippy would appear on the screen and ask, "What would you like to do?" Many users of *Office* 1997 found Clippy annoying. One user complained that "it pops up without warning—even when you don't want it or don't need it" (quoted in Shroyer). Another user explained that Clippy "wouldn't be so obnoxious if you could control it. As is, it moves all over the place and gets in the way" (quoted in Shroyer). Because so many users reacted negatively to Clippy, *PC World* published articles telling users how to turn off Clippy (Li-Ron). In October 1998, Microsoft gave customers the code to remove Clippy from the system (Shroyer).

So what went wrong? Why did the users reject Clippy? The users rejected Clippy, in part, because the software designers assumed that the users couldn't work without supervision. In reality, the users found Clippy's help intrusive. The users knew what they were doing; if they didn't, they would ask for help (Shroyer). The software designers assumed that all users knew less than they did and that all users would welcome help from Clippy. These designers' misguided assumptions frustrated many users.

The software designers at Microsoft learned from their users; in *Office* 2000 and later versions, Clippy was more user friendly. Clippy sat in one corner of the screen, and the user could turn off Clippy by simply clicking. Clippy became the "office assistant," and users could change Clippy to the character they wanted.

Assignment

- How could the designers at Microsoft have better anticipated how users would react to Clippy?
- Find a copy of the most recent version of Microsoft *Word*. Does the software have an office assistant? Is the assistant user friendly? Explain your answer.
- Locate a software program and do the following:
    - Find an example of a function that is automatic—one that the user does not or cannot easily change or turn off.
    - Is that function user friendly? If not, how could the designers have made the function more user friendly?
    - Be prepared to share your findings with your instructor and your class.

TAKING IT INTO THE WORKPLACE
Reader-Friendly Web Sites

Let's assume you're looking for information on the Web. What sites do you prefer? Which sites do you leave without finding the information you need? What makes a site reader friendly? Readers on the Web "want to get to their destination quickly without slogging through a lot of verbiage" (Bradbury). S. Gayle Bradbury, a technical communicator, explains that if you don't give readers the information they need quickly, you'll lose the readers. As Alice Fugate, author of *Writing for Your Web Site: What Works, and What Doesn't,* explains, "far too many Web sites tell you everything except what you want to know." Bradbury suggests that writers should "think like a reader" when writing for a Web site.

According to Ben Miller, a user interface design specialist with Edward Jones, the two fastest-growing groups online are seniors and Baby Boomers (Fugate), with seniors (the over-55 group) comprising the fastest-growing group. Many of these users from both groups are new users of technology; these users are "relatively slow to adopt new technologies . . . They view computers as tools," not toys (Fugate). They want the Web writer to filter out "cyberbells" and whistles. They want pure information that they can access and use efficiently (Fugate).

Assignment

To help you think like seniors (the over-55 crowd; not college seniors!), assume that you are creating a Web site for your city. You want to learn more about how seniors in your city use the Web and what they like and don't like in a Web site. You also want to learn how they navigate a Web site. To gather the information, complete the following tasks:

1. Develop a questionnaire for seniors. In the questionnaire, ask questions about what Web sites they like to use, what Web sites they find frustrating, and how they navigate a Web site. As you design the questionnaire, think about what you like in a Web site and ask whether the seniors like similar characteristics. (To learn about writing questionnaires, see Chapter 5).
2. Give the questionnaire to at least five seniors. You might e-mail the questionnaire to some of the seniors.
3. Be prepared to discuss your findings with the class.

Download a Worksheet
for Writing for Your
Readers online at
www.kendallhunt.com/
technicalcommunication.

EXERCISES

1. Write two paragraphs about one of these topics or about a topic that
you select:

    A car accident in which you were involved

    A grade that you received

    A class that you took

    The relationship between body fat and aerobic exercise

    The relationship between ozone and vehicle emissions

    The production of cheese

    Sanitation standards for a meat-processing plant

    Safety precautions for women walking alone at night

    Write each paragraph for a different reader. The readers should have
different purposes for reading and different levels of knowledge on
the topic. Refer to the "Reader Profile Worksheet" online as you
analyze your readers and determine their needs and expectations.

2. This passage appears in a brochure on property tax appraisals written for homeowners. These homeowners have different educational
backgrounds and levels of expertise about property tax appraisals.
Most of these homeowners are general readers and will read these
paragraphs to help them understand their property appraisal, which
determines its fair market value or what an appraiser believes is the
appropriate sale price for the home.

What Is Fair Market Value?

Section 1.04 of the Texas Property Tax Code defines market value
as follows:

    Market value means the price at which a property would trans-
fer for cash or its equivalent under prevailing market conditions
if

    a.   exposed for sale in the open market with a reasonable
time for the seller to find a purchaser;

    b.   both the seller and the purchaser know of all uses and
purposes to which the property is adapted and for which

    it is capable of being used and of the enforceable
    restrictions on its use; and

  c. both the seller and the purchaser seek to maximize their
    gains and neither is in a position to take advantage of the
    exigencies of the other.

Write a memo to your instructor explaining why this paragraph
about market value doesn't adequately respond to readers' needs and
levels of expertise. Comment on specific language that readers
might not understand.

3. **Web exercise**: Using the Web, find an article on a topic related to
your major field of study. The article should be for expert readers.
You might search using Google or Yahoo! or your college or university
online library services. You can also go to government Web sites
such as the Center for Disease Control, FEMA, EPA, etc. These
Web sites publish technical information or link to scientific journals.
When you have located an article, make sure that you understand
it and then complete the following:

  a. Write a memo to your instructor describing the possible educational
background of primary readers of the article, their level of
technical expertise, and their purpose for reading the article.

  b. Select a 200- to 300-word passage that you find particularly
interesting and rewrite it for general readers.

  c. Attach a copy of the article to your memo and rewritten passage.

4. **Collaborative exercise:** Working with a team, locate two ads for the
same type of product. For example, you might locate ads for coffee
from Starbucks and Folgers, or ads for a hybrid vehicle and a traditional
vehicle. When you have found the ads, complete the following
steps:

  a. Profile the intended readers of each ad. Use the "Reader Profile
Worksheet" online as your guide.

  b. Be prepared to discuss how the documents address the attitudes
and expectations of the intended readers.

View the Reader Profile
Worksheet online at
www.kendallhunt.com/
technicalcommunication.

REAL WORLD EXPERIENCE
Informing Students about Financial Aid

Background

Although this assignment is a team project, your instructor may modify it to be an individual assignment. Your team has received the following assignment from the scholarship and financial aid office at your college or university:

> Our school is losing many good students because they cannot afford the cost of tuition, books, and living expenses. Many of these students are unaware of the scholarships and other forms of financial aid that are available. Students don't know that they can apply for scholarships based only on merit, scholarships based on merit and financial need, and scholarships and grants based solely on financial need. Scholarships and other forms of financial aid help students stay in school. Students, however, may be unaware of these scholarships, think they don't qualify, or think that applying for financial aid is too much trouble.

Assignment

Your assignment is to write a document aimed at students at your college or university informing them that scholarships and financial aid are available to qualifying students. You should determine the most appropriate media for presenting this information to students: Web site, student newspaper, flier, e-mail, or letter. Whatever media you select, your document should motivate students with financial difficulties to come to the scholarship and financial aid office for information.

Download the Reader Profile Worksheet online at www.kendallhunt.com/ technicalcommunication.

Follow these steps to complete the assignment:
1. Complete the "Reader Profile Worksheet" found online.
2. Visit your college or university's Web site or financial aid office to gather information about scholarships and financial aid at your university.
3. Determine the most appropriate format for your readers.
4. Turn the document in to your instructor. Include the "Reader Profile Worksheet" that you completed.

iStockphoto 2008.

Researching Information

Bill Garcia, an engineer for a food company, must determine why water pipes in one of the manufacturing plants are vibrating and recommend a solution. Before making an appointment to go to the plant, Bill thinks about the information he will need to recommend a solution and to write an effective report. He knows he will need to do more than just inspect the system, but with the amount of information available to him, he feels overwhelmed. He's not quite sure where to begin or what sources will best suit his purpose. He needs a plan for selecting appropriate research techniques and sources.

Bill's dilemma is not unusual. You, too, may face a similar dilemma as a student and as a professional. As a student, you might need to answer a more abstract question, such as, "What causes the hiccups?" or "How do greenhouse gases affect the health of people living in heavily populated urban areas?" As a professional, you will likely answer more applied questions, such as Bill's question of "How can we solve the problem of vibrating pipes at our plant?" or "What are the advantages of opening a second plant in Mexico?"

Whether you ask abstract or applied questions, your goal is the same: to gather information for answering a question or solving a problem. In either case, you will need a plan to effectively research your topic, tools for researching secondary sources, and methods for conducting primary research. You will also want techniques for evaluating the sources and the information. This chapter presents guidelines to help you effectively research your topic.

PLAN YOUR RESEARCH

Bill thinks about the purpose of his document. He decides that he is writing primarily to recommend a viable solution to the problem of the vibrating pipes. His solution must be not only cost-effective and feasible, but also well supported with research and testing. Bill also recognizes a long-term goal—to establish a reputation as a problem solver and to establish positive relationships with his supervisor and the plant manager. As a professional, you, too, may conduct research to complete documents or projects. For some documents, you may need only a single fact or figure; while for others, you may need to do research using several information-gathering techniques.

Regardless of whether you are researching a simple or complex topic, you will want to plan your research. Figure 5.1 shows the steps that you should follow when planning your research. As you work through the planning process, you may determine that you have to return to one or more of the steps. The planning process is not linear; obstacles such as lack of information, learning that a source is not credible, and running out of funds can send a writer back to the drawing board.

Select the Right Research Method for Your Topic and Your Readers

Once you have determined the type of information that you need, you can select the appropriate research method for your topic, time, budget, and readers. For example, if you want to determine how *E. coli* got into a meat supply, you would want to inspect the processing plant to test the meat processed at that plant and to interview the managers and workers in the plant. If you wanted to learn how *E. coli* contaminated the U.S. food supply in 2005 and how the processing plants responded, you might search Web sites.

As you plan your research, you will want to consider the most appropriate methods for gathering information. You might begin by asking these questions:

FIGURE 5.1

Planning Your Research

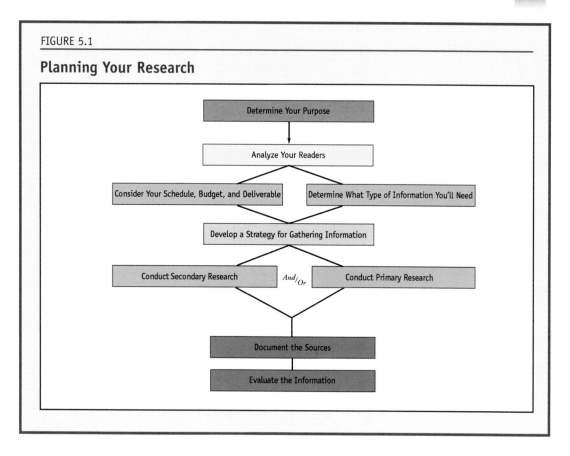

- **What types of tools can you use?** You use tools to help locate information in secondary sources—sometimes called *research media*. For example, will you use online catalogs or databases, or reference works such as encyclopedias, indexes, or abstract services?
- **What research media can you use?** Where can you find the information? For example, will you find reliable, up-to-date information in print sources, online databases, Web sites, discussion groups, or perhaps blogs?
- **What primary research methods can you use?** Should you use primary research methods to gather the information? Should you use primary research methods to verify information from secondary sources?

Use the ideas in the tip box on the next page for selecting and implementing a research plan.

TIPS FOR SELECTING AND IMPLEMENTING YOUR RESEARCH PLAN

- **Determine the appropriate methods for gathering reliable, up-to-date information.** For example, if you want to determine if people are stopping at an intersection, you would observe the intersection at different times of day. However, if you want to know how many accidents have occurred at that intersection, you would search Web sites or government documents. The information you need will determine how you will gather the information.
- **Take detailed notes.** Don't rely on your memory or on haphazard notes. Write down all needed information in a detailed, systematic manner. As you take these notes, include accurate, complete bibliographic information—enough information for your readers to locate the same information.
- **Record your data accurately and thoroughly.** Regardless of the value of the information (or data) that you find, if you don't record it accurately and thoroughly, the information is useless. Write it on paper or enter it online. If you are conducting an interview or making an observation, you might record that interview and video the observation; make sure to get permission before recording an interview or observation. If you are searching Web sites, cut and paste the URLs of the Web sites and save them with your notes. You might consider bookmarking the Web sites so you can easily return to those sites.
- **Record your information in a timely manner.** Don't rely on your memory; instead, record information in a timely manner. For example, if you conduct an interview, record the information during the interview—either in written notes or in a recording. If you find information from a Web site, cut and paste or bookmark the URL immediately; don't wait until you are ready to use that information in the document. You may find that you cannot find the Web site, or you may waste valuable time relocating the Web site.
- **Verify information by using more than one method.** Good researchers rely on more than one or two methods to verify information. Consider driving directions on the Web; these directions are often incorrect. If you want to provide directions for your coworker to a restaurant that you have never visited, you might get the directions online, look at a map, call the restaurant to verify the directions, and try the directions yourself.
- **Be persistent and patient.** You have probably become frustrated trying to find information on the Web or in the library. Even experienced researchers have trouble locating information—or locating *reliable* information. Be persistent and patient, as the information may not be easy to find.

SELECT APPROPRIATE SECONDARY RESEARCH METHODS

Secondary research is gathering information from previously documented research or studies. For example, if you want to research the climate of Mars, you might look at reports written by scientists who analyzed data from the Mars Pathfinder. Even though research in the workplace may rely on primary research, you should conduct secondary research along with primary research.

As a college student, you can find valuable secondary information by searching the Internet or your college or university library. These libraries have tools to help you locate information. They also have media storage where that information is housed or located. Your college or university library gives you access to resources that are not available to the general public; so take advantage of these resources. They will not only provide access to valuable information, they may also save you time. When you use your school's library, talk to reference librarians. They can help you locate the appropriate research tools and teach you how to use them. They are willing to suggest new ways to locate information or to tell you if your library doesn't have the information that you seek. Reference librarians can save you time and frustration.

> *Secondary research involves gathering information from previously documented research or studies.*

In the workplace, you may also find information in your organization's library. Most organizations call this library an *information resource center*. These centers collect information that is vital to the organizations' business. For some organizations, the information resource center has specialists who will do the research for you. For example, you might e-mail a question such as, "How many megawatts of power did Reliant generate between January and March of 2008?" The staff then researches the question and e-mails you the answer. Other organizations have staff that suggests where you can look for the information. If your organization does not have an information resource center, you can conduct research over the Internet or in company archives.

Understanding the Secondary Resources Available to You

You may better understand the wealth of secondary resources by categorizing them into the following types:

Print

Print or paper media still remains a popular medium for producing information. **Print media** includes books, journals, newsletters, trade magazines, reports, product materials, and other forms. Some books and journals are now also available in digital form. To find print media, you will use online catalogs.

Online Databases

Online databases provide access to extensive databases of journal articles, trade magazine articles, conference proceedings, newspapers, government publications, and other documents. Universities and many public libraries subscribe to these databases. At many libraries, you can search for the database by title or by subject. For example, if you are a biology major, you may find up to 27 databases specifically related to biology. Your library probably subscribes to extensive databases such as EBSCOhost, Infotrac, or Lexis-Nexis that provide access to many secondary media. Although you can access free databases online, they are not as extensive as those to which libraries subscribe. Also, free databases often do not access many journals.

Web Sites

Web sites provide access to thousands of pages of information, some of which is reliable, and much of which is not. To find information on the Web, you may be looking for a "needle in a haystack"; therefore, you need a strategy for looking for information on the Web:
- Search with a specific URL.
- Search with a keyword or phrase using a search engine such as Google or Yahoo!
- Search by subject using a search engine.

Online Discussion Groups

You can participate in two types of online discussions: newsgroups and electronic mailing lists, such as Listserv®.
- **Usenet newsgroups** (sometimes called bulletin boards) publish e-mail messages sent by members of a group. Usenet is a "worldwide style of discussion groups," or newsgroups (Hahn). The discussion groups are called newsgroups because the Usenet was originally designed to carry local news between two universities in North Carolina (Hahn). Usenets consist of thousands of newsgroups organized into basic categories such as business, computers, health, science and technology, and socie-

ty and humanities. You can even create your own group. In a news-group, e-mail is stored on a database where members can access it.

• ***Electronic mailing lists*** (like Listserv®) allow members to share infor-mation, to discuss issues, to ask questions, and to get answers. Like newsgroups, electronic mailing lists post e-mail messages sent by mem-bers who subscribe to the group. However, electronic mailing lists send the messages to every person who subscribes to the list. The messages come directly to your computer; you don't have to go get the message. If you subscribe to an electronic mailing list, be wary of information that you receive. If you plan to use the information, know its source.

Personal Publications, Such as Blogs, Podcasts, and Wikis

A ***blog*** is a Web log. Blogs may be personal diaries that are of little use in find-ing reliable information. However, some blogs are published by scientists, engi-neers, programmers, and other professionals; these blogs can help you under-stand topics being discussed in your field. A ***podcast*** is a collection of digital media files that you can download to an MP3 player. The information from blogs and podcasts may not be accurate, reliable, or complete; so be sure to ver-ify any information that you plan to use from these sources. A *wiki* is, in the general sense, a Web page that allows for universal editing. Contributors can access a wiki from any Web browser—adding, deleting, or reorganizing information on the Web page without restrictions. You are likely familiar with "Wikipedia." This online encyclopedia (www.wikipedia.org) is the result of the collaborative efforts of many contributors. Wikipedia has no editor; so the infor-mation may or may not be reliable or accurate.

> *A blog is a Web log. Some blogs are published by scientists, engineers, programmers, and other professionals.*

Using Tools to Find Information in Secondary Sources

To access the information in the various types of sources, you need tools. You have four tools available: online catalogs, indexes, reference works, and abstract services. Because you will use the computer in accessing these sources of information, tips for searching online appear on page 113.

Online Catalogs

You can use an online catalog to search for holdings at a specific library. An ***online catalog*** is a database of books, films, microforms, compact discs, recordings, maps, and other materials. For most college and university libraries, you can search their holdings from your desktop; you don't have to go to the library building to use the catalog. With many online catalogs, you can find out if the source you need is checked out by another library patron. To use the online catalog, follow the instructions on the library's Web site.

Indexes

Indexes are lists of books, newspapers, periodical articles, or other works on a particular subject. Indexes also list the documents available in government documents.

- **Periodical indexes** offer you an excellent source of some of the most current research information. (*Periodicals* are magazines and journals.) Although periodicals offer excellent information, you may have trouble identifying the articles related to your topic. Periodical indexes can make this task much easier. These indexes list articles classified according to title, subject, and/or author. Some indexes are general, covering multiple fields, such as the *Readers' Guide to Periodical Literature*. Some indexes also categorize articles by subject. For example, the following indexes list articles in particular fields or by discipline:
  - *Agricultural Index*
  - *Business and Periodicals Index*
  - *Applied Science and Technology Index*
  - *Engineering Index*

You can also use a directory search engine to access periodicals. Many search engines contain a searchable directory of specialty search engines; you can search using the terms "journals" or "periodicals."

When you need periodicals, ask your reference librarian which indexes will best help you find the information you need. Once you have gathered a list of periodical articles that may have the information you need, you have to locate those articles. Check your library's online catalog to determine if your library carries the periodical that you need. Be sure to check the volume number or years of the periodical; sometimes libraries don't carry all the volumes or may be missing volumes. If your library doesn't have the periodical or volume that you need, you may still get a copy of the article by using the following sources:

- **Interlibrary loan**. Go to the interlibrary loan office at your library. It will help you locate the nearest library that has the article. That library will photocopy the article and mail it to your library. This service can be slow; sometimes it takes two weeks or more to receive an article, and some libraries charge you a small fee for the service.

- **Online document-delivery services**. You access these services on the Web. These services search a database of periodicals. If the service has the article you need, it faxes or e-mails the article to you, usually within an hour. These services may charge a fee; however, if you're in a hurry, the article may be worth the fee.

 TIPS FOR SEARCHING ONLINE FOR INFORMATION

- **Use more than one search engine.** Search engines are not identical; they index only a portion of the information available online. Use more than one search engine to make sure that you get the most reliable, up-to-date information. If you don't find the information you need or if you get limited results, try a different search engine.
- **Use varied keywords or search phrases.** Vary the keywords. For example, if you are looking for information on greenhouse gases, you might search the terms "greenhouse gases," "greenhouse effect," "greenhouse gas emissions," "global warming," or "climate change." By using all of these terms, you will gather more information.
- **Use discipline-specific Web sites when possible.** Once you find a list of links, select the Web sites that are discipline specific. Let's consider the greenhouse gas example. If you search for greenhouse gas emissions, you will find links to information from the Environmental Protection Agency (EPA), Wikipedia, the Department of Energy, and *Science Daily*. The specialized Web sites would be the EPA, Department of Energy, and *Science Daily*, not Wikipedia.
- **Download only what you need.** Many Web sites include graphics, video files, and sound. If you only need the text, download only that information. Graphics, video files, and sound can take up valuable memory on your computer or flash drive.
- **Save or print what you need before the information changes or goes away.** Write down the URL or copy and paste the URL and the date that you accessed the information. Keep an electronic file of the URLs where you have accessed information. You will appreciate this file when you get ready to prepare your works cited or list of references.
- **Consider the ethics of using information from the Internet.** Information, videos, sounds, and graphics that you download from the Internet may be the intellectual property of an author or company. The information may be copyrighted; make sure that you credit the source and/or get permission for using the information, video, sound, or graphic.

TAKING IT INTO THE WORKPLACE
Copyright Laws and Your Research

Passed by the U.S. Congress, the Copyright Act of 1976 and the Digital Millennium Copyright Act of 1998 protect the authors of published and unpublished works. The European Union passed a similar directive, the EU Copyright Directive, in 2001. Under the U.S. laws, the author of a work is entitled to the profits if someone sells or distributes the work, except in cases of "fair use." For example, the information in a Web site may be copyrighted: the text, artwork, music, photographs, and audiovisual materials (including sounds). You can use copyright-protected works without the author's permission if you follow the fair use guidelines established by the Copyright Act of 1976. This law protects you if you use a small portion of an author's work to benefit the public. Fair use allows you to use works for education, research, criticism, news reporting, and scholarship purposes that are nonprofit in nature. To determine fair use, consider these four factors:

- **The purpose of the use.** Is the use for commercial, nonprofit, or educational purposes? If you are using the work for commercial purposes, you may be violating the copyright law. You are responsible for knowing if the work is copyrighted.

- **The nature and purpose of the copyrighted work.** If the information is essential for the good of the public, you may be able to use the copyrighted work without the author's permission.

- **The amount and substantiality of the portion of the work used.** The law doesn't give us strict guidelines, so you must use your judgment to determine how much of a copyrighted work you can use without the author's permission. For example, if you use 400 words of a 1,000-word document, you are not following fair use guidelines. If you use 400 words of a 100,000-word document, you are following fair use guidelines. If you are using even a part of a

iStockphoto 2008.

copyrighted graphic, you must get permission. You can't use a portion of a copyrighted graphic.

- **The effect of the use upon the potential market for or value of the copyrighted work**. If your use of the copyrighted document hurts the author's potential to profit from the work or hurts the potential value of the work, you have violated fair use guidelines.

The copyright laws protect authors even if they haven't registered that work with the U.S. Copyright Office. Copyright exists from the moment an author creates the work. As you write and create documents—online or in print—ask yourself these questions:

- Have you relied on information from copyrighted works that you did not write or create? If so, do you have permission from the author(s) to use the work?
- Have you appropriately cited sources or acknowledged that you have used the work by permission of the author?
- Are you unfairly profiting from the copyrighted work of another author(s)?
- Have you asked for legal advice? If you don't know whether you can legally use a portion of work, ask for advice from legal counsel. If you ask for advice, you may prevent legal action against you or your employer.
- Are you doing the right thing? For more information on ethics and doing the right thing, see Chapter 2.

Assignment

Visit these Web sites to learn more about copyright laws in the U.S.:
- U.S. Copyright Office: www.copyright.gov
- Copyright Clearance Center: www.copyright.com

After visiting these sites, write a memo to your instructor answering these questions:
- How do you know if a document is copyrighted?
- What is the Copyright Clearance Center? How can students and businesses use this center's Web site?
- How do you register a copyright?
- How long is a copyright protected?
- What works are protected by copyright?
- What is not protected by copyright?
- How does copyright affect you when conducting research?

- **Newspaper indexes** will tell you what topics various newspapers have covered. Newspapers contain many types of information, but they don't cover most subjects in depth. Newspapers may summarize some topics, especially local issues and present statistics, trends, and demographic information. Many major newspapers are indexed by subject. Some of the more important newspaper indexes are listed below:
  - The *New York Times*
  - The *Wall Street Journal*

  You can access many newspapers on the Web. However, the print and the online versions may vary.

- **Government document indexes**. The U.S. government may be the largest publisher in the world. The government publishes documents on business, science, the environment, engineering, health, and many other topics. You can find documents published by federal agencies and departments such as the Department of Labor, the Department of Housing and Urban Development, the Environmental Protection Agency, and the Internal Revenue Service. You can find these documents through libraries that are registered repositories for U.S. government documents. Many university and some large public libraries are repositories. They usually have a separate reference librarian and staff for the government documents because these documents are cataloged separately. Some libraries also house publications published by state agencies. These publications can provide information on topics such as food, nutrition, agriculture, animal science, natural resources, and state demographics.

  Government documents may not be listed in online catalogs at your library. Unlike other documents in the library, government documents are classified according to the Superintendent of Documents system (not the Library of Congress system). You might think of government documents as a separate library within the library. You can also find many government documents on the Internet. To locate information in government documents, use the following indexes:
  - *The Monthly Catalog of the United States Government.* Although this catalog isn't technically an index, it provides a roadmap to locating government documents. You can access the

catalog on paper, on CD-ROM, or on the Internet at http://catalog.gpo.gov:80/F.

- *Fedworld Information Network.* This federal Web site has searchable databases and indexes to locate information produced by the federal government. You can find the Web site at http://www.fedworld.gov.

To effectively and efficiently locate government documents on the Web, you can start by going to www.gpoaccess.gov or to www.USA.gov. At www.gpoaccess.gov, you can search for government publications. As you can see in Figure 5.2, page 118, this Web site is easy to use. If you plan to use government publications, you can also talk with a librarian in the government documents section.

Reference Works

Any time you look up a word in a dictionary or look up a topic in an encyclopedia, you are using a ***reference work***. Reference works include

- general dictionaries
- specialized dictionaries
- encyclopedias
- almanacs
- atlases and maps
- writing style manuals
- thesauri
- other reference tools

You can access these reference works in print or online. For example, you might look at the following guides to reference works:

- *Guide to Reference*, published online by the American Library Association, www.guidetoreference.org/HomePage.aspx.. This online guide to reference sources in print and on the Internet is organized by subject.
- *Internet Public Library* at www.ipl.org. At this site, you can navigate to reference works available on the Internet.

You can also go to your library's Web site and navigate to the reference (sometimes called ready reference or quick reference). At this area of your library's Web site, you will find links to free Internet sources.

FIGURE 5.2

Web Site for Locating Government Publications

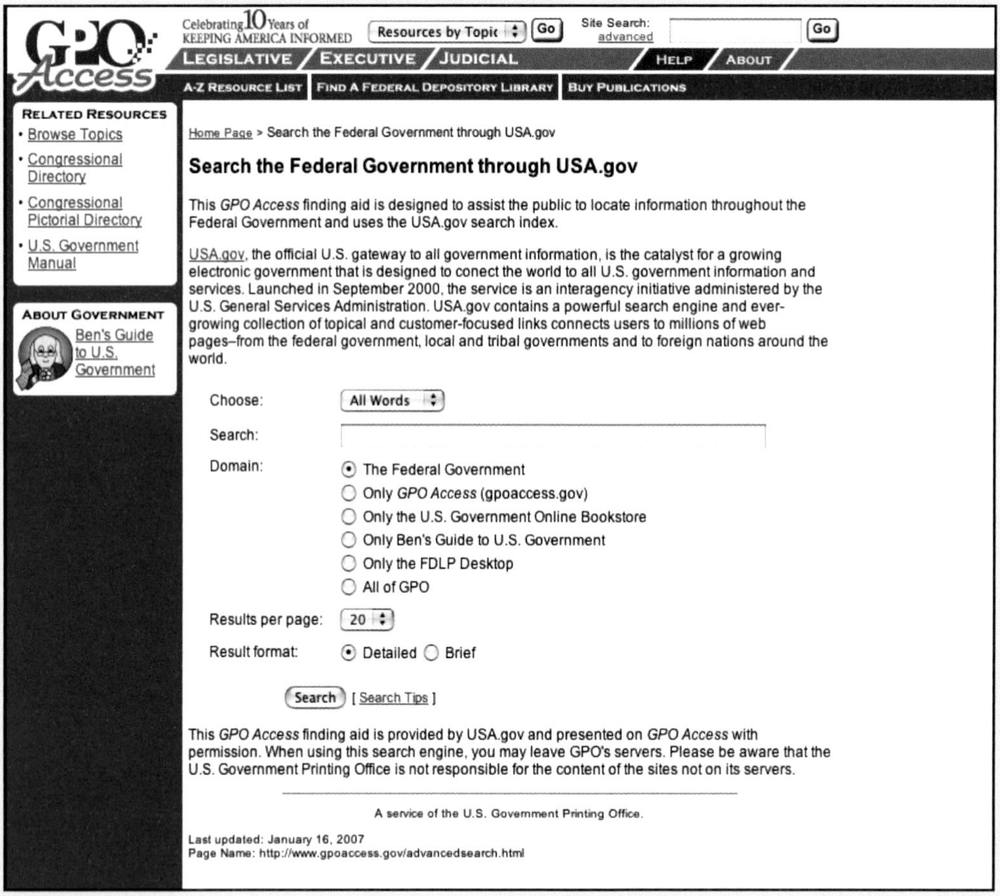

Source: Downloaded from the World Wide Web, November 2008: www.gpoaccess.gov-advancedsearch.html. GPO Access.

Abstract Services

Abstract services go one step beyond indexes. Along with bibliographic information, abstract services include a summary of the article. These abstracts can save you time. By reading the abstract before you locate the article, you can decide whether the article contains the kind of information you are seeking.

Abstract services cover
- a broad field, such as *SciFinder Scholar*, the electronic version of *Chemical Abstracts*
- a specialized field, such as *Abstract in Social Gerontology*, which provides abstracts of articles related to gerontology (the study of aging)

Your library's Web site has links to many abstract services where you can read abstracts and decide whether to get the full-text version of an article.

Evaluate the Information

Once you have gathered information, you must evaluate that information and your sources. Just because the information is published in print or online doesn't mean that it is reliable. Some information may be out-of-date, incomplete, incorrect, misleading, or biased. The source of the information may not be reliable or credible. Use these questions as you evaluate information:
- **Is the information credible and unbiased?** Does the author or source include credentials? Are the credentials credible, relevant, and reliable? Can you find a section such as "About Us"? Does the author or source have a financial or political stake in the information? If the author or source has such a stake, the information may be biased and unreliable.
- **Is the information up-to-date?** You want to base your research and your conclusions on the most current information. You can sacrifice your professional reputation by basing your decisions or your research on out-of-date information.
- **Is the information complete?** Does information seem to be missing? Have the researchers or authors covered all areas of the topic? Is the information sufficiently detailed for your research?
- **Is the information accurate?** The information should be accurate; you can verify accuracy by using more than one source or conducting observations or experiments. You should also make sure that any estimates used in your research are accurate.

You may find information from the Internet to be the most difficult to evaluate. Anyone may publish information on the Internet—and the disreputable sources are just as prominent as the reputable ones. At first glance, disreputable sources may look equally reliable; so you have to learn how to sort the reliable from the unreliable, the reputable from the disreputable. Few, if any, standards regulate publications that appear on the Internet. Many online publications have no overseeing editorial board. Currently, no market forces drive incompetent or unreliable publications off the Internet; so carefully evaluate any online sources that you use in your research. As you evaluate these sources, consider the following guidelines and the questions listed in Figure 5.3:

> *To ensure you have compiled complete, up-to-date information, use print sources, as well as the Internet, to conduct your research. Not all print sources have been transferred to digital records.*

- **The Internet may not contain all the information available about your topic**. Even in this digital age, not all printed records have been transferred to digital storage; you may find important information about your topic in traditional print sources, such as journals, books, and trade publications. Although many journals are now available online, some excellent journals are still only available by subscription or in print. Until all print sources are transferred to digital records, be sure to include print sources as part of your search strategy.
- **No search engine indexes all information on the Internet.** Therefore, use a combination of search engines and online tools.
- **Some online sources are not up-to-date**. You cannot tell whether a site is current by simply looking at the site. You need to check when the site was last updated.
- **Some search engines and subject directories typically index sites that have many links to them** (Lawrence and Giles). Search results, then, will be populated with results from the more popular sites. A site's popularity may have nothing to do with the quality or reliability of its information.

SELECT PRIMARY RESEARCH TOOLS

Primary research (sometimes called original or firsthand research) is gathering information for the first time—not relying on research previously conducted by others. If you interview an expert or conduct an experiment, you are using primary research tools. In this section, you will learn some of the more common primary research tools:

- interviews
- inquiry letters and e-mails
- surveys and questionnaires
- observations and experiments

FIGURE 5.3

Questions for Evaluating Online Sources[1]

Is the Online Source . . .	Questions to Ask	Comments
Written or prepared by reliable authors?	• Who created the Web site or the information? • What are the credentials of the author(s)? • If you don't recognize the name of the author(s), company, or organization, did you find the site through another reliable source? • Does the site contain biographical information on the author(s)? • Can you find references to the author's, company's, or organization's credentials or work?	Anyone, whether qualified or not, can publish information online and all voices appear equal. • Be wary of using information from an author, company, or organization you don't recognize, or from a source that you cannot verify as credible. • If you cannot find who operates a Web site, consider using information from another source.
Published by a reliable group?	• Who is the publisher of the Web site? • Why is the publisher qualified to produce the site? • Is the publisher affiliated with a reputable group or organization? • Does the domain name give you a clue to the publisher and its purpose? For example, epa.gov tells you that the publisher is a government institution. • Does the Web site include contact information for the publisher?	If the Web site is published from a personal account with an Internet service provider, the site may contain unreliable information. If the information looks interesting or valuable, verify that information through other sources. If you cannot verify it, don't use it.
Up-to-date?	• When was the Web site or document created or published? • When was the site or document updated or revised? Has it been updated in the last three months? • Are the links up-to-date? Do the links work? • Is the site "under construction" or only partially complete?	If the Web site has not been updated in the last three months, the information may not be up-to-date or reliable. If the site or item is "under construction," you may want to use another Web site or source. If the links don't work, the site may not contain up-to-date information.
Presented clearly and accurately?	• Is the site well constructed? Does it have a professional appearance and design? • Does the site follow basic rules of grammar, spelling, and punctuation? • Do the authors cite sources? Do they support all claims with appropriate evidence? • Does the information appear biased?	If a site looks unprofessional, the information that it contains may not be reliable. If the site doesn't follow the basic rules of grammar, spelling, and punctuation, the information may not be reliable or accurate. If the authors do not cite sources or support their claims, use other sources. A reliable site • has a professional, well-constructed design • follows the basic conventions of grammar, punctuation, and spelling • supports all claims • documents sources • presents unbiased information or tells you when authors are stating opinions

[1]Some questions adapted from *Web Site Evaluation,* published by the University of North Texas Libraries.

Interviews

You can gather valuable firsthand information from informal and formal interviews. *Interviews* give you the opportunity to gather factual information from an expert, to hear firsthand observations of a situation, and to discover the experiences of the interviewee. For example, you may want to learn how a proposed road expansion will affect small businesses located next to the road; so you might interview some of the business owners. From these interviews, you may learn how the business owners perceive the expansion will affect their businesses both during the construction of the road and after the road is completed. You can conduct interviews in person, by phone, or by e-mail.

Informally Interview Coworkers and Colleagues

Informal interviews may be informal conversations in the office, by telephone, or by e-mail. Informal interviews can be a valuable tool for gathering information. For many writing situations, informal interviews are the primary means of gathering information. Review the tips for conducting successful informal interviews on the next page.

Formally Interview People Whom You Don't Work With or Know Well

Formal interviews are the best choice for gathering accurate and thorough information from people whom you don't work with or know well. Use the tips on page 124 as you prepare for and conduct your formal interviews. Too frequently, beginners go to an interview with only a notepad and pencil—expecting to ad-lib the questions. Although this strategy (or lack of it) may occasionally work for experienced interviewers, most researchers who arrive at an interview without a written list of questions will leave without much of the information they intended to gather.

Inquiry Letters and E-Mails

Instead of an interview, you can send an inquiry letter or e-mail. Many people prefer the ease of an e-mail inquiry; they can simply click on the reply button to respond. Be aware, however, that because many respondents receive junk e-mail, they may ignore your inquiry if they do not know you or recognize your e-mail address. On the other hand, many organizations and experts will respond to inquiry e-mails and send you detailed answers to your questions. If your topic is sensitive or confidential, consider sending a more formal inquiry letter.

Although inquiry letters and e-mails are convenient, they are not as useful as face-to-face interviews for the following reasons:

- In most cases, you cannot ask follow-up questions or ask the respondent to clarify.
- The respondent may misunderstand your question.
- Because the respondent has not agreed to answer your inquiry, he or she may not respond.

Surveys and Questionnaires

Questionnaires, sometimes called surveys, are used for surveying large groups. If you want to gather information from a large group (target population), you can survey a sample of that group. You might use a questionnaire to gather information about a group's preferences, attitudes, or beliefs. For example, if you are planning to change the company's health plan, you might send a questionnaire to the employees to determine their opinions about the current company health plan. You can also use questionnaires to develop profiles about a

---

 TIPS FOR CONDUCTING INFORMAL INTERVIEWS

- **As you plan your research, write down questions that you might ask.**
- **Make a list of all the people who could answer your questions.** From this list, select people whom you work with or know well to interview informally. (Set up formal interviews with people whom you don't work with or know well.)
- **Try to interview these people when they are least busy.** When people are less busy, they may share information more freely and answer your questions more accurately.
- **Be willing to return the favor.** If you are open to answering your coworkers' questions, they, in turn, will be more willing to answer yours.
- **If you use e-mail to interview, ask your interviewees if they would like to have the questions inserted directly into the e-mail message or attached in a file.** You can also simply do both: attach the file and put the questions into the e-mail. The interviewees can then decide which format they prefer.

TIPS FOR EFFECTIVE FORMAL INTERVIEWS

Determine the Purpose of the Interview	**Before the interview** • Identify the purpose of the interview. • Identify the specific information you want to gather. Write down your purpose. For example, *I will interview Richard Hampton, Senior Engineer at PEC, to ask about the effect of pipe vibration on the plant's operation and the operations that trigger the vibrations. I will also ask what he believes is causing the vibration.*
Contact the Interviewee to Set Up the Interview	**When you call or e-mail** • State the purpose of your interview, so the interviewee can tell you whether he or she can give you the needed information. If you need the information by a specific date, tell the interviewee. • Be flexible about the time. As possible, let the interviewee determine the time and date for the interview. • Set the interview at the interviewee's office or at a location most convenient for the interviewee–even if the location is not convenient for you. • Ask permission to record the interview if you plan to use audio- or video-recording equipment. • Offer to submit questions ahead of time, if possible.

Prepare for the Interview	**Before the interview, find out as much about your subject and the interviewee as possible and plan your questions.**
	• Gather background information. Look at secondary sources. You may look unprepared if you ask questions that the professional literature answers.
	• Plan specific, open-ended questions. Avoid questions that can be answered with "yes" or "no."
	  • **Yes/No Question:** *Can greenhouse emissions affect children's health?*
	  • **Open-Ended Question:** *How do greenhouse emissions affect children's health?*
	• Avoid leading, biased questions:
	  • **Leading/Biased:** *Don't you agree that greenhouse emissions are one of the greatest dangers to children's health in large cities?*
	  • **Impartial:** *How do you see greenhouse emissions as affecting children's health in large cities?*
	• Write each question on a separate notecard. You can also write the questions in a file on your laptop or on a notepad. Be sure to use ample white space so you can see the questions and summarize the answers.

Conduct the Interview	**Before the interview**
	• Give yourself plenty of time to get to the interview.
	• Arrive on time (or a little early).
	• Check your video- and audio-recording equipment if you plan to use it.
	
	At the beginning of the interview
	• Thank the interviewee for agreeing to do the interview.
	• State the purpose of the interview.
	• Tell the interviewee how you plan to use the information.
	• Respect cultural differences. Make sure you use the appropriate level of formality, politeness, and other behaviors in the given culture. (See Chapter 4.)
	
	During the interview
	• Let the interviewee do most of the talking.
	• Maintain eye contact. Your interviewee will give you more information if you seem sincerely interested.
	• Use your prepared questions.
	• Ask follow-up questions, such as,
	  • *Can you give me an example?*
	  • *What additional actions would you suggest?*
	• Ask the interviewee to clarify any answer you don't understand by asking additional questions, such as,
	  • *Can you give me an example?*
	  • *Can you simplify?*
	  • *Would you go over that again?*
	• Verbally summarize the interviewee's answers to make sure that you understand.

	• If the interviewee gets off the subject, be prepared to respectfully move the interview back to the intended focus. You might say, *"Thank you for that interesting information. I know our time is short, so let's move on to the next question."*
	 • Take notes only as necessary. Be careful to write only important information. Don't write so much that you can't maintain some eye contact with the interviewee or that the interviewee has to wait while you complete your notes.
	
	 At the end of the interview
	 • Ask the interviewee for final comments.
	   • *Would you like to add any other information?*
	   • *Can you suggest other sources or people whom I should interview?*
	 • Ask permission to quote the interviewee (if you have not already done so).
	 • Ask for the interviewee's correct title and position.
	 • Ask if you may contact the interviewee again if you have questions or need to verify information.
	 • Offer to send the interviewee a copy of the document, so he or she may review the information for accuracy.
	 • Thank the interviewee for his or her time.
	 • Leave promptly.

Follow Up	**After the interview** • Write down the information that you want to remember from the interview.

group. For example, you might gather demographics about consumers as you ask them what product brand they prefer or what newspaper they prefer. Questionnaires, however, have three disadvantages:

- The response rate for questionnaires is poor, usually only 15 to 20 percent.
- You can't guarantee that the respondents are a representative sample, so you must carefully draw conclusions from the information that you gather from a questionnaire. The respondents who choose to respond may have a bias about the subject of the questionnaire (Plumb and Spyridakis).
- You won't know if respondents misinterpret a question because questionnaires don't allow you to follow up and clarify, as you can in an interview (Plumb and Spyridakis).

ETHICS NOTE

Report and Analyze Results Ethically

You must guard against bias when analyzing collected data; you may be tempted to see what you want to see or what your organization wants you to see—not what the data actually shows. As we discussed in Chapter 2, you may face ethical challenges when analyzing and reporting the results of your observations, experiments, or other research. You must analyze and report your results in an honest manner. As you report your results, remember the following:

- Do not mislead your readers.
- Use clear language.
- Include all the information that readers need or have a right to know.

If you decide to use a questionnaire, remember that you are asking the respondents to do you a favor—with little, if any, benefit to them. For this reason, make sure that the questionnaire is as efficient and simple as possible.

Determining the Purpose, Sample Group, and Method

Before you begin preparing the questions for your questionnaire, ask yourself these questions:

- **What is the purpose of this questionnaire?** What do I want to learn?
- **What kinds of questions will I ask?**
- **What will I do with the information that I collect?** (Plumb and Spyridakis)
- **Who is the target group I want to question?**
- **How will I select the target population (the intended respondents)?** Will I select the respondents at random? See Figure 5.4 for questions that you might consider as you select the target group.
- **How many questionnaires will I send?** The more questionnaires you send, the more responses you will receive.
- **How will I administer the questionnaire—by mail, by e-mail, by phone, or in person?** Mailed questionnaires take less time; but they are expensive, and fewer people will respond. You will receive more and quicker responses when you administer the questionnaire by e-mail, by phone, or in person. However, some respondents are uncomfortable, even annoyed, by phone questionnaires. By phone or in person, respondents may respond with less candor than with mail or e-mail questionnaires. If you personally deliver a questionnaire, the response rate may be higher, but the geographic area you can survey is greatly reduced. To guarantee a higher response rate, personally pick up the questionnaires or offer to wait while participants fill out the questionnaire. Tell participants when you will return.

Preparing Effective Questions

Figure 5.5, page 128, illustrates types of questions that you might include in a questionnaire. Both closed-ended and open-ended questions are appropriate. *Closed-ended questions* elicit answers that you can count or quantify. With closed-ended questions, you limit respondents' answers. The respondents select only from the answers you provide. With open-ended questions, respondents answer with their own words. A question-

---

FIGURE 5.4

Selecting a Target Group

To narrow the group you will survey, identify the following characteristics about target group members:
- Do group members need a certain level of education? If so, what level?
- Should group members be a certain age? If so, what age?
- Must group members live in a specific geographic area? If so, where?
- Is the target group best represented by only males, females, or both genders?
- Should group members have a certain level of knowledge about the topic? If so, what knowledge level?
- Must group members be savvy users of a certain type of tool or software? If so, what tool or software?
- Must group members share a particular language? If so, what language?
- Must group members have a certain level of income? If so, what level?

---

naire that uses closed-ended questions might include
- multiple-choice questions
- yes/no questions
- ranking questions that ask respondents to arrange items in order of preference
- ranking questions that ask respondents to rate items on a scale

Because closed-ended or quantitative questions yield totals and percentages, you can use software to read and tabulate the answers. Such questions are particularly valuable when you have a large number of respondents.

Open-ended questions yield answers that you can't easily quantify, but they may give you valuable information that you can't gain through closed-ended questions. Open-ended questions tend to elicit more accurate information because they don't limit the respondents to the writer's suggested answers (Anderson). For example, with the question, "Why didn't you choose to use the Fleet Assistance Program when you bought your 2008 car?" respondents can list whatever reason may come to mind, whether or not the writer of the questionnaire has considered it as a possibility. Figure 5.6A and 5.6B, pages 131-132, show a sample cover letter and accompanying questionnaire. The questionnaire includes both closed-

FIGURE 5.5

Types of Questions Used in Questionnaires

Type of Question	Type of Question	Sample Question
Multiple choice	Respondents select from one or more alternatives	What is your classification? ____ Freshman ____ Junior ____ Sophomore ____ Senior ____ Graduate
Yes or No	Respondents select "yes" or "no"	Would you use the express rail service to commute to your office? ____ Yes ____ No
Likert scale	Respondents rank their responses along a scale. Select an even number of choices. Plumb and Spyridakis recommend including no fewer than 3 and no more than 11 choices.	The online reporting system has made my job easier. Circle your response. Strongly disagree Disagree Agree Strongly agree
Semantic differential scale	Respondents rate their responses along a continuum of opposing concepts. Limit the choices to no fewer than 3 and no more than 11 choices. Use this type of question to measure attitudes and feelings.	Your technical communication class is Easy __ __ __ __ Hard Interesting __ __ __ __ Boring Current __ __ __ __ Out of date
Ranking	Respondents rank (prioritize) their responses.	Rank the importance of the following factors when you purchase a new car. Put a 1 next to the most important factor, a 2 next to the second most important factor, and so on. ____ Fuel economy ____ Safety ratings ____ Price ____ Consumer satisfaction ratings ____ Size ____ Dealership ____ Color ____ Upgrades available
Formal rating scale	Respondents rate an item or quality on a specific scale, usually 1 to 5 or 1 to 10.	How do you rate your satisfaction with your new vehicle? Please circle only one rating with 1 being not satisfied and 5 being satisfied. 1 2 3 4 5 Not satisfied Satisfied
Checklists	Respondents check one response. With an expanded checklist, respondents may check more than one response. (See the sample question to the right.) Plumb and Spyridakis point out that "to ensure that all respondents interpret the questions similarly, the question instructs respondents to 'Check all that apply.' The 'Other' response option is provided in case the researcher has not considered all possible activities."	Which of the following activities describe tasks you complete daily using a computer? (Check all that apply.) ____ Check and send e-mail ____ Surf the Internet ____ Communicate with friends ____ Pay bills ____ Watch videos ____ Get directions ____ Get news and weather ____ Check my stocks ____ Use word-processing software ____ Prepare spreadsheets ____ Other (please explain) _____
Short answer/essay	Respondents answer open-ended questions using phrases or sentences.	How well do you think telecommuting will work in your department? What do you believe are the advantages and disadvantages of telecommuting?

TIPS FOR PREPARING AN EFFECTIVE QUESTIONNAIRE

- **Write unambiguous questions.** If respondents can interpret a question in more than one way, your results will be meaningless.
- **Use simple, plain language.** Plumb and Spyridakis recommend avoiding "technical terms, acronyms, and abstract or ambiguous words."
- **Avoid questions that influence your respondents' answers or that indicate your opinions.** For example, the following questions unnecessarily influence, or lead, readers:
   "Do you think that curbside recycling is an environmentally sound idea?"
   "Do you think that curbside recycling is a waste of the city's valuable tax dollars?"
- **Test your questions on a small group of respondents to make sure the questions are clear and unambiguous before you finalize the questionnaire.** Correct any problems with the questionnaire. If you have time, test the questionnaire a second time. Once you send out the questionnaire, you can't revise it.
- **Ask only the questions you need.** Respect your respondents' time. Respondents often ignore questionnaires that are long or that seem to waste their time.
- **Attach a cover letter.** Clearly and concisely explain the purpose of your questionnaire.
- **Explain the significance of the questionnaire.** Persuade your respondents that their responses will benefit them, their workplace, or their community.
- **Tell the respondents when you will pick up the questionnaire or when they should return it.** Include this information in the cover letter or "cover e-mail." Include a self-addressed, postage-paid envelope for returning the questionnaire (if you're using traditional mail), or use an electronic format. Your respondents will be more likely to complete the survey and return it to you in a timely manner if they can simply drop it in the nearest mailbox or respond to an e-mail.
- **In the cover letter or e-mail, thank respondents for their time.**

ended and open-ended questions. As you write your questionnaire, use the tips for preparing an effective questionaire found above.

Observations and Experiments

Once you have gathered adequate background information, you may decide to observe the problem or situation. In our example at the beginning of the chapter, for instance, Bill might personally observe the vibrating pipes. If you directly observe people and situations as part of your research, remember these guidelines:

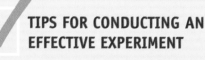

TIPS FOR CONDUCTING AN EFFECTIVE EXPERIMENT

- **Establish a hypothesis.** A *hypothesis* is an assumption. For example, to determine the relationship between calcium supplements in preventing bone density loss (or osteoporosis), you might test the following hypothesis: *Women will lose less bone density when taking 600 milligrams per day of calcium.*

- **Test the hypothesis.** You will need an experimental group and a control group. For example, to test our hypothesis, we would need an experimental group of women that takes 600 milligrams of calcium for one year and a control group that takes no calcium supplements for one year. We would conduct bone density tests on each woman in each group before and after the experiment.

- **Analyze the data.** Once you have all the data from the experiment, you have to understand the data so you can determine whether your hypothesis is true.

- **Report the data.** After you analyze the data, you will report what you have learned. You will learn more about reports in Chapters 17 and 18.

- **Gather background information**. Before you observe, gather as much background information as possible so you don't waste your time.

- **Know what to look for.** Do your homework. Find out what you are looking for and where to find it.

- **Take notes as you observe**. As you observe, answer the questions Who? What? When? Where? Why? How? Record your observations immediately; don't rely on memory.

An *experiment* is a controlled observation where you test a hypothesis. Each field has specific guidelines for designing and conducting experiments. If you decide to conduct an experiment, spend time designing the experiment; you will only have valid results if you have designed an effective and reliable experiment. To learn about experimental design in your field, visit a Web site on research methods used in your field. Follow the tips at left for conducting an effective experiment.

FIGURE 5.6A

Sample Cover Letter for a Questionnaire

January 30, 2008

Department of Linguistics and Technical Communication
University of North Texas
Denton, TX 76203

Name
Title
Company Name
Address

Dear _____ :

I am researching the types of writing done by entry-level employees and the types of software used for that writing. You are receiving this questionnaire because you have hired graduates with a certificate in technical communication from the University of North Texas. Your responses will help us better understand how to prepare graduates for your company. I will use this information in a recommendation report for the faculty and lab staff in the technical communication program.

I am specifically interested in the following information:
- whether entry-level employees are prepared for the types of writing done in the workplace
- the types of writing done by entry-level employees in the workplace
- the types of software used by entry-level employees for writing tasks
- whether our curriculum is preparing entry-level employees for the writing tasks
- whether our labs provide relevant software training.

Please take a few minutes to complete this questionnaire. Your responses will remain anonymous. If possible, return the survey by fax (940-555-5555) or by e-mail (labsurvey@unt.edu). I appreciate you taking time to respond.

Sincerely,

Elizabeth Smidt
Technical Communication Student

FIGURE 5.6B

Sample Questionnaire

Technical Communication Questionnaire

1. Describe the work of your company (manufacturing, service, health care, high tech, etc.)

   _____

2. How many employees work for your company?
   ____ less than 10   ____ 100-200

   ____ 10-25   ____ 200-300

   ____ 25-50   ____ 300-400

   ____ 50-100   ____ more than 400

3. What types of documents do entry-level employees write? Check all that apply.
   _____ e-mail   _____ feasibility reports
   _____ letters and memos   _____ progress/status reports
   _____ PowerPoint presentations   _____ procedures
   _____ proposals   _____ white papers
   _____ definitions   _____ Web pages
   _____ product descriptions/definitions _____ manuals
   _____ other (please explain) _____

4. What types of software do entry-level employees use to prepare documents? Check all that apply.
   _____ word processing   _____ desktop publishing
   _____ Web design and development   _____ graphics
   _____ spreadsheet
   _____ other (please explain) _____

5. How would you rate the preparedness of your entry-level employees for writing tasks (with 1 being not prepared and 5 being highly prepared)?

   1   2   3   4   5

6. How would you characterize the writing skills of your entry-level employees?

   very poor   poor   good   very good   excellent

7. What percentage of writing tasks do entry-level employees perform in a collaborative setting?
   _____ none   _____ 50%-70%
   _____ less than 20%   _____ more than 70%
   _____ 20%-50%

8. Do you provide additional writing training for entry-level employees within the first year of their employment?
   ____ Yes   _____ No

9. Please provide any additional comments that may help us better prepare our graduates.

   _____
   _____

CASE STUDY ANALYSIS
Prominent Doctor Admits to Creating Phoney Research

Background

When his department chair asked him to clear up some research data discrepancies, Dr. Andrew Friedman—a prominent surgeon and researcher at Brigham and Women's Hospital and Harvard Medical School—pulled patient files and frantically wrote in information to support his research findings. His deception was discovered days later when he confessed to colleagues and managers that he had manipulated data. Because his research had been published in some of the nation's top medical journals, he retracted his articles.

As a result of his deceptive research practices, Dr. Friedman was punished with a $10,000 fine, a one-year suspension of his medical license, and exclusion from federally funded research for three years. However, because he was from a family prominent in medicine and research and he had a highly regarded personal reputation that included groundbreaking work, awards, honors, and more than 100 published works, he was hired as a drug consultant with pharmaceutical companies for the next three years. During that time, he also volunteered with the American Red Cross and attended ethics and record-keeping lectures.

Later, he petitioned to have his license reinstated and was hired at Ortho-McNeil Pharmaceuticals as director of women's health care. In that position, he designs and reviews clinical trials on hormonal birth control, including writing the information inserts for products. He also gives lectures and appears as a media expert on hormonal birth control methods.

Dr. Friedman attributed his deceptive research to the professional pressure he was under to publish articles in addition to maintaining his surgical, patient, teaching, and lecture schedule.

Assignment

1. Working with a team, write a procedure for Brigham and Women's Hospital that details a way to ensure research and articles prepared by doctors of the institution are not falsified or plagiarized.

    a. Use information found in this chapter to support your procedure.

    b. Have one student present your team's procedure to the class.

2. Assume that you are the hiring director at Ortho-McNeil Pharmaceuticals. Draft a press release that details why you have faith in your hiring of Dr. Andrew Friedman. Specifically address concerns that patients might have about drug research conducted by Dr. Friedman. Turn in your press release to your instructor.

*Compiled from information downloaded from the World Wide Web; July 10, 2005: http://www.msnbc.msn.com/id/8474936. "Charges of fake research hit new high; Doctors accused of making up data in medical studies." The Associated Press, updated 3:17 p.m. CT. Original article, copyright 2008 The Associated Press. All rights reserved.

EXERCISES

1. **Web exercise:** Use a search engine to find five Web sites about a current topic in your field. For example, if you are majoring in engineering, you might research using ethanol for fueling vehicles. If you're majoring in biology or chemistry, you might look at the impact of decreased acid rain on the water quality of our rivers and lakes. Visit each Web site and do the following:
   - Print the home page of the site.
   - Locate the contact information of the publisher of the site.
   - Write a memo to your instructor evaluating the reliability of these sites. Use the questions in Figure 5.3, page 121, to guide you as you evaluate the sites.

Download a Worksheet for Researching Information online at www.kendallhunt.com/ technicalcommunication.

2. Using the topic you selected for Exercise 1, develop a research strategy for developing that topic into a report:
   - Make sure that you have appropriately narrowed the topic.
   - Prepare a preliminary bibliography of the sources that you might use for your report. Include at least five print sources, five Web sites, and five articles.
   - Determine what primary sources will be appropriate for your topic.
   - Write a memo to your instructor explaining your strategy. Be sure to a list all of your planned secondary sources, with complete bibliographic information.

3. Interview someone who works with a non-profit group that does service (philanthropy) work on your campus or in your community. The purpose of the interview is to learn the following:
   - types of services the group performs
   - purpose of the group
   - qualifications and skills of the volunteers
   - current needs of the group
   - examples of work done by the group

   After the interview, write a memo to your classmates summarizing your interview, and send a copy of your interview questions to your instructor.

4. **Web exercise:** Using search engines, answer the following questions related to your major. Be sure to print a copy of the homepage of any Web site that you use and include complete bibliographic information for each site. Put your answers in a memo to your instructor:

    - What are two professional organizations in your field?
    - What publications do each of the professional organizations publish? (For example, does either of the organizations publish a journal or magazine?)
    - How often do the publications appear—monthly, quarterly, annually?
    - Do the organizations have a discussion group or electronic mailing lists? If so, how do you join? Print a copy of the instructions for joining the discussion group or electronic mailing lists.
    - Do the organizations have a code of ethics? If so, print a copy of the code and attach it to the memo to your instructor.
    - Do the organizations have a student rate for joining?

5. **Collaborative exercise:** Working with a team, find a topic for a report you will write for your classmates. You might consider any of the following topics:

    - A problem on your campus, such as parking, crowded classrooms, etc. (Identify the following: What is the problem? Why is it a problem? Whom does the problem affect? How would you solve the problem?)
    - A problem in your community. You might search the local newspapers to find possible topics. (Identify the following: What is the problem? Why is it a problem? Whom does the problem affect? How would you solve the problem?)
    - An environmental issue. (Identify the following: What is the problem? How does the problem affect the environment? What solutions have been proposed? Which solutions are feasible? What solution would you suggest?)

    Once you have decided on a topic, your team should do the following:

    - Plan a research strategy. E-mail a summary of your plan to your instructor.
    - Select at least one primary research tool for gathering information. Decide who will conduct this research.

- Select appropriate secondary research methods. Use a variety of methods and media and take detailed notes.
- Evaluate the sources using the questions. Send an e-mail to your instructor summarizing your evaluation.
- Prepare a report for your classmates explaining the problem or issue that you researched and your proposed solution. (See Chapters 17 and 18 for information on reports.) Your instructor may want you to present your report orally. If so, see Chapter 20.

REAL WORLD EXPERIENCE
Discovering Job Prospects in Your Field

Background

You will soon be graduating from college and looking for a job in your field. You want to learn more about what to expect when you begin looking for that job. You want to answer the following questions:
- What are the job prospects in my field, internationally and locally?
- What areas of specialization seem to have the best prospects for employment?
- What types of skills and knowledge are employers expecting from new college graduates?
- What annual salary should I expect to receive for an entry-level job?
- What Web sites can I use to look for a job and to post my resume?

Assignment

1. Develop a research plan to answer the questions listed above about employment in your field.
2. Find secondary sources giving you employment and salary information. Include two citations from each of the following sources:
   - government or state publications
   - periodical publications such as trade magazines, journals, or newspapers
   - Web sites
3. Use a primary research tool to gather information to answer the questions. You may use any of the following tools:

- Interview a professional in your field or a job recruiter in your institution's job placement center.
- Survey faculty and employers in your field.
- Send letter or e-mail inquiries to two or more employers or professionals. Be sure to save a copy of the inquiry and the responses. If you don't receive responses, select another tool.

4. Write a memo to your classmates discussing the answers to the questions. Include a works cited page to document your sources. (See Chapter 12 for information on writing a memo.)

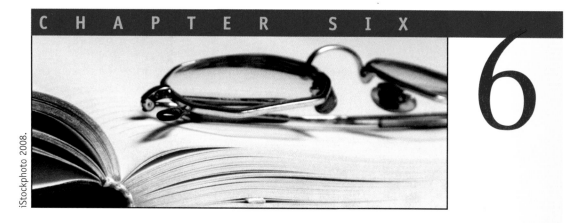

CHAPTER SIX 6

iStockphoto 2008.

Organizing Information for Your Readers

Paul Das sits at his desk looking at the pages of data he has collected for a manual on minimizing pollution from waste disposal sites. Paul works in a research laboratory where he and his coworkers develop technology for cities to use in treating and managing wastewater and solid and hazardous waste. They have a contract with the U.S. Environmental Protection Agency (EPA) to prepare this manual. After eighteen months, they have completed their research on ways to minimize pollution; Paul is ready to begin writing the manual. He considers the ways he can organize the information and realizes that structuring it is more than just preparing an outline. It is a process that goes beyond the data he has collected; it involves understanding the readers, the expectations of his workplace, and his purpose.

In this chapter, you will learn guidelines to help you effectively organize your documents. You will also learn nine possible organizational patterns for the documents that you write.

DECIDE HOW TO ORGANIZE YOUR DOCUMENT[1]

You may have two or more options for organizing information. By recognizing these options, you can select the one that will work best for your readers. Ask yourself the following question to determine the most effec-

[1] Based on a suggested approach to arranging business documents created by Jack Selzer in *Arranging Business Prose.*

tive structure: *Can I group similar information?* When you combine similar (closely related) information, you help your readers to gather the information they need and to receive the intended message of your document. For example, the writers of the document in Figure 6.1 group three areas of similar information: the definition of acid rain in one section, wet deposition in another, and dry deposition in the final section.

To determine how to group similar information and effectively present it to your readers, follow these guidelines:

- Consider your readers.
- Consider your workplace context.
- Use the standard patterns of organization.
- Outline your information.
- Make the organization visible.

FIGURE 6.1

A Document that Groups Similar Information

What is Acid Rain?
"Acid rain" is a broad term referring to a mixture of wet and dry deposition (deposited material) from the atmosphere containing higher than normal amounts of nitric and sulfuric acids. The precursors, or chemical forerunners, of acid rain formation result from both natural sources, such as volcanoes and decaying vegetation, and man-made sources, primarily emissions of sulfur dioxide (SO_2) and nitrogen oxides (NO_x) resulting from fossil fuel combustion. In the United States, roughly 2/3 of all SO_2 and 1/4 of all NO_x come from electric power generation that relies on burning fossil fuels, like coal. Acid rain occurs when these gases react in the atmosphere with water, oxygen, and other chemicals to form various acidic compounds. The result is a mild solution of sulfuric acid and nitric acid. When sulfur dioxide and nitrogen oxides are released from power plants and other sources, prevailing winds blow these compounds across state and national borders, sometimes over hundreds of miles.

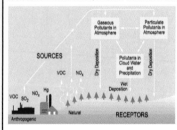

Wet Deposition
Wet deposition refers to acidic rain, fog, and snow. If the acid chemicals in the air are blown into areas where the weather is wet, the acids can fall to the ground in the form of rain, snow, fog, or mist. As this acidic water flows over and through the ground, it affects a variety of plants and animals. The strength of the effects depends on several factors, including how acidic the water is; the chemistry and buffering capacity of the soils involved; and the types of fish, trees, and other living things that rely on the water.

Dry Deposition
In areas where the weather is dry, the acid chemicals may become incorporated into dust or smoke and fall to the ground through dry deposition, sticking to the ground, buildings, homes, cars, and trees. Dry deposited gases and particles can be washed from these surfaces by rainstorms, leading to increased runoff. This runoff water makes the resulting mixture more acidic. About half of the acidity in the atmosphere falls back to earth through dry deposition.

Source: Downloaded from the World Wide Web, November 2008:
www.epa.gov-acidrain-what-index.html.

Consider Your Readers

Even though your information seems to suggest a particular organizational pattern, that pattern may not work for your readers. To analyze your readers' needs and expectations, ask yourself the following questions before you decide how to organize the information in your document:

- *Can I put important information at the beginning of the document?* Most readers prefer to read the important information—or a summary of that information—at the beginning of the document. For instance, readers of proposals like to know at the beginning of the document what you are proposing. They don't want to wait until the middle or the end of the document to learn your proposed solution. Whenever possible, place the most important information or a summary of it at the beginning of the document.

- *Can I order the information from the simplest to the most complex, the easiest to the most difficult, or the most familiar to the least familiar to clarify it for readers?* If you structure the information so that readers begin with what they already know or understand and move to what they don't know or understand, you will help them to read and remember your information.

- *Will readers scan the document or read it selectively?* Readers rarely read documents from beginning to end. Therefore, choose an organization that allows readers to find the information they need without reading the whole document. Make your organization visible to readers by using headings.

- *Can I begin with the least controversial or surprising information and move to the most controversial or surprising?* For most readers, you will want to begin with the information that will least surprise or upset them, especially if you are trying to persuade them to take some action or to adopt your viewpoint. Try to establish common ground with your readers by beginning with information that is not controversial or surprising.

Consider Your Workplace Context

Your manager, your organization's policies, your field, or your coworkers may influence how you organize information. You may not be the sole decision maker in determining how you will organize your documents. Before you decide how to organize a document, consider the conventions and expectations in your workplace. The following questions may help you:

- *How will my manager want me to organize the information?* Your manager may expect you to organize a document in a particular way. If you are in doubt about what he or she expects, discuss your planned organization with your manager or with the person who asked you to write the document.

- *Does my organization have a predetermined organization for similar documents?* For some documents, your company or group will determine the organization before you even begin writing. In these situations, you may not have much control over the organization of sections or the overall document. However, within paragraphs, you can organize sentences logically (Felker et al.). You can find out whether your workplace has a predetermined organization by looking at similar documents, searching the company's internal Web site, asking your manager, or reading the organization's style manual.

Use the Standard Patterns of Organization

You can use any of the following standard patterns for organizing information:

- spatial order
- chronological order
- general-to-specific order
- classification
- partition
- comparison and contrast
- problem and solution
- cause and effect
- order of importance

The first two patterns are sequential: the items that you arrange follow each other in physical location (spatial order) or in time (chronological order). The other seven patterns require you to choose the main point and then group the subpoints in a specific way. You can use more than one pattern in a document. For instance, you can use partition to break a whole into parts and then use spatial order to describe the parts. You may encounter a writing situation where none of the standard patterns are appropriate. In such a situation, try grouping together closely related information. At the end of this chapter, you will learn the definitions of each pattern, see examples of each pattern, and learn tips to help you use each pattern effectively.

Outline Your Information

Once you have analyzed your readers' and your workplace's expectations and have determined the standard pattern(s) you will use, outline your document. An outline will help you identify sections that are organized illogically or that lack information.

An outline also gives you a plan to follow as you write. You may find that you need to change your outline as you write. You may also decide to write the sections of the outline in sequence or to begin with the section that you know the most about or for which you have gathered all the necessary information. You can use two forms of outlines: informal and formal.

An Informal Outline

An *informal outline* may be a list of what you plan to include in a document. Informal outlines don't include sentences or parallel structure. Instead, they may be lists of initial thoughts and pieces of information that you write down but don't organize. These outlines may also be the topics that you plan to discuss in the document in the order you plan to discuss them. Informal outlines may be sufficient for short reports, instructions, and correspondence.

Let's look at how a writer takes the first draft of a document on babesiosis (see Figure 6.2, page 144) and expands the document for the final draft. The writer begins with an informal outline (see Figure 6.3, page 145). You can use an informal outline to create a draft for a formal outline. This draft can help you learn where the information is incomplete or where the organization is illogical. In the informal outline, you group similar information and do not use parallel structure.

Informal outlines may be lists of initial thoughts and pieces of information that you write down but don't organize. These outlines are useful for short reports, instructions, and correspondence.

A Formal Outline

A *formal outline* differs from an informal outline in two ways:

- It shows a more detailed structure for the information.
- It uses numbers and letters.

Because a formal outline establishes a hierarchy among pieces of information, you can easily convert topics and subtopics into headings for a document or into a table of contents for a formal report. As you prepare your outline, determine what format you will use for it; then use parallel structure for the topics and subtopics.

Selecting an Outline Format

If you are writing an outline that you will not show to others, use any format that you can read, understand, and follow. If you will use your outline

---

FIGURE 6.2

A Document Before the Writer Expands the Information

Diagnosing Babesiosis[1]

Babesiosis, also known as Malignant Jaundice, is a tick-transmitted disease of dogs. Babesiosis is caused by a protozoan organism, *Babesia canis*, that enters and destroys the red blood cells. The principal carrier of babesiosis from infected to non-infected dogs is the brown dog tick. The disease can also spread through blood transfusions or, in rare cases, from an infected female dog to her pups before birth.

When a dog becomes infected with the *Babesia* organism, it may become critically sick and die in a few days; or it may become a carrier without showing any signs of the disease. The most specific signs for babesiosis are
- bloody urine
- jaundiced mucous membranes and skin

Other signs include poor appetite, listlessness, fever, weight loss, and pale mucous membranes from anemia; however, these signs also occur with other diseases.

A veterinarian can diagnose babesiosis by finding the microscopic organisms in the red blood cells. If a diagnosis is not possible with this method, the veterinarian must test the blood serum for *Babesia* antibodies. Even if a dog tests positive for babesiosis, the veterinarian cannot treat it because the most effective drugs for treating babesiosis are unavailable in the United States.

[1]Source: Adapted from W. Elmo Crenshaw and Bruce Lawhorn, *Tick-borne Diseases of the Dog*, L-22667, rpt. 10M-7-88 (College Station: Texas Agricultural Extension Service, n.d.).

when collaborating, or if your document requires a table of contents, select one of the formal outline formats:

* topic outline
* sentence outline

FIGURE 6.3

Informal Outline

Ehrlichiosis

* Ehrlichiosis is a frequently diagnosed tick-borne disease
* The symptoms of ehrlichiosis are depression, poor appetite, weight loss, fever, enlarged lymph nodes, pale mucous membranes, and bleeding tendencies
* Ehrlichiosis has either an acute form or a chronic form. Acute form may cause death within a few days of infestation or may last for 3 to 6 weeks. If dog survives acute, the disease may become chronic
* Diagnose ehrlichiosis through finding the rickettsial organism in blood cells or testing blood serum for ehrlichia antibodies

Babesiosis

* Babesiosis is caused by *Babesia canis*, which is transmitted by the brown dog tick
* Babesiosis is treated with drugs not currently available in the U.S.
* Babesiosis symptoms: poor appetite, listlessness, fever, weight loss, and pale mucous membranes from anemia, bloody urine, and jaundiced mucous membranes and skin
* Diagnose babesiosis by finding *Babesia canis* in blood cells or its antibodies in blood serum

Borreliosis

* Borreliosis, or lyme disease, was first diagnosed in 1975
* Borreliosis signs include intermittent lameness in one or more legs from swelling, pain in the joints of the legs and feet, and fever
* Borreliosis usually occurs in dogs less than 4 years old
* Treat borreliosis with antibiotics

Prevention

* Most effective prevention is treatment such as dipping, spraying, or using powders to kill ticks before they can cause the anemia or spread disease
* The lawn where the dog lives should also be sprayed to eliminate ticks
* The pet owner should carefully follow the label instructions for any dip, spray, or powder or have a professional treat the dog

Topic outlines use phrases for the topics and subtopics. For example, a topic outline might use the phrase "signs of ehrlichiosis." *Sentence outlines* use sentences for the topics and subtopics: "A dog with ehrlichiosis may exhibit any one of seven signs."

Once you have determined which format to use, decide how you will number the outline—with a combination of numbers and letters (see Figure 6.4) or with decimals (see Figure 6.5).

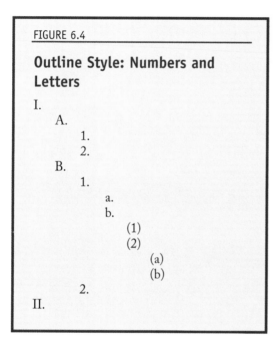

FIGURE 6.4

Outline Style: Numbers and Letters

I.
   A.
      1.
      2.
   B.
      1.
         a.
         b.
            (1)
            (2)
               (a)
               (b)
      2.
II.

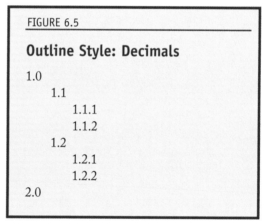

FIGURE 6.5

Outline Style: Decimals

1.0
   1.1
      1.1.1
      1.1.2
   1.2
      1.2.1
      1.2.2
2.0

Using Parallel Structure

The statements of topics and subtopics in a formal outline should be parallel in structure. *Parallel structure* means that the items of the same level should have the same grammatical structure. The items in the following numbered list are *not* parallel:

  A. Logging on to the network
     1. Find the login prompt
     2. Typing your login
     3. Passwords
     4. What you will see if you login correctly

Items 1, 2, 3, and 4 have different grammatical structures. Item 1 begins with a verb; item 2 begins with a different verb form; item 3 begins with

with a noun; and item 4 begins with a pronoun. These differences obscure the writer's reason for listing the items in this sequence.

The items in the following list are parallel:

A. Logging on to the network
    1. Find the login prompt
    2. Type your login
    3. Type your password
    4. Look for the message "Welcome to the network"

Now items 1, 2, 3, and 4 have the same grammatical structure; each item begins with a verb.

Using Your Outline to Analyze Your Organization and Your Information

You can use your formal outline to analyze the information you have gathered and the organization you plan to use. Let's consider the information for the document on tick-borne diseases of dogs (see Figure 6.3, page 145) and put it in a draft for a formal outline (see Figure 6.6, page 148). The draft reveals several problems, which appear in blue:

- **The information is incomplete and inconsistent**. The sections on the three diseases contain different kinds and amounts of information. The sections on ehrlichiosis and babesiosis each have three subsections, but the subsections are different. The section on borreliosis has only two subsections: signs and treatment. If possible, all three sections should have the same subsections to create consistent and complete information.
- **The structure of Section II ("How to prevent tick-borne diseases of dogs") is not logical because it has only one subsection**. Each section of an outline should have at least two subsections.
- **The draft lacks parallel structure**. See the parallelism errors marked on Figure 6.6, page 148.

Figure 6.7, page 149, shows a revised version of the outline. This version eliminates the problems in the first outline. You will see the following improvements noted in blue:

- **The information is complete and consistent**. The sections on the three diseases contain consistent information on signs, diagnosis, and treatment. Ehrlichiosis is the only one of the three diseases that has a

FIGURE 6.6

First Draft of a Formal Outline

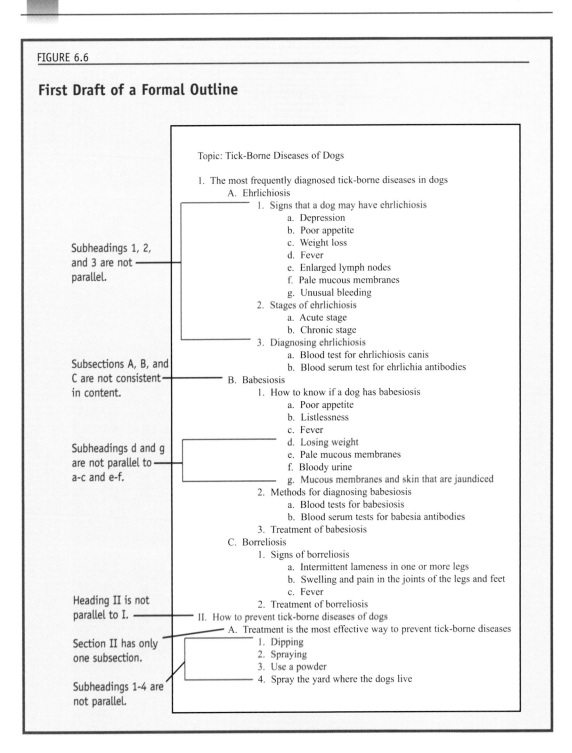

Topic: Tick-Borne Diseases of Dogs

1. The most frequently diagnosed tick-borne diseases in dogs
   A. Ehrlichiosis
      1. Signs that a dog may have ehrlichiosis
         a. Depression
         b. Poor appetite
         c. Weight loss
         d. Fever
         e. Enlarged lymph nodes
         f. Pale mucous membranes
         g. Unusual bleeding
      2. Stages of ehrlichiosis
         a. Acute stage
         b. Chronic stage
      3. Diagnosing ehrlichiosis
         a. Blood test for ehrlichiosis canis
         b. Blood serum test for ehrlichia antibodies
   B. Babesiosis
      1. How to know if a dog has babesiosis
         a. Poor appetite
         b. Listlessness
         c. Fever
         d. Losing weight
         e. Pale mucous membranes
         f. Bloody urine
         g. Mucous membranes and skin that are jaundiced
      2. Methods for diagnosing babesiosis
         a. Blood tests for babesiosis
         b. Blood serum tests for babesia antibodies
      3. Treatment of babesiosis
   C. Borreliosis
      1. Signs of borreliosis
         a. Intermittent lameness in one or more legs
         b. Swelling and pain in the joints of the legs and feet
         c. Fever
      2. Treatment of borreliosis
II. How to prevent tick-borne diseases of dogs
   A. Treatment is the most effective way to prevent tick-borne diseases
      1. Dipping
      2. Spraying
      3. Use a powder
      4. Spray the yard where the dogs live

Subheadings 1, 2, and 3 are not parallel.

Subsections A, B, and C are not consistent in content.

Subheadings d and g are not parallel to a-c and e-f.

Heading II is not parallel to I.

Section II has only one subsection.

Subheadings 1-4 are not parallel.

FIGURE 6.7

Revised Formal Outline

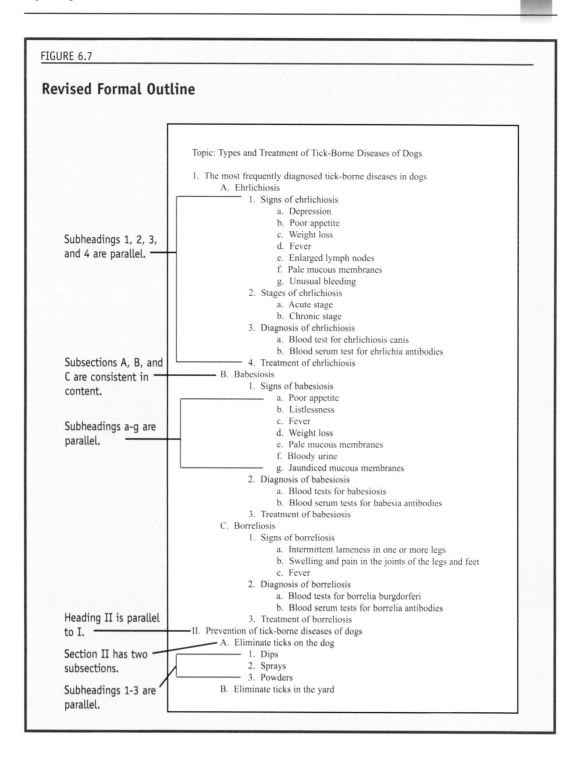

Topic: Types and Treatment of Tick-Borne Diseases of Dogs

1. The most frequently diagnosed tick-borne diseases in dogs
    A. Ehrlichiosis
        1. Signs of ehrlichiosis
            a. Depression
            b. Poor appetite
            c. Weight loss
            d. Fever
            e. Enlarged lymph nodes
            f. Pale mucous membranes
            g. Unusual bleeding
        2. Stages of ehrlichiosis
            a. Acute stage
            b. Chronic stage
        3. Diagnosis of ehrlichiosis
            a. Blood test for ehrlichiosis canis
            b. Blood serum test for ehrlichia antibodies
        4. Treatment of ehrlichiosis
    B. Babesiosis
        1. Signs of babesiosis
            a. Poor appetite
            b. Listlessness
            c. Fever
            d. Weight loss
            e. Pale mucous membranes
            f. Bloody urine
            g. Jaundiced mucous membranes
        2. Diagnosis of babesiosis
            a. Blood tests for babesiosis
            b. Blood serum tests for babesia antibodies
        3. Treatment of babesiosis
    C. Borreliosis
        1. Signs of borreliosis
            a. Intermittent lameness in one or more legs
            b. Swelling and pain in the joints of the legs and feet
            c. Fever
        2. Diagnosis of borreliosis
            a. Blood tests for borrelia burgdorferi
            b. Blood serum tests for borrelia antibodies
        3. Treatment of borreliosis
II. Prevention of tick-borne diseases of dogs
    A. Eliminate ticks on the dog
        1. Dips
        2. Sprays
        3. Powders
    B. Eliminate ticks in the yard

Subheadings 1, 2, 3, and 4 are parallel.

Subsections A, B, and C are consistent in content.

Subheadings a-g are parallel.

Heading II is parallel to I.

Section II has two subsections.

Subheadings 1-3 are parallel.

TAKING IT INTO THE WORKPLACE
Organization Does Make a Difference

Do you spend more time outlining and planning your documents than you do checking the grammar, sentence structure, and spelling of your documents? If you spend more time outlining and planning, you will likely create a more effective document because researchers have found a positive correlation between outlining (and/or other writing plans) and the quality of the document (Perl; Taylor and Beach; Kellogg; Spivey and King). Effective writers prepare outlines and "spend more time on macrowriting issues such as organization" (Baker). Less-effective writers "spend less time on outlining and more time on microwriting issues (sentence structure, grammar, spelling)" (Baker; Hayes and Flower).

Readers process information on two levels:
- microprocessing (focusing on the meaning of individual words and sentences)
- macroprocessing (focusing on the relationship of paragraphs and sections) (Lorch and Lorch). Through macroprocessing, readers create a mental roadmap of a document (Baker).

Readers can more easily remember and process information from documents with an effective macrostructure (or organization) than from documents with an ineffective organization (Lorch and Lorch). An effective organization allows readers to more easily relate word- and sentence-level information to the document as a whole. Without an effective organization, readers relate individual sentences and words—local information—only "to the immediately preceding ones"; the readers don't have a clear organization in which to integrate the local information (Baker; Kintsch).

What does this research mean for your writing? If you want readers to understand your documents and better recall the information, use a logical, clear organization that gives readers a mental roadmap (Baker).

Assignment

Let's test Baker's theory that "outlining does make a difference." For your next assignment in your technical communication class, do the following:
- Prepare an informal or formal outline.

> - Write a document based on the outline.
> - Send an e-mail to your instructor answering these questions:
>   - Did you spend more time on writing the outline and planning or on editing the document?
>   - Did you change the outline after you began writing?
>   - Did the outline help you to write a more reader-focused document? Explain your answer.
>
> Remember to prepare the outline before you write the document.

chronic stage; thus, the section on ehrlichiosis contains a subsection on the stages of the disease.

- **Section II has two subsections, instead of one**.
- **The structure is parallel**.

Make the Organization Visible

You can make the organization visible and easy to follow in these ways:

- **Use headings**. Headings make the organization of your document visible and easy to follow. *Headings* are subtitles within a document. If you have prepared a formal outline, you can use the entries in the outline as headings and subheadings. You will learn how to design and write effective headings in Chapters 7 and 10.
- **Use a detailed, accurate table of contents**. You will use a table of contents for longer documents, such as manuals, formal reports, or proposals. If you use a table of contents, include at least two levels of headings in the document to indicate the organization and scope of a document to help readers who want to determine whether a document contains information they need. You will learn more about tables of contents in Chapter 18.
- **Use topic sentences at the beginning of each paragraph**. A *topic sentence* announces what a paragraph is about; it gives readers a frame of reference (context) for the sentences that follow.
- **Use overviews at the beginnings of documents, chapters, and sections**. *Overviews* are introductory statements that describe what a document is about, what it may be used for, or how it is organized

TIPS FOR ORGANIZING INFORMATION SPATIALLY

- **Describe the object, device, or location as if readers were looking at it.** Tell your reader how you are describing the object with words such as *"beginning at your right."*
- **Give readers a "roadmap" to follow.** Give your readers signposts to help them follow your organization. Include words that help readers follow the organization. For example, use phrases such as *"from the right," "north view,"* or *"left to right."*
- **Use graphics, when possible, to help readers follow your organization.** Graphics help readers follow the spatial organization of your document. For example, imagine trying to follow a description of a physical site without a picture or drawing. If you use graphics, be sure to introduce and explain them.

(Felker et al.). You will learn how to write effective overviews in Chapter 7.

SELECT THE APPROPRIATE STANDARD PATTERN OF ORGANIZATION

The following subsections will help you
- define each pattern
- give examples
- provide tips for using each pattern effectively

These subsections will help you select the patterns most appropriate for your readers, your information, and your purpose.

Spatial

You might use spatial order to describe an object, device, or physical location. You use ***spatial order*** when you include words that indicate location, such as *"east of the location," "to the right,"* or *"below."* When using spatial order, you follow an organizational principle such as left to right, top to bottom, or inside to outside. For example, in Figure 6.8, the writers show how radon gets into a home. They begin at the left and move to the right of the house. The writers use numbers in the list to correspond to the numbers in the graphic. You might use spatial order in instructions, feasibility reports, definitions, or accident reports.

Chronological

When you read information arranged in order of occurrence or sequence, you are reading information arranged ***chronologically***. Chronological order occurs most frequently in instructions, process descriptions, and event descriptions. You use reverse chronological order when you write a resumé. Figure 6.9, page

---

FIGURE 6.8

Information Organized Spatially

How Does Radon Get Into Your Home?

Radon is a radioactive gas. It comes from the natural decay of uranium that is found in nearly all soils. It typically moves up through the ground to the air above and into your home through cracks and other holes in the foundation. Your home traps radon inside, where it can build up. Any home may have a radon problem. This means new and old homes, well-sealed and drafty homes, and homes with or without basements.

> Any home may have a radon problem

Radon from soil gas is the main cause of radon problems. Sometimes radon enters the home through well water (see "Radon in Water" below). In a small number of homes, the building materials can give off radon, too. However, building materials rarely cause radon problems by themselves.

RADON GETS IN THROUGH:
1. Cracks in solid floors
2. Construction joints
3. Cracks in walls
4. Gaps in suspended floors
5. Gaps around service pipes
6. Cavities inside walls
7. The water supply

Nearly 1 out of every 15 homes in the U.S. is estimated to have elevated radon levels. Elevated levels of radon gas have been found in homes in your state. Contact your state radon office for general information about radon in your area. While radon problems may be more common in some areas, any home may have a problem. The only way to know about your home is to test.

Radon can also be a problem in schools and workplaces. Ask your state radon office about radon problems in schools, daycare and childcare facilities, and workplaces in your area.

Source: Downloaded from the World Wide Web, November 2008: www.epa.gov/radon/pubs/citguide.html#howdoes

---

154, demonstrates chronological order to describe the five stages of mitosis in an onion root tip.

General-to-Specific

When you use *general-to-specific* organization, you assume that readers need to understand the general topic before they can understand the specific information. For most information, you will want to "write about the 'big picture' before you describe the parts and pieces that make up the whole" (Felker et al.). Most readers also need the conclusion or recommendations first, so they can interpret and understand the specific information in the context of the

FIGURE 6.9

Information Organized Chronologically

STAGE	CHROMOSOME ACTION	IMAGE
STAGE I	CHROMOSOMES APPEARING (early prophase—DNA condenses and becomes visible)	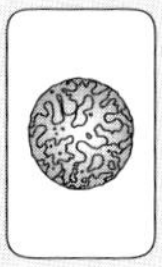
STAGE II	CHROMOSOMES ALIGNING (late prophase—chromosomes migrate toward the center)	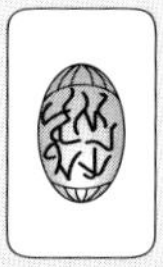
STAGE III	CHROMOSOMES AT MIDREGION (metaphase—chromosomes are aligned in the center of the cell)	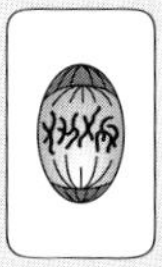
STAGE IV	CHROMOSOMES SEPARATING (anaphase—the chromosome copies begin to move to opposite poles)	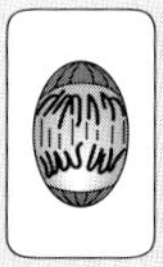
STAGE V	CHROMOSOMES SEPARATED (telophase) and cytoplasm divides (cytokinesis)	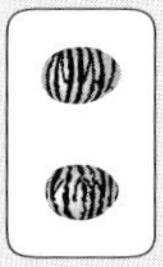

conclusions or recommendations (Samuels). When you solve a problem or make a recommendation, you come to that solution, recommendation, or conclusion last; however, for your readers to understand your work, they need the solution, recommendation, or conclusion first.

TIPS FOR ORGANIZING INFORMATION CHRONOLOGICALLY

- **Use words and phrases that give readers a "mental roadmap" of the chronological sequence.** Use "signposts" such as *"Step 1, Step 2,"* or *"Phase 1, Phase 2,"* etc. You might use an introduction to introduce the major parts of the sequence. Then use the parts as headings to guide your readers.
- **Use graphics when appropriate to illustrate the chronology.** If you use a graphic, be sure to introduce and explain it.
- **Explain the steps, events, or phases when appropriate for your readers and your purpose.** When organizing information chronologically, you are simply telling what happened in the order it happened. You do not have to explain why or how. For some information or readers, you may want to analyze the steps or the events. For example, if you are telling a new cook how to make a cake, you may need to explain some of the steps; you may need to explain what you mean by *"sifting the dry ingredients"* or *"creaming the butter and sugar."* If you are explaining an accident at work, the information may not be useful without your analysis of why the accident happened.

Figure 6.10, page 156, is an excerpt from a document that uses a general-to-specific organization. The document begins with general information on the risk of living with radon and moves to specific (and technical) information about the increased risk if you smoke. You will see this specific information in the table.

Classification

Classification is a means of grouping items into categories. In its simplest form, *classification* is grouping like items into a broad category or group. If you can find items that share common characteristics, you may be able to classify the information into meaningful categories (groups) and subcategories. To classify information, identify the broad group to which that information belongs. For example, Figure 6.11, page 157, classifies food we should eat to be healthy into 6 groups: grains (orange), vegetables (green), fruits (red), oils (yellow), milk (blue), and meat and beans (purple).

FIGURE 6.10

Information Organized from General to Specific

The Risk of Living With Radon

Radon gas decays into radioactive particles that can get trapped in your lungs when you breathe. As they break down further, these particles release small bursts of energy. This can damage lung tissue and lead to lung cancer over the course of your lifetime. Not everyone exposed to elevated levels of radon will develop lung cancer. And the amount of time between exposure and the onset of the disease may be many years.

Like other environmental pollutants, there is some uncertainty about the magnitude of radon health risks. However, we know more about radon risks than risks from most other cancer-causing substances. This is because estimates of radon risks are based on studies of cancer in humans (underground miners).

Smoking combined with radon is an especially serious health risk. Stop smoking and lower your radon level to reduce your lung cancer risk.

Children have been reported to have greater risk than adults of certain types of cancer from radiation, but there are currently no conclusive data on whether children are at greater risk than adults from radon.

> Scientists are more certain about radon risks than from most other cancer-causing substances.

Your chances of getting lung cancer from radon depend mostly on:

- **How much radon is in your home**
- **The amount of time you spend in your home**
- **Whether you are a smoker or have ever smoked**

Radon Risk If You Smoke

Radon Level	If 1,000 people who smoked were exposed to this level over a lifetime*...	The risk of cancer from radon exposure compares to**...	WHAT TO DO: Stop smoking and...
20 pCi/L	About 260 people could get lung cancer	250 times the risk of drowning	Fix your home
10 pCi/L	About 150 people could get lung cancer	200 times the risk of dying in a home fire	Fix your home
8 pCi/L	About 120 people could get lung cancer	30 times the risk of dying in a fall	Fix your home
4 pCi/L	About 62 people could get lung cancer	5 times the risk of dying in a car crash	Fix your home
2 pCi/L	About 32 people could get lung cancer	6 times the risk of dying from poison	Consider fixing between 2 and 4 pCi/L
1.3 pCi/L	About 20 people could get lung cancer	(Average indoor radon level)	(Reducing radon levels below 2 pCi/L is difficult.)
0.4 pCi/L	About 3 people could get lung cancer	(Average outdoor radon level)	

Note: If you are a former smoker, your risk may be lower.
* Lifetime risk of lung cancer deaths from EPA Assessment of Risks from Radon in Homes (EPA 402-R-03-003).
** Comparison data calculated using the Centers for Disease Control and Prevention's 1999-2001 National Center for Injury Prevention and Control Reports.

Source: Downloaded from the World Wide Web, November 2008: http://www.epa.gov/radon/pubs/citguide.html#howdoes

TIPS FOR USING GENERAL-TO-SPECIFIC ORDER

- **State the general information clearly and directly at the beginning of the document, section, or paragraph**. In a longer document, consider putting the general information in a separate section or in an overview that introduces the topic. You may want to indicate to the readers that you will be organizing the information from general to specific. You might say something like, *"We will first define radon and then move to how to test for and eliminate radon."*
- **Use headings to separate the general information from the specific information**. The headings should clearly indicate the general and the specific information; so use informative headings.
- **Use graphics when appropriate to supplement the organization.** You can use diagrams, photographs, drawings, and maps to help readers understand the information.

FIGURE 6.11

Information Organized by Classification

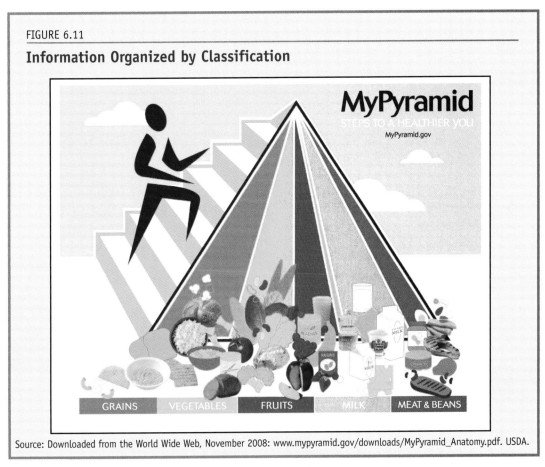

Source: Downloaded from the World Wide Web, November 2008: www.mypyramid.gov/downloads/MyPyramid_Anatomy.pdf. USDA.

For more tips on classifying information, see page 158.

Partition

Partition is the division of an item into its individual parts. For example, Figure 6.12, page 159, illustrates how you might partition the cornea into its five layers: the epithelium, Bowman's membrane, the stroma, Descemet's membrane, and the endothelium.

Comparison and Contrast

You may need to compare or contrast two or more options. When you examine these options, you'll need standards, or criteria, for comparing. For example, if you were deciding which apartment to rent, you might

TIPS FOR CLASSIFYING INFORMATION

- **Make sure each item will fit in only one category.** Categories should not overlap. The classification shown in Figure 6.11, page 157, works because the categories are mutually exclusive.
- **Make sure that each item will fit into a category.** The categories should not overlap. If you have even one item that doesn't fit into a category, you need to add another category or you may have selected a basis for classification that doesn't fit the items. For example, if you were classifying homes, you could classify them by their market value because every home has at least some market value. However, if you tried to classify homes by their exterior building materials of brick, wood, or aluminum siding, it would not work because some homes have a stucco exterior. You would need to add stucco and other building materials to your possible categories.
- **Classify the items in ways suited to your readers and your purpose.** Choose a principle for classifying and dividing that will fit the information and your purpose. When creating the food pyramid, the U.S. Department of Agriculture thought not just about all foods that the readers could eat. Instead, they focused on the foods that the readers should eat within the groups, so they changed the word "fats" to "oils" and made that section of the pyramid the smallest. In fact, in some graphics of the pyramid, the fat category does not have a label.
- **Use graphics when possible to illustrate the categories and subcategories.** A graphic may help your readers understand how you have classified the information. You can use the following types of graphics to illustrate your classifications: diagrams, pictographs, charts, tables, and photographs. You also might consider using color to show the categories, as in Figure 6.11.

compare apartments based on size, cost, amenities, and location. You might even rank these criteria in order of importance to help you select the best apartment for you, your budget, and your needs.

You may use the comparison and contrast pattern in many documents, but especially in feasibility reports. *Feasibility reports* compare two or more options. As you report your comparisons, you can organize by options or by the criteria used to compare the options. Figures 6.13 and 6.14, page 161, illustrate how you might organize a document comparing apartment complexes. Figure 6.13 is organized according to options. Notice that all the information about Sunrise Apartments appears in one section, and all

---

FIGURE 6.12

Information Organized by Partition

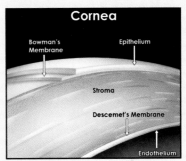

Cornea

Shaped like a dome, the cornea is the transparent layer that covers the front of the eye. It reflects light once it enters the eye. The cornea needs tears to help nourish and clean it from undesirable elements. It also receives nourishment from the aqueous humor, which fills the chamber behind it. Because of the cornea and the lens, we can see objects that are near or far.

Source: Downloaded from the World Wide Web, November 2008: www.lasik.md/img/learnAbout. Used with permission.

The epithelium:
The epithelium, the layer that covers the surface of the cornea, is about 5-6 cell layers thick and filled with tiny nerve endings, making the cornea very sensitive to pain. The epithelium blocks the passage of dust and germs and provides a smooth surface that absorbs oxygen and cell nutrients from tears, and then distributes these nutrients to the rest of the cornea.

Bowman's membrane:
This layer lies beneath the epithelium and is very difficult to penetrate. The difficult access to the Bowman's membrane protects the cornea from injury. But once injured, it resiliently regenerates. It leaves a scar when the injury is deeper. The scar becomes an opaque area, causing the cornea to lose its clarity and luster.

The stroma:
Lying beneath the Bowman, the stroma is the thickest layer. Composed of tiny collagen fibrils that run parallel to each other, this precision formation gives the cornea its clarity, strength, elasticity, and form.

Descemet's membrane:
Lying beneath the stroma, Descemet's membrane is a thin but strong sheet of tissue that acts as protection against infection and injuries. It is composed of collagen fibers (different from those of the stroma) and is made by the endothelial cells that lie below it. Descemet's membrane is regenerated readily after injury.

The endothelium:
This is the extremely thin, innermost layer of the cornea. Endothelial cells are essential in keeping the cornea clear. It pumps the excess fluid out of the stroma, which has the danger of swelling with water. Once endothelium cells are destroyed by disease or trauma, they cannot be recovered. Too much damage to endothelial cells can lead to corneal edema (swelling caused by excess fluid) and blindness ensues, with corneal transplantation the only available therapy.

Source: Downloaded from the World Wide Web, November 2008: http://www.uniteforsight.org/course/cornealdisease.php

TIPS FOR PARTITIONING INFORMATION

- **Choose a principle for partitioning that will meet your readers' needs and your purpose**. For example, if you want to help a technician understand how the parts of a machine function, you might partition them by function.
- **Organize the parts in a way that your readers will find helpful**. You can help readers use and understand your document if you discuss each group of parts in a logical fashion. For example, if you are discussing the parts of the cell, you might also use spatial order, beginning with the parts on the right and moving clockwise to the parts on the left.
- **Use graphics when possible to illustrate the parts**. A graphic will help your readers to understand how you have partitioned the information. For example, the diagram of the cornea in Figure 6.12, page 159, helps readers locate the parts of the cornea and see the location of the layers.

TIPS FOR COMPARISON AND CONTRAST

- **Determine your criteria for comparison or contrast**. Criteria help you compare items equally. To help you interpret criteria, rank items by order of importance.
- **Use comparison/contrast in feasibility reports**. Because feasibility reports compare two or more options, the comparison and contrast pattern can help you organize information so readers can more easily understand and interpret it. (For more information on feasibility reports, see Chapter 18.)

the information about Sunset Apartments appears in another section. This organization has the advantage of presenting the whole picture for each option in one section and emphasizes the options.

In Figure 6.14, the writer has organized according to the criteria used to compare the apartment complexes. The writer groups the information about the distance from campus, the rent per month, and the amenities of both apartments in separate sections. This organization emphasizes the criteria and helps the reader easily compare options, based on individual criteria.

FIGURE 6.13

Comparison and Contrast: Organizing by Options

Sunrise Apartments
- Distance from campus
- Rent per month
- Amenities

Sunset Apartments
- Distance from campus
- Rent per month
- Amenities

FIGURE 6.14

Comparison and Contrast: Organizing by Criteria

Distance from Campus
- Sunrise Apartments
- Sunset Apartments

Rent per Month
- Sunrise Apartments
- Sunset Apartments

Amenities
- Sunrise Apartments
- Sunset Apartments

Problem and Solution

You can use the *problem-and-solution* pattern to explain both actual and proposed solutions. This pattern is frequently used in proposals and progress reports.

The document on biological pollutants in Figure 6.15, page 164, illustrates the problem-and-solution pattern. The writer begins by identifying the problem of biological pollutants and their health effects, then concludes with solutions for reducing these pollutants in and around the home.

Cause and Effect

You can use the *cause-and-effect* pattern to help readers understand the consequences or the causes of a particular action or series of actions. Depending on your purpose, you can move from the cause to the effect or from the effect to the cause. For example, you can talk about a leak (effect) in a steam-generating plant and then explain the cause, sagging pipes. You can also discuss the sagging pipes (cause) and predict the effect (a leak).

Figure 6.16, page 165, presents a document from the Center for Disease Control with a cause-and-effect organization. The document explains the

TIPS FOR USING THE PROBLEM-AND-SOLUTION PATTERN

- **Identify the problem before you begin discussing the solution.** Before your readers can understand and appreciate your solution, they must first know the problem. Emphasize the significance of the problem to your readers and the parts of the problem that your solution addresses.
- **Show how your solution will solve the problem.** Your readers will see the value of your solution only if they understand how it relates to the problem. Give specific details revealing how the solutions will eliminate the problem.
- **Group the stages of your solution into meaningful categories.** If your solution has several stages, you can help your readers by grouping the stages into mutually exclusive, nonoverlapping categories.
- **Give readers ample evidence that your solution will solve the problem.** If you are trying to persuade readers to adopt a solution that you are recommending, give them the evidence and reasoning they need to accept it. Even if the solution is already in place, you may want to persuade readers that it is worthwhile and effective.
- **Use graphics when appropriate to illustrate or clarify the problem or the solution.** Graphics can help you summarize or clarify the problem or the solution.

most common causes of poor oral health and discusses the impact of each condition on the patient's quality of life.

Order of Importance

You can organize information according to its *importance*. You can use either a descending or an ascending order. You can begin with the most important and move to the least important (descending), or you can begin with the least important and move to the most important (ascending). With descending organization, you get the reader's attention with the most important point. You may want to use ascending order when you are trying to persuade a reader who may be hostile or who may disagree with you.

The document shown in Figure 6.17, page 166, moves from the most-important to the least-important question for the readers. It begins with a discussion of radon and its effect on health and ends with less-important information about how to test for radon and eliminate it in the home.

TIPS FOR USING THE CAUSE-AND-EFFECT PATTERN

- **Identify either the cause or the effect near the beginning of the document**. Near the beginning of the document, tell your readers what you are trying to do in the document: to explain the causes of a specific effect or the effects of a specific cause. If your readers know how you have organized the information, they can better understand it.
- **Show how the cause directly relates to the effect or how the effect directly relates to the cause**. Make sure that your readers understand the links between the cause and the effect or between the effect and the cause. Don't expect readers to infer the connection; tell them.
- **Group the causes or effects into logical categories**. Grouping helps readers understand the relationships among causes and effects.
- **Use graphics when appropriate to illustrate or clarify the effect or the cause**. Graphics may help you summarize or clarify the cause and the effect. Graphics are especially effective for showing how the cause and effect relate to each other.

TIPS FOR USING THE ORDER OF IMPORTANCE PATTERN

- **Give readers a context for the information**. State the main point or topic at the beginning.
- **Tell readers how you are organizing the information**. Clearly tell readers that you are beginning with the most important information or that you are ending with the most important information.
- **Tell readers why one point is more or less important than another**. Don't assume that your readers will see the information in the same order of importance as you do.
- **Use graphics when appropriate**. Graphics can help you clarify the information.

FIGURE 6.15

Information Organized from Problem to Solution

The writers identify the problem.

What Are Biological Pollutants?

Biological pollutants are living organisms. They promote poor indoor air quality and may be a major cause of days lost from work or school, and of doctor and hospital visits. Some pollutants can even damage surfaces inside and outside your house. Biological pollutants can travel through the air and are often invisible.

What Can You Do About Biological Pollutants?

Control Water
Fix leaks and seepage. If water is entering the house from the outside, your options range from simple landscaping to extensive excavation and waterproofing.

The writers identify four solutions.

Maintain and Clean All Appliances That Come In Contact With Water
Have major appliances, such as furnaces, heat pumps and central air conditioners, inspected and cleaned regularly by a professional, especially before seasonal use.

Clean Surfaces
Clean moist surfaces, such as showers and kitchen counters. Remove mold from walls, ceilings, floors, and paneling.

Control Dust
Always wash bedding in hot water (at least 130°F) to kill dust mites. Launder bedding at least every 7 to 10 days. Clean rooms and closets well; dust and vacuum often to remove surface dust.

Source: Adapted from U.S. Environmental Protection Agency, 2008.
http://www.cpsc.gov/cpscpub/pubs/425.html.

FIGURE 6.16

Information Organized from Cause to Effect

The writers identify the causes.

Oral Health and Quality of Life

Diseases and disorders that damage the mouth and face can disturb well-being and self-esteem. The effect of oral health and disease on quality of life is a relatively new field of research that examines the functional, psychological, social, and economic consequences of oral disorders. Most of the research has focused on a few conditions: tooth loss, craniofacial birth defects, oral-facial pain, and oral cancer. The impact of oral health on an individual's quality of life reflects complex social norms and cultural values, beliefs, and traditions. There is a long tradition of determining character on the basis of facial and head shapes. Although cultures differ in detail, there appear to be over-all consistencies in the judgment of facial beauty and deformity that are learned early in life. Faces judged ugly have been associated with defects in character, intelligence, and morals.

The Impact of Craniofacial-Oral-Dental Conditions on Quality of Life

The writers identify the effects.

Missing teeth
People who have many missing teeth face a diminished quality of life. Not only do they have to limit food choices because of chewing problems, which may result in nutritionally poor diets, but many feel a degree of embarrassment and self-consciousness that limits social interaction and communication.

Craniofacial birth defects
Children with cleft lip or cleft palate experience not only problems with eating, breathing, and speaking, but also have difficulties adjusting socially, which affects their learning and behavior. The tendency to "judge a book by its cover" persists in the world today and accounts for many of the psychosocial problems of persons affected by craniofacial birth defects.

Oral-facial pain
The craniofacial region is rich in nerve endings sensitive to painful stimuli, so it is not surprising that oral-facial pain, especially chronic pain conditions where the cause is not understood and control is inadequate, severely affects quality of life. Conditions such as temporomandibular (jaw joint) disorders, trigeminal neuralgia, and postherpetic neuralgia (chronic pain following an attack of shingles affecting facial nerves) can disrupt vital functions such as chewing, swallowing, and sleep; interfere with normal activities at home or work; and lead to social withdrawal and depression.

Oral Cancer
Surgical treatment for oral cancer may result in permanent disfigurement as well as functional limitations affecting speaking and eating. Given the poor prognosis for oral cancer (the five-year survival rate is only 52 percent), it is not surprising that depression is common in these patients.

Source: Downloaded from the World Wide Web, November 2008: www.cdc.gov-oralhealth-publications-factsheets-sgr2000_fs5.htm.jpg

FIGURE 6.17

Information Organized from Most Important to Least Important

The writers begin with the most-important infor-mation.

Radon is estimated to cause thousands of lung cancer deaths in the U.S. each year.

Overview

Radon is a cancer-causing, radioactive gas.
Radon is estimated to cause many thousands of deaths each year. That's because when you breathe air containing radon, you can get lung cancer. In fact, the Surgeon General has warned that radon is the second leading cause of lung cancer in the United States today. Only smoking causes more lung cancer deaths. **If you smoke and your home has high radon levels, your risk of lung cancer is especially high.**

Radon can be found all over the U.S.
Radon comes from the natural (radioactive) breakdown of uranium in soil, rock and water and gets into the air you breathe. Radon can be found all over the U.S. It can get into any type of building - homes, offices, and schools - and result in a high indoor radon level. But you and your family are most likely to get your greatest exposure at home, where you spend most of your time.

You should test for radon.
Testing is the only way to know if you and your family are at risk from radon. EPA and the Surgeon General recommend testing all homes below the third floor for radon. EPA also recommends testing in schools.

The writers end with the least-important infor-mation.

You can fix a radon problem.
Radon reduction systems work and they are inexpensive. Some radon reduction systems can lower radon levels in your home by up to 99%. Even very high levels can be reduced to acceptable levels.

Source: Adapted from *A Citizen's Guide to Radon: The Guide to Protecting Yourself and Your Family from Radon.* U.S. Environmental Projection Agency, 2007.
http://www.epa.gov/radon/pubs/citguide.html#overview

CASE STUDY ANALYSIS
2004 Presidential Candidates Use Online Sources

Background

In 2004, presidential election campaigns were changed forever by using online sources to make information immediate and accessible to citizens across the nation. Web pages, meetup services, and blogs allowed campaign headquarters to post news stories, posters, videos, letters, donation requests, and more online, so workers at the grassroots level could access the information. The 2004 campaign effectively showed that using online sources could make a virtually unknown candidate a household name in a short time; the impact of the shift in strategy has been likened to the Nixon/Kennedy televised debate.

With this new venue, campaigns changed how they marketed the 2004 candidates; the information was managed and controlled after it was disseminated. Traditional marketing sources—such as television, radio, print, and mail—create an awareness and desire for a product through branding and consistent use of typography, design, color, and wording. However, once information is posted online, it becomes "owned" by the people who use it; candidates cannot control how people at the grassroots level change or use the information. This online format results in an "over-the-back fence" feel to discussions carried on in coffee shops, town halls, and blog pages across the country.

Assignment

1. Imagine you are creating information that will be downloaded from a presidential candidate's Web site. What would you do to ensure that your candidate's political profile remains consistent and accurate after your information is posted and disseminated? Write a memo using a problem-and-solution pattern to answer the question.
2. Write a memo to your manager explaining how information organized for the Web is different from information organized for print.

EXERCISES

1. Select an appropriate criterion that you can use to compare or contrast two or more items or groups of items. Some suggested items appear below. Prepare a formal outline. Using your outline, write a brief document in which you compare or contrast the items.
   - apartment complexes near your campus
   - dorms on your campus
   - coffee shops near your campus
   - grocery stores in your community
   - state parks in your home state

2. Partition a single object into its parts. Use one of the items listed below, or select one of your own. Specify the specific brand or model of the object. For example, you will not partition a generic printer; instead, you might partition a Hewlett Packard Officejet Pro L7680 printer. Partition the object into at least two levels of parts—main parts and subparts. For instance, you might partition the printer into the control panel and the paper tray. Once you have partitioned the object, create a diagram (or locate a diagram from the manufacturer's literature on the Web) showing the partitions, as in Figure 6.8, pages 153.
   - lawnmower
   - bicycle
   - refrigerator
   - tools, equipment, or instruments that you use in your major field of study
   - auditorium
   - microwave

3. Using the partitions and the diagram that you created in Exercise 2, write a description of the parts of the item you diagrammed. Use spatial order to describe the parts.

4. Determine the most effective principle of classification or division for a topic and specific readers. Select one of the following topic-and-reader combinations or create a combination of your own. Prepare a formal outline (as in Figure 6.7, page 149), a diagram, or

a description of your classification or division. Remember that the categories of the classification or division must not overlap.

- Types of financial aid available to students at your college or university

  Readers: students or parents
- Restaurants

  Readers: visitors to your community
- Types of jobs available in your major field after graduation

  Readers: graduating college seniors

5. **Web exercise**: Prepare a document that explains the cause of some situation or event. Address your document to a specific reader, such as a friend, family member, or classmate. Search the Web for information on your topic. Be sure to document information that you use from the Web. Select one of the following topics, or choose one of your own:

- comets
- global warming
- hurricanes
- eclipses of the sun or moon
- tsunamis

6. Prepare a document that explains the effects of a situation or event for a specific reader, such as a friend, family member, or classmate. Select one of the following topics, or choose one of your own:

- writing skills and your job
- recycling
- lower or higher interest rates on the local economy
- mercury levels in water and food

7. Select a problem in your community, on your campus, or at your workplace, and determine an appropriate solution. Prepare a formal outline of the problem and your proposed solution. Your outline should have at least two levels. Using the outline, write a letter detailing the problem and solution to the appropriate community leader, to your local or campus newspaper, or to the appropriate person on campus or at your workplace. (For information on letters, see Chapter 12.)

8. Assume that a friend has asked you to describe a process or procedure. Using chronological order, write a description of the process or procedure for your friend. You can use one of the processes or procedures below or select one of your own:
   - how to create a slideshow using Microsoft *PowerPoint*®
   - how to create a wiki
   - how to use a tool or equipment in a chemistry or biology lab
   - how to set the security alarm in your apartment or home
   - how to create and edit a graphic using graphics software (you must select a specific software)

9. **Collaborative exercise**: You and your team will select a topic below, or choose one of your own, and write a memo to your instructor or to your classmates. Use the order-of-importance pattern to organize your memo. (For information on memos, see Chapter 12.)
   - reasons pilots should or should not carry firearms
   - reasons your college should or should not offer more courses online
   - reasons your college should add more parking
   - potential effects of global warming on your community

10. Interview (in person or by e-mail) a person in your field to find out how his or her workplace or manager influences the way he or she organizes documents. You should ask the following questions; you may also add questions of your own. (For information on setting up and conducting interviews, see Chapter 5.)
    - Does your organization or field have a predetermined organization for some of the documents that you write? If so, what types of documents? How does the organization compare with the guidelines that you learned in college?
    - Does your organization use templates that give you a specific organization for a document? If so, which documents?
    - How frequently does your manager expect you to organize a document in a particular way? What do you do when you disagree with the organization that he or she expects?
    - If your workplace or manager does not have a required organization, how do you decide to organize the document? What if you

have never written such a document? Where do you find out about the organization? From a company archive? From a Web site? From a coworker?

Write a memo to your instructor summarizing the information that you gathered from the interview. (For information on memos, see Chapter 12.)

REAL WORLD EXPERIENCE
Working with a Team to Illustrate the Standard Patterns of Organization

You will find the standard patterns of organization used in many technical documents. You will work with a team to find documents that illustrate the patterns of organization.

Working with your team, complete the following for each standard pattern of organization:

• Find at least two documents that illustrate each of the standard patterns of organization.
  - You might search the Web or look at documents from businesses in your community or offices on your campus.
  - Make sure that the documents follow the tips for that pattern.
  - Make a copy of each document.
• Prepare a presentation that will define and illustrate the organization. You might consider using a slideshow (see Chapter 20). Your presentation should
  - define each pattern
  - explain how each document illustrates the pattern
  - include appropriate graphics
  - cite the source of the document
• Be prepared to deliver your presentation to your classmates.

iStockphoto 2008.

Writing Easy-To-Read Documents

When you begin your work in business or industry, you will want to make sure that your sentences and paragraphs are clear, concise, and reader-focused and that your writing says what you intend. Clear, concise writing will help you sound like a professional, make a positive impression, and gain the respect of your managers and your coworkers. Unclear or poorly organized writing will make a negative impression. This chapter gives you guidelines for making your writing easy to read—from the sentence to the document level.

MAKE ACTORS THE SUBJECTS OF YOUR SENTENCES[1]

In most of your writing, you will tell a story involving actors and actions. In even the most convoluted writing, you usually can find actors acting. Let's look at a passage in which the writer buries the actors and actions:

> It is Sabrina's proposal for the adoption of spreadsheet software by the accounting department. This software provides assistance in the preparation of the annual budget.

[1] Williams, Joseph.

Who are the real actors? In the first sentence, "Sabrina" and "the accounting department" are the actors, although the sentence does not describe their actions. In the second sentence, the actor seems to be "software"; but because of the sentence structure, we can't be sure whether the software or the accounting department is preparing the budget.

What actions does the writer of that passage mention? In the first sentence, the actions appear in the nouns "proposal" and "adoption." Again, in the second sentence, the actions appear in nouns: "assistance" and "preparation." The writer buries the action in nouns, instead of using verbs.

Let's look at a revised version of the passage:

> Sabrina proposed that the accounting department adopt spreadsheet software. The accounting department personnel can use this software when they prepare the annual budget.

In this revised passage, the actors ("Sabrina" and "the accounting department personnel") are the subjects, and the actions ("proposed," "adopt," "use," and "prepare") appear in verbs. The revised passage illustrates two powerful style principles:
- Make actors the subjects of your sentences.
- Put the action in verbs.

Most readers expect the actor to appear in the subject. Think of the subject as a fixed location (see Figure 7.1). In that location, readers expect to find an actor. (An *actor* is a noun—a person, a place, or a thing.)

Use People as Subjects Whenever Possible

Make people the subjects of your sentences whenever possible. For some sentences, you may not be able to make a person the subject; for example, if you are describing a piece of equipment, you might make the equipment the actor; or you might want to make your company the actor. When you use people as actors, you lessen the distance between you and the readers and eliminate ambiguity. Consider this passage:

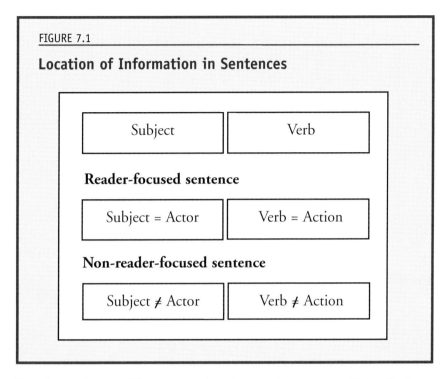

FIGURE 7.1

Location of Information in Sentences

Subject	Verb

Reader-focused sentence

Subject = Actor	Verb = Action

Non-reader-focused sentence

Subject ≠ Actor	Verb ≠ Action

People ≠ subjects: The prescription drug card program will be eliminated, effective at the end of the year. Beginning January 1, prescriptions that are purchased from local pharmacies may be filed and will be paid at 80% after a $200 deductible. This means that all prescriptions purchased from local pharmacies must be paid in full and the receipts saved. A claim form and your receipts should then be mailed in for reimbursement.

Readers of this passage must infer who is purchasing, eliminating, filing, and so on. They could confuse the actions that they are to perform (purchasing, filing, paying, saving, and mailing) with the actions that the pharmacy or insurance company will perform (eliminating, paying, and reimbursing). Let's look at a revision in which people are the subjects:

People = subjects: You will begin using a new prescription drug program on January 1. Instead of using your prescription drug card, you may file with us any prescription that you purchase from local pharmacies. We will pay 80% of the cost of your prescriptions after you meet the $200 deductible. To take advantage of this new program, follow these four steps:

1. Pay for all prescriptions purchased from local pharmacies.
2. Save the receipts.
3. Complete the attached claim form.
4. Mail the completed form in the envelope provided, so we can reimburse you.

In this version, each sentence has people as actors; the passage now clearly explains what readers must do and what the company will do.

USE THE ACTIVE VOICE WHEN APPROPRIATE

Let's look again at the unrevised prescription passage. The object of the action is the subject of these sentences because the writer uses passive-voice verbs instead of active-voice verbs. In *passive voice*, the subject is acted upon. In *active voice*, the subject performs the action.

When you use the passive voice, readers must search for the actors or assume who is doing the acting. When you use the active voice, the actor is the subject of the sentence, as your readers expect. Notice how the active voice changes the two passive-voice sentences:

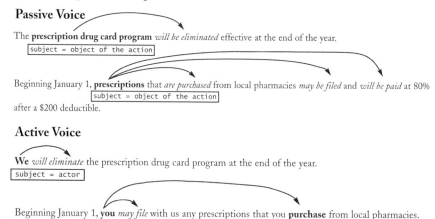

Passive Voice

The **prescription drug card program** *will be eliminated* effective at the end of the year.
subject = object of the action

Beginning January 1, **prescriptions** that *are purchased* from local pharmacies *may be filed* and *will be paid* at 80%
subject = object of the action
after a $200 deductible.

Active Voice

We *will eliminate* the prescription drug card program at the end of the year.
subject = actor

Beginning January 1, **you** *may file* with us any prescriptions that you **purchase** from local pharmacies.
subject = actor

In the active-voice sentences, the actors—"we" and "you"—are the subjects. Active-voice sentences have these characteristics:

• They tell the readers who is performing the action.
• They use fewer words than passive-voice sentences.

Passive-voice sentences have these characteristics:

- The actor and the subject are not the same. The actor may be missing or may appear in a prepositional phrase beginning with "by."
- The verb consists of a form of the verb "to be" plus the past participle of the main verb. (The verb "to be" has eight forms: is, am, are, was, were, be, being, and been.)
- The object of the action appears in the subject.

Let's look at another passive-voice sentence. In this sentence, the subject ("proper procedures") and the actor ("the front-desk staff") aren't the same. The actor appears in a prepositional phrase. The verb ("be learned") consists of the verb "be" plus the past participle "learn."

Passive Voice

In this active-voice sentence, the subject ("front-desk staff") is also the actor. The verb does not have a form of "to be." This and other active-voice sentences have these characteristics:

- The actor and the subject are the same.
- The verb does not consist of a form of "to be" plus the past participle of the main verb.
- The object appears after the verb.

Active Voice

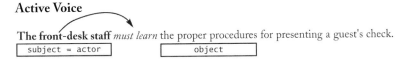

The active voice makes the actor the subject, so readers can read sentences quickly and understand them easily.

When to Use the Passive Voice

You may have to choose between the active and the passive voice. Let's look at an example:

Passive voice   During the last six months, more than 3,000 **coats** *were distributed* to children across the city.

TIPS FOR CHANGING A PASSIVE-VOICE SENTENCE INTO AN ACTIVE-VOICE SENTENCE

- Identify the actor.
- Put the actor in the subject.
- Follow the actor with the action (the verb).
- Follow the action (the verb) with the object or the receiver of the action.

In this example, the more important information is not who distributed the coats, but instead, who received the coats. The writers of these sentences use the passive voice because they wanted to focus on the children (the object) rather than on the many people (the actors) who distributed the coats.

Before you decide to use the passive voice, answer these questions. If you answer "no" to the first two questions or "yes" to the third question, the passive voice may be appropriate:

1. Do your readers need to know who is acting? Is the actor important?
2. Do you know who is acting?
3. Do you want to focus attention on the object rather than on the actor? (We discuss this situation later in this chapter in the context of using the old/new pattern.)

PUT THE ACTION IN VERBS

Readers expect to find the action of a sentence expressed in verbs. Many writers, however, bury the action in nouns, as in this sentence:

Action in noun   The **police** *are conducting* an investigation of the robbery that occurred this morning.

You must look beyond the verb "are conducting" to find the action "investigation." The sentence names this action in the noun "investigation," instead of using the verb "investigate." The verb "are conducting" doesn't give you the information you need. To improve this sentence, we can eliminate the verb "are conducting" and put the primary action in the verb "investigate":

Action in verb   The **police** *are investigating* the robbery that occurred this morning.

In this revised sentence, you don't have to search for the primary action; it appears in the verb where you expect to find it. The revised sentence is also shorter and more direct.

Identifying Sentences Where the Verb Does Not Express the Action

Most sentences in which the verb is not expressing the action have one or both of these characteristics:

- A noun expresses the primary action of the sentence.
- The verb of the sentence is a form of "to be."

If a noun expresses the primary action, you may be able to identify that noun from its suffix: -tion, -ment, -ion, -ance, -ence, or -ery. However, the noun may not end in a suffix because some verbs (for example, "hope," "result," and "change") do not change form when used as nouns.

Let's look at some sentences in which nouns express the primary action; when revised, these sentences are clearer and more concise:

Action in noun   Her **discovery** of the missing bolts *happened* on Friday while she was cleaning the lab.

Action in verb   **She** *discovered* the missing bolts on Friday while she was cleaning the lab.

The second sentence is more direct and reader-focused because the action appears in the verb "discovered." The sentence also now has an actor in the subject.

Let's consider another example. In this example, the verb is a form of "to be"; and the action of the sentence appears in the noun "discussion." Notice that the noun ends in "-sion." When the verb "discussed" express-es the action, the sentence is more effective and reader-focused. This revised sentence focuses on the actor, which now appears in the subject.

Action in noun   There *was* a **discussion** of the zoning ordinances by the city council.

Action in verb   **The city council** *discussed* the zoning ordinances.

Keep the Actor and the Action Together

Once you have successfully expressed the action in a verb, keep that verb and the actor or subject together in the sentence. If several words separate the subject from the verb, readers may have to reread the sentence. Consider this example:

Actor and action separated	Our **branch managers**, because they have insufficient managerial experience or management training, *cannot motivate* unproductive employees.

By the time you finally read the action ("cannot motivate"), you may have forgotten the actor ("branch managers") because nine words separate the actor from the action. The sentence is easier to understand when the actor and the action are together:

Actor and action together	Because they have insufficient managerial experience or management training, **branch managers** *cannot motivate* unproductive employees.

EMPHASIZE THE IMPORTANT INFORMATION IN YOUR SENTENCES

Readers can read your documents more quickly if you emphasize the important information in your sentences. You can emphasize the most important information by doing the following:
- putting the more important information at the end of the sentence
- putting unfamiliar technical terms at the end of the sentence

Put the More Important Information at the End

The natural stress point of most sentences is the end. When reading a sentence aloud, you tend to raise the pitch of your voice near the end and stress the last few words (Williams, Joseph). When writing, you can take advantage of this natural stress point to emphasize important information. In the examples that follow, the more-important information appears in bold and the less-important in italics.

You have not sent us your **December progress report**, *according to our records*.

Our tests show that reliability **increased by 15 percent**, *for example*.

In both of these sentences, unimportant phrases appear at the end. By moving these phrases to the beginning of the sentences, you emphasize the more-important information:

According to our records, you have not mailed us **your December progress report**.

For example, our tests show that reliability **increased by 15 percent**.

Put Unfamiliar Technical Terms at the End

Readers will better understand unfamiliar technical terms if you put them at the end of sentences (Williams, Joseph). If you put them near the beginning, readers don't have a context for understanding unfamiliar terms. Consider these sentences:

Unfamiliar technical terms at the beginning	**Fast-twitch fibers and slow-twitch fibers** are two basic types of muscle fibers.
Unfamiliar technical terms at the end	Muscles have two types of fibers: **fast twitch and slow twitch.**

In the first sentence, the unfamiliar terms appear at the beginning, and the familiar term "muscle" appears at the end. When we put the familiar term "muscles" at the beginning, the sentence gives readers a context for the unfamiliar terms "fast twitch" and "slow twitch."

WRITING EFFECTIVE PARAGRAPHS

When you are writing, you understand how your sentences relate to each other. You also know the topic of each paragraph. However, your readers probably don't. To guide them through your paragraphs and help them understand how each sentence relates to the topic, do the following:

- Begin with an effective topic sentence.
- Tie the remaining sentences together by putting old information near the beginning of the sentences.
- Use transitions, keywords, and pronouns.

> *Effective paragraphs have a topic sentence, put old information near the beginning of sentences to tie ideas together, and use transitions, keywords, and pronouns to link information.*

Begin the Paragraph with a Topic Sentence

To understand the sentences in a paragraph, readers need a frame of reference. Consider the following paragraph. What is the topic?

> The idea is to latch every hook on each of the two tracks into a hollow on the opposite track. The latching mechanism, called the slide, is just a collection of wedges. As the slide moves up, the two teeth strips must enter at a specific angle. As the strips move through the slide, the slide's inclined edges push the teeth toward each other. The strips are offset from each other, so each hollow settles onto a hook in sequence. For this to work properly, each tooth must be exactly the same size and shape, and they all must be perfectly positioned on the track. This would be all but impossible without modern manufacturing technology.[2]

The paragraph is about some type of machine, yet you must guess what type of machine. Many of you may have guessed the correct machine, but some of you may still have doubts. Now, let's look at the paragraph with a topic sentence.

> A zipper is made up of dozens of teeth, each of which combines a hook and a hollow. The idea is to latch every hook on each of the two tracks into a hollow on the opposite track. The latching mechanism, called the slide, is just a collection of wedges. As the slide moves up, the two teeth strips must enter at a specific angle. As the strips move through the slide, the slide's inclined edges push the teeth toward each other.

[2] Reprinted courtesy of HowStuffWorks.com

The strips are offset from each other, so each hollow settles onto a hook in sequence. For this to work properly, each tooth must be exactly the same size and shape, and they all must be perfectly positioned on the track. This would be all but impossible without modern manufacturing technology.[3]

Once you know the machine is a zipper, the paragraph is much easier to understand.

A *topic sentence* announces what a paragraph is about; it provides the frame of reference that readers need to establish meaning in a paragraph. The topic sentence gives readers a key to unlock the meaning of the sentences that follow.

Put Old Information Near the Beginning of Sentences

Effective paragraphs guide readers with key words that help them understand how the sentences relate. These paragraphs follow a pattern where the old information appears before the new information. *Old information* has previously appeared in that paragraph—not in a previous paragraph. *New information* has not yet appeared in the paragraph. In the following paragraph, the old information appears at the beginning of the second through fifth sentences. However, the third and seventh sentences begin with new information. The old information appears in boldface.

[1]When your muscles contract, the muscles' thick and thin filaments do the actual work. [2]**Thick filaments** are made of a protein called myosin. [3]At the molecular level, a **thick filament** is a shaft of myosin molecules arranged in a cylinder. [4]**Thin filaments** are made of another protein called actin. [5]**The thin filaments** look like two strands of pearls twisted around each other. [6]During **contraction**, the thick myosin filaments grab on to the thin actin filaments by forming crossbridges. [7]Making the sarcomere shorter, **the thick filaments** pull the thin filaments past them. [8]In a **muscle** fiber, the signal for contraction is synchronized over

[3] Reprinted courtesy of HowStuffWorks.com

the entire fiber so that all of the myofibrils that make up the sarcomere shorten simultaneously.[4]

In sentences 1, 2, 3, and 4, you can easily see how the sentences relate. However, the old/new pattern breaks down in sentence 3 and then again in sentence 7 because the writer introduces new information ("At the molecular level" and "Making the sarcomere shorter") at the beginning of the sentences. Let's look at a visual representation of the information in the paragraph (See Figure 7.2). The new information appears in red. In sentence 3, notice that the word "molecular" does not appear in sentences 1 or 2. Likewise, in sentence 7, the word "sarcomere" does not appear in sentences 1, 2, 3, 4, 5, or 6. Therefore, these words are new information to the reader—information that has not yet been introduced in the paragraph.

If we revise the paragraph to put the old information before the new information in sentences 3 and 7, the paragraph is easier to read.

[1]When your muscles contract, the muscles' thick and thin filaments do the actual work. [2]**Thick filaments** are made of a protein called myosin. [3]A **thick filament** is a shaft of myosin molecules arranged in a cylinder. [4]**Thin filaments** are made of another protein called actin. [5]**The thin filaments** look like two strands of pearls twisted around each other. [6]During **contraction**, the thick myosin filaments grab on to the thin actin filaments by forming crossbridges. [7] **The thick filaments** pull the thin filaments past them, making the sarcomere shorter. [8]In a **muscle** fiber, the signal for contraction is synchronized over the entire fiber so that all of the myofibrils that comprise the sarcomere shorten simultaneously.[5]

Use Topics to Tie Sentences Together

You can also relate sentences by topic. The first sentence in a paragraph introduces the topic. The second sentence comments on that topic, and

---

[4] Reprinted courtesy of HowStuffWorks.com
[5] Reprinted courtesy of HowStuffWorks.com. Adapted.

FIGURE 7.2

Understanding the Old/New Pattern (new information appears in red)

Sentence	First part of the sentence	Second part of the sentence
1	When your muscles contract	the muscles' thick and thin filaments do the actual work.
2	Thick filaments	are made of a protein called myosin.
3	At the molecular level,	a thick filament is a shaft of myosin molecules arranged in a cylinder.
4	Thin filaments	are made of another protein called actin.
5	The thin filaments	look like two strands of pearls twisted around each other.
6	During contraction,	the thick myosin filaments grab on to the thin actin filaments by forming crossbridges.
7	Making the sarcomere shorter,	the thick filaments pull the thin filaments past them.
8	In a muscle fiber,	the signal for contraction is synchronized over the entire fiber so that all of the myofibrils that comprise the sarcomere shorten simultaneously.

the third, fourth, fifth, and subsequent sentences comment further on that topic. In a well-written paragraph, the writer signals the topic by mentioning it in the subject position of each sentence.

In the following paragraph, the subject of each sentence appears in boldface, but the topic shifts from sentence to sentence:

Shifts in topic

Your **Personal Identification Number (PIN)** should arrive within three days after the card. The **bank** currently does not have the software to allow cardholders to personalize their PINs. The **company** selected a four-digit number for your PIN.

"Your Personal Identification Number (PIN)," the topic of the first sentence of the paragraph, appears as the subject, and the remainder of the sentence provides information about the topic. The second sentence, however, does not comment on that topic, but instead, introduces a new topic ("the bank") as the subject. The third sentence introduces another new topic ("the company"). The paragraph would be more effective with consistent topics in the subjects of each sentence:

Revised

Your **Personal Identification Number (PIN)** should arrive within three days after the card. Your **PIN** is a four-digit number that the bank has selected for you. Your **PIN** cannot be personalized because the bank currently does not have the software to allow cardholders to personalize their PINs.

In the revised version, the common topic appears as the subject of each sentence; and readers can easily relate the old information to the new information in each sentence. The writer uses the passive voice in the third sentence to place the common topic in the subject position. This writer appropriately uses the passive voice to tie together related information. Sometimes, you can only put a common topic into the subject position by using the passive voice.

Use Transitions

Transitions are words, phrases, and even sentences that connect one idea or one sentence to another. Transitions indicate relationships of time, cause and effect, space, addition, comparison, and contrast. Figure 7.3 lists common transitions that writers use to connect ideas.

When using transitions to tie your sentences together, put the transition at or near the beginning of the sentence. When you put a transition after

---

FIGURE 7.3

Common Transitions

Time	before, while, during, after, next, later, first, second, then, subsequently, the next day, meanwhile, now
Cause and Effect	because, therefore, since, thus, consequently, due to, if...then, so
Place	below, above, inside, outside, behind, at the next level, internally, externally
Addition	furthermore, in addition, also, moreover, and
Comparison	likewise, as, like, similarly, not only...but also
Contrast	conversely, on the other hand, unlike, although, however, yet, nevertheless, but

the verb or near the end of the sentence, you weaken the effect of the transition and frequently create an awkward sentence, as in this example:

Awkward

The results of the flight tests concerned the company executives. The executives asked the research and development team to retest the new plane, **therefore**.

Because the transition ("therefore") occurs at the end of the sentence, you do not see the cause-and-effect relationship until you reach the end of the sentence. With the transition at the end, the sentence also fails to stress the more important information. The transition is more effective at the beginning of the sentence:

Revised

The results of the flight tests concerned the company executives. **Therefore**, the executives asked the research and development team to retest the new plane.

Although transition words can help readers to see relationships, use these words in conjunction with the other techniques presented in this chapter to tie your sentences together. Transition words alone aren't enough to tie all your sentences together. If you find that you are using transition words in sentence after sentence, revise your paragraphs to eliminate some of the transition words and make sure that you are using the old/new pattern.

Repeat or Restate Key Words or Phrases

You can tie sentences together by repeating or restating key words or phrases to help readers remember information and to understand the point you are making.

Repeating Key Words and Phrases

You can repeat key words or phrases to tie sentences together. However, avoid overusing this technique. The following sentences effectively repeat key words:

> The accident on the space station **depleted** the oxygen **supply**. Because of the **depleted supply**, the crew had to limit its physical activities.

> The city council **recommended** that the city redraw the lines of the districts. This **recommendation** upset many citizens.

In the first example, the sentence repeats two words from the first sentence, "depleted" and "supply," to tie the sentences together.

In the second example, the verb "recommended" in the first sentence is echoed in the second sentence by the noun "recommendation." The repetition allows the writer to put old information at the beginning of the sentence.

Restating with Pronouns

You can use pronouns to refer to nouns that appear in a previous sentence. Pronouns not only help tie your sentences together, but also avoid the monotony that results when the same noun appears several times in a sentence or paragraph.

Muscles have two types of fibers: fast twitch and slow twitch. Fast-twitch fibers can contract faster. **These** fibers also have greater anaerobic capacity.

Restating with Summary Words

You can tie sentences together by using words that summarize ideas or information presented earlier. **Summary words** allow you to restate information, usually in just a few words. In the example below, the subject of the second sentence ("this new equipment") summarizes the equipment mentioned in the first sentence:

The computer lab purchased six new laser printers and twelve new personal computers. **This new equipment** will allow the lab to better serve the computer science students.

The summary words concisely restate the information from the previous sentence and put that old information at the beginning of the new sentence.

WRITING COHERENT DOCUMENTS

Once you have written your sentences and individual paragraphs, you need to turn to the overall document. You want to consider how the overall document flows from one topic to another. The following tools can help your readers navigate through your document:
- order of the paragraphs
- overviews
- headings
- lists

Organizing a Coherent Document

Have you ever tried to read a document where the paragraphs were organized illogically? Did you have trouble locating the information you needed? Documents with an illogical organization make the reader's task difficult—whether that task is to follow a procedure, to make a decision, or to gather information (Felker et al.). A logical organization helps readers remember, understand, and follow a document (Duin, "How People Read") As an example, consider the document in Figure 7.4. This document, from a manual on minimizing pollution, organizes the paragraphs as follows:

Paragraph 1	Introduction to the use of chemicals to stabilize or destroy waste
Paragraph 2	More information about using chemicals to stabilize or destroy waste
Paragraph 3	The process of chemical fixation
Paragraph 4	An alternative to applying chemicals to in situ landfills
Paragraph 5	Firms that provide the chemicals to stabilize the landfills
Paragraph 6	Problems with chemical fixation in the landfills
Paragraph 7	Problem landfills and chemical fixation

As these paragraphs are organized, readers have trouble understanding how one paragraph relates to the next for the following reasons:

- Information about in situ landfills appears in paragraphs 4 and 6. This information should appear either in one paragraph or in two successive paragraphs.
- The paragraphs do not follow a logical order.
- The document does not have headings that help readers locate the major topics.
- The document lacks an overview describing its organization.

Let's see what happens when we revise this document. Examine the revised text in Figure 7.5, page 192.

Paragraph 1	Introduction to the use of chemicals to stabilize or destroy waste
Paragraph 2	The process of chemical fixation
Paragraph 3	Problems with chemical fixation in the landfills
Paragraph 4	An alternative to applying chemicals to in situ landfills
Paragraph 5	Stabilized waste materials in in situ landfills

FIGURE 7.4

Information Organized Illogically

Chemical Fixation

[1] The application of chemicals to destroy or stabilize hazardous materials and potential pollutants has been a common practice for many years, particularly for industrial wastes. Generally, chemical treatment is quite waste-specific. Thus, an effective system in one case may be ineffective or totally inapplicable in another.

[2] Since the 1970s, several processes involving chemicals have been developed which may be effective on a broader range of wastes. Some of these newer processes are more effective on liquids and thin sludges while others function best with heavier sludges and solids. These processes rely on the reactions of such materials as Portland cement, lime, and common silicates to encapsulate, solidify, or cement waste material.

[3] Each of these processes involves the mixing of a chemical agent such as cement, lime, or silicates with the waste material. With liquid water, the agent absorbs the waste. With solids, the agent coats the surface of the solids to cement them together. With sludges, the agent absorbs and cements the waste. Some of these processes rely mainly on the ability of the chemical system to insulate each particle of pollutant from adjacent leaching fluids; others rely on the formation of a relatively impermeable mass to exclude leaching fluids from passing through the waste.

[4] An alternative to the application of chemical fixation agents to the in situ landfill is to use these agents for stabilizing waste materials. These waste materials can then serve as cover for a problem landfill. After proper processing, these waste materials can be spread, graded, and thereafter cemented into a stable, relatively impermeable cover.

[5] The earliest commercially prominent stabilization system was a process offered by Chemfix for applying to hazardous liquids and sludges. Now, stabilization processes are offered by other firms such as the Environmental Technology Corporation, IU Conversion Systems, Inc., and the Dravo Corporation. The latter two firms primarily offer systems for stabilizing sulfur dioxide scrubber sludge.

[6] We have included information on stabilizing waste materials with chemical agents because, in particular instances, the process is a viable means for controlling potential pollutants. However, this process is not feasible for in situ landfill problems because the success of the system depends on the intimate mixing of the chemical agents and the material to be stabilized; without this mixing, the municipal refuse cannot be coated and encapsulated, and the normal landfill processes of degradation and leaching cannot occur. To ensure the mixing of the chemicals with the refuse would require excavating the entire landfill, which provides little advantage over excavating and relocating the landfill to an environmentally sound site.

[7] An ideal situation would be a problem landfill located near a source of chemically stabilized waste material. The material would be readily available for applying to the landfill as a cap. The chemically stabilized material would then be applied at an approximate compacted thickness of 0.6 m, with appropriate drainage swales to remove surface water.

Source: Tolman, Andrews L., et al. *Guidance Manual for Minimizing Pollution for Waste Disposal Sites.* EPA-600/2-78-142. (Washington: GPO, Aug. 1978) 52.

FIGURE 7.5

Document with a Logical, Reader-Focused Organization

Chemical Fixation

Brief overview

[1] The application of chemicals to destroy or stabilize hazardous materials and potential pollutants has been a common practice for many years, particularly for industrial wastes. Generally, chemical treatment is quite waste-specific. Thus, an effective system in one case may be ineffective or totally inapplicable in another. For example, applying chlorine to destroy cyanide and applying lime to precipitate and insolubilize flourides are standard but waste-specific processes. Since the 1970s, several firms have developed chemical fixation processes that may be effective on a broader range of wastes.[1] In this section, we will discuss how chemical agents destroy or stabilize waste and when chemical agents are ineffective.

Heading

How Chemical Agents Destroy or Stabilize Waste

[2] To destroy or stabilize waste, each of these chemical fixation processes mixes a chemical agent such as Portland cement, lime, or silicates with the waste material. With liquids, the chemical agent absorbs the waste. With solids, the agent coats the surface of the solids to cement them together. With sludges, the agent absorbs and cements the waste. Some of these processes rely mainly on the ability of the chemical system to insulate each particle of pollutant from adjacent leaching liquids; others rely on the formation of a relatively impermeable mass to exclude leaching fluids from passing through the waste.

Heading

Information on existing landfills is in successive paragraphs 3 & 4.

Similar information is now grouped in paragraphs 4 & 5.

When Chemical Agents are Ineffective

[3] Although chemical fixation is a viable means for controlling potential pollutants, this process is not feasible for existing landfill problems because the success of the process depends on the intimate mixing of the chemical agents and the material to be stabilized. Without this mixing, the municipal refuse cannot be coated and encapsulated, and the normal landfill processes of degradation and leaching cannot occur. To ensure the mixing of the chemicals with the refuse would require excavating the entire landfill, providing little advantage over excavating and relocating the landfill to an environmentally sound site.

[4] An alternative to applying chemical fixation agents to an existing landfill is to use these agents for stabilizing waste materials not currently in a landfill. These waste materials can then serve as cover for a problem landfill. After proper processing, these waste materials can be spread, graded, and thereafter cemented into a stable, relatively impermeable cover.

[5] When such waste material serves as cover, the problem landfill must be near a source of chemically stabilized waste material. The material would be readily available for applying to the landfill as a cap. The chemically stabilized material could then be applied at an approximate compacted thickness of 0.6 m, with appropriate drainage swales to remove surface water.

[1] Chemfix offered the earliest commercially prominent stabilization system for hazardous liquids and sludges. Now, other firms such as the Environmental Technology Corporation, IU Conversion Systems, Inc., and the Dravo Corporation offer stabilization processes. The latter two firms offer systems for stabilizing sulfur dioxide scrubber sludge.

Source: Adapted from Andrews L. Tolman, Antonio P. Ballestero Jr., William W. Beck Jr., and Grover H. Emrich, *Guidance Manual for Minimizing Pollution for Waste Disposal Sites,* EPA-600/2-78-142 (Washington: GPO, Aug. 1978) 52.

In this version, closely related information is grouped into five, instead of seven, paragraphs. For example, the information on existing landfills now appears in successive paragraphs. The document begins by briefly introducing chemical fixation, moves to more specific information about how chemicals stabilize or destroy hazardous waste, and ends with even more specific information about when chemical fixation is ineffective. The document also gives readers a brief overview at the beginning and includes informative headings that allow readers to locate information.

Using Overviews to Introduce Readers to Your Document

Your readers want to know what they are reading. At the beginning of a document and at the beginning of each section, readers want to know what is coming. You can help them by using overviews and topic sentences.

Use Overviews to Introduce Readers to Your Document

Overviews are introductory statements that describe what a document is about, what it may be used for, or how it is organized (Felker et al.). Overviews preview the topic and the overall organization of information in your document. Overviews appear immediately before the text that they describe or summarize. That text might be a short document, a specific section of a document, or an individual chapter or section. For some short documents, such as memos, e-mail, and letters, overviews may not be necessary. Instead, a topic sentence at the beginning can function as an overview. You can use overviews in three ways (Felker et al.):

- to point out the type of information presented in the document
- to identify the specific sections included in the document
- to give instructions about how to use the document

Overviews that Introduce the Types of Information in the Document

Consider the overview in Figure 7.6, page 194, from the *Earthquake Safety Guide for Homeowners* published by the Federal Emergency Management Agency (FEMA). This overview tells readers about the two types of information available in the guide: common weaknesses that can result in a home being damaged by earthquakes and steps homeowners can take to correct those weaknesses. With this overview, readers can quickly decide whether the document contains the information they need.

FIGURE 7.6

An Overview that Introduces the Types of Information in the Document

Earthquakes, especially major ones, are dangerous, inevitable, and a fact of life in some parts of the United States. Sooner or later another "big one" will occur. Earthquakes:

- Occur without warning
- Can be deadly and extremely destructive
- Can occur at any time

As a current or potential owner of a home, you should be very concerned about the potential danger to not only yourselves and your loved ones, but also to your property.

The major threats posed by earthquakes are bodily injuries and property damage, which can be considerable and even catastrophic. Most of the property damage caused by earthquakes ends up being handled and paid for by the homeowner.

- In a 2000 study titled *HAZUS 99: Average Annual Earthquake Losses for the United States,* FEMA estimated U.S. losses from earthquakes at $4.4 billion per year.
- Large earthquakes in or near major urban centers will disrupt the local economy and can disrupt the economy of an entire state.

However, proper earthquake preparation of your home can:

- Save lives
- Reduce injuries
- Reduce property damage

As a homeowner, you can **significantly reduce** damage to your home by fixing a number of known and common weaknesses.

The overview tells readers right up front what types of information they will find in the documents.

This booklet is a good start to begin strengthening your home against earthquake damage. It describes:

- Common weaknesses that can result in your home being damaged by earthquakes, and
- Steps you can take to correct these weaknesses.

Source: FEMA, *Earthquake Safety Guide for Homeowners.* FEMA 530, September 2005. Page 1.

Overviews that Identify the Specific Sections of the Text

You can also write overviews that identify the sections of the text and describe the information that they contain. These overviews particularly help readers by explaining the organization of the text that follows and by providing signposts to that organization (Felker et al.). Figure 7.7 includes signposts in the overview to *Electric and Magnetic Fields.* Such overviews are especially useful in long documents, in documents divided into sections, and in documents written for more than one type of reader (Felker et al.).

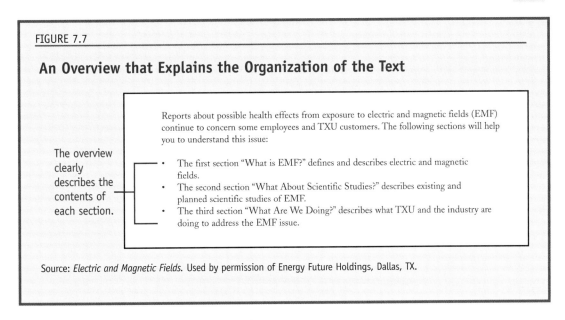

FIGURE 7.7

An Overview that Explains the Organization of the Text

The overview clearly describes the contents of each section.

Reports about possible health effects from exposure to electric and magnetic fields (EMF) continue to concern some employees and TXU customers. The following sections will help you to understand this issue:

- The first section "What is EMF?" defines and describes electric and magnetic fields.
- The second section "What About Scientific Studies?" describes existing and planned scientific studies of EMF.
- The third section "What Are We Doing?" describes what TXU and the industry are doing to address the EMF issue.

Source: *Electric and Magnetic Fields.* Used by permission of Energy Future Holdings, Dallas, TX.

Overviews that Tell Readers How to Use the Document

In some overviews, you may want to tell readers how to use the document. This type of overview combines signposts with instructions for using the document. Figure 7.8, page 196, shows an overview with instructions for a software manual. In the first paragraph, the writer lists the manual's three major sections and then tells readers how to use each section. This type of overview clearly tells readers "what the document is about, who should use the different parts of it, and when they should use it" (Felker et al.).

Use Headings to Show the Organization of Your Document

Headings are subtitles within a document. If you have prepared a formal outline, you can use the entries in the outline as headings and subheadings in your document.

Headings accomplish the following:
- indicate the organization and scope of a document to help readers who want to determine whether a document contains information they need
- help readers locate specific information
- give readers cues to the information contained in specific sections

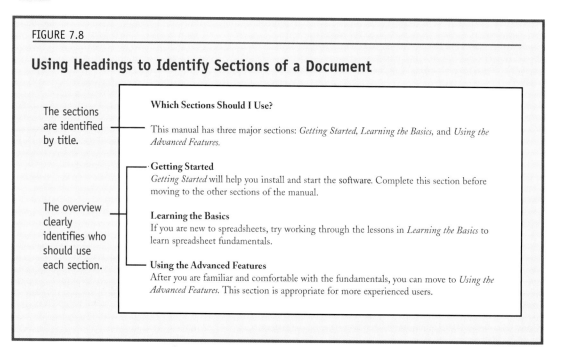

Figures 7.9 and 7.10, pages 198-199, show how adding headings can help reveal content. The document shown in Figure 7.9 has no design elements other than the title and paragraph breaks. It doesn't include headings or lists to help readers locate information, so readers can't read selectively or easily locate specific information. Readers who are interested in specific information must read until they find that information. For example, if you want to know how to treat canine epilepsy, you must read six paragraphs before finding that information. In contrast, in Figure 7.10, if you want to locate that same information, you simply scan the document until you see the heading "Treatment of Canine Epilepsy."

Readers process a document from top to bottom unless the writer gives them clues about the organization (Duin, "How People Read"). Without these clues, readers may assume that information near the beginning is more important than information near the end, and they may read the information near the beginning of a document especially closely, recalling it better than they recall information near the end. If readers are likely to need information that appears in the middle or near the end, you can give clues through the document design to help readers find that information.

The format of the document in Figure 7.10 gives readers clues through headings and lists. The headings visually categorize information so readers don't have to process the document from top to bottom. Instead, readers can scan through the document for the information that interests them. The format of the document in Figure 7.9 gives readers no clues about content and forces them to process the document in a top-down manner.

Make Your Headings Informative

Effective headings inform readers, giving them enough information to decide whether a section contains the information that they need. Compare the headings in the left and right columns:

Introduction	How to Use This Manual
Body	Description of the Rehabilitation Clinic
Part 2	How to Set Up the Scanner
Donation Limitations	How Often Can You Give Blood?

The headings in the left column give few, if any, clues about the information in the sections that they introduce. "Body" doesn't give readers any clues to the information they will find if they turn to that section. "Body" could refer to a human body, a body of people, or the body (main part) of the document. Because the headings in the left column give limited, if any, clues about the information contained in a section, they can't help readers locate specific information. In contrast, the headings in the right column describe the information that readers will find in each section.

FIGURE 7.9

A Document with No Headings or Design Elements

Canine Epilepsy

Epilepsy is a disorder characterized by recurrent seizures. Seizures, also known as fits or convulsions, occur when an area of nerve cells in the brain becomes overexcitable. This area is often called a seizure focus. The mechanism responsible for developing this focus is unknown.

A dog can inherit or acquire canine epilepsy. Inherited epilepsy affects about 1% of the canine population. Breeds which may inherit epilepsy include the beagle, Belgian shepherd, German shepherd, dachshund, and keeshond. Researchers also suspect a genetic factor in the following breeds: cocker spaniel, collie, golden retriever, Labrador retriever, Irish setter, miniature schnauzer, poodle, Saint Bernard, Siberian husky, and wire-haired fox terrier.

Acquired epilepsy may occur months to years after an injury or illness that causes brain damage. In many cases, the dog is completely normal except for occasional seizures. Causes of acquired epilepsy include trauma, infection, poisons, hypoxia (lack of oxygen), and low blood sugar concentrations.

A dog with inherited epilepsy has generalized seizures that affect its entire brain and body. The dog usually falls on its side and displays paddling motions with all four limbs. During or immediately after the seizures, the dog may also exhibit loss of consciousness (i.e., the dog will not respond when you call its name), excessive drooling, and urinating or passing of feces. The seizure of inherited epilepsy usually occurs between the ages of 1 and 3 years. Seizures that occur before 6 months or after 5 years of age probably result from acquired epilepsy.

A dog with acquired epilepsy has partial seizures. A partial seizure affects only one part of the body, and the dog may not lose consciousness. During a partial seizure, the dog may exhibit turning of the head to one side, muscular contractions of one or both legs on the same side of the body, or bending of the body to one side. These signs are localizing because they help to determine the location of the seizure focus in the brain. The localizing sign may occur only briefly, after which the seizure becomes generalized. If the seizure becomes generalized, you or your veterinarian may have difficulty distinguishing between acquired and inherited epilepsy. The first seizure may occur at any age.

You can treat epilepsy by giving anticonvulsant medication orally several times a day. This treatment is effective in 60 to 70% of epileptic dogs. Unfortunately, the medication will not completely eliminate the seizures. Instead, the medication reduces the frequency, severity, and duration of the seizures. Most veterinarians recommend that dogs receive the anticonvulsant medication when the seizures occur more often than once every 6 weeks or when severe clusters of seizures occur more often than once every 2 months. To successfully treat epilepsy, you must consistently give the medication as directed by the veterinarian and continue the medication without interruption. If you discontinue the medication, status epilepticus could occur, resulting in the dog's death. Status epilepticus is a series of seizures without periods of consciousness. If this condition occurs, contact a veterinarian immediately.

Source: Adapted from S. Dru Forrester and Bruce Lawhorn, *Canine Epilepsy* (College Station: Texas Agricultural Extension Service, n.d.).

FIGURE 7.10

A Document Broken Up with Headings and Design Elements

Canine Epilepsy

Epilepsy is a disorder characterized by recurrent seizures. Seizures, also known as fits or convulsions, occur when an area of nerve cells in the brain becomes overexcitable. This area is often called a seizure focus. The mechanism responsible for developing this focus is unknown.

Types of Canine Epilepsy

A dog can inherit or acquire canine epilepsy. Inherited epilepsy affects about 1% of the canine population. Breeds which may inherit epilepsy include the beagle, Belgian shepherd, German shepherd, dachshund, and keeshond. Researchers also suspect a genetic factor in the following breeds: cocker spaniel, collie, golden retriever, Labrador retriever, Irish setter, miniature schnauzer, poodle, Saint Bernard, Siberian husky, and wire-haired fox terrier.

Acquired epilepsy may occur months to years after an injury or illness that causes brain damage. In many cases, the dog is completely normal except for occasional seizures. Causes of acquired epilepsy include trauma, infection, poisons, hypoxia (lack of oxygen), and low blood sugar concentrations.

Characteristics of Inherited Epilepsy

A dog with inherited epilepsy has generalized seizures that affect its entire brain and body. The dog usually falls on its side and displays paddling motions with all four limbs. During or immediately after the seizures, the dog may also exhibit some or all of the following signs:
- loss of consciousness (i.e., the dog will not respond when you call its name)
- excessive drooling
- urinating or passing of feces.

The seizure usually lasts no longer than 1 or 2 minutes. The first seizure of inherited epilepsy usually occurs between the ages of 1 and 3 years. Seizures that occur before 6 months or after 5 years of age probably result from acquired epilepsy.

Characteristics of Acquired Epilepsy

A dog with acquired epilepsy has partial seizures. A partial seizure affects only one part of the body, and the dog may not lose consciousness. During a partial seizure, the dog may exhibit one or more of the following localized signs:
- turning of the head to one side
- muscular contractions of one or both legs on the same side of the body
- bending of the body to one side.

These signs are localizing because they help to determine the location of the seizure focus in the brain. The localizing sign may occur only briefly, after which the seizure becomes generalized. If the seizure becomes generalized, you or your veterinarian may have difficulty distinguishing between acquired and inherited epilepsy. The first seizure may occur at any age.

Treatment of Canine Epilepsy

You can treat epilepsy by giving anticonvulsant medication orally several times a day. This treatment is effective in 60 to 70% of epileptic dogs. Unfortunately, the medication will not completely eliminate the seizures. Instead, the medication reduces the frequency, severity, and duration of the seizures. Most veterinarians recommend that dogs receive the anticonvulsant medication when the seizures occur more often than once every 6 weeks or when severe clusters of seizures occur more often than once every 2 months.

To successfully treat epilepsy, you must
- consistently give the medication as directed by the veterinarian
- continue the medication without interruption.

If you discontinue the medication, status epilepticus could occur, resulting in the dog's death. Status epilepticus is a series of seizures without periods of consciousness. If this condition occurs, contact a veterinarian immediately.

Source: Adapted from S. Dru Forrester and Bruce Lawhorn, *Canine Epilepsy* (College Station: Texas Agricultural Extension Service, n.d.).

TIPS FOR WRITING EFFECTIVE HEADINGS

- **Identify the primary topic of the section**. The heading should clearly identify the information that follows. For example, in Figure 7.10, page 199, the heading "Characteristics of Inherited Epilepsy" clearly identifies the characteristics of only inherited epilepsy.
- **Use parallel headings**. Let's look again at Figure 7.10, page 199. Each of the headings has the same grammatical structure: a noun combined with a prepositional phrase. These headings, therefore, are parallel.
- **Use key words to tell readers the topic of a section**. Often writers use key words such as "Budget" as headings. In many instances, these key words adequately inform readers of the information that follows. If you use single-word headings, make sure the headings clearly identify the information that follows.
- **Use questions to tell readers the topic of a section and to create a friendly tone for less-formal documents and for non-expert readers** (Felker et al.). You can draw readers into your document and add a friendly tone by using question-style headings. For example, on a university Web site, you might find a heading such as "How do you pay your bill online?" or "How do you drop a class?" The pronoun "you" adds a personal, informal tone to the headings, yet the headings are still clear and informative.
- **Use "how-to" headings for instructions**. "How-to" headings help readers locate the tasks or procedures they need or want. For example, in a manual for a fax machine you might see a heading such as "How to use a shared phone line" or "How to set up the fax header."

Using Lists Effectively

Lists allow you to break up long sentences and to emphasize important information. Let's look at two examples from the document on canine epilepsy in Figures 7.9 and 7.10. The original document contains the following sentence:

> During a partial seizure, the dog may exhibit turning of the head to one side, muscular contractions of one or both legs on the same side of the body, or bending of the body to one side.

We cannot easily follow this sentence because of the long series of signs. The sentence becomes much easier to read when the series appears in a list:

During a partial seizure, the dog may exhibit one or more of the following signs:

- turning of the head to one side
- muscular contractions of one or both legs on the same side of the body
- bending of the body to one side

With the list, we can easily follow the sentence because the arrangement on the page directs our eyes to each of the signs.

TIPS FOR WRITING EFFECTIVE LISTS

- **Use a number when the list has an implied sequence or hierarchy.** For example, if you are writing instructions, use numbers to indicate the sequence of the steps. If you have a subordinate (sub) list, use lowercase letters or bullets to indicate that sub list:
  1. List item
     a. Sub item
     b. Sub item
  2. List item
     a. Sub item
     b. Sub item
- **Use a symbol, usually a bullet, when you do not have an implied sequence or hierarchy**. In many word-processing programs, you can find several styles for bullets; you can even customize bullets. For most business and technical documents, use a round or square bullet. If you have a sub list, use dashes or hollow bullets:
  - List item
    - Sub item
    - Sub item
  - List item
    - Sub item
    - Sub item
- **Align the list correctly.** Look at the lists in Figure 7.11, page 203. In the left column, the lines of text align below the bullet or number. However, in the right column, the lines of text align below the text, and the bullet or number stands out. The right column is easier to follow.

- **Break up long lists.** Most readers can only remember about 7 items; so if your list has more than 7 or 8 items, break the list into two or more smaller lists.
- **Make your lists parallel.** Let's look again at Figure 7.10, page 199. Notice the lists of characteristics of acquired epilepsy. Each item in the list has the same grammatical structure: a noun combined with a prepositional phrase. These list items, therefore, are parallel.
- **Punctuate the lead-in correctly.** Although standards vary among organizations, the most commonly used lead-in is a complete clause followed by a colon, as in the following example:

    > To log in to the network, complete these steps:

    If you cannot use a complete clause, do not use the colon—or any punctuation—after the lead-in:

    > To successfully treat epilepsy, you must
    > - consistently give the medication as directed by the veterinarian
    > - continue the medication without interruption

- **Use correct punctuation and capitalization in your lists.** The rules for punctuating lists vary, so you should find out what your organization prefers. If your organization does not have a preference, follow these guidelines for punctuating lists:
    - *If the list items are phrases,* begin each list item with a lowercase letter and do not use a period or comma at the end of the item. You do not need the period or comma because the white space after the last list item signals the end of the list.
    - *If the list items are complete sentences,* begin each list item with an uppercase letter and end the list with a period.
    - *If the list items are phrases followed by complete sentences,* begin each list item with an uppercase letter and end the initial phrase with a period. Begin the complete sentence with an uppercase letter and end the sentence with a period. You can use italics and/or bold to emphasize the phrase, as in the example below.
    - *If the list includes two types of grammatical structures—such as phrases and complete sentences,* begin each list item with an uppercase letter and end each item with a period, as in the example below.
        - *Increased fuel economy.* The new hybrid design will increase fuel economy by 15%.
        - *Increased cargo room.*
        - *Improved navigation systems.* The newly designed vehicle includes a built-in compass and a GPS navigation system.

FIGURE 7.11

Aligning Lists Correctly

Incorrectly Aligned	Correctly Aligned
1. Text	1. Text
• sub text	• sub text
• sub text	• sub text
2. Text 2	2. Text 2
• sub text 2	• sub text 2
• sub text 2	• sub text 2

CASE STUDY ANALYSIS
Bad Writing Costs State Governments a Bundle

Background

Poor writing costs taxpayers big money. A 2005 report released by the National Commission on Writing says that poor writing skills cost state governments nearly $250 million annually in employee training programs to improve writing skills.

In a survey conducted by the Commission, human resource managers listed clear writing as one of the most important skills needed by state government employees. However, only 33 percent of managers at the state level reported that their clerical- and support-level employees had adequate writing skills. Consequently, many employees at the state level must take remedial writing classes, at a cost of as much as $400 per employee.

The consequences of poor writing skills cost taxpayers money because of
• slower response times
• time-consuming and costly rewrites and printing

- time lost in reading and rereading unclear documents
- additional hires required to answer citizens' questions about unclear documents.

Assignment

1. Find an example of a poorly written document distributed by your state or local government.
2. Highlight sentences and paragraphs that are not clear and that do not follow the guidelines presented in this chapter.
3. Revise the document. Make sure your revised version follows the guidelines for clear writing in this chapter.
4. Hand in both the original document and your revised document.

*Compiled from information downloaded from the World Wide Web:
http://www.usatoday.com/news/washington/2005-07-04-employees-lack-skills_x.htm. "Report: State employees' lack of writing skills cost nearly $250M," Copyright 2005, The Associated Press.

EXERCISES

1. Rewrite these sentences to make the actor the subject.
   a. Attempts were made by the engineering staff to complete the project.
   b. The addition of the new building will allow 20 new businesses to relocate downtown.
   c. There have been threats from other landowners nearby due to the trash and mud in the streets caused by unauthorized employees using this lot.
   d. There is no alternative for us except to recall the faulty software.
   e. Working together will ultimately allow us to finish the new product by the deadline.
   f. It is extraordinarily important that we turn in our application by the required deadline.

Download a Worksheet for Writing Easy-to-Read Documents online at www.kendallhunt.com/ technicalcommunication.

2. Change these sentences from the passive to the active voice. Follow the guidelines in "Tips for Changing a Passive-Voice Sentence into an Active-Voice Sentence" on page 178.
   a. The merit raises were approved by the board of directors.
   b. Dues must be paid within 90 days or in three installments by all members.
   c. Please e-mail us soon as a shipment date is known.
   d. Parking on-site is allowed for your plant employees.
   e. Reduction in the size of the laser beam has been accomplished.
   f. The ethics training procedure was completed by the new employee.

3. Rewrite these sentences to put the action in the verb.
   a. We must provide support for the three candidates from our district who are running for president, vice president, and senator.
   b. Our expectation was to begin the interview at 10:00 p.m.
   c. The detectives are conducting an investigation of the burglary that occurred this morning.
   d. Failure to pay the fine within 15 days may result in a warrant being issued for your arrest.
   e. The team has taken into consideration the results of the usability tests.

f. To ensure the safety of all visitors and employees, we will conduct a test of the new evacuation procedures on May 16.

4. Rewrite these sentences to put the more-important information and technical terms near the end of the sentence.
   a. The outcome of the election changed because of some unfortunate comments about the city manager, according to the press release.
   b. Business owners in the historic district are suing the developer for faulty foundations and poor street drainage, however.
   c. Water pipes in the older condominiums may freeze and burst when the temperature drops below 20° F.
   d. Myosin and actin are tiny filaments inside your muscles.
   e. Creatine phosphate is broken down by the muscles.
   f. Epithelial cells slide over the cornea and patch the injury if your cornea is scratched by dirt.

5. For each sentence listed below, complete the following:
   • Identify the problems in the sentence.
   • Rewrite the sentence to eliminate the problems.

   a. Because the technician did not have the parts, there was a delay in the repair of my laptop until the next day.
   b. Because the team did not have much time to complete the testing, it was surprising that it was able to submit the report by the deadline.
   c. It is our intention to begin the project when the new equipment arrives.
   d. After conducting a reinvestigation of the new employee's expense report, it was determined that the accounting division could provide reimbursement for the expenses.
   e. The opening of an international branch will increase our budget for the new year, however.
   f. Our international sales offices, which are located in Germany, China, Canada, and Brazil, have increased sales by 15 percent.
   g. There has been a decrease in the number of infants killed because of the redesigned car seats.

h. Updates to your account can be accessed on the Web in five business days.

i. The updates will be ready in one month after the staff has conducted the installation, according to the press release.

6. Complete the following for each paragraph:
   - Underline the old information.
   - Make sure the paragraph has an effective topic sentence.
   - Rewrite the paragraph so that the old information appears before the new information.
   - Make sure that the paragraph follows the guidelines presented in this chapter for writing effective sentences and paragraphs.

Paragraph 1

Almost everyone has had the unpleasant experience of being bitten by a mosquito. Severe skin irritation can occur through an allergic reaction to the mosquito's saliva; this is what causes the red bump and itching. But a more serious consequence may be transmission of certain serious diseases, such as malaria, dengue fever, and several forms of encephalitis from mosquito bites. Not only can mosquitoes carry diseases that afflict humans, but they also can transmit several diseases and parasites that dogs and horses are very susceptible to. These include dog heartworms and Eastern Equine Encephalitis. There are about 200 different species of mosquitoes in the United States, all of which live in specific habitats, exhibit unique behaviors, and bite different types of animals. Despite these differences, all mosquitoes share some common traits, such as a four-stage life cycle. After the female mosquito obtains a blood meal (male mosquitoes do not bite), she lays her eggs directly on the surface of stagnant water, in a depression, or on the edge of a container where rainwater may collect and flood the eggs. The eggs hatch and a mosquito larva, or "wiggler," emerges. Living in the water, the larva feeds and develops into the third stage of the life cycle, called a pupa, or "tumbler." The pupa also lives in the water, but no longer feeds. Finally, the mosquito emerges from the pupa case and the water as a fully developed adult, ready to bite.

Paragraph 2

There are many rules regarding the presentation of the evidence in both civil and criminal trials which the judge must apply in deciding evidence which may be admitted and the form and manner in which it may be admitted. These rules are complicated and not easily understood by the layperson. They have been developed through hundreds of years of experience so that we may have fair and orderly trials. When a question is asked which either attorney believes is in violation of these rules, they have a right to object. Therefore, no juror must allow himself or herself to be prejudiced for or against one side or the other on account of objections made by an attorney to the introduction of evidence. A ruling by the judge does not mean that he or she is taking sides. He or she is merely deciding that the law does or does not allow the question or the form of the question to be answered. At times, the jury will be excused from the courtroom or the judge will speak to the lawyers beyond the hearing of the jury while questions of law are being discussed. The law provides that certain matters of law be discussed out of the presence of the jury. When a trial is interrupted for these reasons, you should not feel that your time is being wasted or that information is being withheld from you. Oftentimes, these hearings expedite the case in the long run.[6]

Paragraph 3

Insect repellents are available in various forms and concentrations. Aerosol and spray products are intended for skin applications as well as for treating clothing. Liquid, cream, lotion, spray, and stick products enable direct skin application. Products with a low concentration of the active ingredient may be appropriate for situations where exposure to insects is minimal. A higher concentration of the active ingredient may be useful in highly infested areas, or with insect species that are more difficult to repel. And where appropriate, consider nonchemical ways to deter biting insects—screens, netting, long sleeves, and slacks.[7]

[6] Adapted from jury instructions from Collin County Courts.
http://www.texasjudge.com/juryduty/index.html.
[7] Adapted from the U.S. Environmental Protection Agency, Office of Pesticide Programs, *Using Insect Repellants Safely.* 25. Nov. 2001.
http://www.epa.gov/pesticides/citizens/inspectrp.htm.

Paragraph 4

Bovine spongiform encephalopathy (BSE) is a chronic, degenerative disorder affecting the central nervous system of cattle. It is commonly called Mad Cow Disease. Evidence suggests that certain contaminated cattle feed ingredients are the source of BSE infection in cattle. The process that leads to the contaminated feed starts when livestock already harboring the BSE agent are slaughtered. After cows and sheep are killed, the edible parts are removed. The inedible remnants are taken to a special plant where they undergo a process called rendering. Two major products are created:

- fat, which is used in an array of products (such as soap, lipstick, linoleum, and glue)
- meat-and-bone meal, a powdery, high-protein supplement that is often processed into animal feed

Although the animal remnants are "cooked" at high temperatures during the rendering process, the BSE agent, if present, is able to survive. When this contaminated meat-and-bone meal is fed to cattle as a protein supplement, the BSE agent can be passed on to many new cattle. It is believed that this is how BSE spread through the U.K. cattle herds. Currently, no test can reliably detect BSE in live cattle. A diagnosis is confirmed by examining brain tissue after death.[8]

7. **Web exercise**: Find two documents on the Web. One document should include overviews and headings; the other document should not include overviews and headings. Print a copy of the documents and complete the following:

   a. Highlight the overview(s).

   b. Highlight the headings.

   c. Write a memo to your instructor explaining how the overviews and headings make the document easier to read.

   d. Write headings for the document that does not include headings or overviews.

---

[8] Adapted from Bren, Linda, "Trying to Keep 'Mad Cow Disease' Out of U.S. Herds." *FDA Consumer Magazine,* U.S. Food and Drug Administration, March-April 2001. http://www.fda.gov/fdad/features/2001/201_cow.html.

8. Using the "Tips for Writing Effective Lists," pages 201-202, rewrite the following sentences into a list:

a. Compost is organic material that can be used as a soil amendment or as a medium to grow plants. Mature compost is a stable material with a content called humus that is dark brown or black and has a soil-like, earthy smell. It is created by combining organic wastes (e.g., yard trimmings, food wastes, manures) in proper ratios into piles, rows, or vessels; adding bulking agents (e.g., wood chips) as necessary to accelerate the breakdown of organic materials; and allowing the finished material to fully stabilize and mature through a curing process.[9]

b. Before installing your virtual private network (VPN), you will begin by importing your browser favorites. If you use *Explorer*® at home, you have an "import" choice under the file menu that opens a wizard. If you use another browser, you can save these favorites one at a time by opening the document you sent from work, clicking each favorite in your home computer browser and adding it to your favorites list instead of importing the entire list.

c. You'll need to apply in person if you are applying for a U.S. passport for the first time: if your expired U.S. passport is not in your possession; if your previous U.S. passport has expired and was issued more than 15 years ago; if your previous U.S. passport was issued when you were under age 16; or if your currently valid U.S. passport has been lost or stolen.[10]

REAL WORLD EXPERIENCE
How Are Your Revising and Proofreading Skills?

You probably feel relieved when you write the last word of a document. However, when you write that last word, you still aren't finished—that is, you're not finished if you want to ensure that the document is coherent and error free. You still need to revise and proofread it. If you have time, ask your classmates, friends, or coworkers to help you revise and proofread your documents. Most writing professionals agree that revising and proof-

[9] Adapted from the Environmental Protection Agency 2008. http://www.epa.gov/epaoswer/non-hw/composting/basic.htm
[10] U.S. Department of State, 2008.

reading your own work "is not the best way to go" (Hansen, J.). However, often you'll write under deadlines or won't be able to find someone to help you revise and proofread your document. When you revise (sometimes called edit) your own documents, James B. Hansen, author of *Editing Your Own Writing*, suggests these tips:

- **Use a style sheet or style guide.** If you, your team, or your company does not have a style sheet or guide, start by creating your own, or use an established style guide, such as *The Chicago Manual of Style*.
- **Create a checklist.** Checklists can help you locate your particular writing problems. For example, if you tend to misuse the comma, a checklist can help you isolate comma errors.
- **Wait a day or two before revising and proofreading.** If possible, don't revise or proofread your document immediately after you write it (Hansen, J.). By waiting, you can perceive what the document actually says rather than what you intend it to say.
- **Revise and proofread longer documents on paper.** If possible, revise and proofread your documents, especially those over two pages, on paper instead of on screen. "Reading speed can drop as much as 30 percent on a computer screen" (Hansen, J.; Gomes; Krull and Hurford). You are also more prone to overlook problems and errors on a screen.
- **Read the document more than one time.**

Assignment

Using a document that you are writing for your technical communication class or for another class you are currently taking, follow the tips above and revise and proofread that document before you turn it in to your instructor. You can also use the "Worksheet for Writing Easy-to-Read Documents" online at www.kendallhunt.com/technicalcommunication.

Download a Worksheet for Writing Easy-to-Read Documents online at www.kendallhunt.com/technicalcommunication.

---

[9] Gomes, Lee. "Advanced Computer Screens Have Age-Old Rival." San Jose Mecury News 21 Feb. 1994.

CHAPTER EIGHT

8

iStockphoto 2008.

Using Reader-Focused Language

You read technical documents to complete a task, to gather information, or, perhaps, to answer a question. In short, you read to take care of business as quickly and efficiently as possible. Like you, your readers will use your technical documents for reasons similar to those listed above. Therefore, you should use language that is reader-focused. Reader-focused language is

- specific and clear
- concise
- simple
- positive when possible
- inoffensive
- sensitive to readers of other cultures and languages

This chapter will help you to use language that will help readers take care of business.

USE SPECIFIC AND CLEAR LANGUAGE

To understand your documents and respond appropriately, readers need specific and clear language. Without such language, they may misunderstand or misinterpret what you write. When you use specific, clear language, you convey your intended meaning precisely.

Using Specific Language

Specific language is clear and precise; vague language is often unclear and is always imprecise. Specific language clarifies the meaning and eliminates questions that readers may ask.

Vague   A computer in the lab isn't working properly.

Specific   In the usability lab, the monitor on computer 16 is flickering erratically.

After reading the vague sentence, a reader might wonder: Which computer? Which lab? What is wrong with the computer? The specific sentence answers all those questions. It identifies the lab where the computer is located, the specific computer, and the problem.

To make your language specific, you can include examples and details. You can use "such as" or "for example" to introduce the examples. Consider the following:

Vague   For its mission, the relief organization needs food and supplies.

Specific   For its mission to the area damaged by hurricane Gustav, the relief organization needs food and supplies such as
- canned milk
- bottled water in one-gallon plastic containers
- canned meat such as tuna, chicken, or ham spread
- ready-to-use baby formula and disposable diapers
- gauze, bandages, and rubbing alcohol

The vague sentence is imprecise; it does not tell readers what specific food and supplies are needed. The specific sentence tells readers clearly what items the relief effort needs.

Consider another example.

Vague	KH&S will replace the masonry frames with the correct metal frames. We must have these frames soon, so we can complete the first floor of both buildings on schedule.
Specific	KH&S will replace the masonry frames with the correct metal frames. To complete the first floor of both buildings on schedule, KH&S must deliver the frames by 11/26/2008.

The vague sentence includes the word "soon" instead of a date. "Soon" could mean by tomorrow, in a month, or in three months. When you don't specify a date, readers can insert their own meaning of "soon." To ensure that readers understand what is meant by "soon," the writer should use a date, as in the specific example. Notice how the writer has placed the date at the end of the sentence to emphasize its importance.

Using Clear Language

Your readers need clear language to take care of business. Clear language conveys only one meaning. Unclear language results from
- misplaced modifiers
- dangling modifiers
- stacked nouns
- faulty word choice
- inconsistent technical terminology

Eliminate Misplaced Modifiers

Modifiers are words, phrases, or clauses that refer to other elements in the sentence. These elements are called referents. The location of a modifier may determine the meaning of a sentence. Let's consider the following example:

> While in **San Francisco**, our manager suggested to the vice president that we register for the class.
> *Meaning:* The manager is in San Francisco.

Our manager suggested to the vice president in **San Francisco** that we register for the class.
Meaning: The vice president is in San Francisco.

Our manager suggested to the vice president that we register in **San Francisco** for the class.
Meaning: We should register in San Francisco.

Our manager suggested to the vice president that we register for the **San Francisco** class.
Meaning: The class is in San Francisco.

By simply moving the location of "San Francisco," we change the meaning of the sentence.

In your sentences, make sure that the modifiers are "pinned" next to the words to which they refer. In the following example, the words "chubby" and "overweight" modify "veterinarian" but are meant to modify "my dog."

Chubby and overweight, the veterinarian says my dog needs a new diet and more exercise.

When we move the modifiers, the sentence reads correctly because the adjectives "chubby" and "overweight" occur next to the words "my dog," not "veterinarian":

The veterinarian says my **chubby and overweight** dog needs a new diet and more exercise.

Let's look at another example. Here, the misplaced modifier occurs because the writer has incorrectly placed the modifier "growing in the sterile solution" next to "the lab technicians" instead of "bacteria," which is actually what is growing in the solution—not the lab technicians.

Misplaced   **Growing in the sterile solution**, the lab technicians observed the bacteria.

Correct	The lab technicians observed the bacteria **growing in the sterile solution.**

Eliminate Dangling Modifiers

Dangling modifiers occur when the modifier does not have the correct referent in the sentence, as in the following example:

Dangling	**Trying to put out the fire**, the fire extinguisher broke.

In this sentence, the writer has not identified who is putting out the fire. To eliminate the dangling modifier, rewrite the sentence to add the person in either the main clause or in the modifier:

Correct	Trying to put out the fire, I broke the fire extinguisher.
Correct	As I was trying to put out the fire, the fire extinguisher broke.

> *The location of a modifier may determine the meaning of a sentence. To eliminate ambiguity, make sure your modifiers are "pinned" next to the words to which they refer.*

You can also create dangling modifiers when switching from the indicative mood (a statement of fact) to the imperative mood (a command or request, usually with an understood "you" subject). To identify dangling modifiers in those situations, you can often look for a passive-voice construction following the dangling modifier, as in the following example:

Dangling	**To link to other Web sites and topics**, the green keywords should be clicked on.
Correct	To link to other Web sites and topics, click on the green keywords.

To correct the dangling modifier, the writer simply puts the passive voice ("be clicked") into active voice with an understood referent—in this case, "you."

Eliminate Stacked Nouns

You can create ambiguity when you use two or more nouns to modify another noun. When you use a stack of nouns as a modifier, you can create hard-to-read and often ambiguous passages. Consider this example:

> **Stacked nouns** The consultant suggested the manager allow time for a **fitness center member evaluation**.

The sentence is ambiguous because of the stacked nouns: Is the consultant suggesting that the manager evaluate the members or the fitness center? Or is the consultant suggesting that the members evaluate the fitness center? To eliminate the ambiguous language, unstack the nouns:

> **Correct** The consultant suggested that the manager allow time to evaluate the members of the fitness center.

> **Correct** The consultant suggested that the manager allow time for the members to evaluate the fitness center.

You occasionally will use one noun to modify another noun, as in **fitness** center, **space** shuttle, **software** manual, or **school** superintendent. However, whenever possible, avoid using a noun to modify a noun.

Eliminate Faulty Word Choice

Let's look at another ambiguous sentence. In this one, the ambiguity occurs because of the word choice:

> **Ambiguous** We were **held up** at the bank.

> **Clear** We were delayed at the bank.

The ambiguous sentence has two possible meanings: We were either delayed, or we were robbed. To eliminate the ambiguity, you can select the verb "delay," which can have only one meaning.

Let's look at another example. In this case, the ambiguity occurs because the word "tragedy" refers either to the type of play or to the quality of the children's performance.

Ambiguous	The seventh graders will be presenting Shakespeare's "Hamlet" in the school auditorium on Friday at 8:00 p.m. Parents are invited to attend this **tragedy**.
Clear	The seventh graders will be presenting Shakespeare's "Hamlet" in the school auditorium on Friday at 7:00 p.m. They invite all parents to attend the performance.

In the clear example, the writer uses "perform-ance," which can have only one meaning in the sentence. Carefully select the words in your sentences to make sure readers will understand them in only one way.

Use Technical Terminology Consistently and Appropriately

Readers of technical documents expect writers to consistently use the words that refer to technical concepts, instructions, and equip-ment. For instance, readers of a computer manual may be confused if the writer uses the words "screen" and "monitor" interchangeably. You should pick one term or the other and use it consistently. When you are writing instruc-tions or describing equipment, consistent lan-guage is especially important. For example, if a software manual explains how to select "typefaces," but the software itself uses the word "fonts," readers who don't know that typefaces and fonts are synonyms may be con-fused.

Technical terminology—or jargon—is the specialized vocabulary of a particular field, profession, or workplace. For instance, profes-

TIPS FOR USING TECHNICAL TERMINOLOGY

- **Use technical terminology only if your readers have detailed knowl-edge of the topic.** You can assume, for example, that an engineer knows what the abbreviation "psi" (pounds per square inch) means; however, someone without an engineering background may not understand the abbreviation.
- **Use technical terminology for expert readers**. Expert readers expect techni-cal terminology.
- **If your readers have casual or little knowledge of your topic or of your field, avoid using technical terminol-ogy that they won't understand**. If you cannot avoid using such terminology, define the words that you use. Consider whether you can use a more familiar term to refer to a concept, instruction, piece of equipment, and so on.
- **Use technical terminology consistently**. Once you have selected the term you will use in the document, use that term consistently throughout the document.

sionals in the restaurant business may use the term "back server" for employees who clear guests' tables during and after meals. Horticulturalists may use the scientific rather than the common name for plants—for example, referring to a pecan tree as *Carya illinoinensis*, or a daylily as *Hemerocallis*. Technical terminology offers a concise way to convey technical information. However, readers who are unfamiliar with the technical terms may find them confusing or may misinterpret them. You can solve this problem for some readers by defining technical terms in parentheses the first time you use them or by defining them in a glossary. However, if your readers will not read your document from beginning to end, they may not see your parenthetical definition. If your readers may begin reading at any point in your document, consider defining terms in a glossary or using nontechnical terms.

How will you know whether to use technical terminology in your documents? Ask yourself these questions:

- Will all of my readers understand the technical terminology and abbreviations? Will they misunderstand or be confused by any of the technical terms and abbreviations? (For example, ATM to most of us means "automated teller machine." However, ATM can also refer to "asynchronous transfer mode"; so if you use the abbreviation ATM to mean asynchronous transfer mode and your readers are not familiar with the abbreviation, they may think you are referring to an automated teller machine.)
- How can I help readers who don't understand the technical terminology and abbreviations? Will I define terminology and explain abbreviations in parentheses the first time I use them, or will I refer readers to a separate glossary for technical terminology?
- Will my readers expect me to use technical terminology and abbreviations?

USE ONLY THE WORDS YOUR READERS NEED

Readers want to read your documents without wading through unnecessary words. Therefore, use only the words needed to help your readers understand the information. You can write concisely if you

- eliminate redundancy
- eliminate unnecessary words

Eliminating Redundancy

Redundancy occurs when you use
- doubled words
- redundant modifiers

Your language is redundant when you use words or phrases that unnecessarily repeat the meaning of other words in the sentence.

Doubled Words

Redundancies can occur in pairs—two words that have the same meaning combined by "and" —as in the following example:

Redundant	Please give our proposal your **thought and consideration** because the proposed relocation can **help and benefit** the engineering division to better serve the southern region.
Concise	Please consider our proposal because the proposed solution can help the engineering division better serve the southern region.

Each pair of doubled words ("thought and consideration" and "help and benefit") uses two words when one word will do. The words "thought" and "consideration" have a similar meaning as do "help" and "benefit." To be concise, use only one of the words in each pair. Figure 8.1, page 222, lists common doubled words. When you see these doubled words in your writing, revise to use only one word in the pair and delete the word "and."

Words That Imply Other Words

Redundancy also results from words that imply other words in the sentence; these *redundant modifiers* repeat all or part of the meaning of other words in a sentence. (*Modifiers* are a word or a group of words that describe, limit, or qualify another word.) Examples of redundant modifiers include the following (the redundant words appear in bold type):

Redundant	**end** result, **very** unique, **absolutely** free, **completely** eliminate

---

FIGURE 8.1

Common Doubled Words

advice and counsel	fair and equitable	null and void
agreeable and satisfactory	fair and reasonable	opinion and belief
any and all	first and foremost	prompt and immediate
assist and help	full and complete	thought and consideration
basic and fundamental	help and benefit	true and accurate
due and payable	help and cooperation	
each and every	hope and trust	

For more examples, see Figure 8.2.

The sentences below illustrate the positive effect of eliminating words that imply other words:

Redundant   The proposed budget cuts will not affect the **final** outcome of our current projects or our **future** plans for improving the street drainage.

Concise   The proposed budget cuts will not affect the outcome of our current projects or our plans for improving the street drainage.

In the redundant sentence, the modifier "final" is unnecessary because it repeats the meaning of "outcome" (an outcome is always final), and the modifier "future" is unnecessary because it repeats the meaning of "plans" (all plans involve the future). Let's look at another example:

Redundant   In the geology lab, the students analyzed a rock that was pink **in color**, cylindrical **in shape**, and 61 pounds **in weight**.

Concise   In the geology lab, the students analyzed a pink, cylindrical rock that weighed 61 pounds.

FIGURE 8.2

Common Redundant Modifiers

Instead of These Words . . .	Use These Concise Alternatives
absolutely essential	essential
absolutely free	free
anticipate in advance	anticipate
circle around	circle
consensus of opinion	consensus
continue on	continue
decrease down	decrease
end result	result
final outcome	outcome
free gift	gift
future plans	plans
green (red, black, etc.) in color	green (red, black, etc.)
human volunteer	volunteer
mail out	mail
past history	history
past memories	memories
past experience	experience
personal opinion	opinion
reduce down	reduce
repeat again	repeat
quite unique	unique
rarely ever	rarely
round (square, oval, etc.) in shape	round (square, oval, etc.)
return back to	return
seldom ever	seldom
small (large, medium, etc.) in size	small (large, medium, etc.)
true facts	facts
twenty (ten, two, etc.) in number	twenty (ten, two, etc.)
very latest	latest
very unique	unique

The prepositional phrases in the redundant sentence repeat the meaning of "pink," "cylindrical," and "61 pounds": pink is a color, cylindrical is a shape, and 61 pounds is a weight.

Eliminating Unnecessary Words

Readers want to read technical documents as quickly as possible, so eliminate any words not absolutely necessary to convey your meaning and purpose. Figure 8.3 lists wordy phrases and suggests concise alternatives.

At first, you may not notice the wordy phrases that appear in your writing because you have used them for a long time or have read them in many documents. Thus, you may have to make a special effort to spot these phrases and then to replace them with more concise, effective words.

You can also simply eliminate some wordy phrases instead of replacing them. Figure 8.4, page 227, lists several expressions that you usually can eliminate. Let's look at the effect of wordy phrases on sentence length and clarity. In each of the concise versions, the sentences are clearer and more concise.

Wordy	**As a matter of fact, there is** an old warehouse that the emergency relief groups can use to house the hurricane victims **at this point in time**.
Concise	The emergency relief groups can now use the old warehouse to house the hurricane victims.
Wordy	**It should be pointed out that** there are three candidates whom our organization **without further delay** will endorse, **despite the fact that we are not in a position** to contribute any money to their campaigns.
Concise	Our organization will now endorse three candidates although we cannot contribute any money to their campaigns.

Wordy	**It should be noted that** mercury levels in the river have increased this year, and, **in accordance with** your request, our department will **take into consideration** whether **it is essential** to conduct a study **with regard** to possible sources of this pollution.
Concise	Mercury levels in the river have increased this year; as you requested, our department will consider whether to study possible sources of this pollution.

FIGURE 8.3

Common Wordy Phrases

Instead of These Wordy Phrases . . .	Use These Concise Alternatives
a limited number	a few (or the specific number)
a majority of	most (or the specific number)
a number of	many (or the specific number)
at a later time (date)	later
at the conclusion of	after, following
at this point and time	now, currently
by means of	by
concerning the matter of	about
conduct an investigation	investigate
conduct a study	study
despite the fact that	although, even though
due to the fact that	because
have the ability to	can
have the capability to	can
has the capacity for	can
has the opportunity to	can
in accordance with your request	as you requested
in connection with	about, concerning
in order that	so that
in order to	to
in reference to	about
in regard to	about

FIGURE 8.3 CONTINUED

Instead of These Wordy Phrases . . .	Use These Concise Alternatives
in the event that	if
in the near future	soon (or give a specific time or date)
in this day and age	today, now
in view of the fact that	because
is able to	can
is in a position to	can
it is crucial that	must, should
it is important that	must, should
it is incumbent upon	must, should
it is my (our) understanding that	I (we) understand that
it is necessary that	must should
it is my (our) recommendation that	I (we) recommend
it is possible that	may, might, can, could
make reference to	refer to (or referred)
not withstanding the fact that	although
on a weekly (daily, monthly, yearly) basis	weekly (daily, monthly, yearly)
prior to	about
relative to	to
subsequent to	after
take into consideration	consider
there is a chance that	may, might, can, could
there is a need for	must
until such time as	until
we are not in a position	we cannot
will you be kind enough to	please
with reference to	about
with regard to	about
with respect to	about

USE SIMPLE WORDS

When you want to impress your reader, resist the temptation to use words that you don't normally use—words that you rarely use when talking.

FIGURE 8.4

Commonly Used Phrases that You Can Usually Delete

as a matter of fact	it is interesting to note that
I believe	it should be noted that
I hope	it should be pointed out that
in my opinion	thanking you in advance
in other words	the fact that
I should point out that	there are
I think	there is
it is essential	to the extent that
it is evident	

Many writers believe that they will impress readers by using "fancy," less-familiar words, such as those listed in Figure 8.5, page 228. In technical communication you are more likely to impress readers not with fancy words, but with simple, clear, everyday words.

Your readers will prefer words that are more familiar. Consider these examples:

Less-familiar words   **Pursuant to our conversation** on December 15

Simple, familiar words   **As we discussed** on December 15

Less-familiar words   Your firm's response **purports** to explain why you deviated from the grant guidelines.

Simple, familiar words   Your firm **attempts** to explain why you deviated from the grant guidelines.

Less-familiar words   The accounting office will **endeavor** to **procure** our **compensation** checks.

Simple, familiar words   The accounting office will **try** to **find** our **paychecks**.

When using technical documents, readers usually need to gather information quickly and effortlessly. Documents that contain fancy, less-familiar words take longer to read than documents that include simple, familiar words.

FIGURE 8.5

Commonly Used "Fancy" Words

Instead of These Fancy Words . . .	Use These Simple Words
accumulate	gather
apparent	clear
ascertain	learn, find out
cognizant	know
commence	begin, start
commitment	promise
deem	consider
endeavor	try
facilitate	help, ease
herewith is	here is
indebtedness	debt
initiate	begin
locality	place
optimum	best, most
proceed	go
procure	buy, get
pursuant to	as
subsequent to	after, next, later
sufficient	enough
terminate	end
utilize	use

USE POSITIVE LANGUAGE WHEN POSSIBLE

Whenever possible, tell readers what something is, instead of what it is not. Similarly, readers would rather be told what to do rather than what not to do. Readers comprehend positive language more easily and quickly than negative language. The presence of several negative constructions in a sentence or paragraph slows the pace of reading because readers have to work harder to gather the information and meaning. Figure 8.6, page 230, lists some negative phrases and their positive counterparts; you'll discover more as you write.

As the following examples show, positive language leads to clearer and often to more-concise sentences:

Negative   **Do not discontinue taking** the medicine until **none** of the medicine is left.

Positive   **Continue taking** the medicine until it is **all** gone.
　　　　　　　　　OR
Take all the medicine.

Negative   Fourteen team members were **not** absent.

Positive   Fourteen team members **were present**.

Negative   Six of the twenty team members did **not attend** the meeting.

Positive   Fourteen of the team members **were present**.

Negative   Even though the area was experiencing severe thunderstorms, the planes **were not late**.

Positive   Even though the area was experiencing severe thunderstorms, the planes **arrived on time**.

FIGURE 8.6

Examples of Making Negative Language Positive

Instead of Saying What Something Is Not . . .	Say What It is
did not succeed	failed
not many	few
not all	most
not on time	late, delayed
not late, not delayed	on time
not continue	discontinue
not discontinue	continue
not efficient	inefficient
not sad	happy
not accurate	inaccurate
not approve	disapprove
not disapprove	approve
not now	later
not familiar	unfamiliar
not absent	present

USE INOFFENSIVE LANGUAGE

Communication is in large part the perception that your language creates. When you use offensive language, your readers perceive that your attitude is offensive—even if you do not intend it to be offensive. When you avoid offensive language, you think about your readers and help to break down stereotypes and incorrect perceptions.

Use Nonsexist Language

Sexist language incorrectly favors one gender over another. For example, if you refer to firefighters as "firemen," then you are favoring male firefighters and leaving out female firefighters. This section describes several ways to eliminate sexist language. When using nonsexist language, you may find personal pronouns especially troublesome. Many writers use "he/she"

TIPS FOR USING NONSEXIST LANGUAGE

- **Replace gender-specific nouns with non-gender-specific nouns when referring to job functions or occupations.** Gender-specific nouns exclude one gender. For example, instead of "chairman" or "chairwoman," use "chair" or "chairperson." Instead of "waitress," use "server."
- **Use plural nouns to eliminate gender-specific pronouns**. The following examples illustrate how using plural nouns results in nonsexist language:

Sexist	**Each employee** should maintain **his** equipment and uniforms.
Nonsexist	**Employees** should maintain **their** equipment and uniforms.
Sexist	**Each teacher** must pass the qualifying test before **she** can receive a contract.
Nonsexist	**Teachers** must pass the qualifying test before **they** receive a contract.

- **To eliminate gender-specific pronouns, use "you" and "your" or the understood "you."** Often you can avoid sexist language by using second-person pronouns ("you," "your") or the understood "you." Use second-person pronouns to address your readers directly, as in the following examples:

Sexist	The user should read the troubleshooting section of the manual before **she** calls the help line.
Nonsexist	**You** should read the troubleshooting section of the manual before **you** call the help line.
	OR
	Read the troubleshooting section of the manual before **you** call the help line. [understood "you"]

- **Use "he or she" or "she or he" when you must use gender-specific pronouns.** Whenever possible, use plural nouns or second-person pronouns to avoid gender-specific pronouns. However, when you can't avoid gender-specific pronouns, use "he or she" or "she or he." Even though these constructions are awkward, they are clear and inoffensive.

TIPS FOR FOCUSING ON THE PERSON, NOT THE DISABILITY

These tips are adapted from Kathie Snow's article, "People First Language."[1]

- **Use language that focuses on the person not on the person's disability**. For example, we should say "people with disabilities" not "the handicapped or disabled." "Handicapped" calls to mind negative images. Similarly, we should say "She has a brain injury" rather than "She is brain damaged."
- **Refer to the person first, then the disability**. For example, write "the child with autism" instead of "the autistic child."
- **Use positive, accurate language when referring to persons with disabilities.** For example, rather than writing "She is emotionally disturbed" write "She has a mental-health condition."

[1]Snow, www.disabilityisnatural.com. Used with permission.

or "s/he" to eliminate nonsexist language, but these expressions are awkward, especially when they appear several times in a paragraph. The tips on page 231 will help you to avoid he/she and s/he constructions and sexist language.

Use Inoffensive Language when Referring to Persons with Disabilities

You may need to write for persons with disabilities—persons who have a physical, sensory, emotional, or mental impairment. When referring to persons with disabilities, focus on the person, not the disability. In her article "People First Language," Kathie Snow writes that "the words used to describe a person have a powerful impact on the person's self-image."

CONSIDER YOUR READERS' CULTURE AND LANGUAGE

With computer networks and the global marketplace, many companies are doing business with people and companies abroad. When communicating with people from other countries, companies encounter two problems: cultural differences and language interference (Mirshafiei). For example, miscommunication often occurs when people in the United States communicate with people whose cultures value "detailed, subjective analyses" and "philosophical argumentation" (Mirshafiei). For example, in Middle Eastern cultures, writers often use what people in the United States regard as overstatement and exaggeration; these writers are "highly rhetorical and use a highly complex and decorative language" that people in the United States often find bewildering (Mirshafiei). Likewise, many Middle Eastern readers may not understand American writers' tendency for directness and individualism. Such misunderstanding and miscommunication can result when readers and writers don't understand the cultures that drive and dictate communication styles.

Japanese writers often use "telepathic communication," an indirect communication style that avoids direct confrontation. Telepathic communication allows writers to imply conflicting opinions and keep the communication smooth (Mirshafiei). When the Japanese communicate with people in cultures that expect and value direct, not vague, language, miscommunication may occur. For example, communication problems may occur between people in the United States and people in Japan because the telepathic communication style conflicts with the direct style expected by readers in the United States.

As you communicate with international readers, remember that their culture shapes their communication style, just as your culture shapes yours. A communication style that is unfamiliar to you is not inherently wrong or right; it is only different from the style to which you are accustomed. Your international readers may have as much difficulty with your communication style as you might have with theirs. Do your best to minimize this difference and to eliminate possible miscommunication by learning as much as you can about your readers' expectations and culture.

Consider How Your Language Differs from Your Readers' Language

Along with cultural differences, consider language interference. Your international readers may not understand the idioms and technical or workplace language that you might use with readers from your country or your workplace. *Idioms* are expressions whose meanings are different from the literal or standard meanings of the words they contain. Examples of idioms in the United States include "put up with," "turn over a new leaf," and "know the ropes." International readers can't understand idioms logically; international readers must memorize what they mean. For example, you might say to your roommate: "Let's run down to McDonald's and grab a burger." You don't literally mean that the two of you should run to the restaurant and snatch a sandwich from the server. Instead, you mean, "Let's get in the car, drive to McDonald's, and buy something to eat."

When you write for international readers, consider the words and phrases you use so that your readers will understand what you mean. Even international readers who speak English may not understand some of the

TIPS FOR WRITING FOR INTERNATIONAL READERS

- **Avoid *idioms*—expressions whose meaning is different from the standard or literal meaning of the words they contain.** Most U.S. readers, for example, would realize that "dig their heels in" means that people stubbornly refuse to change their positions. International readers, however, would be more likely to interpret the expression literally and think that people dig holes in the ground with their feet.
- **Use workplace and technical language with which international readers will be familiar or comfortable.** Avoid using any terminology that is likely to be unfamiliar to your readers. For example, when writing for readers in England and Europe, use metric measurements, such as kilometers rather than miles.
- **Avoid *localisms*—phrases familiar only to people living in a specific area.** For example, many people in the southern part of the United States use the phrase "fixin' to" as in "I am fixin' to eat lunch." This phrase sounds odd to people not from the southern U.S. and certainly would sound odd to international readers.
- **Avoid brand names.** Brand names are another type of localism. Many people in the United States mistakenly use the brand name "Kleenex" to refer to all facial tissues, and the brand name "Coke" to refer to all soft drinks. Such brand names may baffle some international readers.
- **Avoid metaphors and allusions.** Metaphors and allusions may refer to or imply a concept or information that is familiar to readers in the United States, but is probably unfamiliar to international readers. For example, you might write the following: "After the heated debate, John was smoldering for weeks." Some international readers may not understand that "smoldering" is a metaphor for angry.
- **If you must use expressions and terminology that international readers may not understand, explain what the language means to avoid misinterpretation or misunderstanding.** Try to find and use the corresponding word or phrase in the native language of your readers. You can also have the document you are writing translated into your readers' native language. If you select this option, be sure the translator knows the readers' language, country, and culture well enough to translate idioms and workplace and technical language correctly, not just literally.
- **Write simple, clear, complete sentences.** Divide long sentences into two or more shorter sentences. International readers or readers who aren't native speakers of your language will comprehend information more easily in short sentences than in long ones.
- **Use a version of Simplified Technical English.** Because nonnative speakers of English will likely read your technical documents, consider using a version of Simplified Technical English. To learn more about Simplified Technical English, see "Taking It Into the Workplace," pages 236-237.

expressions you use. For instance, in the United States, people say "line up" outside a theater box office, but people in England say "queue up." In everyday speech, you use many words and phrases that international readers may not understand. Likewise, you might use workplace language or technical terminology that is clear to others in your workplace but is baffling to international readers—especially to people who have never been to your country or to your workplace. For example, the expression "boot up" (referring to a computer turning on and starting the operating system) may be unclear and even humorous to some international readers.

When writing for international readers, try to find out as much as possible about their knowledge of your language and your country and about their language and customs. Follow the tips for writing for international readers on the previous page.

Writing Your Document for Translation and Localization

You may write a document or Web site that will be translated into another language. To make your document easy to translate, follow these techniques:

- Use relatively short sentences.
- Use the simplest verb form.
- Use consistent terminology.
- Avoid abbreviations and acronyms when possible.
- Use the active voice.
- Use simple language.
- Use words that have only one meaning.
- Do not use metaphors, localisms, or idioms (including brand names).

Even when a document's words have been translated word for word, the document still may be ineffective. The document may need to be adapted to fit the economic, technical, and/or marketing realities of a particular country or region. For example, a U.S. company that produces heart treatment equipment "introduced its products with a cartoon of happily smiling hearts" (Klein). Although these smiling hearts worked for U.S. readers, German readers were offended by the light treatment of heart disease. Readers in different locales may have different rules, data, and cultural expectations.

TIPS FOR LOCALIZING DOCUMENTS

- **Avoid country-specific information when possible.** For example, 800 numbers are country-specific to the United States.
- **Use the date and address formats appropriate for the locale.** For example, some countries use the 24-hour clock. So you might use multiple formats such as 2:00 p.m./14:00.
- **Know the format for numerical values in the locale.** For example, two million dollars is $2,000,000.00 in English, $2.000.000,00 in Spanish, and $2 000 000,00 in French.
- **Use international symbols when possible.**
- **Use graphics when possible.**

TAKING IT INTO THE WORKPLACE
Writing in a Global Workplace

Developed primarily for nonnative speakers of English, Simplified Technical English gives writers a basic set of grammar rules, style points, and vocabulary words. The objective of Simplified Technical English is "clear, unambiguous writing" (Boeing). Simplified Technical English was developed by the Aerospace and Defense Industries Association of Europe (ASD). Other versions of simplified languages include the following: Attempto Controlled English, Global English, and the U.S. government's Plain Language specification.

The ASD Simplified Technical English was first published in 1986 as *The AECMA Simplified English (E) Guide.* The guide sets up the following characteristics for Simplified Technical English:
- simplified grammar and style rules
- a limited set of approved vocabulary with restricted meanings; each word has a limited number of clearly defined meanings and a limited number of parts of speech (Boeing)
- guidelines for adding new technical words to the approved vocabulary

The ASD Simplified Technical English also requires writers to
- use the active voice
- use articles (such as "a" and "the") wherever possible
- use simple verb tenses
- use language consistently
- avoid lengthy compound words
- use relatively short sentences

Although ASD Simplified Technical English was developed for the aerospace industry, companies in other industries have modified it or produced their own version.

Boeing has adopted ASD Simplified Technical English for its technical documentation and has also modified this version for more general types of technical communication. Boeing has also developed a Simplified Language Checker that can be modified for more general types of technical communication. You can learn more about Boeing's Simplified English Checker on the Web site www.boeing.com.

Assignment

Search the Web for at least one company (other than Boeing) that uses a version of Simplified Technical English. When you find a company, provide the following information in an e-mail to your instructor:
- name of the company
- the features of that version of Simplified Technical English
- Web site address

Source: Downloaded from the World Wide Web, 2008: http://www.boeing.com/phantom/sechecker/se.html. Adapted from Boeing, *What is Simplified English?*

CASE STUDY ANALYSIS
Why Clear Language Matters

Background

In May 1996, SabreTech mechanics loaded five cardboard boxes of old oxygen generators and three tires into the forward cargo hold of ValuJet Flight 592. These oxygen generators had come to the end of their licensed lifetime; they had "expired." ValuJet had provided SabreTech with a seven-step process for removing the generators; the second step of this process instructed the workers as follows: "If generator has not been expended, install shipping cap on firing pins" (Stimpson; Langewiesche). This instruction required 72 SabreTech workers to distinguish between generators that were "expired"—meaning most of the ones they were removing—and generators that were not "expended"— meaning many of the same ones, loaded, and ready to fire.

Some of these mechanics were temporary employees. For some of these mechanics, Spanish was their native language. Working under a tight schedule, the mechanics did not clearly distinguish between "expired" and "expended" generators. In reality, many of the generators being removed and then loaded onto ValuJet 592 were expired but were not expended; that is, they could still be fired. Most of the oxygen generators should have been destroyed, or at the very least, had the shipping caps placed on the pins. With five boxes filled with generators that could explode, ValuJet 592 took off and within six minutes crashed into Florida's Everglades Holiday Park. Two pilots, three flight attendants, and 105 passengers died in the crash.

The engineers who wrote the instructions clearly understood the distinction between "expired" and "expended." This distinction, however, was not clear to all the mechanics.

Assignment

Write a memo to your instructor answering the following questions:
- How could the company have prevented the fatal misunderstanding?
- How could the company have distinguished between "expired" and "expended"?
- What is an example of other disasters or problems created because of unclear language? Document your sources.

EXERCISES

1. Rewrite these sentences, substituting specific language for vague language. Add details and examples to make the language specific.
   a. The dryers in the laundry room are acting funny.
   b. The results of the survey will be available soon.
   c. Please turn in your project ASAP.
   d. The profits of our foreign offices decreased significantly.

Download a Worksheet for Using Reader-Focused Language online at www.kendallhunt.com/technicalcommunication.

2. Rewrite these sentences to eliminate misplaced modifiers.
   a. A brochure about the scholarship program is enclosed with the application that gives complete details.
   b. The study suggests that we should continue the recycling program to the city and the council.
   c. The police shot the protesters with guns.
   d. The technician banged angrily on the flashlight in the laboratory that was dimly lighted.
   e. As a soccer mom, Mrs. Garcia's van was always full of soccer players.
   f. After returning to his office, Patrick's phone rang.
   g. Like Jessica, Sarah's automobile insurance went up after her speeding ticket.

3. Rewrite these sentences to eliminate dangling modifiers.
   a. After six months as an exchange student in Italy, the United States was a wonderful sight.
   b. Running to the meeting, the cover for her flash drive was lost.
   c. During discussions with the architect, it was determined that the building needed a new elevator shaft.
   d. After testing the new airbag, the new design was approved for delivery to the factory.
   e. At the age of 18, my parents bought me a new car.
   f. The software was upgraded in time for the December meeting by working overtime.

4. Rewrite these sentences to eliminate stacked nouns.
   a. The school district technology innovation committee meeting will begin at 6:00 p.m. in the district technology center.

b. The protocol changes list states that investigation modifications occurred after the study enrollment period.

c. The teacher recommended that the students allow time for a writing center student analysis of their research papers.

d. The human resources benefit study task force will recommend a revised employee tuition reimbursement policy.

5. Rewrite these sentences to eliminate faulty word choices.

a. After the town hall meeting, the citizens were revolting.

b. The teacher was mad.

c. The operators were held up in the briefing room.

d. The tailor pressed his suit in court.

e. At the end of the hearing, most of the opponents of the curfew were gone.

6. Revise these sentences to eliminate redundant words and phrases.

a. This important and significant network upgrade should help each and every employee to work more efficiently and effectively.

b. When you complete the programming, we will document the software fully and completely.

c. To enhance the marketing of our new line of computers, we are offering free gifts to the first 100 buyers.

d. The pipe to the main generator rarely ever leaks.

e. The designer plans to paint the auditorium walls green in color and to repeat the color again in the foyer.

f. We have a very unique opportunity to see the future plans for the training center.

g. It is my personal opinion that we should again repeat the study to determine the very latest user preferences.

7. Rewrite these sentences to eliminate unnecessary words and to condense wordy phrases. Correct any other style errors that you find in the sentences.

a. There are several scientists who are uneasy about the end results of the water quality tests.

b. Preparedness for an attack involving biological agents is complicated by the large number of potential agents (most of which are rarely encountered naturally) and the fact that these possible potential agents sometimes have long and lengthy incubation periods.

  c. It is important that we turn in our applications by the required deadline.

  d. As a matter of fact, we will be offering the vaccines again next month.

  e. You should check the roof for damage on a yearly basis.

  f. The majority of the citizens support the tax rebate, but have not taken into consideration the cost to educational programs.

  g. With reference to the revised design, it is possible that not withstanding the fact that we have a good design, we should still conduct a discussion on ways to improve it.

  h. It is interesting to note that at the conclusion of the council meeting, most of the opponents of the hands-free cell phone policy had left.

  i. Until such time as the new health plan goes into effect, all employees should continue to file claims according to the current plan without further delay.

  j. It is my understanding that the lower electricity rates will begin in August.

8. Revise these sentences, replacing fancy words with simple words. Correct any other style errors that you find in the sentences.

  a. The new tuition plan will commence for the fall semester of next year.

  b. During this time of heightened national security alerts, bomb threats are proliferating nationwide.

  c. To obtain optimum performance from your vehicle, you should endeavor to follow the maintenance program furnished in the owner's manual.

  d. We are cognizant of the fact that you are attempting to facilitate our reimbursement for the damaged equipment that we purchased.

  e. We will attempt to ascertain the opinions of the students through a survey.

9. Change the negative words to positive words in these sentences.

  a. You should not treat any customer or coworker unprofessionally or discourteously.

b. Since the construction team did not know of the crack in the wall before they began their work, they could not repair the wall according to their original estimate.

c. Not many of the racers finished the triathlon because of the extreme heat and humidity.

d. Even though the plane was delayed because of thunderstorms, we were not late to the meeting.

e. Only ten percent of the students were not absent.

10. These sentences contain offensive language. Revise the sentences, replacing the offensive language with inoffensive language. You may change singular nouns to plural nouns when appropriate.

a. Before the plane made an emergency landing, the stewardess checked the children's seatbelts.

b. Each student should discuss his degree plan with his adviser at least two years before his planned graduation.

c. The network operator should read instructions before she installs the updates.

d. Many shoes and equipment contain man-made materials.

e. The ramp will help handicapped students.

f. The school district has a program for autistic children.

11. Assume that international readers or non-native speakers of English will read these sentences. Eliminate any language that these readers may not understand.

a. Before the weekly meeting, our manager told us to stick to our guns when answering questions about what our employees need to complete the project.

b. After the midterm exam, we walked to our apartment and crashed.

c. The victims of the hurricane need medical supplies such as Band-Aids®, Kleenex®, and alcohol.

d. After the meeting, the team needed to get their bearings.

e. An iPod® was given to each member who completed the questionnaire.

EXERCISES

iStockphoto 2008.

Building Persuasive Arguments

At various points in your life, you have experienced someone trying to persuade you to do something. In other instances, you have been the one trying to persuade someone. You might have been simply persuading someone to go out to eat with you or to help you with a volunteer project. You might have prepared a resumé and cover letter to persuade an employer to interview you. We use persuasion in our everyday communication. **Persuasion** is the process of convincing others to act in a certain way or to accept a viewpoint.

In the workplace, you will use persuasion both informally and formally. You may use persuasion for a situation as informal as convincing your manager to buy additional office supplies or as formal as convincing a client to hire your organization for some kind of service. You may also use persuasion to bring a group to consensus. Whether the persuasion is formal or informal, you will use the same process and techniques for building persuasive arguments.

In this chapter, you will learn how to build a persuasive argument and to use techniques to help you deliver that argument effectively. At the end of this chapter, you will study samples of effective written persuasion. In the workplace, you will also be called on to deliver an effective verbal argument. This chapter will help you to build the argument, and Chapter 20 will help you to deliver that argument.

IDENTIFY YOUR DESIRED OUTCOME

In building any argument, you begin with identifying the desired outcome: what you want to achieve. After you have determined this outcome, you need to consider how your audience may react.

Consider How Your Audience May React

Your audience may react positively to your desired outcome. However, your audience may also resist or reject that outcome. You need to anticipate how your audience may react, so you can prepare an argument that addresses those potential reactions. Consider these guidelines:

- **Expect some members of the audience to resist or to disagree.** Regardless of the logic or the worth of your desired outcome, some audience members may not agree with it. For example, you may try to persuade your manager that your division needs an additional employee to share the workload because you and your coworkers have been working 50 hours a week for the past six months. You know that your coworkers agree and that you can document the overtime. However, your manager may disagree because the company does not have the money for a new hire or your manager may not see a problem with working overtime. You should be prepared for your audience to see your desired outcome from a viewpoint different than yours.

- **Be prepared for resistance and for other ideas.** Prepare as though your audience will resist your desired outcome or may offer another idea. Be prepared to listen, to consider the idea, and to be flexible. You don't have to accept a new idea. However, if the idea is reasonable and you can accept it, be flexible. Don't be defensive if some or all of your audience disagrees. Instead, listen sincerely.

- **Tailor your arguments to the audience.** Before you prepare your argument, take time to research your audience and to put yourself in your audience members' "shoes": if you were in their place, how would you see the desired outcome? Let's consider your request for another employee in your division. From your vantage point, your desired outcome is ideal. Now, look at the request from your manager's vantage point: How much will it cost? Where will the funds come from? Can you propose a more cost-effective way to lessen the workload on employees of the division? For you to persuade your manager, you need to consider these questions and then tailor your argument to answer these questions.

- **Make sure that your desired outcome is reasonable and appropriate**. You have a responsibility to offer reasonable arguments. For example, let's again consider your request for a new employee in your division. You know that a more-experienced employee will be a good fit, but experienced employees come at a higher salary than less-experienced ones. You also know that the company is experiencing some cutbacks in other divisions. Should you ask for the more-experienced employee and a higher salary knowing that an employee with less experience could do the job? You will need to weigh each of these options in light of the company's situation and your manager's viewpoint. When you develop arguments, make sure that they are reasonable, and then be willing to adjust your desired outcome when appropriate.

When you know the desired outcome and have considered how your audience may react, you can begin the process of building an effective argument. Figure 9.1, page 246, illustrates the process of building a persuasive argument.

> *Anticipate how your audience may react, so you can prepare an effective argument that addresses those potential reactions.*

CONSIDER THE CONSTRAINTS THAT MAY IMPACT YOUR ARGUMENT

Once you have determined your desired outcome and anticipated how your audience may react, you are ready to consider the constraints that may impact your argument. *Constraints* are anything that may restrict or limit the actions of others. For example, if you plan to propose hiring an employee for your division and your company has a hiring freeze, the hiring freeze is a constraint that impacts your desired outcome. As you prepare your arguments, consider these constraints:

- the audience's situation
- your workplace

Audience Constraints

Your audience members will view your argument from their personal situation. Your audience members may react positively or negatively to your argu-

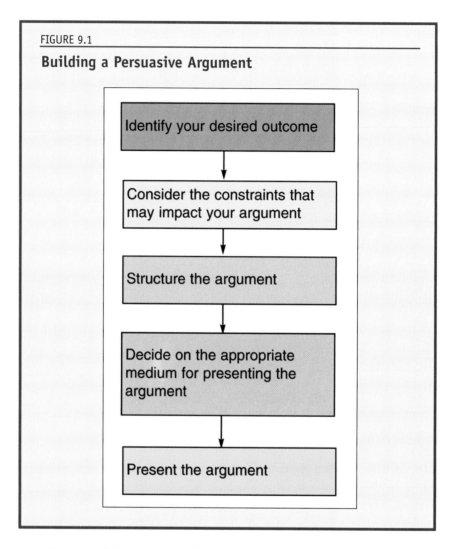

FIGURE 9.1

Building a Persuasive Argument

Identify your desired outcome

Consider the constraints that may impact your argument

Structure the argument

Decide on the appropriate medium for presenting the argument

Present the argument

ment because of their situation. As you prepare your argument, consider how the following constraints may influence your audience members' reactions.

Their Place in the Organizational Hierarchy

Whether your audience members are internal or external to your organization, their place in the organizational hierarchy may influence how they react to your organization. If you are writing for *internal audience members* who are your subordinates, they may feel pressured to agree with your argument simply because of their place in the hierarchy. If you are trying to persuade in this context, you want to assure your audience members that you want their honest reactions. Make your audience feel comfortable disagreeing. If you cannot change your argument, consider sending a directive (see Chapter 17 for information on directives). Include persuasive language to help readers understand why the directive is important to them. If the audience members are above you in the hierarchy, determine the level of formality and the amount of detail expected. Your argument will be more persuasive if you follow the expected conventions. If your internal audience members are your coworkers, your job will be easier in some ways because you will understand their vantage point.

If you are writing for ***external audience members***, consider how their place in the organizational hierarchy may influence how they react to your argument. For example, these audience members may not have the authority to accept your argument or they may not feel comfortable accepting your argument without seeking the counsel of others in their organization.

Your Relationship with the Audience

As you consider the constraints of your audience's relationship with you, begin by asking yourself: Do I have a relationship with the audience? If not, consider how that lack of relationship may impact how the audience members will react to your argument. If they do have a relationship with you or your organization, is it a positive or a negative one?

When audience members don't know you or your organization, they may resist your argument simply because they don't know if they can trust you. You will need to persuade them that you are a qualified professional and that your organization is of high quality. If the audience members have a negative impression of you or your organization, they may not have confidence that your argument is valid or they may believe that they cannot trust you or your organization. To persuade these audience members, spend some time rebuilding that relationship and assuring them that they can trust you.

The Audience's Familiarity with the Topic

Your audience members may resist or reject your argument because they are not familiar with the topic. Often when we encounter a new topic, we initially resist or reject it simply because we don't understand it. Members of your audience may react similarly if they aren't familiar with the topic. Find out what your audience knows about the topic. If they are unfamiliar with it, consider helping the audience understand the topic as part of your persuasive appeal.

Workplace Constraints

The following workplace constraints may impact both you and your audience:
- time and budget
- legal constraints
- ethical constraints
- political constraints

Time and Budget

Regardless of the merit of your desired outcome or how persuasively you argue for that outcome, time and budget may limit or prevent it. Either you will not have the time and money for the outcome, or your audience members' time and budget will not allow it. Time and especially budget are powerful constraints in the workplace. In some cases, you may be able to adjust your desired outcome to fit your budget and time or that of your audience. In other cases, you may have an excellent outcome, but you may not have the time to prepare an effective argument—especially if you will present the argument in writing. Try preparing a schedule working backwards from the deadline. With this schedule, you can determine if you have the time and resources to present an effective argument; you can also determine if your audience has the time and/or resources to implement your proposed outcome.

Legal Constraints

Before you present your argument, make sure that you are abiding by all laws. If you are unsure if the desired outcome is legal, contact your organization's legal counsel. If your organization doesn't have such counsel, contact an attorney outside the organization. *When in doubt about the legality of the desired outcome, don't present it.*

Ethical Constraints

As with any decision in the workplace, you should consider whether your desired outcome and your means of achieving that outcome are ethical.

Follow the code of conduct for your profession and the ethics guidelines of your organization. If your profession or organization does not have established guidelines, follow those in Chapter 2. You have a responsibility to all involved to do the right thing.

Political Constraints

Consider the political climate of your workplace. For example, if you know that your coworkers and your manager disagree with your idea or that others have proposed the same idea and it was rejected, consider revising or simply not proposing that idea. You will gain credibility with your managers and your coworkers when you do your homework before presenting an argument.

Determine if your desired outcome will in all likelihood be rejected. You don't want to lose credibility or to waste your time (or that of your coworkers and manager) by proposing something that you know will be rejected. However, if by not presenting or by changing an argument you violate legal or ethical guidelines, you must present your argument as you intended (see Chapter 2 for information on legal and ethical guidelines).

STRUCTURE A PERSUASIVE ARGUMENT

A persuasive argument has three elements:
- the claim: your desired outcome
- the evidence: what supports your claim
- the appeal: why the audience should accept your claim

The *claim* is the outcome that you want your audience to accept. For example, your claim

TIPS FOR COMMUNICATING PERSUASIVELY TO INTERCULTURAL AUDIENCES

- **Understand how an audience's culture influences what makes an argument persuasive.** For example, in Western cultures, we place the claim at the beginning of an argument. However, in Eastern cultures, the claim usually appears later in the document and may be implied rather than stated (Tebeaux and Driskill).
- **Spend time learning about the audience's culture and business customs.** When you know about the audience's culture and business customs, you can better structure a persuasive argument.
- **Show respect for the audience's culture and customs.** You show respect when you realize that your customs are not the only way to do business.
- **Construct your argument so that it translates into the audience's cultural frame of reference.** If you violate that frame of reference, you may offend the audience. An offended audience is less likely to accept your argument.
- **Review the guidelines for writing to intercultural readers in Chapter 4.**

may be that your company should move to a 4-day work week. You want your audience (company executives) to implement this idea.

The *evidence* is the information used to support your claim. This evidence might be facts, statistics, examples, or testimonials. Let's consider your proposal for a 4-day work week. Your evidence might include the following:

- The company would save on janitorial services and energy costs.
- Company X and Company Y have moved to a 4-day work week and have seen less absenteeism because employees schedule medical and personal appointments on their extra day off.
- Employees would spend less on gas and would have less wear-and-tear on their vehicles.
- Employees would drive to work one fewer day a week, reducing harmful emissions that cause ozone pollution.
- In a survey of employees last month, 83 percent of employees said they would prefer working a 4-day schedule. In that same survey, employees were asked if they would prefer working a 10-hour day if they only worked 4 days per week; 87 percent said yes.

The *appeal* explains why the audience should accept the claim. Evidence alone may not be enough to persuade your audience members to accept your desired outcome or to change their minds. You must also appeal to the audience members' needs and values (Rottenberg). Let's consider some possible appeals for our proposal for a 4-day work week:

- The company will save money (reduced janitorial services and energy costs).
- The company may be more productive (less absenteeism).
- The company would be more environmentally responsible (reduced harmful emissions that cause ozone pollution).
- A majority of the employees will be happier (survey results).
- Other companies have found this work week successful (Company X and Company Y).

These appeals link directly to the evidence, as seen in parentheses next to each appeal.

Types of Evidence

Evidence is a powerful piece of your argument. Without evidence, you will rarely persuade your audience to agree with your claim. Most audiences will also expect your evidence to be credible. The tips on page 252 will help you to present persuasive evidence to support your claims.

Your arguments will be most effective when you use the following types of persuasive evidence:
- facts
- statistics
- examples
- expert testimony

Facts

Facts are evidence that you or your audience can verify. A fact can be observed, demonstrated, or measured. For example, the following statement demonstrates a factual statement. An audience can verify both of these statements.

Factual evidence	Hybrid vehicles use less gasoline.
Factual evidence	With a 4-day work week, employees would spend less on gas and would have less wear-and-tear on their vehicles.

Statistics

Statistics, or numerical data, make your arguments highly persuasive. For example, look at the following statements.

Statistical evidence	In a survey of employees last month, 83 percent said they would prefer working a 4-day week. In that same survey, 87 percent said they would prefer working a 10-hour day if they only worked 4 days per week.

In these statements, the numerical information (83 percent and 87 percent) is powerful evidence to support the claim to move to a 4-day work week. When possible, use numerical data to support your argument; however, when you use such data, make sure that readers can understand it and that it is accurate and ethical.

TIPS FOR PRESENTING PERSUASIVE EVIDENCE

- **Present credible, verifiable evidence**. Audiences expect credible evidence to support claims. The evidence should directly support your claim, and the audience should be able to verify that evidence. If you present evidence that can't be verified, you will quickly lose credibility with your audience.
- **Focus on the best evidence**. You may be tempted to present every possible piece of evidence, thinking that you will persuade your audience with the quantity of evidence. However, your argument will be much more convincing if you select the strongest evidence and use it to support your claim. As one professional put it, "When presenting your argument, less is more" (Sims, W.).
- **Cite your sources**. Tell your audience where you got your evidence. If you present your argument in writing, include a list of references and/or a footnote. If you present your argument orally, tell your audience where you got your evidence and/or prepare a handout listing the sources.

Examples

Examples help audiences to visualize abstract information, to remember your arguments, and to identify with your argument. The following example helps the audience to visualize what a 4-day work week could do for an employee and for the company.

Example   Rachel Olsin takes off an average of four hours a week to take her 92-year-old parents to the doctor. As the only living child, Rachel is the sole caretaker and the only person who can take them to the doctor. If she was able to work 4 days a week, she could schedule the appointments on her extra day off and lessen her absenteeism.

Examples, especially when coupled with statistics, make powerful arguments.

Expert Testimony

You are probably familiar with *expert testimony* from television shows and movies where an expert gives testimony at trial. If unbiased, the testimony from an expert lends credibility to your argument. However, if the expert is not unbiased, the testimony can damage your argument and your credibility. If you decide to use expert testimony, make sure that you select experts with credentials that the audience will recognize as valid. Let's look at an example:

Expert Testimony   Vicki Peake, CEO of DesignAds, reports that employees have reduced their absenteeism by 25 percent since her company moved to a 4-day work week in 2007.

This expert testimony is credible because the writer includes the credentials of the expert, and the expert has credentials that most audiences would recognize as valid.

Types of Appeals

You can use the following appeals to convince your audience to accept your claim:
- appeal to shared goals and values
- appeal to common sense
- appeal by recognizing the opposing viewpoint/evidence
- appeal to the audiences' emotion

Appeal to Shared Goals and Values

Our goals are shaped by our values. *Values* are characteristics that we live by (integrity, honesty, loyalty, friendship, fairness, etc.). Even though these values and goals do not constitute evidence, they do affect how audiences react to arguments. If, for example, an audience values fairness and integrity, you can use that value when you present your evidence. By identifying these shared goals and values, you can structure a more effective argument. These shared goals and values are especially effective when you are trying to build consensus.

For example, if you are trying to convince employees that a 4-day work week is a good idea, you could appeal to the common goal of spending less money on gas and increasing time for personal activities. Most employees, will share these goals: saving money on gas and increasing the time available for personal activities. Now, consider that same argument to company executives. You might appeal to the common goal of cost savings: executives will share your goal of saving the company money, so any successful argument for them must appeal to this shared goal.

Appeal to Common Sense

For some arguments, you may not have evidence to support your claim, but you know the claim is sound. For these arguments, you can appeal to your audience's common sense. **Common sense** is a judgment based on something that you or others have perceived. For example, common sense dictates that you register for classes early because you will have a better chance of getting the classes you want or that you arrive at the theatre early to get a better seat. In our argument for a 4-day work week, we might use the following appeal to common sense:

> A 4-day work week makes sense because gasoline prices have increased by almost 25 percent.

If your audience members share your common sense argument, you will communicate to them more persuasively. However, you cannot rely solely on common sense appeals. You also need appropriate evidence.

Appeal by Recognizing the Opposing Viewpoint or Evidence

When you want to gain credibility with your audience, recognize the opposing viewpoint or evidence. Many arguments fail because the person presenting the argument does not recognize and appropriately respond to opposing viewpoints and evidence. For example, let's again consider the argument for a 4-day work week. An opposing viewpoint might be that employees will be less productive. You might recognize this viewpoint as follows:

> You may be concerned that with fewer workdays, employees will be less productive. I, too, was concerned with this potential problem. The CEO of Design Ads and of Pinnacle report that when their company moved to this shorter work week, employees were as productive as they were in the traditional 5-day work week. These executives report that in some departments, employees were more productive.

When faced with these opposing viewpoints and evidence, you can address them by following these guidelines.

- **Respectfully recognize the opposing viewpoint and evidence**. Be careful to treat opposing viewpoints in a professional, objective manner.

Avoid subjective statements such as "This argument has no merit" or "I am not sure how anyone could believe this evidence."

- **Focus on the viewpoint, not on the person holding that viewpoint**. You want to keep the goodwill not only of the person(s) with the opposing viewpoint, but also with the others in the audience. If you appear to attack that person(s) or to respond disrespectfully, you will lose much of your power to communicate persuasively.
- **Address the merits of the opposing viewpoint or evidence**. You want to keep the goodwill of your audience. When you recognize the merits of opposing viewpoints, you gain the trust and goodwill of your audience members—especially if they hold those viewpoints.
- **Explain the merits of your argument**. Focus on the merits of your argument, not on subjective statements such as, "This method is so much better." Instead, demonstrate why the method is better.

Appeal to Emotion

Emotional appeals can be effective when used responsibly. You use emotional appeals responsibly when you combine them with appeals to shared goals and values and with sound evidence. An example of an emotional appeal appears in Figure 9.2, page 257. In this request for donations to sponsor a child, the writers appeal to audience members' emotions with the photo, specific information about the child, and the second-person pronouns "you" and "your." The writers combine the emotional appeal with an appeal to common goals and values ("so he can grow up to be a healthy and productive adult").

DECIDE ON THE APPROPRIATE MEDIUM FOR PRESENTING YOUR ARGUMENT

Once you have structured your argument, you are ready to present it. You want to begin by deciding the best medium for that presentation. Will you present it orally? Will you present it in writing? If so, what format will you use? Once you have determined the medium, follow the tips on page 256 to present an effective argument.

SAMPLE DOCUMENTS

The following sample documents illustrate how writers have crafted persuasive arguments appropriate for the situation and the audience.

Figure 9.3, page 258, presents a page from a Web site for a community-minded construction company. Figure 9.4, page 260, presents pages from a brochure for consumers.

TIPS FOR PRESENTING YOUR ARGUMENT PERSUASIVELY

- **Present yourself as a professional, reasonable person.** Many arguments are won or lost based on the persona of the presenter. If you appear to be a professional, reasonable person, your audience will be more likely to listen to your arguments. In most cases, you will not win an argument solely on your persona; however, if you present yourself as someone unreasonable and/or unprofessional, you will rarely win your argument.
- **Acknowledge the contributions, knowledge, and ideas of others.** If others have helped you to develop your argument or idea, acknowledge their contributions. When appropriate, acknowledge the contributions, knowledge, and ideas of your audience. When you make these acknowledgements, you add power to your persuasion.
- **Don't oversell.** You can oversell in these ways: by using unsupported subjective language, by including too much evidence, or by continuing to push your idea when your audience members are not interested or will not change their minds. If you find that you can't get your audience to agree, try your argument at a later time or, if appropriate, drop the argument.
- **Don't be defensive.** When someone in your audience disagrees with you, respectfully consider his or her arguments. Have an open mind. If you get defensive, your audience may assume that you don't have the evidence to support your argument or that your argument is not sound.
- **Present effectively designed documents and graphics.** When you present documents and graphics that follow the guidelines in Chapters 10 and 11, you add to your professional persona. Effectively designed documents and graphics add credibility to your argument. Likewise, if the documents appear sloppy and poorly designed, you appear to lack credibility.

FIGURE 9.2

Example of an Emotional Appeal

Aid for Children International

Name: Brian

Birthday: August 2, 2003

Age: 7

Gender: Boy

Country: Uganda

Favorite Activity: Soccer

Favorite Subject: Reading

Brian lives with his parents. His father sometimes works as a laborer. His mother makes clothes when she can buy fabric. Brian helps at home by carrying water and running errands. He has two brothers and three sisters. Brian enjoys playing soccer and reading. He attends primary school when he doesn't have to work at home. His performance is satisfactory.

Your financial support will help Brian to receive the clothing, food, and education he needs to develop to his full potential.

Sponsor now

FIGURE 9.3

Persuading Customers and Potential Employees about a Company's Commitment to Supporting the Community

The company states its claim: "At the heart of everything we do, we are people serving people."

The company includes evidence through examples.

The photographs give the company a persona of "real" people like those in their communities.

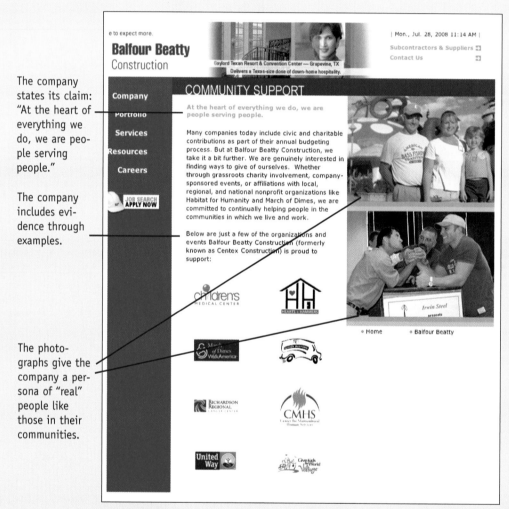

FIGURE 9.4

Persuading a Consumer

The claim is implied in the title.

> # An oven for *your* time.
>
> A meal can say many things. It says welcome home, thank you, or relax. It says let's celebrate. And it always says you care.
>
> Meals can make friends, nurture relationships, and create memories that last forever. But at today's busy pace – with our lives overflowing with daily tasks – finding time to serve great meals is a real challenge.
>
> To help you meet this challenge, TurboChef introduces the oven that *gives you the freedom to be the cook you want to be more often.*

The claim is stated in the last line of the first page: "gives you the freedom to be the cook you want to be more often."

The writers use design and graphics to make the persuasion more effective.

turbochef.com 866.54ENJOY

Source: *TurboChef 30 Double Wall Oven Guide*: www.turbochef.com/residential/shared/pdf/ TurboChef-30DoubleWall-Oven.pdf, page 1. Used with permission of TurboChef.

FIGURE 9.4 CONTINUED

Our oven is designed around *you*.

TurboChef is dedicated to providing a real solution for your lifestyle. From the fashionable retro-modern styling to the easy-to-use controls to the amazing cooking performance, our oven gives you more freedom.

Introducing the TurboChef 30″ Speedcook Ovens – Single and Double Wall.

Imagine cooking with a touch of magic. Steamed asparagus – perfectly al dente – in 45 seconds. A rack of lamb in 4 minutes. A 12 pound Thanksgiving turkey in 42 minutes. From family favorites to a gourmet dinner, our oven enables you to cook fresh, delicious food up to 15 times faster than conventional methods with the flavor and quality endorsed by four-star chefs. Meats are caramelized, moist and tender. Baked goods are golden and flaky. And, vegetables come out crisp, succulent and nutritious.

Whether it's dinner on a busy Wednesday night, a dinner party for friends on Saturday night, or a once-a-year special occasion, *the TurboChef oven gives you the freedom to be the cook you want to be more often.*

turbochef.com   866.54EN701

The writers use statistics and examples as evidence.

The writers use the appeal to shared goals and values: the audience is busy but wants to serve great meals.

Source: *TurboChef 30 Double Wall Oven Guide*: www.turbochef.com/residential/shared/pdf/TurboChef-30DoubleWall-Oven.pdf, page 3. Used with permission of TurboChef.

TAKING IT INTO THE WORKPLACE
Persuasion and Leadership

Persuasion is an important tool in every professional's toolbox. How you use that tool is important for your success in leading others in the workplace. As Lois Zachary, author of *Rekindling the Art of Persuasion*, explains,

> We live in a world where the command and control model of persuasion just doesn't work anymore. Achievement, money, and status no longer persuade but actually raise the level of skepticism. Positional power is slowly being replaced by personal power and new strategies are being used to connect and reconnect with the people with whom we do business. Getting someone to 'buy in' to something and commit to it is a more effective approach than authority-based management.

If you want to persuade your coworkers, your managers, or your clients to "buy in" to your products and ideas, you have to convince them to take part in your vision (Zachery). You want to effectively lead them to not just go along with your ideas or product, but to want to be a part.

In the workplace, you may lead a team of coworkers or you may manage a large team. As part of your leadership, you will need to persuade others. Use the techniques presented in this chapter for developing effective arguments and for presenting those arguments.

Assignment

Interview a professional in your field (see Chapter 5 for information on interviews). You may conduct your interview in person, by phone, or by e-mail. At the interview, ask the professional about the following:
- how he or she uses persuasion in informal meetings to convince others to accept his or her ideas
- in these settings, what techniques has he or she found successful and unsuccessful

E-mail a summary of your interview to your instructor.

CASE STUDY ANALYSIS
Persuading a City Council

Background

In 1992, the New York City Council voted to approve a controversial garbage disposal plan by a solid majority—with votes to spare. Interestingly, only one month prior to the vote, the measure was considered dead. What persuaded the council was "education," according to Bill Lynch, Jr., the political lobbyist and mastermind of then-Mayor David N. Dinkins.

City council members are reluctant to approve unpopular or unpleasant measures. In this case, garbage incineration was not popular. To make matters more complicated, this garbage plan would have ramifications far into the future and left the door open for additional incinerators to be built. No council members wanted to be responsible for putting an incinerator in their constituents' "backyard."

The mayor and his associates began a system of educating the council. They used a variety of methods to persuade council members to vote in favor of the plan: phone calls, discussions, action on local problems, and rewrites of the plan. Ultimately, the mayor dropped two incinerators and an ash land fill from the plan and agreed to include more future recycling.

Mayor Dinkins and Mr. Lynch were pleased by the vote. Their success resulted from persuasion rather than strong-arm tactics.

Assignment

1. Pretend you are Bill Lynch, Jr. Write a memo to the city council members, persuading them to support the mayor's garbage plan. Remember to consider your readers' resistance to the plan, the constraints they face, and the relationship the mayor's office has with the council members. Turn your memo in to your instructor. For evidence to support your arguments, research the topic using techniques that you learned in Chapter 5.

2. The mayor's garbage plan included building a new trash incinerator in the Brooklyn Navy Yard and adding an ash fill in Staten Island. The plan also seemed to side-step recycling efforts and left the door open to building additional incinerators, which environmentalists, city council members, and their constituents found unappealing. Pretend you are Bill Lynch, Jr. Write a memo to the city council members, persuading them to support the mayor's garbage plan. Remember to consider your reader's resistance to the plan, the constraints they face, and the relationship the mayor's office has with the council members. Turn your memo in to your instructor.

Download a Worksheet for Communicating Persuasively online at www.kendallhunt.com/technicalcommunication.

EXERCISES

1. **Web exercise**: Visit the Web site of a major company such as Dow Chemicals (www.dow.com), Exxon (www.exxonmobil.com), or GE (www.GE.com). Look at how the company presents information about its commitment to its communities and the environment.

   a. Identify the stated or implied claim about the company's commitment to the environment and the community. If the claim is implied, what words and graphics does the site use to make the claim?

   b. Identify the types of evidence that the company uses. Consider both the words and the graphics.

   c. Identify the appeals. Consider both the words and the graphics.

   d. Write a memo to your instructor. In your memo, use the information in Exercises 1a-1c to identify whether the company presented an effective argument. (For information on memos, see Chapter 12.)

2. **Web exercise**: Visit the Web sites or offices of two local or regional non-profit organizations such as the Society for the Prevention of Cruelty to Animals (SPCA) or Habitat for Humanity. Look at how the organizations persuade people to donate and to volunteer.

   a. Identify the appeals. Consider both the words and the graphics.

   b. Identify the evidence. Consider both the words and the graphics.

   c. Write a memo to your instructor summarizing the information from 2a-2b and stating whether or not you believe the organizations' persuasion was effective. Explain your answer.

3. Find two brochures for the same type of product. The purpose of these brochures should be to persuade a consumer to purchase the product.

   a. Write a memo to your instructor comparing and contrasting the persuasive evidence and appeals used in the two brochures. (For information on comparison and contrast, see Chapter 6.)

   b. Attach a copy of the two brochures to your memo.

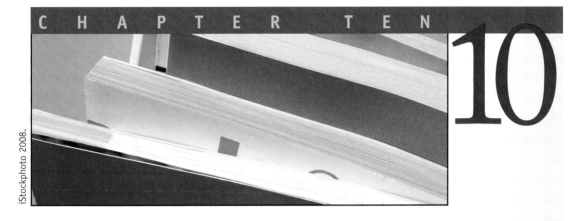

CHAPTER TEN

10

iStockphoto 2008.

Designing Reader-Focused Documents

Effective design is more than simply making a document look "pretty." An effectively designed document follows principles to help you reach the following goals:

- to make a good impression on your readers and to leave the readers with a good impression of you and your organization
- to help readers locate the information they need
- to show readers how the document is organized so they can better use and understand the information
- to help readers remember the information

In this chapter, you will learn the principles of effective design. You will also learn to apply those principles when designing the whole document and the individual pages.

PRINCIPLES OF EFFECTIVE DESIGN

In *The Non-Designer's Design Book*, Robin Williams suggests the following four basic principles for designing effective documents:

- contrast
- repetition
- alignment
- proximity

Williams adds that good design is easy if you
- learn these principles
- "recognize when you're not using them"
- apply these principles

Contrast

You create **contrast** when two elements on a page are different. For example, when you put white text on a black background, you create contrast through color. To use contrast effectively, make sure that the elements are clearly different—not just sort of different (Robin Williams). You cannot create contrast when the items are only slightly different. For example, you cannot create contrast by making one element dark blue and one element black; the colors are too similar. You can create contrast with elements such as type size, white (blank) space, alignment, and color.

Contrast has two purposes:
- **It creates interest on the page**. When a page has visual interest, readers are more likely to pay attention to the page and to read it.
- **It shows how you are organizing the information**. Contrast can direct readers through the organization of a document.

Figure 10.1 demonstrates contrast in three ways:
- The white text contrasts against the green background.
- The blue of the girl's hat and shirt stands out against the background and the green space below.
- The dark green "Texas Parks and Wildlife" logo contrasts with the lighter green background.

Repetition

You use the principle of **repetition** when you repeat some aspect of the design throughout the document (Robin Williams). You might repeat a color, a graphic element, a logo, or an icon. You can also create repetition by using the same layout and location for similar items. You might consider repetition as being consistent. For example, the first-level headings in this book repeat the same format; the repeated format distinguishes the first-level headings from the second-level headings or from the paragraphs. The purpose of repetition is to
- unify a document
- add visual interest

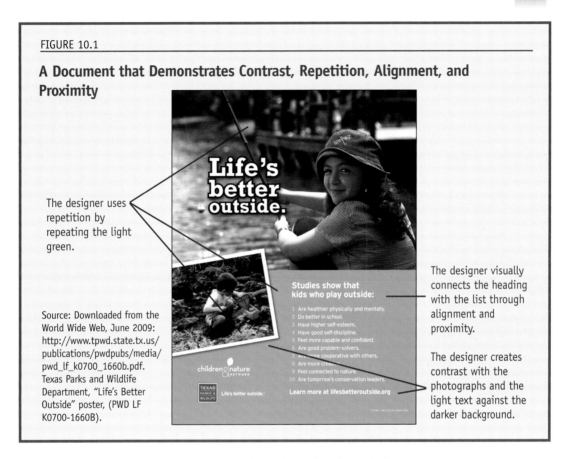

FIGURE 10.1

A Document that Demonstrates Contrast, Repetition, Alignment, and Proximity

The designer uses repetition by repeating the light green.

Source: Downloaded from the World Wide Web, June 2009: http://www.tpwd.state.tx.us/publications/pwdpubs/media/pwd_lf_k0700_1660b.pdf. Texas Parks and Wildlife Department, "Life's Better Outside" poster, (PWD LF K0700-1660B).

The designer visually connects the heading with the list through alignment and proximity.

The designer creates contrast with the photographs and the light text against the darker background.

Figure 10.1 demonstrates repetition through color. The light green appears three times: the background for the list, the foliage in the smaller photograph, and the grass in the upper left-hand corner of the larger photograph (look at the callouts on the figure).

Alignment

The principle of ***alignment*** helps unify the elements on a page. As Robin Williams explains, every element on a page should visually connect with something else on the page. When you follow the principle of alignment, you don't simply place items wherever there happens to be room. Instead, you align every item with the edge of another item on the page. For example, in Figure 10.1 above, the heading, "Studies show that kids who play outside," aligns with the list and the statement that begins "Learn more." This alignment visually connects the text and allows the readers to scan the list. If the features were placed wherever the designer had space, the readers would not know if the elements were related.

Proximity

You use the principle of ***proximity*** when you group related items together. When related, or like items, are closely grouped together, these items appear as a cohesive group instead of as a bunch of unrelated items (Williams, Robin). If items are not related, they should not be in proximity. The purpose of proximity is to visually organize information or graphic elements on a page. For example, in Figure 10.1, page 267, the list and its related heading are close together. The proximity implies that the heading and list are related.

PLAN YOUR DOCUMENT DESIGN

Before you begin designing the document, consider your readers' needs and expectations and your resources.

Consider Your Readers' Needs and Expectations

Think about your readers and how they will use your document:

- **What design elements will help readers fulfill their purpose for reading your document?** Readers have a goal when reading your document. They may be reading to answer a question, to gather information, to complete a task or procedure, or to learn how to do something. For example, if your reader is in a hotel room and the fire alarm goes off, what type of format will help the reader to quickly and efficiently know what to do?
- **Where will readers use the document?** Think about where your readers will use or read the document. For example, if you are writing a manual for an automobile mechanic, you would want the pages to be made of durable material and to be large enough to see while working on the automobile. You would not want the type of manual that a car owner would get—a manual that easily fits into the glove compartment.
- **Do readers have expectations about quality?** Some readers expect certain material to have high-resolution color graphics, as in a proposal for a multi-million dollar project. However, if you were proposing to buy a new copier for your office, your readers probably would not expect high-resolution graphics.

- **Do readers have expectations for the presentation of the information?** Your readers may expect information to be presented in a particular format. For example, if you are preparing an owner's manual for a car, owners would expect that manual to fit easily into the glove compartment or console of the car.

Consider Your Resources

If you wait until you have written one or more drafts, you may not have the resources to incorporate the design elements you want, or you may have to unnecessarily reformat the text to fit your design. Consider these resources before you begin designing your document:

- **Budget:** How much money do you have for the document? Do you have the budget to have the document professionally printed or professionally designed? For example, many companies budget thousands of dollars for a manual that accompanies a product, but not for a manual that is used in-house.
- **Time:** How much time do you have to create the document? Some designs can take hours to create. If you plan to use a professional print shop or designer, you need to consider how much time will be required for production.
- **Equipment:** Some designs require graphics or desktop-publishing software. Do you have that software and do you know how to use it? Some designs also require specific printers. For example, a basic laser printer can produce a fine black-and-white document, but a high-resolution color document requires a more expensive printer.

DESIGN THE WHOLE DOCUMENT

Before you begin writing and designing individual pages, think about how you will design the whole document. Decide on the

- page size
- paper
- binding
- tools to help readers locate information

For example, if you know that your readers will use the document in a confined space where they will use liquids, consider laminating the pages and using a smaller page size (such as 6 x 9 inches instead of the standard 8½ x 11 inches). If, for instance, you are designing a document that UPS employees will carry in their delivery vans, the standard $8\frac{1}{2}$ x 11 inch page would be awkward whereas a $4\frac{1}{2}$ x 5 inch page would better meet employees' needs. The following sections provide guidelines for designing elements that will motivate readers to use your documents.

Page Size

Think about what page size will best meet the needs of your readers and their environment. You might be tempted to use $8\frac{1}{2}$ x 11 inch pages because you are accustomed to seeing and using that size for documents you write in school. Your term papers and reports are mostly $8\frac{1}{2}$ x 11 inches; this page size is also the default size for word-processing software and basic printers. Because we are so accustomed to seeing and using this page size, we often forget to consider what page size will best meet the needs of our readers and help us achieve our purpose. However, if you are designing a poster that shows servers in a restaurant the procedure to follow when a customer is choking, an $8\frac{1}{2}$ x 11 inch page would be too small; instead, you might use a 22 x 28 inch poster. Conversely, if you are preparing a proposal for a client, the client would probably expect you to use a standard $8\frac{1}{2}$ x 11 inch page.

Paper

You will want to consider the best type of paper for your readers' purpose and environment. Paper comes in different weights, brightness, and coatings. You will want to select the appropriate weight, brightness, and coating for your budget and your readers: the higher the weight and the brightness, the higher the quality (and the more expensive). Follow the tips on page 272 for selecting paper.

Binding

For long documents, several types of binding are available (see Figure 10.2).
- **Loose-leaf binders**: In these binders, users can open the rings and remove or insert pages. This type of binding is especially effective if you or your readers will be updating a document. This binding is excellent when the document needs to lie flat or stay open to a specific page.
- **Wire or plastic spiral binding**: With this type of binding, users cannot remove or insert pages. However, this binding is less expensive than

---

FIGURE 10.2

Types of Binding

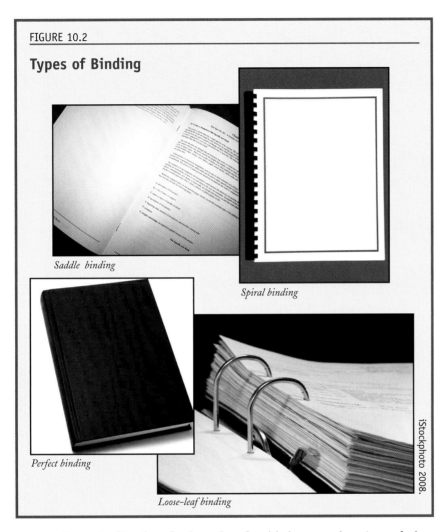

Saddle binding

Spiral binding

Perfect binding

Loose-leaf binding

iStockphoto 2008.

---

most loose-leaf binders. It also is less durable because the wire and plastic coils can be easily crushed and broken. This binding is excellent when the document needs to lie flat or stay open to a specific page.

- **Saddle binding**: Saddle binding uses large staples to bind the pages. This binding is not practical for long documents, but works well with small booklets and brochures.
- **Perfect binding**: Perfect binding is how most books are bound: the pages are glued together along the spine. This type of binding produces a formal appearance to your documents, but pages can easily fall out and the document does not lie flat.

TIPS FOR SELECTING PAPER

- **Select a lower weight and lower brightness for most internal, everyday documents.** If you are printing a memo to your coworkers, you can use a lower-weight paper. This paper generally has a brightness in the 80s. This paper is the type available in most photocopy machines.
- **Select a 20- or 30-pound bond paper for external and formal internal documents.** This weight paper gives your documents a more professional feel. On a more practical note, it doesn't tear as easily. If you select a lighter-weight paper, the document may look flimsy and could tear. If you print the document on both sides and use a lower-weight paper, the print will bleed through the paper.
- **Use paper with a high brightness number for external and formal internal documents, especially documents with color graphics.** The brightness of paper is expressed on a number scale from 1 to 100. You can find the brightness number on the paper packaging. High-quality paper has a brightness number in the 90s. Some paper manufacturers simply use terms such as "bright white" or "ultra bright" instead of a number. When you use paper with a lower brightness, colors are darker and less vibrant. When you use paper with a higher brightness, colored graphics are more vibrant.
- **If you will be printing on both sides of the paper, make sure that the printing will not bleed through from one side to the other.** Selecting at least a 20- or 30-pound paper will ensure that the printing will not bleed through.
- **Use coated paper to increase the durability of the paper and the resolution of the print.** Coated paper can be more expensive, but print resolution is much higher on the paper. Coated paper also creates a sharper, more professional look. If you use coated paper, use an off-white or ivory paper to decrease glare.
- **For formal documents, select a white, ivory, or off-white paper.** Select colored paper (other than the colors above) only when appropriate for the tone, formality, subject, and readers. For example, if you are writing a proposal to a client, select a white, ivory, or off-white paper. However, if you are inviting that same client to a Mardi Gras celebration of your winning proposal, you might use light-purple paper with green accents. Regardless of the color, make sure that your readers can easily read the print.

Locating Tools

To help readers locate information in your documents, use the following tools:

- tabs
- divider pages
- headers and footers
- page numbers
- headings
- color
- icons

Examine the design of a Reader-Focused Document in the Interactive Student Analysis online at www.kendallhunt.com/technicalcommunication.

Tabs

Frequently used in manuals, procedures, and proposals, tabs allow readers to quickly and easily locate sections, chapters, and divisions. Tabs are often color-coded and follow the design principle of contrast because they exceed the paper margins so readers can quickly go to specific sections of a document. If you decide to use tabs in your documents, select tabs with a professional appearance and print a shortened version of the chapter or division title on the tab. Figure 10.3, page 274, shows an example of tabs.

Divider Pages

Divider pages help readers locate chapters and major sections. **Divider pages** appear before chapters or sections. These pages list the title of the section or chapter and often a brief list of the contents. They are often a different color than the pages of the chapters or sections.

Headers and Footers

Headers and footers help readers locate specific pages and navigate through each page to locate specific information. A **header** is a word or phrase that you put at the top of each page to identify a document or a specific section of a document. A **footer** serves the same purpose, but appears at the bottom of each page. Headers and footers can include information such as the chapter title, the most recent first-level heading, the page number, and the book title; they may also include identifying information about the author or the organization. Figure 10.4, page 275, shows a header and footer.

Locating tools make your document more usable by helping readers quickly scan and locate the information they need.

FIGURE 10.3

A Document with Tabs

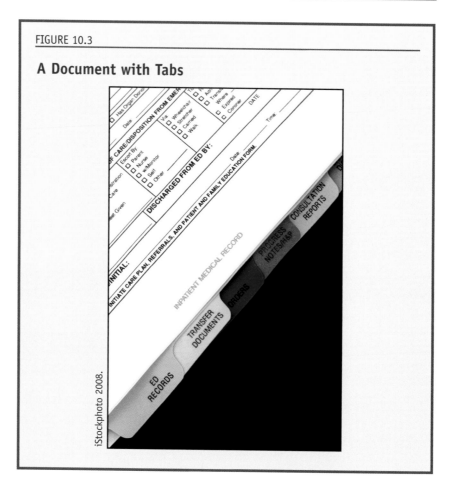

iStockphoto 2008.

Page Numbers

Page numbers help your readers find pages, locate information, and use an index. You should position page numbers so readers can easily see and use them. Follow the tips for numbering pages on the next page.

Headings

Headings are subtitles that show readers how you have grouped the information in your document. When information is grouped, or "chunked," readers can easily locate the information they need. In the absence of headings, readers read down from the top of a page until they find the information they want. With headings, readers can scan your document to locate the information they need.

FIGURE 10.4

Headers and Footers

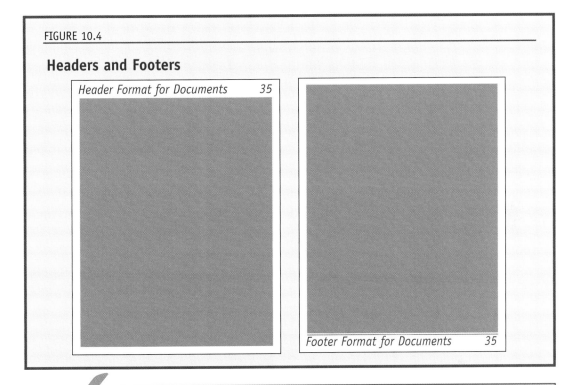

Header Format for Documents   35

Footer Format for Documents   35

TIPS FOR NUMBERING PAGES

- **For one-sided pages, put page numbers in the top or bottom corner.** Be consistent: always place the page number in the same place on every page. If you include a header or footer with the page numbers, use the same style of header or footer on every page. (See Figure 10.4.)
- **For two-sided pages, put the page numbers in the outside top or bottom corners of the pages.** If you put the page numbers in the center of the page, readers may not see them as they thumb through your document to locate a specific page.
- **Do not number the first page of most documents.**
- **For longer and more complex documents, use a different number sequencing for the front matter.** Use lowercase Roman numerals (i, iii, iv, etc.) for the front matter. Use Arabic numerals (1, 2, 3, etc.) for the body. Chapter 18 will give you more information on numbering the front matter.
- **For documents that will be updated, number the pages by section.** By numbering each section separately, you will only have to reprint the sections that change. If you number by section, the page numbers should have two parts: section number – page number. That is, for section 2, page 4, the page number would read 2–4.

Figure 10.5 shows three ways to position headings. As you write your headings, follow the tips for creating effective headings found on the next page.

Color

Color is one of the most effective tools for organizing information because it can be the most dominant visual element in a document. You can use color to emphasize important information and to help readers locate information. Because color can create contrast, consider combining color with these locating tools to make them stand out:

* headings
* tabs
* divider pages
* headers and footers
* page numbers

Figure 10.7, page 279, illustrates how color highlights information. In this page from a cell phone user guide, color highlights the header, the page numbers, the first-level headings, and the features. Follow the tips on page 278 for using color.

Icons

Icons are graphics that symbolize an action or a concept. You see and probably use icons daily on your cell phone or your computer. Icons operate on the

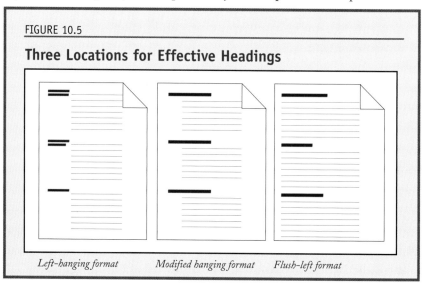

FIGURE 10.5

Three Locations for Effective Headings

Left-hanging format *Modified hanging format* *Flush-left format*

✓ TIPS FOR CREATING EFFECTIVE HEADINGS

- **Position headings flush with the left margin**. Readers can more easily locate a heading if you align it flush against the left margin. Do not center headings because readers can't locate centered headings as quickly as flush-left headings; centered headings violate the principle of alignment.
- **Use no more than four levels of headings in most documents**. Too many levels can clutter a document and confuse readers.
- **Use more lines of space above your headings than after your headings**. For instance, in a double-spaced document, use three lines of space above the heading and two lines of space after. In a single-spaced document, use two lines of space above and one line after.
- **Include at least two lines of text below a heading**. Don't leave a heading floating at the bottom of a page. Include at least two lines of text below a heading. Headings without these lines appear to "float" at the bottom of a page (see Figure 10.6).
- **Use different type sizes and styles (such as boldface or color) to make headings stand out**. Readers associate size of type with importance of information (White). If the text is in 12-point black type, you can make the heading stand out by using 14-point type in another color. Don't underline the heading: underlining distorts the descenders of some letters such as *p* and *y*. Whatever technique you select to make the headings stand out, use that technique consistently. If you make one first-level heading 14-point blue type, make all first-level headings 14-point blue type.

FIGURE 10.6

Appropriate Spacing for Headings

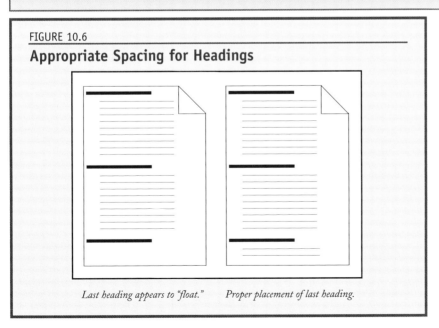

Last heading appears to "float."   *Proper placement of last heading.*

✓ TIPS FOR USING COLOR

- **Use the same color throughout the document for the same type of information.** For example, if you use green for the first-level headings in the first chapter, use green for the first-level headings in all chapters. Using random colors for the same type of information serves only to decorate rather than to help readers locate information.
- **Use color along with other devices, such as white space, boldface, or type size.** Color-blind readers need clues other than color to help them locate information. If you indicate a locating device using only color, some readers may not be able to use that device.
- **Use colors primarily to communicate, not to decorate.** Color attracts the reader's eye. If the color merely decorates, it may distract readers from the information you are trying to communicate.
- **Consider your readers' culture when selecting colors.** Some colors have different meanings in different contexts (Horton, "The Almost Universal Language"). For example, when U.S. readers see instructions printed in red, they may associate the red with danger. However, when they see red while driving, they know to stop. In business, red is often associated with power. In other contexts, black connotes formality and power. In the U.S. and other Western cultures, people associate black with death and mourning; but in China, people associate white with death. As you select colors for your document, consider what the color may indicate to your readers—especially those readers from other cultures.

design principle of repetition. For example, when you see the scissor icon in a software program, you know it indicates "to cut." Figure 10.8 shows some commonly used media and publishing icons. When users see these icons, they know what media devices are available.

DESIGN THE PAGE

As readers move beyond external packaging, they look at the page to determine what—if anything—to read. To plan the page, you want to explore how and where you will place the text and graphics. To help you get ideas, you might look at layout books such as *The Layout Index* and *The Idea Index* by Jim Krause. These books display hundreds of "idea-starters" for designing brochures, posters, Web sites, and other documents. As you are designing the page, consider the following elements:

- layout
- white space
- type
- margins
- line spacing

Designing the Layout

To create the most effective layout for your information, your purpose, and your readers, explore various folds and layouts. Many writers simply select the types of layouts they are familiar with or that will fit on $8\frac{1}{2}$ x 11 inch paper. However, these layouts may not fit the

---

FIGURE 10.7

Effective Color Use

Retrieving Your Messages

Retrieving Your Voice Mail Messages

You can retrieve your messages from home or from afar using any touch-tone phone. You may access your voice mail in the following three ways:

- Dialing star (*) 98
- Dialing your digital phone telephone number
- Dialing the voice mail access number 214-555-0211

Retrieving Your Voice Mail at Home Using Your Digital Phone

1. Lift the handset of your phone.
2. Listen for a stutter tone. If you hear this tone, you have voice mail.
3. Dial star (*) 98 or your 10-digit home phone number.
4. Wait for your voice mail to answer. The voice mail system will tell you how many new messages you have.
5. Press the "1" key to listen to your messages.

Retrieving Your Messages

Retrieving Your Voice Mail from Afar by Dialing Your Home Phone Number

1. Using any touch-tone phone, dial your 10-digit home phone number.
2. Wait for the voice mail system to answer. You will hear your voice mail greeting.
3. Press the star (*) key.
4. Enter your PIN followed by the pound (#) key. The voice mail system will tell you how many new messages you have.
5. Press the "1" key to listen to your messages.

Retrieving Your Voice Mail from Afar by Dialing Your Voice Mail Access Number

1. Using any touch-tone phone, dial the voice mail access number 214-555-0211.
2. Enter your 10-digit home phone number.
3. Press the pound (#) key. The voice mail system will tell you how many new messages you have.
4. Press the "1" key to listen to your messages.

Source: Design courtesy of Rachel Oslin Bradford

FIGURE 10.8

Media and Publishing Icons

iStockphoto 2008.

needs of your readers or your information. To design the layout, use the following techniques and tools:

- thumbnail sketches
- prototypes
- style sheets
- styles

Thumbnail Sketches

Thumbnail sketches are rough drawings of possible page layouts (see Figure 10.9). Roger Parker and Patrick Berry recommend sketching your initial page layout ideas: "Try out a variety of ideas. When you finish one sketch, begin another.... Don't bother with excessive detail—use thin lines for text, thick lines or block lettering for headlines [headings], and happy faces for art or photographs. Even simple representations such as these will give you a sense of which arrangements work and which don't." As you sketch pages, you might also consider different types of page layouts and sizes. Figure 10.10 presents some of the common layouts (sometimes called grids).

Prototypes

Prototypes are similar to thumbnail sketches, except prototypes are the actual size of the design. For example, if you plan to design a bifold booklet with a page height of $8^1/_2$ inches and a width of $6^1/_2$ inches, you would take standard paper ($8^1/_2$ x 11 inches) and fold the document to create the bifold

FIGURE 10.9

Thumbnail Sketch Designs

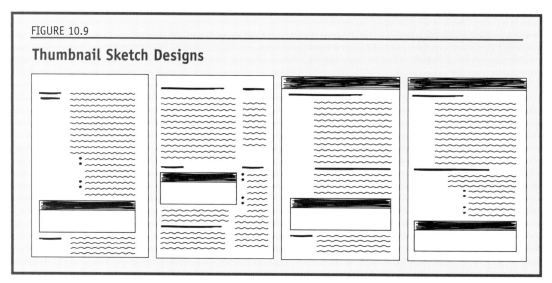

booklet. You would then simply sketch where the headings, text, and graphics would appear.

Style Sheets

Style sheets are a tool to help writers and designers maintain consistency throughout a document. A style sheet might include language choices such as those discussed in Chapter 8; it also can serve as a plan for designing a document. Figure 10.11, page 282, shows a simple style sheet for the design elements of a software manual written by a team of students. The team e-mailed the style sheet to each team member and also put it in a folder on the

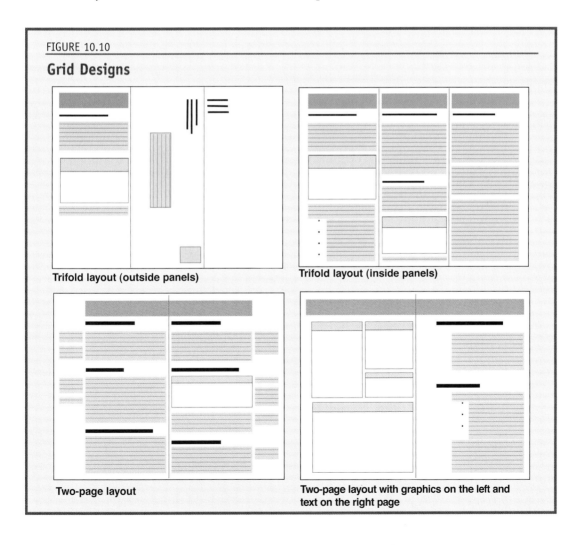

FIGURE 10.10

Grid Designs

Trifold layout (outside panels)

Trifold layout (inside panels)

Two-page layout

Two-page layout with graphics on the left and text on the right page

FIGURE 10.11

Sample Style Sheet

Team 3
Style Sheet for Software Manual

Page Elements	Page size	9" x 6"
	Margins	¾"
	Layout	two uneven columns, the left column for marginal comments and headings
		left column 1½"
		right column 3"
	Spacing	single-space for text
		double-space between paragraphs
	Visual aids	no captions for screen captures placed at point of reference
	Headings	left-hanging
Type Elements	Typeface	Bookman
	Size	12 point for text
		14 point for headings
		12 point for subheadings
	Style	bold for subheadings
	Color	RGB 44 18 238 for major headings

local area network in their computer lab. A style sheet can help you maintain consistency. When you are working on a long document, a style sheet can help you remember the design decisions that you made at the start of the project. Style sheets can also help you create a consistent appearance for similar documents or for all of the documents written for an organization.

Many organizations have their own style guides or specific design requirements for their documents. For example, a telecommunications company in the U.S. requires the company logo always to appear in the same typeface, type size, and color. This company also has specific page layout requirements for business letters and reports. As you prepare documents for your organization, find out whether it has a style sheet or specific design requirements.

When you are working as part of a team, a style sheet helps you and other team members to use the same design elements and to format consistently. Keep the style sheet simple, so team members can easily follow it. The style sheet should list at least these design elements:
- typefaces
- type sizes

- margins
- heading style
- line spacing

If team members follow the guidelines spelled out in the style sheet, the team can easily combine each member's section to create one document.

You also can use preformatted templates available in most word-processing or desktop-publishing software. These templates provide page layouts that you can use for many documents. If you decide to use a preformatted template, you should consider the possible problems. Preformatted templates often
- don't fit the purpose of your document
- don't follow good design principles
- omit key conventional elements required in some documents
- are commonplace (many writers use these templates, so your document could look like many other documents)
- include inappropriate graphics or design elements

TIPS FOR PLANNING THE PAGE LAYOUT

- **Make thumbnail sketches.** Thumbnail sketches help you to see multiple possibilities for the page layout.
- **Create a prototype page.** A prototype page helps you to determine the appropriate size and layout for your readers and your purpose.
- **Create a style sheet or use your organization's style sheet.** A style sheet helps you to create a consistent design. If you're working with a team, a style sheet saves you time.
- **Use the styles function.** Styles can save you time and help you to use the design elements consistently.
- **If you're working with a team, create a style sheet before you begin writing.** Make sure each team member has a copy of the style sheet and the styles.
- **If you select a preformatted template, make sure it follows good design principles.** If it doesn't, consider creating your own template.

Styles

Consider using the styles tool of your word-processing software. Using styles ensures a consistent format for headings, bullets, and text. Figure 10.12 shows the styles tool in Microsoft *Word*® and how you can modify the preformatted styles. By using the styles tool, you save time and achieve a consistent look in one document or among several documents. When you create a style for headings, for example, you don't have to format a heading each time you type one. Instead, you type a heading and then apply the style by putting the cursor in the heading text and selecting the style for that heading using the styles tool.

White Space

Readers look for relationships among the elements on a page—text, graphics, and headings. These elements should look as though they belong together; otherwise, readers may be confused. To create a unified layout, frame the elements on the page with white space (Lay; Yeo). **White space** is the space on the page not occupied with text or graphics. As Robin Williams explains, many beginning designers tend to "be afraid of white space." They feel they must fill every space on the page. However, effec-

FIGURE 10.12

Using the Styles Tool

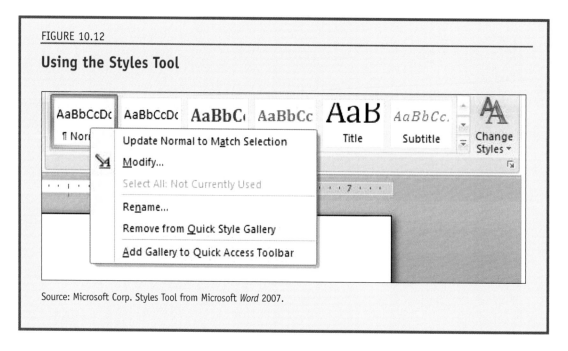

Source: Microsoft Corp. Styles Tool from Microsoft *Word* 2007.

tive designers use lots of white space. White space allows you to follow two of the design principles: proximity and contrast. When related items are in proximity, you create white space that helps to show readers what items are related. Without white space, readers cannot easily see which items are related. White space also helps to create contrast. For example, Figure 10.13, page 286, has excellent contrast around the Bent Tree Press advertisement.

Type

You may see type referred to as "font." When you go to select type for your documents, you have hundreds of choices. Yet many of these choices are not appropriate for your purpose or your readers. As you select the appropriate type for your documents, you should consider the following:

- typeface
- size
- style
- case

The type that you select can "help or hinder" the readability of your document (Parker and Berry).

Typeface

A *typeface* is a set of letterforms (letters, numbers, punctuation marks, and other symbols) with the same design. In word-processing software, typefaces are often referred to as "fonts." You can select from different groupings of typefaces: serif and sans-serif (see Figure 10.15, page 287). A *serif* is a short stroke or line at the top or bot-

---

✓ TIPS FOR USING WHITE SPACE

- **Push related elements together with white space**. White space helps you to follow the principle of proximity because this space "pushes" page elements together (Lay), helping readers to see what elements belong together. For example, when you leave more white space before than after a heading, you help readers clearly see what text the heading describes.

- **Use white space to surround any elements that you want to emphasize**. By surrounding elements with white space, you emphasize them. For example, surrounding your company's logo with white space will draw the reader's eye to the logo. (See Figure 10.13, page 286.)

- **Set off elements such as headings, bullets, and graphics with white space**. The white space increases their visibility and allows readers to locate them easily. For example, notice that white space surrounds the bullets in this textbook. The text aligns to the right of the bullet not under the bullet; and the white space highlights the bullet. If the text aligned under the bullet (see Figure 10.14, page 286), the displayed list would be less visible on the page or screen.

FIGURE 10.13

Brochure Demonstrating Effective Use of White Space

The white space contrasts with the words, "Bent Tree Press."

White (or empty) space highlights the company's logo.

FIGURE 10.14

Framing Elements with White Space

Incorrect

Correct

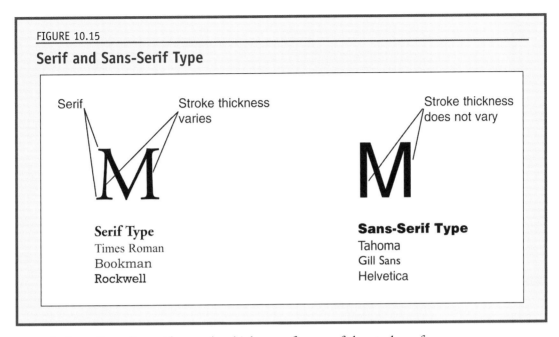

FIGURE 10.15

Serif and Sans-Serif Type

tom of a letterform. In serif type, the thickness of some of the strokes of a letterform may vary, helping readers to distinguish the shapes of different letterforms. In a sans-serif typeface—type without serifs (*sans* means "without" in French)—no small strokes project from the top or bottom of the letterform, and, generally, the thickness of the lines of a letterform are uniform. As you select type for your documents, follow the suggestions found in the tip box on page 288.

Type Size

Type size is measured in points; 72 points equal 1 inch. Most word-processing and desktop-publishing programs allow you to adjust the type size up to 72 points. When deciding what type size to use in your documents, follow these guidelines:

- **For text, use 10-, 11-, or 12-point type.**
- **For headings**, **use a type size 2 to 4 points larger than the text**. For example, if the text is in 12-point, use a 14- to 16-point type for the headings.
- **For footnotes, use 8- to 10-point type**.
- **For slides, use 24- to 36-point type.** See Chapter 20 for more information on slides.

TIPS FOR SELECTING TYPEFACES

- **Use serif typefaces for body text**. Serifs guide readers' eyes from letter to letter; the serifs help readers see the text "in terms of words and sentences instead of as individual letters" (Parker and Berry). Serifs and the variations in stroke thickness help readers distinguish between letters with similar shapes (such as the numeral "1" and a capital "I") and to recognize the shapes of all letters.
- **Use sans-serif typefaces for titles and headings**. Sans-serif type is difficult to read in long blocks of text and in small sizes, but small amounts of it can "add impact to a document," especially when white space surrounds the elements (Parker and Berry). You can effectively use sans-serif type for titles and headings; however, you can use the same serif type that you are using for the text. The key is contrast: if you want to use a different typeface for the titles and headings, the type should be noticeably different from the text. For example, if your text is in Times Roman (a serif typeface), don't select Garamond—a similar serif typeface—for the headings. Instead, select a sans-serif type, such as Tahoma or Ariel.
- **Limit the number of typefaces in your documents to two: a serif typeface for the text and a sans-serif typeface for the titles and headings**. You can use a different, perhaps more-decorative typeface for title pages, chapter titles, covers, or divider pages. Otherwise, use no more than two typefaces.
- **Select a typeface that is easy to read**. Script and decorative typefaces are not appropriate for text or headings in technical documents. These typefaces may be hard to read and appear unprofessional. You might use these typefaces in logos, party invitations, title pages, divider pages, or covers. If you decide to use a script font, use a large size. The smaller the script font, the harder it is to read.
- **Use a sans-serif typeface for reverse type**. When you place white (or light) type on a dark background, you are using reverse type. Reverse type should always be sans-serif type.
- **Use a sans-serif typeface for text that will appear on screen or online**. Sans-serif typefaces are easier to read on screen.

Appropriate Type Styles

You can modify the appearance of the type to create different looks in your document. With word-processing software, you can vary the style (often referred to as "effects") of a typeface. For example, commonly used styles are boldface, italics, underlining, shadowing, outlining, and reversed type. Some styles can improve your documents' appearance by providing "visual relief in an otherwise uniform page of text" (Felker et al.). However, some styles make type unreadable, create an unprofessional appearance, clutter the page, and fail to focus readers' attention on what you intend to communicate.

Case

Text in upper- and lowercase letters is easier to read than text in all uppercase letters. Lowercase letters take up less space, so readers can "take in more words as they scan a line of text"; lowercase letters give each word a distinct shape (Benson). Shape helps readers distinguish letters and identify words (Felker et al.). As Figure 10.16, page 290, shows, words set in uppercase letters have the same basic shape or outline, but words set in lowercase or in both upper- and lowercase letters have different shapes. The uniform shape of words set in uppercase letters slows readers' ability to recognize each word.

Margins

When setting up the margins for your pages, you want to consider two items:
- Justification
- Margins

Justification

Justification refers to the alignment of text along the left and right margins. You are familiar with alignment through the alignment icons in the toolbar of word-processing software (see Figure 10.17, page 290). When lines of text are of different lengths and do not

TIPS FOR SELECTING TYPE STYLES AND CASE

- **Use boldface type to add emphasis.** Boldface type increases the visibility of headings and individual words and phrases. Use boldface type for headings and, sparingly, to emphasize individual words in blocks of text; do not use it for entire paragraphs or for more than two or three lines of type. In online documents, designers recommend that you use boldface type only for headings (Yeo).
- **Use italics to add emphasis.** Italics can effectively emphasize individual words and short phrases, though less dramatically than boldface. Use italic type for isolated words and short phrases, such as for non-English words, not for entire paragraphs or large blocks of text.
- **Use reverse type sparingly.** If you use reverse type, use a sans-serif typeface in a relatively large size.
- **Do not use outlined or shadowed type.** These type styles can "seriously hinder legibility" and are especially hard to read in small sizes and in uppercase letters (Parker and Berry). These type styles make your documents appear unprofessional.
- **Do not underline.** Underlining interferes with readers' ability to recognize the shapes of some letters. It can distort letters with descenders—*y, j, p, q, g*—and marks of punctuation such as commas and semicolons. Instead of underlining, use boldface, italics, or color.

align on the right, the text is unjustified, or ragged. When lines of text are equal in length, the text is justified and the right margin is even (see Figure 10.18, below). Unjustified right margins are easier to read (Benson). When line lengths vary, readers' eyes can move more easily from one line to the next. When text is justified, the lines all look the same.

FIGURE 10.16

Type Case and Readability

Because readers recognize words by their shape, text in

UPPERCASE LETTERS ARE HARDER TO RECOGNIZE.

FIGURE 10.17

Alignment Icons in Word-Processing Toolbar

Source: Microsoft Corp. Microsoft *Word 2007* alignment icons.

FIGURE 10.18

Justification

Spacing between words is inconsistent.

Lines are the same length.

Justified Text

Justified text gives documents a formal look; but it is harder to read than unjustified text, and the inconsistent spacing between words may bother your readers. Unjustified text gives documents a more open look. The unequal line lengths of unjustified text help readers to move smoothly from line to line and eliminate the inconsistent spacing associated with justified text.

Spacing between words is consistent.

Line length varies.

Unjustified Right Margin (Ragged)

Justified text gives documents a formal look; but it is harder to read than unjustified text, and the inconsistent spacing between words may bother your readers. Unjustified text gives documents a more open look. The unequal line lengths of unjustified text help readers to move smoothly from line to line and eliminate the inconsistent spacing associated with justified text.

Readers can't easily distinguish one line from the next and may find their eyes moving to the wrong line as they read down a page or screen. You will see justified type in many books and some formal documents.

The space between words in justified text is inconsistent. From one line to the next, the space between words may vary so that all lines will align evenly on the right side of the page or screen. This inconsistent spacing can slow reading and make readers wonder whether a word is missing. In the justified example in Figure 10.18, notice the inconsistent spacing between words and the uniformity in line lengths. In the unjustified example, notice the uniform spacing between words that results when the right margin is unjustified.

Margins

For most documents, you will use a 1-inch top, bottom, right, and left margin. If you plan to bind the document, use a 1½-inch left margin. You also want to use consistent margins:

- **Use the same top, bottom, left, and right margins on each page**. For example, if you use a 1-inch margin for one section of the document, use a 1-inch margin for every section.
- **Use consistent paragraph indents and spacing between columns, within lists, and before and after headings**. For example, if you use a 3-space indent for the first paragraph, use the same indent for all paragraphs.

Line Spacing

Line spacing refers to the space between lines of type and the space between text and graphics. For most technical documents, you will select single spacing, one-and-a-half spacing, or double spacing.

- **Use single spacing for e-mail, memos, letters, and most manuals.**
- **Use single, one-and-a-half, or double spacing for reports and proposals.** If you cannot determine the spacing appropriate for a report or proposal, look at other reports or proposals prepared by your coworkers. If you do not have such a document, use single spacing.

TAKING IT INTO THE WORKPLACE
Using Color to Structure Information

Color is an important design tool, yet inexperienced designers often think primarily about what and how much color to use, rather than how color can "enhance and clarify" the information (Wilson). When used as a key to information structure, color can help readers handle more information and process it more efficiently (Horton, "Overcoming Chromophobia," *Illustrating Computer Documentation*).

- **Color can help readers group objects, "taking precedence over other visual" cues** (Keyes). Readers group by color before they group by shape, size, or other design elements (Keyes; Horton, "Overcoming Chromophobia," *Illustrating Computer Documentation*; Martinez and Block).
- **Color grabs a "reader's attention first, before the reader has understood the surrounding informational context**—where it is in the hierarchy, what type of information it is, or its relation to other text" (Keyes). Readers perceive a color element independently of its surrounding text.
- **Color creates a separate "visual plane" that differentiates and consolidates visual information** (Keyes). For example, readers might separate type in color from type not in color. This separation helps readers to scan documents and to see the organization of the information.
- **Multiple colors distract readers because each color group forms a separate category that competes for the reader's attention** (Krull and Rubens). When selecting color, "less is definitely more" (Keyes).

When selecting color for your documents,
- determine what you want to emphasize. Remember that readers will perceive information in color as more important than information not in color.
- use different shades of one color rather than several different colors. For more information on color, see Jim Krause's *Color Index* or Leatrice Eiseman's *Messages & Meanings: A Pantone Color Resource.*

Assignment

Visit a local business or non-profit organization and gather a color document that the business or organization produced.
- Ask an employee, manager, or owner the following questions:
  - How much did you spend to produce the document?
  - Why did you select the color(s) used?
- Write a memo to your instructor
  - summarizing what you learned about the cost of producing the document in color and why the organization selected the color(s) used.
  - explaining whether the color is used effectively to structure and emphasize information and if the color detracts or enforces the message.

SAMPLE DOCUMENT DESIGNS

Figures 10.19, 10.20, and 10.21, pages 293-296, illustrate how writers have effectively designed documents.

FIGURE 10.19

An Effectively Designed Document

The designer effectively uses contrast to highlight the company's creed.

The designer has effectively aligned the text along the left margin and has used an unjustified right margin. The designer also correctly aligns the text after the bullets in the boxes.

The designer uses the principle of proximity through the boxes; however, the boxes would be more effective with upper- and lowercase type for the words "service," "talent," and "choices."

The designer uses a sans-serif typeface for the reverse type.

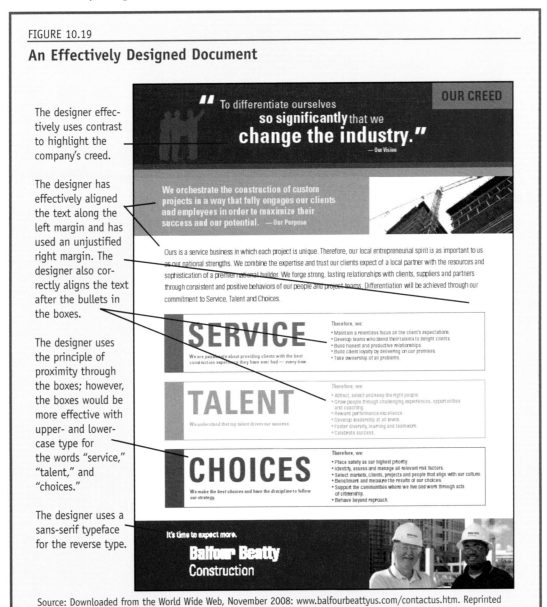

Source: Downloaded from the World Wide Web, November 2008: www.balfourbeattyus.com/contactus.htm. Reprinted courtesy of Balfour Beatty Construction Company.

FIGURE 10.20

Sample Document Design

The designer uses a sans-serif typeface for the headings.

The designer creates repetition by repeating the red twice: in the flowers and in the red background on the facing page.

FIGURE 10.20 CONTINUED

The designer correctly selected a sans serif typeface because the text is white on a dark background (reverse type). For reverse type, sans serif typefaces are more readable than serif typefaces.

The designer creates contrast by using white (blank) space.

The designer creates effective alignment by aligning the heading with the text, the "Texas Parks and Wildlife" logo, and the publication information.

Source: Downloaded from the World Wide Web, June 2009:
http://www.tpwd.state.tx.us/publications/pwdpubs/media/pwd_bk_e0100_003_01_09.pdf. Texas Parks and Wildlife Department, 2008 Annual Report, pp 2-3, (PWD BK E0100 003).

FIGURE 10.21

An Effectively Designed Document

Modified hanging heads help readers locate information.

Chapter 3
Keeping the Participants Interested

Once you have your equipment in place and you have planned and rehearsed your presentation, you're almost ready. However, you can improve your chances for a successful presentation by understanding ways to keep your particpants interested (and awake). In other words, put yourself in your participants' seat! This chapter will present some strategies to help you keep participants interested:

The designers effectively use color to emphasize the headings. Notice that a color-blind reader could easily locate the headings because they appear in large, boldface type as well as color.

- Give participants only the information they need.
- Anticipate participants' needs and questions.
- Provide participants with a "road map" and examples.
- Help participants enjoy your presentation.

Strategy 1: Give Participants Only the Information They Need

Keep your presentation short and simple. Participants want to hear only the information they need and no more. As you prepare for your presentation, consider the following:

The designers have used a hanging indent for the bulleted lists. The text aligns after the bullets.

- **Listening to information takes twice as long as reading that same information.** Thus, if you can read 10 pages in 8 minutes, your participants can comprehend the same information in about 16 minutes.
- **Condense your presentation into a few points.** Don't try to give participants every bit of information you have about a topic or all the tiny details. Instead, select the key points and present those. If necessary, you can refer your participants to the quick reference cards or to other printed handouts.
- **Plan the presentation to take slightly less than the allotted time.** Look for ways to tighten your presentation, so you have time for the participants to ask questions. Your participants will prefer a presentation that is a couple of minutes short rather than a presentation that exceeds the allotted time.

Strategy 2: Anticipate Participants' Needs and Questions

As you are preparing and even as you are speaking, think about what participants already know and what they will want to know about the topic.

The designers have used a sans-serif typeface for the headings and a serif typeface for the text.

- **Customize your presentation according to what you know about your participants.** You will always begin your presentation with the same databases: Project Description, Project Contacts, and Specifications.

Making Dynamic Presentations   32

Source: Reprinted courtesy of Balfour Beatty Construction Company.

CASE STUDY ANALYSIS
Bad Design Makes Its Mark on History: The Palm Beach Ballot and The 2000 Presidential Election

Background

If you think design doesn't matter, consider the effect it had on the 2000 U.S. presidential election. In 2000, the U.S. Census Bureau estimated that more than 27 percent of the Palm Beach, Florida, population was age 55 or older. As a result, election officials in that county decided to make the ballot "easier to read" for its elderly voters. Rather than placing all the candidates on one page, as was typically done, officials decided to use a facing-page layout, which would allow the typeface to be larger. Unfortunately, it also positioned the punch holes in a single line down the center of the ballot, which proved to be confusing for voters (see Figure 10.22, page 298). Consequently, many people ended up inadvertently voting for more than one candidate, and, as a result, their ballots were disqualified. The county had to throw out 4.1 percent of its election ballots for overvoting, as compared to its historical average of about 1 percent disqualified ballots.

Assignment

1. Write a letter to your state election officials, detailing the elements of good design that should be used in their ballot design. Include supporting evidence for your suggestions. E-mail a copy of your letter to your instructor.
2. Redesign the Palm Beach, Florida, 2000 election ballot. Design a ballot that is easy to use and maintains readability for elderly voters. Turn your design in to your instructor.

FIGURE 10.22

Ballot for 2000 Presidential Election

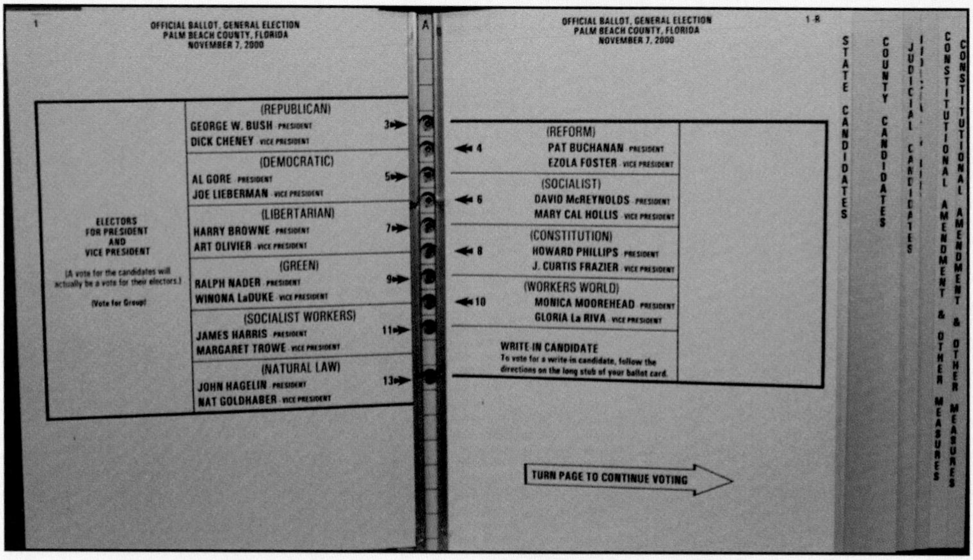

© AP/Wide World Photos. Used by permission.

EXERCISES

1. **Web exercise:** Find a one- or two-page ineffectively designed document on the Web.
   - Write a memo to your instructor explaining the problems with the design. In your memo, comment on how you could improve the design by using the principles of contrast, repetition, alignment, and proximity. With your memo, include a printout of the document. For information on writing memos, see Chapter 12.
   - Redesign the ineffective document that you found. Correct the design problems that you identified in your memo.

Download a Worksheet for Designing Reader-Focused Documents online at www.kendallhunt.com/ technicalcommunication.

2. Write a memo to your instructor evaluating the design of the document "How Do I Choose an Insect Repellent?" (see Figure 10.23, page 300). In your memo, comment on how effectively the document
   - follows effective design principles
   - uses type
   - uses white space
   - uses color

   For information on writing memos, see Chapter 12.

3a. Write a memo to your instructor evaluating the design of the information guide for Enchanted Rock State Natural Area in Figure 10.24, page 301. You will find an electronic copy of this document on the Web site for this book. In your memo, comment on how effectively the document:
   - follows the design principles
   - uses type
   - uses white space
   - uses color

3b. Assume that you work for Enchanted Rock State Natural Area. Your manager has asked you to create a document that campers can carry to the primitive campsites. This document will be used in addition to the current document. Your document should meet these specifications:
   - fits easily into a backpack or pocket

- includes safety information and park regulations
- includes at least two color graphics

4. Write a memo to your classmates telling them how to create a template in the word-processing software that you use on your campus. For information on writing memos, see Chapter 12.

FIGURE 10.23

Document for Exercise 2

HOW DO I CHOOSE AN INSECT REPELLENT?

| 1- 2 hours | 2-4 hours | 5 – 8 hours |

ON SKIN

MOSQUITOES

Protection varies by species of mosquito.

Most mosquitoes that transmit diseases in the US bite from dusk-dawn.

Choose the appropriate repellent for the length of time you'll be outdoors. Reapply according to product instructions

<10% DEET <10% picaridin	~15% DEET ~15% picaridin/KBR 3023 ~30% oil of lemon eucalyptus/PMD	~20%-50% DEET

TICKS

Generally, repellent with 20 – 50% DEET is recommended to protect against tick bites.

Other factors affecting efficacy include: individual chemistry, sweat, numbers of bugs. Apply creams and lotions 15 to 20 minutes before going outdoors.

In areas where both mosquitoes and ticks are a concern, repellents with 20 – 50% DEET may offer best, well-rounded protection.

The American Academy of Pediatrics has recommended that repellents containing up to 30% DEET can be used on children over 2 months of age.

The repellents shown here meet CDC's standard of having EPA registration and strong performance in peer-reviewed, scientific studies. They reflect products currently available in the U.S.

ON CLOTHING AND GEAR

Permethrin

Permethrin treatment of clothing and equipment can provide protection against mosquitoes and ticks through multiple washings. Follow label instructions.

Source: Downloaded from the World Wide Web, November 2008: www.cdc.gov/ncidod/dvbid/westnile/resources/repellent%20 timeline%20poster041207.pdf.

FIGURE 10.24

Document for Exercise 3

General Information

You are about to enter a primitive hiking and camping area, an experience far removed from the everyday "civilized" life to which we have become accustomed. The country through which you will walk is being allowed to revert to its natural condition; the incursions of man will be kept to a minimum. Visitors are encouraged to maximize their experience in this natural setting by closely examining and experiencing the sounds, smells and the feel of nature. In so doing, the trail before you will not only be more enjoyable but inspirational as well. Future generations will be able to enjoy and experience the primitive backcountry area only if today's visitors protect and care for their heritage. The satisfaction and achievement of traveling through and camping in a primitive area will be complete only if you, the user, leave no sign of your visit ... no perceptible traces. Help preserve the harmony and nature of the backcountry.

Climbing

1. The use of pitons is prohibited. Climb "clean" and preserve the resource for others to enjoy.
2. Leather-soled shoes are not recommended for climbing and hiking on the rock due to the slippery nature of the rock surface.
3. Visitors exploring Enchanted Rock Fissure should use CAUTION. This 1,000-foot-long talus fissure contains over 20 entrances with tight passages, wet, slippery surfaces, numerous steep inclines and hazardous vertical drops. It requires some skill and climbing ability to experience safety. To insure safety, carry at least one light source and wear loose, protective clothing and proper foot gear such as climbing boots or rubber-soled shoes.
4. Keep it safe. Nearest medical facilities are in Fredericksburg, and most parts of the park are accessible only by foot. Should **gency arise, contact the headqua** (830) 685-3636 or 911.

Enchanted Rock State Natural Area
16710 Ranch Road 965
Fredericksburg, TX 78624
(830) 685-3636

www.tpwd.state.tx.us

4200 Smith School Road
Austin, TX 78744

TEXAS PARKS AND WILDLIFE

Enchanted Rock

STATE NATURAL AREA

Primitive Campsites (Backpacking)

Overnight camping is permitted only in the designated primitive camping area. CAMP WITHIN CAMPING AREA BOUNDARIES. Do not rearrange the landscape. DO NOT CONSTRUCT YOUR CAMPSITE. Never cut branches, saplings or trees. As much as possible, please avoid disturbing the groundcover and topsoil. Locate your camp in order to take advantage of natural drainage and topography. Bedsites or tents should be pitched on naturally flat ground. Do not dig or level an area for a bedsite. Please remove all traces of your camp when you leave.

Be advised that the river and low-lying areas are subject to flash flooding.

Campfires are not permitted in the primitive area due to the possibility of wildfires.
COOKING SHALL BE DONE ONLY ON CONTAINERIZED FUEL STOVES. NO GROUNDFIRES.

All garbage and litter (including cigarette butts) shall be packed out of the area for disposal in trash receptacles at the trailheads. Burying garbage is not permitted.

Self-composting toilet facilities are located adjacent to the trail at the designated camping areas. DO NOT THROW ANY NON-BIODEGRADABLE WASTE OR SMOKING MATERIALS INTO THE COMPOSTING TANK.

Vernal Pools

A Threatened Natural Resource

The "islands" of vegetation on the bare granite summit of Enchanted Rock are some of the most ecologically significant and severely threatened features of this state natural area. Known as soil islands, weather pits, gammas or vernal pools, these patches of vegetation on bare rock develop in depressions formed by weathering over thousands of years.

The depressions shelter an assemblage of plants and animals uniquely adapted to a harsh environment. In fact, by studying weather pits, ecologists learn: (1) how plants and animals colonize a newly formed habitat; (2) how those organisms modify their environment and help develop soils; and, (3) how plant and animal community structure and composition change over time.

At Enchanted Rock you can see the progressive development from bare rock-bottom pits, to annual plant establishment, to miniature prairies with grasses like little bluestem and even trees like live oak. Vernal pools also support an interesting species of invertebrate, the fairy shrimp. These tiny animals survive total desiccation as fertilized eggs, and hatch into larvae and grow into adults each time water collects after sufficient rainfall.

Because of the fragile nature of weather pits, Texas Parks and Wildlife Department wants to make sure visitors are aware of their significance. Too often they are perceived as nothing more than convenient rest areas or even "bathrooms" for people and pets, and are subject to trampling, littering and other forms of waste. PLEASE refrain from entering or allowing pets in weather pits under any condition. Enjoy and observe these special features only from their granite margins – STAY ON THE ROCK. Thank you for protecting an important part of the Enchanted Rock experience.

Source: Downloaded from the World Wide Web, June 2009: http://www.tpwd.state.tx.us/publications/pwdpubs/media/ pwd_br_p4507_0119l.pdf. Texas Parks and Wildlife Department, Enchanted Rock State Natural Area, (PWD BR P4507-119L).

REAL WORLD EXPERIENCE
Designing Documents for a Non-profit Organization

Working with a team assigned by your instructor, redesign or create new documents for a non-profit organization on your campus or in your community. Follow these steps:

Step 1: Organize the team
- Select a team leader to serve as managing editor of the project. The managing editor is responsible for communicating with your instructor, handing in the final documents, assigning tasks when necessary, and proofreading the final documents.
- Exchange telephone numbers and/or e-mail addresses.

Step 2: Locate a non-profit organization and find the documents that you will redesign or create
- Ask the non-profit organization if they have documents that you can update or redesign. You might begin by visiting non-profit organizations in your community or on your campus.
- Make copies of the documents for each team member.
- If the organization doesn't have documents, determine what documents they need written and designed. Gather any needed information.

Step 3: Plan the design
- Create thumbnail sketches or a prototype for possible page designs.
- When you have decided on a sketch or prototype, determine the page size appropriate for your design.
- Create a style sheet for redesigning the documents.

Step 4: Write and revise the document
- If you are creating new documents, organize and write the documents.
- If you are revising existing documents, make sure the information is up to date and accurate.
- Make sure the documents follow the guidelines in Chapters 6, 7, and 8.

Step 5: Redesign (or create) the documents
- Implement the designs.
- Hand in your documents to your instructor. Attach copies of the original designs.

iStockphoto 2008.

CHAPTER ELEVEN

11

Creating Visual Information

You've probably heard the expression "a picture is worth a thousand words." You can often convey information in technical documents more effectively and efficiently with pictures than with words. Pictures, or graphics, can explain abstract ideas or summarize concepts that may be difficult for readers to understand with only words. Visual information can help you achieve your purpose in a number of ways:

- Visual information—graphics—can support and supplement the text. Graphics are especially helpful to readers who are unfamiliar with concepts or who want to gather information at a glance.
- Graphics can summarize the information in the text and present the information in a different way to help readers understand it.
- Graphics can help readers understand how something works or how to do something.
- Graphics can present some types of information more quickly and efficiently than words. For instance, a map can more efficiently convey the locations of coral reefs than can words. Pictures are also more effective for showing readers emergency information. In a fire evacuation, a building diagram with arrows marking the exits is more effective than a paragraph describing the location of the exits.

Your readers are bombarded with visual information through television, advertisements, and the Web. Because readers are so accustomed to receiving information visually, they will often respond best to documents that use not only words, but also graphics to convey information.

In most documents, you can't rely solely on graphics to communicate information. For example, in Figure 11.1, the writers rely heavily on pictures to tell readers how to tie a Windsor knot. Brief step-by-step instructions for these procedures are included next to the graphics to provide additional information. The graphics alone convey the primary message, but the written instructions provide more detailed information.

To balance the visual and textual information, consider the needs of your readers and what you want to communicate. This chapter will help you choose the most appropriate graphics for readers and strike the proper balance between visual and textual information.

WHY USE VISUAL INFORMATION?

Effective visual information can help you convey part or all of a message. Visual information helps you do the following:
- show how to follow instructions or visualize a process
- show what something looks like
- show and summarize relationships among data
- emphasize and reinforce information
- show how something is organized
- simplify complex concepts, discussions, processes, or data
- add visual interest

Show How to Follow Instructions or Visualize a Process

Graphics are excellent tools for giving readers instructions or helping them visualize a process. Instructions without graphics are often hard to follow. For example, imagine trying to learn to tie a Windsor knot for the first time without a visual demonstration or graphics. For someone who has never tied a tie, the process must be visualized. Without graphics, most of us would not be able to tie the knot; however, with drawings, as in Figure 11.1, you can learn to tie the knot and visualize the process.

FIGURE 11.1

Graphics Help Readers Visualize a Process

The Half Windsor

YOUR MIRROR REFLECTION

1
Start with wide end of tie on your right
and extend a foot below narrow end.

2
Cross wide end over narrow end and
back underneath.

3
Bring up and turn down through loop.

4
Pass wide end around front from
left to right.

5
Then, up through the loop, and...

6
...down through knot in front.
Tighten carefully and draw up to collar.

Source: Taken from the *How to Tie a Tie* brochure, distributed by The Men's Wearhouse, Inc.
Permission granted by Kim Owens via email June 5, 2009.

Show What Something Looks Like

Graphics, such as photographs and drawings, are excellent tools for helping readers to see what something looks like. Often, you can help readers visualize a concept, theory, or object by including a graphic. For example, the photograph in Figure 11.2 shows readers what a coral reef looks like.

Show and Summarize Relationships among Data

For some readers, you may want to display numerical data or show how one set of data relates to another. Perhaps you want to show the results of a laboratory test, a survey, a trend, or changes over time. Graphics help readers quickly see the relationships among data.

You can use several types of graphics to show relationships among numerical data. Figure 11.3 is a line graph that compares the change in gas prices from May 2007 to May 2008, to the change in prices during the same period the following year. The same numbers presented in a paragraph would

FIGURE 11.2

Photograph that Shows What Something Looks Like

iStockphoto 2008.

not adequately convey the scope of the change. Figure 11.4 is a table that shows a relationship between mined acres of land and reclaimed acres of land. The table also summarizes by including the totals of these areas.

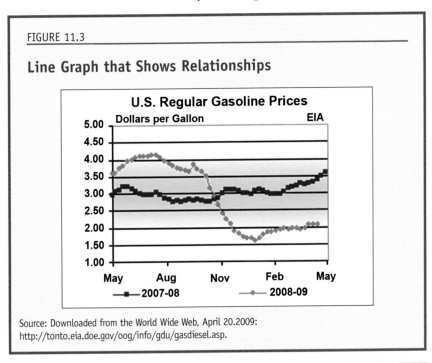

FIGURE 11.3

Line Graph that Shows Relationships

Source: Downloaded from the World Wide Web, April 20.2009: http://tonto.eia.doe.gov/oog/info/gdu/gasdiesel.asp.

FIGURE 11.4

Table that Shows Relationships

Mining and Reclamation

Land Mined and Reclaimed	Oh My Mine	Darling Mine	Clementine Mine	Total (in acres)
Mined in 2007	210	745	643	1,598
Mined Since 1990	13,465	20,442	13,456	47,363
Reclaimed in 2007	268	1875	897	3,040
Reclaimed Since 1990	15,601	21,465	14,575	51,641

Emphasize and Reinforce Information

You can use any type of graphic to emphasize information presented in the text. Your choice will depend on your objectives and the information you want to emphasize or reinforce. For example, if you want to emphasize the findings of a series of tests on airbags, you might first discuss the data in a paragraph and then present the data in a horizontal bar graph or a line graph to visually reinforce the discussion. You might also display the data in a table and then reinforce the data in a bar graph or line graph. Let's look at a specific example: to reinforce the concept that investing even small amounts of money over time can help individuals reach their savings goals, an investment company might use the table in Figure 11.5, below.

Show How Something is Organized

Readers may need to know how something is organized, but they may have trouble understanding textual descriptions of an organizational structure. Graphics can make the organization clear. For example, the Web site for the National Credit Union Administration includes a chart showing how the administration is organized (see Figure 11.6). This chart quickly identifies the three offices that the board and the chair oversee: the Office of Inspector General, the Executive Director, and the General Counsel. The chart also shows that the Office of the Executive Director

FIGURE 11.5

Table that Reinforces Information

When you contribute monthly[1]	When you contribute annually	In 5 years, you could have[2]	In 10 years, you could have	In 20 years, you could have	In 30 years, you could have
50	600	3,698	9,208	29,647	75,015
100	1,200	7,397	18,417	59,295	150,030
150	1,800	11,095	27,625	88,942	225,044

[1] All amounts are in dollars
[2] Assumes an 8% annual return, compounded monthly

oversees two types of offices: central and regional. Under regional offices, the chart includes a U.S. map showing the regions. Notice also how the chart is designed to fit on the screen of a small, as well as a large, monitor. The chart uses color to show that all the central offices belong in the same group and are equal in the hierarchy. Traditional organizational charts use a horizontal layout to show equal rank; however, the chart wouldn't fit on a screen if the central offices appeared horizontally rather than vertically.

FIGURE 11.6

Chart that Shows How Something is Organized

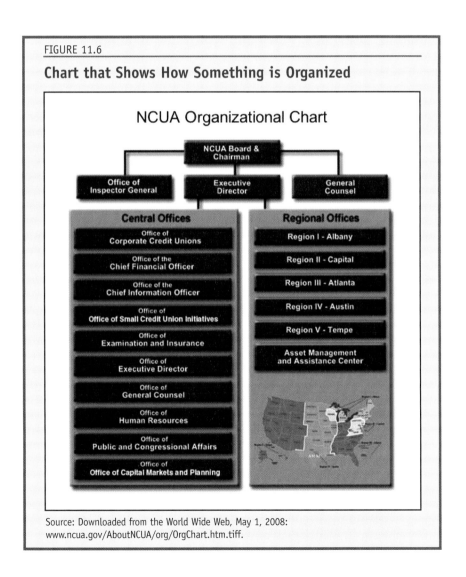

Source: Downloaded from the World Wide Web, May 1, 2008:
www.ncua.gov/AboutNCUA/org/OrgChart.htm.tiff.

Simplify Complex Concepts, Discussions, Processes, or Data

Readers may have difficulty understanding and analyzing complex information presented in words. When complex information is presented visually, readers can more quickly and easily understand that information. For example, readers need a graphic to understand how sea level projections for the future compare with sea levels of the past. Figure 11.7 simplifies information that would be difficult to follow without a graphic. The graph makes the information easily recognizable as an upward trend.

Add Visual Interest to a Document

When used appropriately, graphics such as photographs and pictographs make your documents more visually appealing. For example, when scientists discuss volcanic eruptions, they use graphics—especially photographs—to show the eruptions and lava flows. To add interest to their dis-

FIGURE 11.7

Graph that Simplifies Complex Information

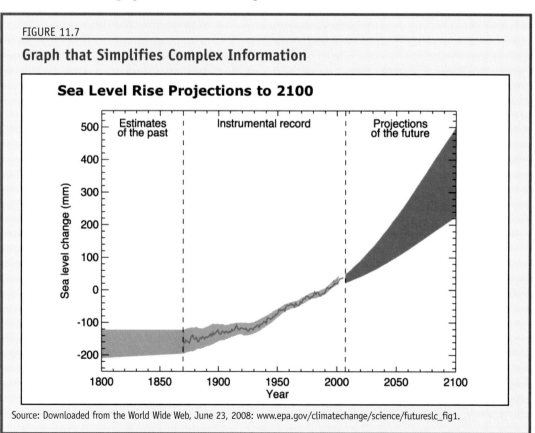

cussion of these volcanoes, the scientists might include a photograph of a volcano erupting like the one in Figure 11.8, below.

Some graphics not only add interest but also convey important information. For example, Figure 11.9, page 312, compares life spans of various animals. This pictograph clearly adds interest to the document, but it also informs.

PLAN YOUR VISUAL INFORMATION

To plan the visual information in your documents, consider three elements: your purpose, your readers, and the most effective format for the information. As you plan, answer these questions:

FIGURE 11.8

Photograph that Adds Interest

iStockphoto 2008.

- **Will visual information help you to achieve your purpose?** For example, if your purpose is to show what something looks like, a visual is essential to achieving your purpose.
- **Who are my readers and will they need or expect information to be presented visually?** For example, expert readers may expect more detailed graphics such as numerical tables or complex drawings, whereas general readers may prefer a less-detailed graphic where the key points are extracted and easily viewed.
- **What types of graphics are appropriate for the information and the readers?** For example, how will your readers' culture or language affect the types of graphics you select? Can you best present the information

FIGURE 11.9

Graphic that Adds Visual Interest

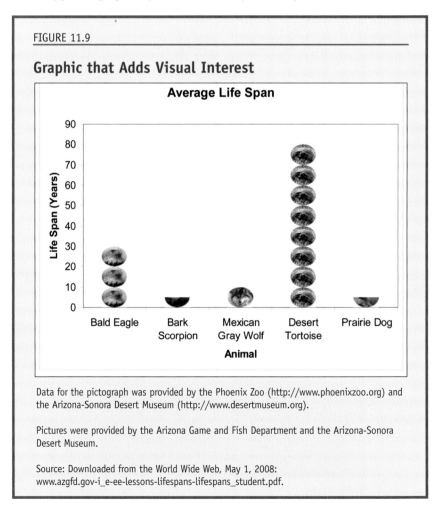

Data for the pictograph was provided by the Phoenix Zoo (http://www.phoenixzoo.org) and the Arizona-Sonora Desert Museum (http://www.desertmuseum.org).

Pictures were provided by the Arizona Game and Fish Department and the Arizona-Sonora Desert Museum.

Source: Downloaded from the World Wide Web, May 1, 2008: www.azgfd.gov-i_e-ee-lessons-lifespans-lifespans_student.pdf.

in a drawing or a photograph? Will the readers' level of knowledge about the topic affect the types of graphics you use?

Plan for the graphics early in the writing and designing process. Think about the graphics as you decide what information to include in the document. If you wait too long, you may not have the time or resources to create the graphics you need, or you may find that adding graphics will require reformatting the document.

SELECT THE APPROPRIATE GRAPHIC

To determine the appropriate graphic, consider what you are trying to illustrate with the graphic. Ask yourself, "What is the purpose of the graphic?" You can use graphics for the following purposes:
- to illustrate quantitative information
- to show relationships
- to illustrate instructions and processes
- to show what something looks like

Figure 11.10, page 314, summarizes the types of graphics and the most appropriate information for each type.

Using Graphics to Illustrate Quantitative Information

You will primarily use the following graphics to illustrate quantitative information: bar graphs, line graphs, pictographs, pie charts, and tables.

Bar Graphs

Bar graphs compare and show relationships among numerical data. Bar graphs display approximate values, not specific values and allow readers to see relationships and trends at a glance. You can orient bar graphs vertically or horizontally; you can use single bars, stacked bars, or multiple bars. For tips on creating bar graphs, see page 316. Bar graphs are excellent choices when you want to do the following:
- **Compare values**. You can use a simple bar graph to compare values. For example, Figure 11.11, page 315, shows the percentage of revenue possible from five fundraising programs.

FIGURE 11.10

Selecting the Appropriate Graphic

Purpose of the Graphic	Type of Graphic	Best Use of the Graphic
Illustrate quantitative (numerical) information	Bar graphs	• show comparisons of approximate values • show relationships among data
	Line graphs	• show trends (changes) over time, cost, or other variables
	Pie charts	• show the relationship of the parts to a whole
	Pictographs	• use icons or pictures to depict statistical information for general readers
	Tables	• summarize and categorize large amounts of numerical information
	Combinations of graphics—especially bar and line graphs	• summarize complex data for expert readers
Show relationships of qualitative (not numerical) information	Organizational charts	• show the hierarchy in an organization or company
	Diagrams	• show a sequence of events
	Tables	• show relationships and summarize data
Show instructions and processes	Flow charts	• explain a process or a sequence of events or steps
	Tables	• organize information; show quantities; indicate troubleshooting and frequently-asked questions with answers
	Line drawings	• show a realistic, but simplified view of what something looks like
	Diagrams	• demonstrate how to do something or show where something is located to complete a task or understand a process

FIGURE 11.10 CONTINUED

Purpose of the Graphic	Type of Graphic	Best Use of the Graphic
Show what something looks like 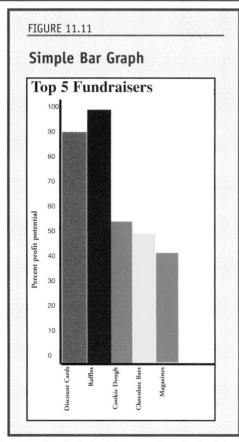	Line drawings	• show a representation of what something will or does look like
	Photographs	• give a realistic picture; show exactly what something looks like
	Maps	• show where something is located
	Screen shots	• show what appears on a computer monitor

iStockphoto 2008.

FIGURE 11.11

Simple Bar Graph

Top 5 Fundraisers

Percent profit potential

- Discount Cards — 90
- Raffles — 99
- Cookie Dough — 54
- Chocolate Bars — 49
- Magazines — 41

TIPS FOR CREATING BAR GRAPHS

- **Use an appropriate scale**. Extend the longest bar nearly to the end of its parallel axis, as in Figure 11.11, page 315. Make sure the scale appropriately and ethically conveys the differences in values. See "Presenting Visual Information Ethically," page 345, for more information on appropriate scale.
- **Begin the scale at 0 if possible to ensure that bars accurately represent values**.
- **Make all the bars the same width**—unless you overlap them.
- **Make the space consistent between bars**.
- **Label the bars**. Label each bar at its base. For multiple- or divided-bar graphs, you can include a key to indicate what the bars or divisions represent, as in Figure 11.12.
- **Put tick (or hash) marks at regular intervals on the appropriate axis**. The tick marks should indicate quantities, such as percentages or amounts of money.
- **Use a different pattern or color for each bar in a divided or multiple bar graph**.
- **Cite the source of your data below the graph**. If you do not generate the data yourself, cite the source of the information. If you use a graph from another source, get written permission for using that graph and cite the source below the graph.

- **Show the values that make up a total**. For example, Figure 11.12 shows the types of retirement investments by age of the investor. Each bar is subdivided to show the percentage invested in stocks, bonds, and cash by age group.
- **Show two or more relationships**. You can show two or more relationships with a multiple-bar graph. The multiple bar graph in Figure 11.13A shows where U.S. citizens get information about scientific issues.
- **Show trends over time**. Figure 11.13B shows how U.S. citizens have changed where they seek scientific information since 2001. For example, the bar graph shows that in 2001, the number of U.S. citizens using books as a source of information about specific scientific issues was considerably higher than it was in 2004 or 2006.
- **Show positive and negative values**. With some data, you may need to indicate both positive and negative values. You can do that with a deviation bar graph like the one in Figure 11.14, page 319.

Line Graphs

Line graphs show relationships among data with more precision than bar graphs. Like bar graphs, line graphs use a horizontal and a vertical axis; but line graphs use lines and sometimes bands instead of bars to indicate relationships. Bar graphs emphasize quantity, and line graphs emphasize changes and trends. Line graphs are especially effective when you want to do either of the following:

FIGURE 11.12

Divided Bar Graph

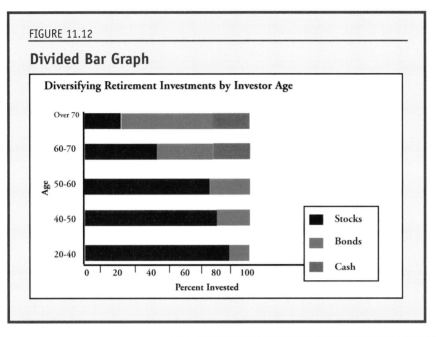

FIGURE 11.13A & B

Multiple-Bar Graph

A

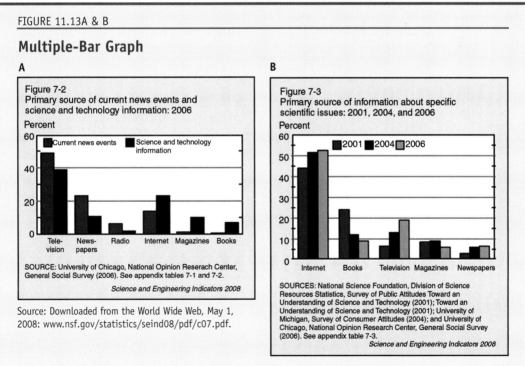

Figure 7-2
Primary source of current news events and science and technology information: 2006

SOURCE: University of Chicago, National Opinion Reserach Center, General Social Survey (2006). See appendix tables 7-1 and 7-2.

Science and Engineering Indicators 2008

Source: Downloaded from the World Wide Web, May 1, 2008: www.nsf.gov/statistics/seind08/pdf/c07.pdf.

B

Figure 7-3
Primary source of information about specific scientific issues: 2001, 2004, and 2006

SOURCES: National Science Foundation, Division of Science Resources Statistics, Survey of Public Attitudes Toward an Understanding of Science and Technology (2001); Toward an Understanding of Science and Technology (2001); University of Michigan, Survey of Consumer Attitudes (2004); and University of Chicago, National Opinion Research Center, General Social Survey (2006). See appendix table 7-3.

Science and Engineering Indicators 2008

- **Show changes in quantity over time**. Figures 11.15A and B show that tables can present the same numerical data as line graphs; but, as Figure 11.15A shows, readers can identify trends from a line graph more easily than from a table of numbers. The table in Figure 11.15B contains the same numerical data as the line graph in Figure 11.15A, but the trends are much easier to spot in the line graph.
- **Compare several variables simultaneously**. You can use three or four lines on a line graph to compare several variables at once.

TIPS FOR CREATING LINE GRAPHS

- **When time is a variable, put it on the horizontal axis**.
- **Place tick marks at regular intervals on each axis**. Use the appropriate scale for each interval. Generally, make tick marks short; longer tick marks add clutter.
- **Use grid lines when readers need to see exact quantities**.
- **Begin the vertical axis with zero**. If it doesn't begin with zero, use breaks to show your readers that the axis begins some place other than zero.
- **Label each axis**. Most readers prefer labels centered along each axis.
- **Make the lines distinct** with color or symbols.

Grid lines

- **Cite the source for any data you did not generate**.

FIGURE 11.14

Deviation Bar Graph

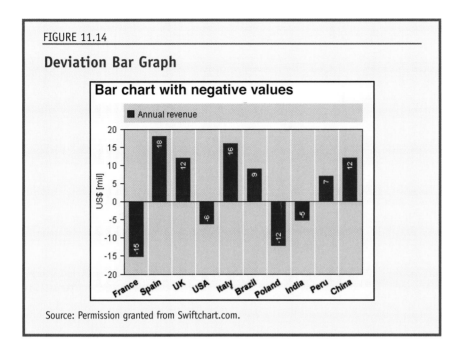

Source: Permission granted from Swiftchart.com.

FIGURE 11.15A & B

The Same Data Presented in Different Graphics

A

FDA/FSIS Food Safety Survey. Eating Potentially Risky Foods: Percent of US Population Who Ate Each Food

B

FDA/FSIS Food Safety Survey. Eating Potentially Risky Foods: Percent of US Population Who Ate Each Food

	Raw Clam	Raw Oyster	Raw Fish	Raw Egg	Steak Tartar	Pink Hamburger
1993	8 percent	16 percent	9 percent	53 percent	6 percent	24 percent
1998	6/*4 percent	12/*8 percent	10 percent	39 percent	3/*4 percent	16 percent
2001	6 percent	12 percent	15 percent	42 percent	4 percent	16 percent

* Note: For three of the foods--raw clams, raw oysters, and steak tartar--there are two numbers reported for 1998. This is because the word ings of these questions changed between 1993 and 2001 and were asked both ways in 1998. The first number is the 1993-1998 comparison; the second number is the 1998-2001 comparison.

Source: Downloaded from the World Wide Web, May 1, 2008: www.cfsan.fda.gov/~dms/fssurvey.html. Chart 2: FDA/FSIS Food Safety Survey. "Eating Potentially Risky Foods: Percent of US Population Who Ate Each Food."

Pictographs

Pictographs are similar to bar graphs but use pictures or drawings instead of bars to depict statistical information. For example, in Figure 11.16, a segmented gas pump replaces bars. As in bar graphs, the measurements in pictographs may not be as exact as those in line graphs. Many pictographs lack a visible vertical or horizontal axis or tick marks, as in Figure 11.16 below. Pictographs make your document visually interesting. As you create pictographs, follow the tips on page 321.

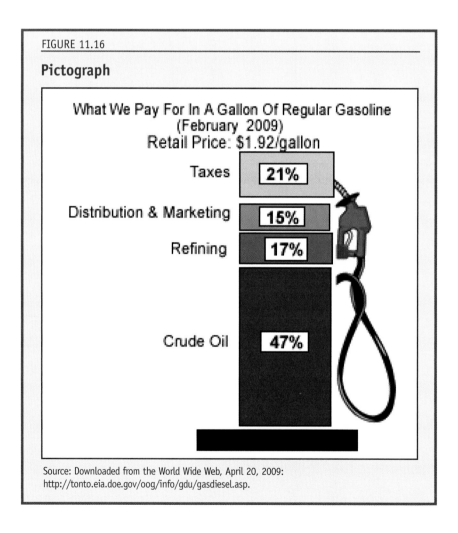

FIGURE 11.16

Pictograph

What We Pay For In A Gallon Of Regular Gasoline
(February 2009)
Retail Price: $1.92/gallon

Taxes — 21%

Distribution & Marketing — 15%

Refining — 17%

Crude Oil — 47%

Source: Downloaded from the World Wide Web, April 20, 2009:
http://tonto.eia.doe.gov/oog/info/gdu/gasdiesel.asp.

TIPS FOR CREATING PICTOGRAPHS

- **Use pictures and drawings that are meaningful and appropriate to the readers, the tone, and the purpose.** Create pictographs that fit your purpose and your readers' expectations. Pictographs should fit the tone of your document. For example, if you are preparing a proposal to the U.S. Department of Defense for building a new aircraft carrier, a pictograph would be inappropriate not only for the purpose, but also for the formal tone of the document.
- **Use drawings rather than photographs.** Photographs contain too much detail and are too realistic for most pictographs.
- **Use color to enhance pictographs.** Color makes pictographs more visually interesting.
- **Label pictographs.** Even though pictographs are less exact than other graphics, they are not merely decorative. Pictographs need appropriate, readable labels, as in Figure 11.16, page 320.
- **Use pictographs primarily for general readers.** If you want to use pictographs for experts or decision-makers, use them only in oral presentations or for less-formal situations. Make sure that the pictograph is appropriate for the occasion of the oral presentation.
- **Cite the source for any data you did not generate.**

Pie Charts

Pie charts are circles divided into wedges—like pieces of pie. Each wedge represents a part of the whole. The pie chart shown in Figure 11.17, page 322, demonstrates how a county spends tax dollars. You can effectively use pie charts to support oral presentations—especially for general readers— and to summarize information in a spreadsheet.

You can easily create pie charts with graphics software. You can make them three-dimensional, or you can rotate them for a professional look. As you create pie charts, follow the tips on page 323.

Tables

Tables present quantitative (numerical) information arranged in columns and rows. With tables, you can present dense quantitative information in a format that readers can quickly read and understand. To create a table, put the information into vertical columns topped with appropriate headings, as in the table shown in Figure 11.18, below. As you create tables, follow the tips on page 324.

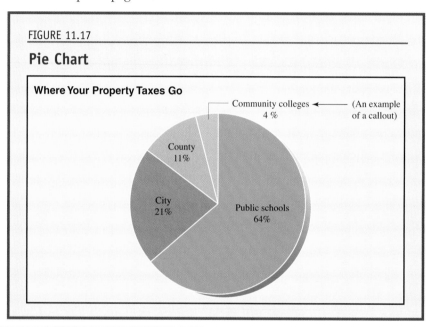

FIGURE 11.17

Pie Chart

Where Your Property Taxes Go

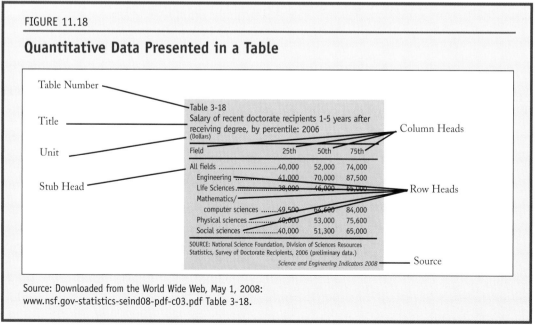

FIGURE 11.18

Quantitative Data Presented in a Table

Table 3-18
Salary of recent doctorate recipients 1-5 years after
receiving degree, by percentile: 2006
(Dollars)

Field	25th	50th	75th
All fields	40,000	52,000	74,000
Engineering	41,000	70,000	87,500
Life Sciences	38,000	46,000	65,000
Mathematics/ computer sciences	49,500	64,600	84,000
Physical sciences	40,000	53,000	75,600
Social sciences	40,000	51,300	65,000

SOURCE: National Science Foundation, Division of Sciences Resources
Statistics, Survey of Doctorate Recipients, 2006 (preliminary data.)

Science and Engineering Indicators 2008

Source: Downloaded from the World Wide Web, May 1, 2008:
www.nsf.gov-statistics-seind08-pdf-c03.pdf Table 3-18.

TIPS FOR CREATING PIE CHARTS

- **Label each wedge and place the labels inside the wedge if possible.** If the labels won't fit inside the wedge, use *callouts*—lines drawn out to each label. Depending on the graphics software, you may be able to pull out or explode some of the small wedges so that the labels will fit inside them.
- **Place the labels horizontally inside the wedge, not diagonally.** In Figure 11.17, page 322, notice how the words "Public schools," "City," and "County" are placed horizontally inside the wedges.
- **Sequence the wedges from the largest to the smallest**. Place the largest wedge in the 12-o'clock position, and move to the smallest wedge as you work around the "clock," as in Figure 11.17.
- **Make sure that the wedges add up to 100**.
- **Use a contrasting color to emphasize one section**. For example, in Figure 11.17, the "Public schools" wedge contrasts with the lighter-colored wedges.
- **Use color to make pie charts visually interesting and to differentiate wedges**. For less-formal documents, you can use photographs and drawings to make the pie chart more visually interesting.
- **Cite the source for any data you did not generate.**

Combined Graphics

You may have two purposes for presenting visual information; for example, you may want to contrast the information while also showing the trends. One type of graphic may not accomplish both of your goals, so you may want to use two types of graphics together. If you decide to use this approach, make sure

- your readers can understand both graphics
- the graphic remains uncluttered
- the relationship between the graphics is clear
- labels are included to clearly identify the information

Figure 11.19, page 325, shows a graphic that combines a line graph with a bar graph.

TIPS FOR CREATING TABLES

- **Use software to create tables.** Word-processing or spreadsheet software can create professional-looking tables and save time by totaling (summing) your data.
- **Put the title and table number at the top.** Readers view tables from the top down.
- **Label the stub, column heads, and row heads to orient your readers**. See Figure 11.18, page 322, to locate the stub, column heads, and row heads.
- **Label all units of measure.** In Figure 11.18, the unit (dollars) is labeled in the title. You can also label the units in the column and row heads. If all of the data is in the same unit of measure, use the title to label the unit. If the data in the columns varies, label the unit in the column heads. For example, in a table summarizing data related to reclaiming land from mining operations, the columns would have different units:

Amount spent reclaiming (in millions of dollars)	**Total area reclaimed (in acres)**

- **Use "X," "NA," or a dash to indicate omitted data or data not available.** "NA" indicates "not available."
- **Align the numbers and words correctly**. Vertically align columns of numerical data at the right or on the decimal points. Align words to the left.

Right-aligned column	**Decimal-aligned column**
7	0.23
11,890	203.78
789	33.90
8,900	3.00

- **Use horizontal rules to separate the column heads from the data.** Rules help readers locate column heads. In a simple table (like Figure 11.19), use few rules; too many rules can make the table hard to read.
- **Use shading to help readers distinguish column subheads.** Graphics and word-processing software offer options for shading or coloring columns and rows.
- **Use footnotes for information you did not generate.** Use letters instead of numbers for table footnotes if numbers could confuse readers (14^b instead of 14^2). Readers can mistake footnote numbers for mathematical notation. For example, 14^2 could mean "14 squared" rather than "footnote 2."
- **Check the data**. Make sure the data is accurate. Double-check your math and make sure you have entered the correct data.

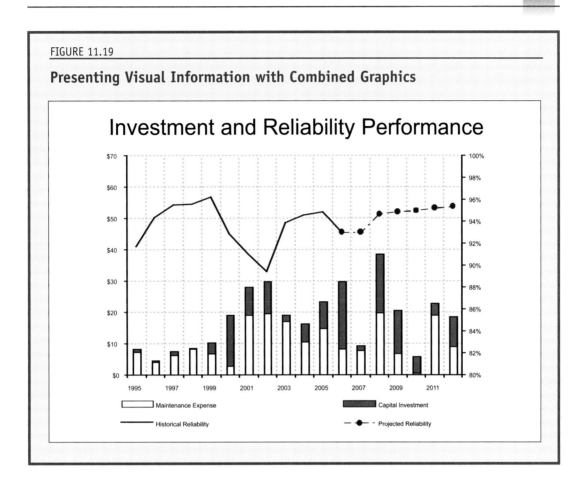

FIGURE 11.19

Presenting Visual Information with Combined Graphics

Investment and Reliability Performance

Maintenance Expense · Capital Investment · Historical Reliability · Projected Reliability

Use Graphics to Illustrate Qualitative Information

When you want to illustrate qualitative—non-numerical—information, you will use primarily organizational charts, diagrams, and tables.

Organizational Charts

Organizational charts are an efficient and clear way to show how something is organized and to help readers understand logical relationships. You are probably most familiar with organizational charts for a company or an organization. You can also use organizational charts to show the divisions of a system. Organizational charts are important because without these charts, people may not understand the lines of responsibility in an organization.

The chart reproduced in Figure 11.20 below shows the organization and lines of responsibility for a university. The person or department with the most responsibility is at the top, and those with the least responsibility are at the bottom. Figure 11.20 indicates that the Board of Regents and President have the highest responsibility and that the heads of Payroll, Public Relations, and the College of Sciences, for example, have less responsibility. Note the use of color to help readers distinguish the various divisions. The chart also uses titles instead of names so it will be relevant longer.

Diagrams

Diagrams are an excellent choice for showing relationships or a sequence of events or actions. For example, the diagrams in Figure 11.21, page 327, show the relationship between beach erosion and seawalls. As you create diagrams, follow the tips on page 327.

Tables without Numbers

Although writers most frequently use tables to display numerical data, tables are also effective for presenting information in words. Using a table, you can summarize quantitative information and show relationships among this information. Confronted with paragraphs of text, readers have to keep on reading until they find the information they need; in contrast, a table lets readers locate key words and information quickly. The table in Figure 11.22 summarizes the maintenance schedule for a car. To create an effective table, follow the tips on page 324.

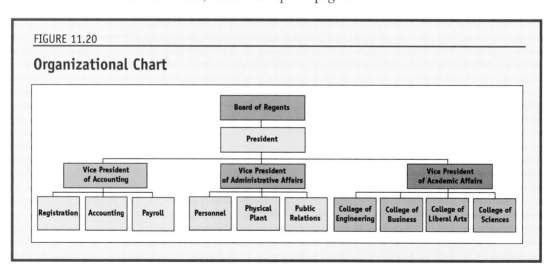

FIGURE 11.20

Organizational Chart

TIPS FOR CREATING DIAGRAMS

- **Sketch rough drafts of the diagram.** Try several drafts to determine exactly what you want in the diagram. Diagrams take a lot of time to create, so before you begin the final draft, make sure you have a clear idea of what information you want the diagram to communicate.
- **Label the diagram and explain the process.** Labels should be easy to read. You can place explanations in the diagram if they won't interfere or cause clutter. If the explanations will be more confusing than helpful, you can explain the diagram in a paragraph that precedes or follows it.
- **Use graphics or word-processing software to produce the diagram.**
- **Cite the source of any data that you did not generate.**

FIGURE 11.21

Diagrams

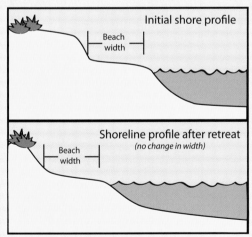

A beach undergoing net long-term retreat will maintain its natural width.

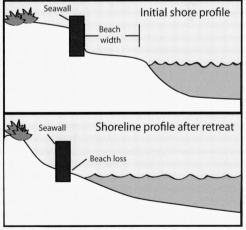

Beach loss eventually occurs in front of a seawall for a beach experiencing net long-term retreat.

Source: Downloaded from the World Wide Web, May 1, 2008:
www.mothernature-hawaii.com/images/beach%20erosion_diagram.

Use Graphics to Show Instructions and Processes

Graphics can help you illustrate a process or show how to do something. The graphics most commonly used for showing instructions and processes are flow charts and line drawings.

Flow Charts

You can use flow charts to explain a process or to show a sequence of steps or events. Flow charts are especially useful for explaining a complex process that has conditional (if/then) steps. Flow charts generally work

FIGURE 11.22

Table without Numbers

Service and Miles (Kilometers)	25,000 (41,500)	50,000 (83,000)	75,000 (125,000)	100,000 (166,000)	125,000 (207,500)	150,000 (240,000)
Inspect fuel system for damage or leaks.	•	•	•	•	•	•
Inspect exhaust system for loose or damaged components.	•	•	•	•	•	•
Replace engine air cleaner filter. See *Engine Air Cleaner/Filter on page 5-21.*		•		•		
Change automatic transmission fluid and filter (severe service only). *See footnote (n).*		•		•		•
CTS-V Only: Change 6-speed manual transmission fluids (severe service only). *See footnotes (l) and (m).*		•		•		
CTS-V Only: Change hydraulic clutch fluid (severe service only). *See footnote (l).*	•	•	•	•	•	•
CTS-V Only: Change rear axle fluid (severe service only). *See footnotes (l) and (m).*		•		•		•
CTS-V Only: Change brake fluid (severe service only). *See footnote (l).*	•	•	•	•	•	•

Source: Cadillac Motor Car Division, www.cadillac.com. *CTS Owner's Manual,* page 6-6.
Used with permission.

best for processes that have a definite beginning and a definite end (use diagrams for ongoing processes, such as recycling).

Flow charts usually consist of circles, rectangles, diamonds, and other geometric shapes that indicate the steps of a process or event. In some fields, various geometric shapes have specific meanings; and people in those fields understand that certain shapes represent specific outcomes and events. If the shapes that you use in your flow chart have specific meanings, make sure your readers understand what the shapes represent, or use a key (see Figure 11.23, page 330).

Line Drawings

Line drawings are excellent graphics for instructing and showing readers what something looks like. Many writers select drawings instead of photographs to help readers see how something is put together. Drawings can emphasize important details or parts that are not apparent in a photograph. Drawings also allow you to explode (make larger) a particular detail. Figures 11.24–11.26, pages 331-332, present drawings that help readers see details.

TIPS FOR CREATING FLOW CHARTS

- **Sketch a rough draft of the flow chart.** By creating a rough draft, you can make sure that the labels will fit inside the geometric shapes and that the chart is accurate.
- **Put all labels identifying a step or event inside the geometric shapes.** Labels can distract and possibly confuse readers if you place them outside the shapes. Make sure that the shapes are large enough to contain the labels, as in Figure 11.23, page 330.
- **Sequence the shapes so that the action flows from left to right or from top to bottom.** When the action flows from left to right and takes more than one line, begin the next line at the left margin.
- **Cite your sources if you did not generate the information.**

Use Graphics to Show What Something Looks Like

Graphics are essential for showing what something looks like. You will most commonly use photographs, line drawings, maps, and screenshots for this purpose. (For information on creating line drawings, see page 330.)

Photographs

Photographs are excellent graphics when you want to do any of the following:

- **Show what something looks like.** Often, words are not enough to help readers know what something looks like, especially if they have never

TIPS FOR CREATING LINE DRAWINGS

- **Give your drawings a professional look**. Try using software to create your drawings. If you plan to use software, allow time to learn the software. If you plan to create your drawings by hand, make sure that you have the tools and training to create drawings with a professional appearance.
- **Render your drawing from the same angle that readers will have when they work with or observe the object in the drawing**. Figures 11.24–11.26, pages 331-332, illustrate three angles or vantage points: a cross section, an exploded view, and a cutaway.
- **When appropriate, make the feature or detail that you want to emphasize larger than it really is**. The large size helps emphasize the feature or detail, as in Figures 11.24, page 331, and 11.27 on page 333.

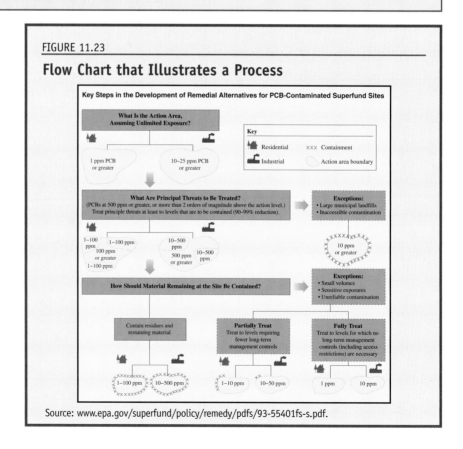

FIGURE 11.23

Flow Chart that Illustrates a Process

Source: www.epa.gov/superfund/policy/remedy/pdfs/93-55401fs-s.pdf.

FIGURE 11.24

Drawing Showing a Cross Section

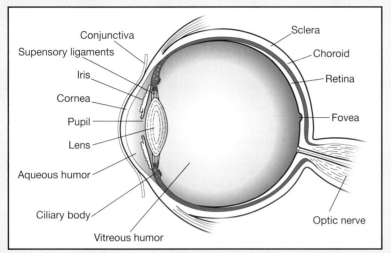

Conjunctiva

Supensory ligaments

Iris

Cornea

Pupil

Lens

Aqueous humor

Ciliary body

Vitreous humor

Sclera

Choroid

Retina

Fovea

Optic nerve

Source: Downloaded from the World Wide Web, November 2008: www.imagedatabase. benttreepress.com/db.bioart.asp. Bent Tree Press Biology Lab Database.

FIGURE 11.25

Drawing Showing an Exploded View

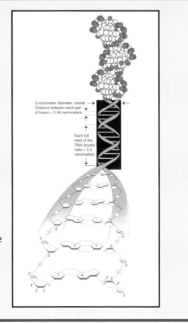

2-nanometer diameter, overall
Distance between each pair of bases = 0.34 nanometers

Each full twist of the DNA double helix = 3.4 nanometers

Source: Downloaded from the World Wide Web, November 2008: www.imagedatabase.benttreepress. com/db.bioart.asp. Bent Tree Press Biology Lab Database.

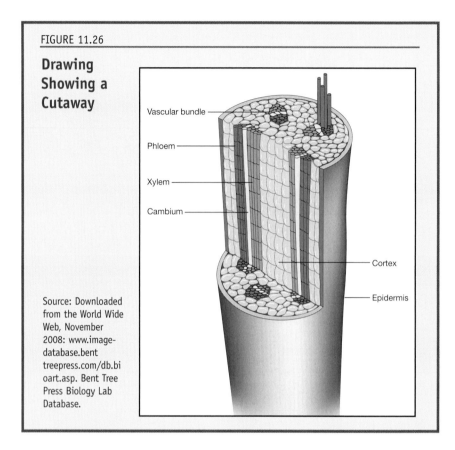

FIGURE 11.26

Drawing Showing a Cutaway

Vascular bundle

Phloem

Xylem

Cambium

Cortex

Epidermis

seen what you are describing. For example, if you are describing a piece of equipment that readers may need for an experiment, a photograph is an effective tool. If you are cataloging various types of sea life, you might use photographs (see Figure 11.27, page 333). Guides to various types of sea life in oceans around the world contain pictures of each type of sea life discussed. Photographs are the only practical means of helping readers to recognize the sea life.

- **Show where something is located**. You may want to show readers where something is located on a machine, a piece of equipment, and so on. For example, a car manufacturer uses a photograph with labels to help readers locate and identify the lights, gauges, and warning indicators (see Figure 11.28, page 333).
- **Show how something is done**. For instance, you might use a photograph to show readers how to plant seeds (see Figure 11.29, page 334).

FIGURE 11.27

Photograph that Shows What Something Looks Like

iStockphoto 2008.

FIGURE 11.28

Photograph that Shows Where Something is Located

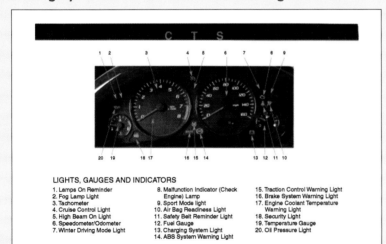

LIGHTS, GAUGES AND INDICATORS

1. Lamps On Reminder
2. Fog Lamp Light
3. Tachometer
4. Cruise Control Light
5. High Beam On Light
6. Speedometer/Odometer
7. Winter Driving Mode Light
8. Malfunction Indicator (Check Engine) Lamp
9. Sport Mode light
10. Air Bag Readiness Light
11. Safety Belt Reminder Light
12. Fuel Gauge
13. Charging System Light
14. ABS System Warning Light
15. Traction Control Warning Light
16. Brake System Warning Light
17. Engine Coolant Temperature Warning Light
18. Security Light
19. Temperature Gauge
20. Oil Pressure Light

Source: Cadillac Motor Car Division, www.cadillac.com. *CTS Owner's Manual.*
Used with permission.

FIGURE 11.29

Photograph of How Something is Done

iStockphoto 2008.

FIGURE 11.30

Map

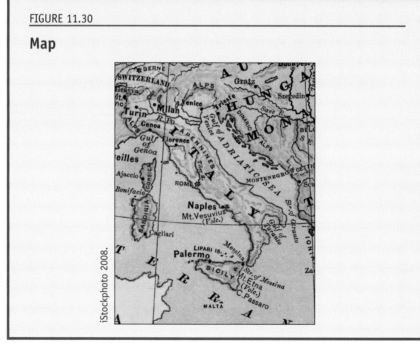

iStockphoto 2008.

TIPS FOR CREATING EFFECTIVE PHOTOGRAPHS

- **Eliminate unnecessary detail and clutter**. Show only what you want readers to see.
- **Use an appropriate angle**. Take photographs from the angle at which readers will actually view the object.
- **Crop the photo to focus on the information that you want the reader to see**. For example, see Figure 11.31, below. In the photo on the left, the flower is off center and too small; however, in the photo on the right, the focus is on the flower because it is larger and framed within the image boundaries.
- **Use software to edit your photographs**. You can use software to size your photographs or to eliminate distracting elements from your photographs.
- **Do not unethically manipulate photographs**. You can ethically crop a photo to eliminate excess background or to draw attention to a particular detail; however, if you airbrush a photograph to take out essential detail, you are unethically manipulating that photograph.
- **Cite the source of your photograph and get permission to use it if you did not take the photograph**.

FIGURE 11.31

Cropping Photos Effectively

iStockphoto 2008.

Maps

You can easily create maps with graphics software or find ready-to-use maps. When you create maps, use conventional colors such as blue for water. Figure 11.30, page 334, shows a simple map.

Screen Shots

A ***screen shot*** is a picture—"snapshot"—of what appears on a computer monitor (screen). Screen shots help readers who are using computer software or who are learning to use that software. The screen shot shows readers what the screen looks like as they use the software. To create a screen shot, use the Print Screen function of your computer or graphics software. Figure 11.32 shows a screen shot.

GIVE YOUR GRAPHICS A PROFESSIONAL APPEARANCE

When you have determined where visual information will help you achieve your purpose and you have determined the most appropriate graphic, you will want to give the graphics a professional appearance. You will create a professional appearance by doing the following:

- use simple, uncluttered graphics
- integrate the graphics into the text
- use software and downloadable graphics when possible and appropriate

FIGURE 11.32

Screen Shot

Source: Downloaded from the World Wide Web, June 23, 2008: www.irs.gov.

TAKING IT INTO THE WORKPLACE
Using Visual Information to Communicate with Intercultural Readers

Companies are increasingly using graphics because of the globalization of markets and the more widely used graphical user interfaces (Bosley). Graphics have these advantages when you are communicating with intercultural readers:

- Graphics can "fit into space too small for text" (Bosley) and can "reduce the size and number of editions [versions] of documents" (Horton, "The Almost Universal Language").
- Graphics can help a reader learn because when combined with text they are clearer than text alone. Readers find it "easier to see and understand than to see, translate, and then understand" (Horton, "The Almost Universal Language").
- Graphics can improve reader comprehension (Horton, "The Almost Universal Language").
- Graphics can replace some technical terms that readers can't easily understand (Bosley).

iStockphoto 2008.

Assignment

Find a graphic that an international company has used for its readers. The company must market its products or services internationally. After you find the graphic, answer these questions in an e-mail to your instructor:

- Does the graphic have a neutral look? Explain your answer.
- Is the graphic simple? Does it use too many words? Explain your answer.
- Does the graphic make the concept, process, instruction, etc., easier to understand? If so, how? If not, how would you improve the graphic?

Use Simple, Uncluttered Graphics

Your graphics will be effective if your readers can quickly and easily understand them. Cluttered graphics have too much information for the space or contain unnecessary detail. Your graphics will also appear cluttered if you try to crowd them into a tight space. To create simple, uncluttered graphics, do the following:

- **Include only the information your readers need**. Don't clutter graphics with unnecessary information or visual details.
- **Create two or more graphics if you have too much information for one graphic**. If you use two graphics, each one should serve a purpose and enhance your document.
- **Use diagrams and drawings to eliminate unnecessary detail**. Photographs often have too much detail and clutter.
- **Exclude distracting visual information in photographs**. Compare the photographs in Figure 11.33A and B on the next page. Photograph A is ineffective because it includes distracting information. Photograph B is more effective because the distracting information is eliminated.
- **Don't crowd graphics into tight spaces.**

Integrate Graphics into the Text

When you have designed and created your graphic, integrate it into the text of your document. To effectively integrate graphics into the text, do the following:

- Give each graphic a number and a title.
- Introduce and refer to each graphic by number.
- Tell readers what is important about each graphic.
- Place each graphic as close as possible to its text discussion.

Give Each Graphic a Number and a Title

Include a number (such as "Table 2.15" or "Figure 3.4") and a title with every graphic. Numbers help readers locate graphics. Titles, or captions, identify the information in the graphic. The titles should be brief yet informative phrases that describe the content of the graphic. Compare these titles:

Vague title   Figure 6. A Figure Showing Inflation
Specific title   Figure 6. A Comparison of Inflation Rates from 1998 to 2008

The vague title needlessly repeats "Figure." The specific title uses "Comparison" to indicate what the graphic shows about inflation. The specific title also identifies the years that the graphic covers and gives

FIGURE 11.33A & B

Comparing Backgrounds in Photographs

A

iStockphoto 2008.

B

iStockphoto 2008.

readers the information they need to locate a specific graphic and identify the information in it. When numbering your graphics, follow these guidelines:

- **Number graphics consecutively within each document**.
- **If you divide your document into chapters or sections, give each figure a two-part number: chapter number first and figure number second**. For example, you would number the graphics in Chapter 2 as 2.1, 2.2, 2.3, and so on; and those in Chapter 3 as 3.1, 3.2, 3.3, and so on.
- **Put the number and title where the reader would begin viewing the graphic**. For example, readers generally read bar graphs from the bottom

to the top, so the number and title should appear at the bottom. However, readers read tables from the top to the bottom, so the number and title should appear at the top.

- **Number tables separately from the rest of the graphics**. All graphics except tables are called figures.

Introduce and Refer to Each Graphic by Number

Introduce and refer to graphics by number. Many readers will only know that you want them to look at a graphic or what you expect them to learn from the graphic if you refer them to it. You can introduce and refer to graphics in one or two sentences or in a parenthetical reference:

One-sentence introduction	As Figure 17 illustrates, heart disease kills more U.S. women than the next four causes of death combined.
Two-sentence introduction	Heart disease kills more U.S. women than the next four causes of death combined. Figure 17 shows the five leading causes of death among U.S. women.
Parenthetical reference	Heart disease kills more U.S. women than the next four causes of death combined (see Figure 17).

In each example, the writer introduces the figure by number and by content.

Sometimes you will want to tell readers how to use or read a graphic or give them information they need to understand a graphic. The following introduction tells readers when and how to use a table:

> If you receive an error message when installing the software, use Table 16. Read down the first column of Table 16 until you find the error message that you received. When you find the message, go to the second column, labeled "What to do when you receive this message."

Tell Readers What is Important about Each Graphic

Briefly explain the purpose of each graphic, or tell readers what they should notice in each one. Readers may not draw the conclusions you drew, so state those conclusions to make sure that readers understand the purpose and meaning of the graphic.

For example, the writers of a scientific paper on women and cardiovascular diseases wanted readers to understand the urgency of studying these diseases specifically in women; so they wrote the following explanation of two bar graphs, one showing the causes of death among U.S. women and the other showing the causes among U.S. men:

> Once a neglected field of research, cardiovascular diseases in women have rapidly become a major topic of scientific investigation. In 2006, cardiovascular disease killed more U.S. women than U.S. men and was the leading cause of death among women. Cardiovascular disease kills more U.S. women than the next four causes of death combined.[1]

The writers clearly state two important pieces of information that they want readers to understand after reading the bar graphs: cardiovascular diseases kill more U.S. women than men, and the diseases kill more U.S. women than the next four causes of death combined.

Place Each Graphic as Close as Possible to its Text Discussion

Graphics are most effective when they appear either on the same page as the text that refers to them or on a facing page. Readers may ignore a graphic if they have to flip from the discussion to hunt for the graphic elsewhere in the document. If you cannot avoid placing a graphic some distance away from its text discussion, refer to the graphic and tell readers where to find it. For example, if a graphic appears in an appendix, you might write the following:

> A large-scale map of the Bobwhite Quail habitat in Louisiana appears in Figure 26 in Appendix C (see page 51).

If you want readers to take another look at a graphic that you discussed earlier in your document, you might write the following:

> The nonspinning portion of the Galileo orbiter, discussed earlier, provides a stable base for four remote sensing instruments (see Figure 3.2, page 120, for a diagram of the orbiter).

Use Software and Downloadable Graphics

When possible, use software to produce professional, high-quality graphics and images. You can use the following types of software to produce graphics:

---

[1] Adapted from Beil, Laura. "Change of Heart: New Insights Gained as Cardiovascular Research Shifts More to Women," *Dallas Morning News* (6 Feb. 1995): 6D.

Evaluate the design of a graphic in the Interactive Student Analysis online at www.kendallhunt.com/technicalcommunication.

- Graphics software, such as Adobe *Illustrator*® or Adobe *Photoshop*®, allows you to create and edit flow charts, diagrams, drawings, and organizational charts.
- Photography editing software allows you to crop and edit digital photographs.
- Presentation software, such as Microsoft *PowerPoint*®, allows you to create slides.
- Spreadsheet software, such as Microsoft *Excel*®, makes it easy to enter data that you can use to create tables, charts, and graphs.
- Desktop publishing software, such as Adobe *InDesign*® or *QuarkXPress*®, allows you to easily integrate graphics into the text.

You can download many ready-to-use graphics. Hundreds of Web sites offer downloadable clip art, photographs, and images. Some graphics are available at no cost, while others may cost hundreds of dollars. For example, you can find downloadable clip art at www.clipart.com and at the Microsoft Web site. You can find downloadable photographs through a search on Google. If you download an image, make sure you do the following:

- Get permission to use the graphic. Even if the graphic is free, you must have permission to use it. If you cannot get permission, don't use it.
- Follow the copyright law (see Chapter 2).

USE COLOR TO ENHANCE AND CLARIFY YOUR GRAPHICS

Color can be a powerful tool. However, according to Roger Parker and Patrick Berry, "the first question to ask yourself when considering color is not how to use it but whether to use it at all." As Parker and Berry explain, most documents can benefit from color only if it is applied correctly. As you consider whether to use color, ask yourself these questions:

- **Can you afford to use color?** Color printing costs are higher than black-and-white printing costs. If color will enhance your graphics, make sure that you have money in your budget for it.
- **Will color enhance the graphic or add to its impact?** Some graphics will not lose much impact if you use black and white. For example, most line drawings don't need color.
- **Will my document compete with color documents?** If your readers expect color graphics, their absence could become a liability (Parker and Berry). However, color does not compensate for a poorly designed, inaccurate, or unclear graphic.

TIPS FOR USING COLOR TO ENHANCE AND CLARIFY GRAPHICS

- **Don't overuse color**. (Parker and Berry) If you use too many colors in a document or even on one page, you won't impress your readers, and you may even confuse them. Make sure that each color has a distinct purpose.
- **Choose colors that will give your documents a unified look**. You can use the color wheel (see Figure 11.34.). When selecting colors for a document, pick corresponding colors—three or four adjacent colors on the wheel (Parker and Berry). For example, you might select blue, green, and yellow.

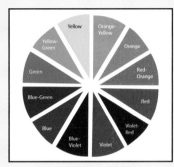

Figure 11.34 The Color Wheel

- **Choose a triad of colors to create contrast in your graphics**. A triad of colors is three colors that are relatively equidistant from each other on the color wheel—such as red, blue, and yellow (Parker and Berry).
- **Choose colors that stand out against the background to create effective contrast**. For example, don't use a shade of red on a red background; readers can't easily distinguish the various shades of red (see Figure 11.35). However, do use contrasting colors such as black and red.

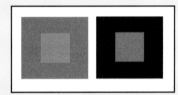

Figure 11.35 Using Color to Create Contrast

- **If your readers associate a color with a particular meaning, use that color as your readers expect**. For example, U.S. readers associate red with danger or warning and yellow with caution. However, for readers outside the United States, these colors have different meanings.

Figure 11.36 Bright Colors Make Objects Look Larger

- **Use bright colors to make objects look bigger**. For example, look at the stars in Figure 11.36. The stars are the same size; however, the yellow star looks larger than the blue star.
- **Make sure the text stands out from the background**. If you have a dark background, make the text white (see Figure 11.37).

Figure 11.37 Make Sure Text Stands Out

CONSIDER WHETHER INTERNATIONAL READERS WILL USE THE GRAPHICS

Visual language, like verbal language, differs from nation to nation and culture to culture. A graphic that is effective for U.S. coworkers may not be appropriate for international readers. Consider these examples:

- British software designers used a wise old owl as an icon for a help file. The designers assumed that this image would work in the international community. However, in Hispanic countries, the owl symbolizes evil. In India, if someone calls you an owl, it means you are crazy (Barthon).
- People in some African countries expect the labels on food products to picture exactly what is in the jar or can: "WYSIWYG" (What You See Is What You Get). In that context, consider how consumers in Africa reacted to jars of baby food with a baby on the label. The baby food manufacturer should have used a picture of carrots—not a picture of a baby (Barthon).

 TIPS FOR CREATING GRAPHICS FOR INTERNATIONAL READERS

- **Give graphics a neutral look** (Horton, "The Almost Universal Language"). For example, use a simple line drawing of a hand; the hand shouldn't appear to be masculine or feminine. Use outlines or neutral drawings, such as stick figures, to represent people (Bosley).
- **Use simple graphics**. Eliminate unnecessary details (Bosley; Horton, "The Almost Universal Language").
- **Use only colors that will imply the correct impression and/or meaning to your intercultural readers**. Colors have symbolic meanings and these meanings vary among cultures. For example, in Japan, blue symbolizes "villainy" whereas in Arabic countries, blue symbolizes virtue, faith, and truth. Bosley suggests using black and white, or gray and white for international graphics; but Horton suggests that "color can prove especially valuable" if the designer carefully considers symbolic meanings when selecting colors.
- **Avoid culture-specific language and symbols.** For example, don't use a red, octagonal shape to indicate "Stop"; not all countries understand that shape to mean "stop" (Bosley).
- **Consider the reading direction of your readers** (Horton, "The Almost Universal Language"). In some countries, readers read from left to right and clockwise. In other countries, readers read graphics from right to left in a counterclockwise direction (Bosley). Horton suggests designing intercultural graphics that readers can read from top to bottom; the graphic might also include an arrow to direct readers.

In both examples, the writers and designers assumed that intercultural readers would interpret the pictures as most U.S. readers would interpret them.

Before you put graphics into a final document, consider how readers in other countries and cultures will "read" them. When feasible, ask people familiar with your readers' culture to look at the graphics you plan to use. These people can help you predict the success of a graphic or can suggest changes that will make it better suited for international readers. As you design graphics for international readers, follow the tips on page 344.

ETHICS NOTE

Presenting Visual Information Ethically

You have a responsibility to present visual information ethically. The visual information that you include should be accurate, complete, and honest. To ensure that you use visual information ethically, do the following:

- **Present an accurate representation of the information**. For example, don't use a table to hide an unfavorable data point when that same data point would stand out in a bar or line graph.
- **Edit photographs ethically**. Airbrushing is ethical and legitimate when used to highlight essential or important information in a picture; airbrushing becomes unethical when it removes information from a photograph to deceive or mislead viewers.
- **Use an accurate scale**. As you create graphics requiring scales, make sure that the graphics accurately and honestly present data and differences among the data. The scale you select will affect how readers perceive your data. If you use inappropriate scales, you will exaggerate the differences in data when differences are minor; or you will make differences in data look small, even though they really are great. Figure 11.38A, page 346, uses an inaccurate scale. The difference between Burgert, the most recommended, and Raign, the least recommended, is only 1.34 percent. The bar graph, however, makes the difference between Burgert and Raign look quite dramatic—certainly more than 1.34 percent. The scale misleads and distorts. The graph needs a smaller scale and size to more accurately represent the difference between Burgert and other builders (see Figure 11.38B, page 346).
- **Begin the axis at zero**. If you can't practically begin at zero, clearly indicate that the axis does not begin at zero.
- **Don't leave out relevant information**. For example, if you have values that you can't explain, don't leave them out.
- **Cite the source of any data and/or information that you did not generate**.

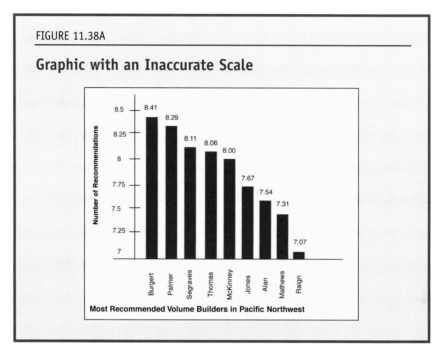

FIGURE 11.38A

Graphic with an Inaccurate Scale

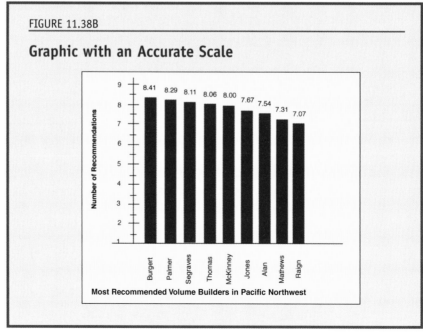

FIGURE 11.38B

Graphic with an Accurate Scale

CASE STUDY ANALYSIS
Forest Service Accused of Using Misleading Photos to Promote Forest Management

Background

In 2005, the United States Forest Service printed 15,000 copies of a brochure entitled "Forests With a Future" to promote old-growth forest management (see Figure 11.39, page 348). The cost was $23,000 for printing and production, paid to a private public relations firm. The brochure featured a series of six photographs, dated from 1909 to 1989, showing a thickening progression of growth in what was portrayed as the Sierra Nevada Forest.

Shortly after the brochure was released, it came under attack from Chad Hanson, Director of the John Muir Project in Cedar Ridge, California, and Timothy Ingalsbee of the Western Fire Ecology Center in Eugene, Oregon. These organizations work to preserve public forestlands. Hanson said he recognized the photographs in the brochure from a similar publication released in Montana. In fact, the photographs were not taken in the Sierra Nevada Forest, but in the Bitterroot National Forest in Montana. Hanson also closely examined the series of six photographs and realized that the 1909 photograph showed an area of forest that had been logged; piles of slash and stumps can be seen in the background. The Forest Service had depicted the lightly forested area in the 1909 photograph as "proof" that forests have grown increasingly thick over time, resulting in an increased threat of wild fire.

After receiving complaints about the brochure, Forest Service spokesperson, Matt Mathes, defended the photographs; he said the Montana forest photographs appeared in the Sierra Nevada brochure because "it is difficult to find a good series of repeat photographs of the same place over almost 100 years." He said the photographs were used only to show a progression of growth through the years. "Our goal here," Mathes said, "was to … increase the clarity and understandability of our message. We needed to be accurate, but not necessarily precise to the 99th degree."

Assignment

1. Pretend you are a member of Congress writing to Matt Mathes. Detail why the photographs, as they appear in the brochure, are misleading. Explain why technical communication must be precise to the "100th degree." Provide a plan for using the photographs in a non-misleading way in a revised brochure. Turn your letter in to your instructor. (For information on writing letters, see Chapter 12.)

2. Write a memo to your instructor recommending how organizations, such as the Forest Service, can ensure that old photographs are used accurately. Recommend ways to archive and label photographs. Defend your ideas in a short memo to your instructor, stating how your recommendations might have prevented the Forest Service from using the misleading photographs in its brochure. (For information on writing memos, see Chapter 12.)

Compiled from information downloaded from the World Wide Web, April 28, 2008: http://www.msnbc.msn.com/id/4722630/ "Forest Service Uses Misleading Photo." The Associated Press, Monday, April 12, 2004.

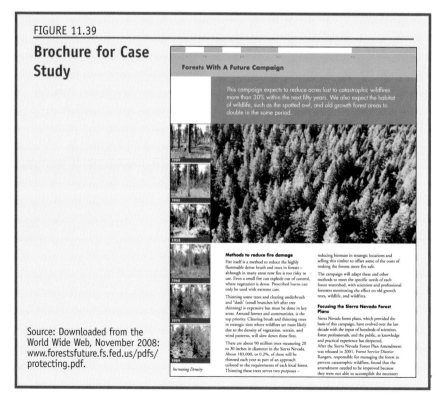

FIGURE 11.39

Brochure for Case Study

Source: Downloaded from the World Wide Web, November 2008: www.forestsfuture.fs.fed.us/pdfs/protecting.pdf.

EXERCISES

1. Visit the computer labs on your campus. Find out the following:
    a. What graphics software is available to students, and what types of graphics can you create using the software?
    b. What drawing software is available to students, and what graphics can you create using the software?
    c. What desktop-publishing software is available to students?

    If your campus doesn't have labs or graphics and drawing software, visit a computer or office supply store (in person or online) to gather information about graphics software. After you have gathered information on the software, create a table summarizing the information you gathered. Write a memo to your instructor reporting what you found. Incorporate the table into your memo. (For information on writing memos, see Chapter 12.)

2. Examine several technical publications or journals in your field.
    a. Find examples of effective and ineffective photographs, diagrams, and drawings.
    b. Write a memo to your instructor analyzing the photographs, diagrams, and drawings that you selected. (For information on writing memos, see Chapter 12.)
    c. Include a copy of the graphics at the appropriate place in the memo. Number the graphics and give them titles.

3. Create a diagram that you might use in a manual for a piece of equipment. You might select a household appliance, a piece of equipment used in your field, or a piece of equipment for which you are writing instructions in your technical communication class. Label the appropriate parts of the equipment, and give your diagram a number and a descriptive title.

4. Create an organizational chart for an organization or company. The chart must include at least four levels of management, divisions, or departments. Use one of these organizations:
    a. a civic organization
    b. the company for which you work

c. a campus or student organization, such as a service organization
d. a city department, such as the fire or police department

5. Prepare a flow chart illustrating one of these processes. Include readable labels, a descriptive title, and a number.
   a. How to apply for a passport
   b. How to apply for financial aid at your college or university
   c. How to explain a process or procedure common in your field or at your place of employment
   d. How to apply for graduation at your college or university

6. **Web exercise**: Find a graphic on the Web that you consider unethical.
   a. Write a paragraph explaining why you consider the graphic unethical.
   b. Print a copy of the graphic.
   c. Revise the graphic so it presents the information ethically.
   d. Be prepared to show your revised graphic in class.

7. The table in Figure 11.40 on the following page gives salary and unemployment information for science and engineering graduates 1 to 5 years after graduating from college. Study this data and create the following:
   a. two different graphics that compare salary information for bachelor's, master's, and doctorate graduates
   b. two different graphics that compare the unemployment rates for each of the fields listed in the column heads
   c. a graphic that compares the involuntary out-of-field rate to the unemployment rate by degree and by field

8. **Collaborative exercise:** Figure 11.41 on the following page contains information that would be more effective if presented with graphics. You and your team should complete the following:
   a. Determine what types of graphics will best convey the information.
   b. Determine what colors would add contrast.
   c. Revise the document to include two or more graphics.
   d. Select the best layout for the information.

FIGURE 11.40

Table for Exercise 7

Table 3-15
Labor market indicators for recent S&E degree recipients 1–5 years after receiving degree, by field: 2003
(Percent)

Indicator	All S&E fields	Computer/ mathematical sciences	Life sciences	Physical sciences	Social sciences	Engineering
Unemployment rate						
Bachelor's	4.7	4.5	4.1	4.0	5.1	4.4
Master's	4.4	5.4	2.9	2.6	4.6	4.5
Doctorate	2.8	2.1	4.6	1.1	1.9	3.3
Involuntary out-of-field rate						
Bachelor's	11.5	9.2	10.9	9.4	15.7	3.6
Master's	5.5	3.4	3.0	6.4	9.5	2.9
Doctorate	2.9	3.0	1.4	4.1	4.0	2.5
Average salary ($)						
Bachelor's	40,900	49,600	34,300	37,500	35,400	53,500
Master's	55,200	65,100	45,000	45,900	43,600	67,600
Doctorate	60,300	65,200	48,500	61,800	59,600	74,100

NOTE: Average salary rounded to nearest $100.

SOURCE: National Science Foundation, Division of Science Resources Statistics. Scientists and Engineers Statistical Data System (SESTAT). 2003, http://sestat.nsf.gov.

Science and Engineering Indicators 2008

Source: Downloaded from the World Wide Web, May 1, 2008: http//www.nsf.gov/statistics/seind08/tables.htm.

FIGURE 11.41

Document for Exercise 8

Backpack-related injuries in children

Overloaded backpacks used by children have received a lot of attention from parents, doctors, school administrators and the media in the past several years. According to the U.S. Consumer Product Safety Commission there were more than 21,000 backpack-related injuries treated at hospital emergency rooms, doctors' offices, and clinics in the year 2003. Injuries ranged from contusions, to sprains and strains to the back and shoulder, and fractures.

"Back pain in children is not so uncommon anymore," according to John Purvis, MD, pediatric orthopaedic surgeon. "Orthopaedic surgeons nationwide have seen an increase in children visiting their offices complaining of back and shoulder pain. If a child complains of back pain, parents should consider that it might be due to the backpack or perhaps something more serious. Back pain that persistently limits a child's activities, requires medication or alters sleep patterns warrants investigation."

The American Academy of Orthopaedic Surgeons recommends that a child's backpack should weigh no more than 15 to 20 percent of the child's body weight. This figure may vary, however, depending on the child's body strength and fitness.

While some experts disagree on whether heavy backpacks are the source of back pain in children, most agree that using good judgment when wearing one will reduce the risk of backpack-related injuries. It is important to partner with your child on the selection, packing and caring of the backpack.

Warning signs a backpack is too heavy

- Change in posture when wearing the backpack
- Struggling when putting on or taking off the backpack
- Pain when wearing the backpack
- Tingling or numbness
- Red marks

Source: Downloaded from the World Wide Web: www.nsc.org/resources/factsheets/hl/backpack_safety. aspx by National Safety Council. Copyright © 2008 by National Safety Council. Reprinted by permission.

iStockphoto 2008.

CHAPTER TWELVE   12

Writing Reader-Focused Letters, Memos, & E-Mail

Like most professionals, you will use letters, memos, and e-mail to correspond with coworkers and people outside your organization. Memos, letters, and e-mail are the everyday communication tools of the workplace. In some workplace situations, your coworkers, clients, and others will know you only through your e-mail; they may not actually talk to you in person. Especially in a virtual workplace, others will develop impressions of you through your e-mail. You will want to think about how your correspondence represents you and your organization.

DETERMINE THE OBJECTIVES OF YOUR LETTER, MEMO, OR E-MAIL

Before writing a letter, memo, or e-mail, decide what you want it to accomplish. Do you want readers to take a particular action after reading your correspondence? Do you want readers to give you information? Do you want to inform readers about good or, perhaps, bad news? Your correspondence may also have the objective of maintaining or establishing a positive relationship with the reader.

As you consider your correspondence, ask yourself these questions:

- **What is the purpose of the correspondence? What do you expect it to accomplish?** Your correspondence frequently will have more than one objective. For example, the primary objective of the letter shown in Figure 12.1, page 355, is to inform the reader of the new customer comment cards.

The secondary objective is to mend a damaged relationship with the reader and ensure the continued business and goodwill of that reader.

- **What action, if any, do you expect readers to take after reading the correspondence?** If you decide that you want readers to do something, clearly and directly state what you want them to do. Much correspondence is ineffective because the writer doesn't clearly and directly state what the reader should do. Many writers assume that readers will know what to do. If you assume incorrectly, the reader may not do what you expect.

- **What do you expect readers to know after reading the correspondence?** If you don't clearly understand and clearly state all that you wish to convey, your readers will miss your intentions. Try informally listing what you want your readers to know. This list can also help you spot information that is irrelevant or unclear.

FIND OUT ABOUT YOUR READERS AND HOW THEY WILL PERCEIVE YOUR MESSAGE

Effective letters, memos, and e-mail have the following characteristics (see Figure 12.1):

- **They are reader-focused.** When your correspondence focuses on the reader, it contains all the information that the reader needs to understand the message. It doesn't contain more than is needed, and it doesn't leave out information about why you sent the correspondence. It also respects the reader's time.

- **They are helpful.** When your correspondence is helpful, it anticipates and answers your reader's questions.

- **They are tactful and professional.** When your correspondence is tactful and professional, it is courteous and, when possible, positive. It maintains or gains the reader's goodwill by using a professional tone.

What can you do to ensure that your correspondence focuses on the reader and is helpful, tactful, and professional? Find out as much as possible about your readers. For much of the correspondence that you will write, this task will be relatively simple because you will know the readers. However, sometimes you will be writing to readers whom you don't know. Use the questions on page 356 to gather information about your readers,

---

FIGURE 12.1

Sample Letter Demonstrating the Characteristics of Effective Correspondence

Computers on Wheels

February 6, 2009

This letter focuses on the reader by acknowledging the writer's conversation with the reader.

Mrs. Wanda Perrill
902 Indian Creek
St. Paul, Minnesota 57904

Dear Mrs. Perrill:

After our conversation last month about the quality of service in your home, we created a yellow "Customer Care Card." To improve our service to you, your computer team will now leave this card on each visit. The purpose of the card is to solicit your comments about our service on a regular basis. These cards resulted directly from our conversation—thank you for the suggestion.

The writer anticipates the reader's questions.

The comment cards will help us to maintain and monitor the quality of service that we provide in your home. They also are part of a new incentive program for our employees, so please take a moment after each maintenance visit to fill out the postage-paid card and drop it in the mail. Your comments will help us to provide the service you need and expect.

This letter is tactful and courteous.

Thank you, Mrs. Perrill, for your comments that led to these new cards. We appreciate the confidence that you have placed in Computers on Wheels.

Sincerely,

Bob Congrove

Bob Congrove
Owner

"Serving you so you can work at home."

302 Macarthur, Suite 204 • St. Paul, Minnesota 57904

especially those readers you don't know. (You can find additional information about writing for your readers in Chapter 4.)

- **Who will read the correspondence? Will more than one person read it?** If you will have more than one reader, prepare to meet the needs and expectations of all your readers. If their needs and expectations vary substantially, consider writing separately to each person or group.

- **What are the readers' positions and responsibilities? How might their positions and responsibilities affect how they perceive your message?** If you know readers' positions and responsibilities in the organization, you can better determine what they know about you, your responsibilities, and, possibly, about the subject of your message. This information about your readers can also help you anticipate how they will perceive and react to your message. Suppose the purpose of your memo is to inform readers about the company's new travel policy. Under this new policy, your readers will no longer receive a corporate credit card to pay for their travel expenses. Instead, they will use their own credit cards or cash, and the company will reimburse them. Because your readers travel extensively, you know that this information may not please them. With this information, you can address their concerns and possible questions in your memo.

- **If your readers are external, what is their relationship to you and your organization? How will this relationship affect how they perceive your message?** Find out as much as possible about past interactions between the readers, their organization, and your organization. This information will help you understand how readers may perceive you and your organization; and it will help guide you in selecting the appropriate information and language.

- **What do your readers know about the subject of the correspondence?** If you can find out what your readers know about the subject of your correspondence, you'll be more likely to include the appropriate amount of background and detail.

USE THE APPROPRIATE FORMAT

Letters, memos, and most e-mail have basic formats that are appropriate in any workplace setting. As you determine the appropriate format, consider whether the reader is internal or external and whether your communication needs to be formal or informal. A letter is more formal than a memo or e-mail.

You will write letters primarily to communicate with people outside your organization. You might also use letters inside your organization to handle confidential matters, such as personnel and salary issues, or when you are sending more formal communications. You will write memos to communicate with people within your organization. You might use a memo for routine correspondence about a new procedure or for an informal report.

You can use e-mail to correspond with people inside or outside your organization. E-mail is especially effective for taking care of routine business and for working with people in different time zones. However, e-mail is not appropriate for confidential or sensitive business; you cannot assume that e-mail is private. You should always assume that your e-mail may be forwarded to someone whom you did not intend to read the correspondence.

> *Consider whether your reader is internal or external and whether your communication needs to be formal or informal. Letters are the most formal; e-mail is the least formal.*

Letters

The three basic formats for letters are block style (see Figures 12.2 and 12.3, pages 359-360), modified block style (see Figures 12.4 and 12.5, pages 361-362), and AMS simplified style (see Figure 12.6, page 363). Figure 12.7A, page 364, describes the elements of a letter and Figure 12.7B, page 365, gives an example.

The AMS simplified style omits the salutation, complimentary closing, and signature. This format is useful when you don't know the reader's name or which courtesy title (Ms., Mrs., Mr., Dr., Rev.) to use. However, the AMS simplified style may strike some readers as impersonal. Therefore, whenever possible, take the time to find the name and title of the reader. The three formats differ in the following ways:

- **Position of the date, the complimentary closing, and the signature block**. In letters set up in block or AMS simplified formats, place these elements flush against the left margin. In letters set up in modified block format, indent these elements from one-half to two-thirds of the width of the page. Be sure to indent all three elements the same distance from the left margin, so they align on the page (see Figures 12.4 and 12.5, pages 361-362).

- **Paragraph indentation**. Indenting paragraphs is optional in the modified block format. Do not indent paragraphs in block or AMS simplified formats.
- **Use of salutation and complimentary closing**. Omit these elements from letters in AMS simplified format.

Using Word-Processing Letter Templates

You can select from **templates**—predesigned letter formats—in your word-processing software. These templates provide fields for you to insert the heading information, the date, your name, etc. Some templates provide decorative elements that may not be appropriate for professional correspondence. Unless the context calls for a more informal, decorative template, select a simple, conservative format. If you cannot find an appropriate format, don't use a template; work from a blank page.

Memos

Even though most companies use e-mail for internal correspondence and for many informal reports, they may use memos when they want a more formal approach. You may frequently use memos for informal reports such as a trip report or a recommendation report (See Chapter 17, Writing Reader-Focused Informal Reports.) Like letters, memos have a conventional format. Unlike letters, memos do not include a salutation, complimentary closing, or writer's signature. You will print most memos on plain paper, not letterhead. Figure 12.8, page 367, presents three typical memo formats.

Using Memo Templates

You can select from **templates**—predesigned memo formats—in your word-processing software; or your organization may have predesigned (and expected) templates for memos. If your organization has such a template, use that template. If you decide to use a template from your word-processing software, select a simple, conservative format.

FIGURE 12.2

Format for Block Style on Letterhead

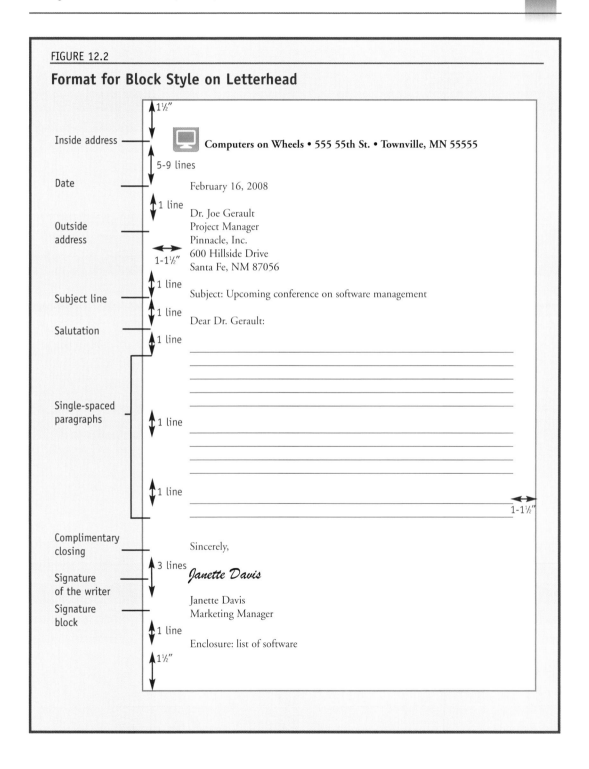

FIGURE 12.3

Format for Block Style without Letterhead

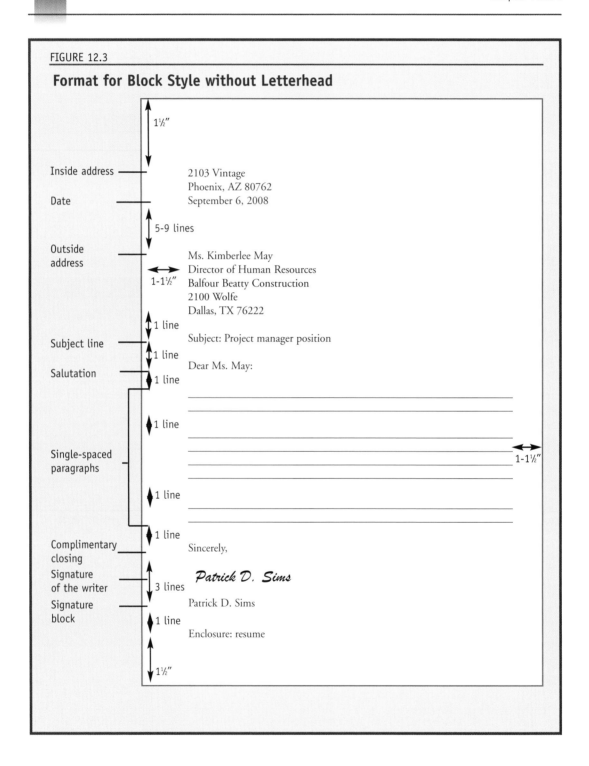

FIGURE 12.4

Format for Modified Block Style on Letterhead

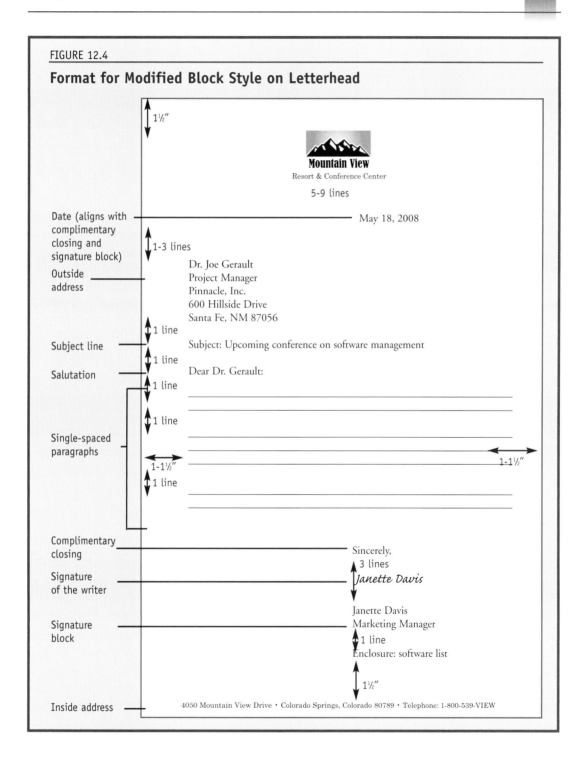

FIGURE 12.5

Format for Modified Block Style without Letterhead

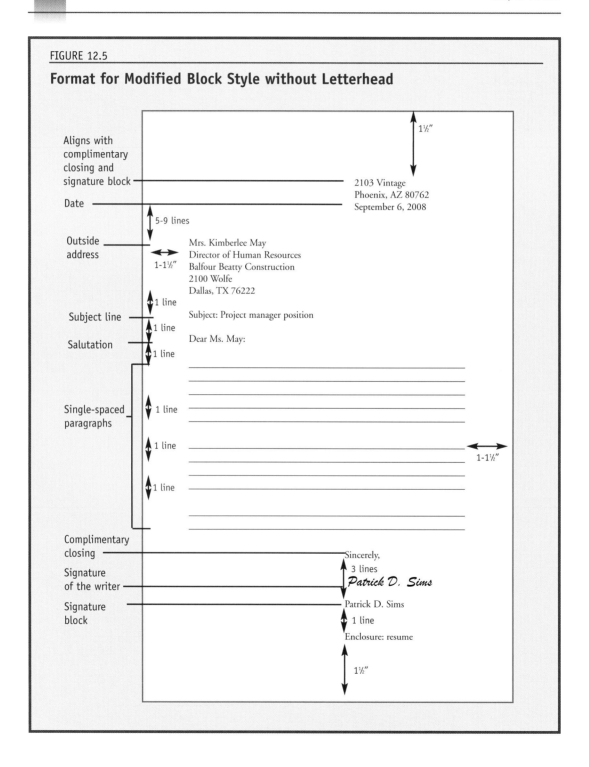

FIGURE 12.6

Format for AMS[1] Simplified Style

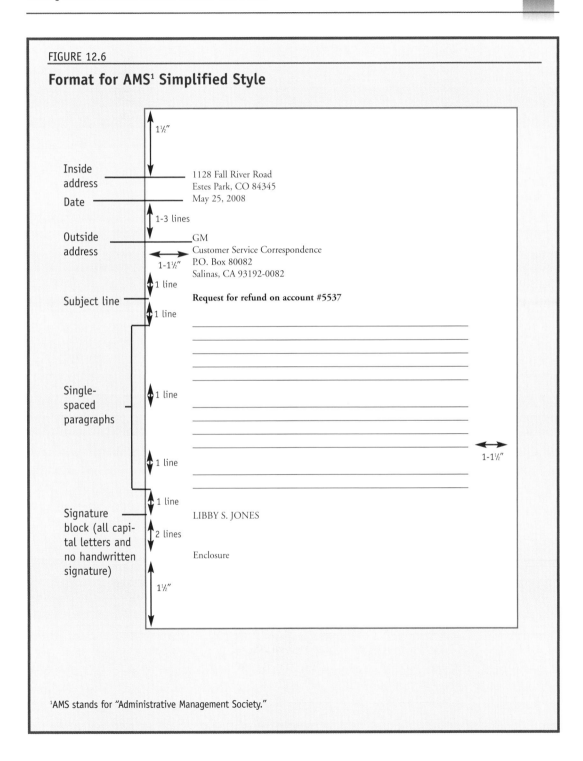

[1]AMS stands for "Administrative Management Society."

FIGURE 12.7A

Elements of a Letter

1. **Inside address**
   The heading consists of your organization's address or your personal address. Most organizations have pre-printed (or digital) letterhead that includes the organization's logo and address. If you are not using letterhead, include your address without your name as the heading.
2. **Date**
3. **Outside address**
   The inside address is that of the person who will receive your letter. The inside address has these elements:
   - Name and position. If that person has a professional title, include that title: for example, if you are writing to a physician, use Dr. John Smith, Director of Medical Services
   - Organization
   - Street address or Post Office box
   - City, state, and ZIP code
4. **Subject or reference line**
   A *subject line* tells readers what the letter is about. A ***reference line*** refers readers to the date of previous correspondence or to the project, order, or account mentioned in the letter. Subject lines begin with "subject" and reference lines with "re." These lines are optional in the block and modified block styles; a subject line is required for the AMS simplified style.
5. **Salutation (a greeting)**
   Use "Dear" followed by the reader's name (or official title if you don't know the reader's name) and a colon (not a comma):
   - Dear Mr. Sampson:
   - Dear Personnel Director:
   Always use a nonsexist salutation. When you don't know the reader's name or gender "Dear Sir" or "Dear Madam" is inappropriate. Instead, use the AMS simplified style and omit the salutation (see Figure 12.6, page 363), or use the reader's title in the salutation. Do not use "To whom it may concern" because it is vague and impersonal.
6. **Body**
   Generally, the body consists of at least three paragraphs. (See pages 368-369 for information on organizing the body of a letter.)
7. **Complimentary closing**
   Select from traditional closings such as the following:
   - Sincerely,
   - Best wishes,
   - Warm regards,
   - Cordially,
   - Best,
8. **Signature**
   Type your full name and your position. Sign your name above your printed name and position.
9. **Enclosure line**
   If you will be enclosing documents along with the letter, include an enclosure line one line below your printed title. The enclosure line indicates the number of enclosures (if more than one). Some writers also identify the enclosure.
   - Enclosures (2): Proposal
                    Drawings of Proposed Renovation
10. **Copy line**
    If you are sending copies of the letter to others, include their names in a copy line. Use the lowercase "c" for copy, followed by a colon and the name, and possibly position, of those receiving copies.
    - c: Norma Rowland, Director of Sales

FIGURE 12.7B

The Elements of a Letter: An Example

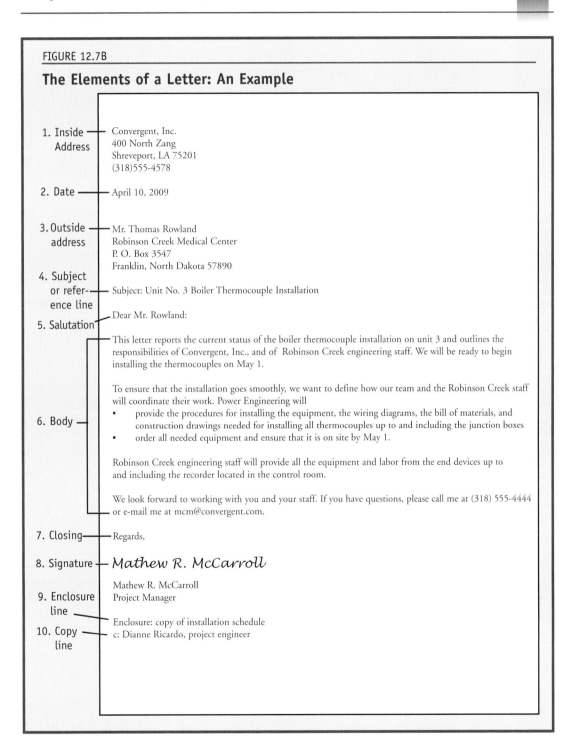

1. Inside Address

Convergent, Inc.
400 North Zang
Shreveport, LA 75201
(318)555-4578

2. Date

April 10, 2009

3. Outside address

Mr. Thomas Rowland
Robinson Creek Medical Center
P. O. Box 3547
Franklin, North Dakota 57890

4. Subject or refer-ence line

Subject: Unit No. 3 Boiler Thermocouple Installation

5. Salutation

Dear Mr. Rowland:

6. Body

This letter reports the current status of the boiler thermocouple installation on unit 3 and outlines the responsibilities of Convergent, Inc., and of Robinson Creek engineering staff. We will be ready to begin installing the thermocouples on May 1.

To ensure that the installation goes smoothly, we want to define how our team and the Robinson Creek staff will coordinate their work. Power Engineering will
- provide the procedures for installing the equipment, the wiring diagrams, the bill of materials, and construction drawings needed for installing all thermocouples up to and including the junction boxes
- order all needed equipment and ensure that it is on site by May 1.

Robinson Creek engineering staff will provide all the equipment and labor from the end devices up to and including the recorder located in the control room.

We look forward to working with you and your staff. If you have questions, please call me at (318) 555-4444 or e-mail me at mcm@convergent.com.

7. Closing

Regards,

8. Signature

Mathew R. McCarroll

Mathew R. McCarroll
Project Manager

9. Enclosure line

Enclosure: copy of installation schedule

10. Copy line

c: Dianne Ricardo, project engineer

E-Mail

You will use e-mail for both interoffice and external communications. Although e-mail software will format the headings of your e-mail, you will want to add some additional elements and follow proper e-mail *netiquette* ("etiquette on a network"). Your workplace e-mail should include the following:

- **Informative, specific subject line**
  The subject line should give readers specific information about the purpose of your e-mail. Readers use the subject line to decide whether to open an e-mail. If the subject line isn't informative, specific, or clear, many readers may ignore your e-mail. Always include specific information in the subject line; don't simply leave it blank.

- **Optional greeting**
  You may include an optional greeting, such as the following:

    Dear Jim:
    Jim,
    Dear Mr. Jackson:

- **Signature block**
  Sometimes called a "signature" in e-mail software, this element includes the following information:

    Your name and position
    Company
    Location (optional)
    Phone number
    Fax number (optional)
    E-mail address

  With most e-mail software, you can create an electronic signature that can be easily inserted into your messages. The software allows you to customize the signature. For example, some e-mail software allows you to insert your business card into an e-mail. For e-mail correspondence with coworkers, you may want to leave the signature block off, and simply end with your name.

Because e-mail is perceived as less formal, you may be tempted to abandon some conventions expected in workplace correspondence. When writing e-mail, remember to follow these rules:

- **Follow the rules of capitalization**. Don't use all uppercase letters or all lowercase letters.

FIGURE 12.8

Typical Memo Formats

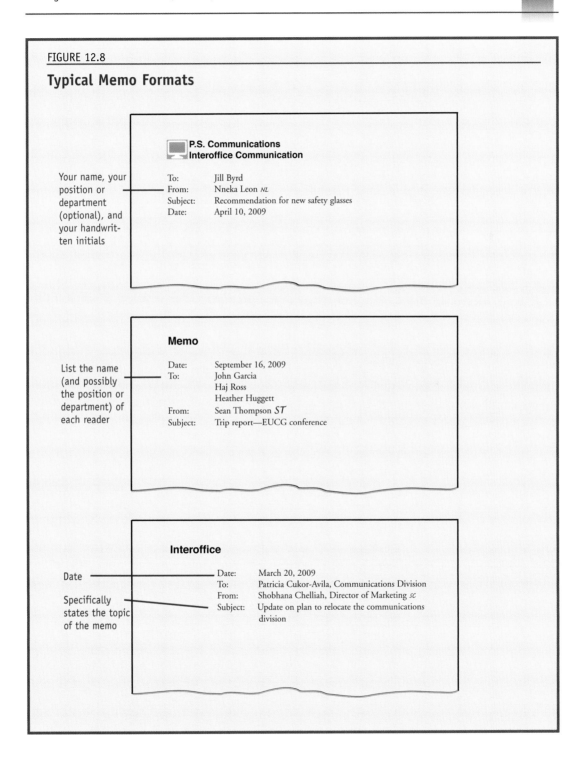

P.S. Communications
Interoffice Communication

Your name, your position or department (optional), and your handwritten initials

To: Jill Byrd
From: Nneka Leon *NL*
Subject: Recommendation for new safety glasses
Date: April 10, 2009

Memo

List the name (and possibly the position or department) of each reader

Date: September 16, 2009
To: John Garcia
           Haj Ross
           Heather Huggett
From: Sean Thompson *ST*
Subject: Trip report—EUCG conference

Interoffice

Date

Specifically states the topic of the memo

Date: March 20, 2009
To: Patricia Cukor-Avila, Communications Division
From: Shobhana Chelliah, Director of Marketing *sc*
Subject: Update on plan to relocate the communications
           division

- **Don't use abbreviations**. You probably use abbreviations, or shortcuts, when sending a text message or communicating via Instant Messenger. However, these abbreviations and shortcuts are not appropriate for workplace correspondence.
- **Limit the e-mail to business issues**. When using e-mail to conduct business, focus on the business and avoid including personal issues.

For more information on following proper e-mail etiquette, see "Taking It Into the Workplace" on page 378.

ORGANIZE YOUR CORRESPONDENCE TO MEET READERS' NEEDS

You may present the main message of your correspondence either directly or indirectly (Dragga).

The Direct Approach

For most of your letters, memos, and e-mail, you will use the direct approach. This approach helps readers find the purpose of your correspondence quickly. The direct approach has three sections, as follows:

In the first paragraph, present the main message
- Tell readers why you are writing.

In the middle paragraph(s), explain the main message
- Explain the main message (your purpose for sending the correspondence) presented in the first paragraph.
- Present necessary details about the main message.

In the final paragraph, close the correspondence
- Tell readers if and when they or you will act next.
- Tell readers, if necessary, what you are requesting or expecting them to do.
- Tell readers how to contact you (your phone number, fax number, e-mail address, or mailing address). You can simply tell readers where or how you would prefer to be contacted. For example, if you prefer that readers e-mail you, give your e-mail address. If you prefer they call you on your cell phone, include only your cell phone number.

You can adapt the direct approach for most correspondence situations. The letter shown in Figure 12.9, page 370, illustrates the direct approach.

The Indirect Approach

When you use the indirect approach, you delay or buffer the the main message until you have graciously opened the letter and explained the message. The indirect approach has three sections, as follows:

In the first paragraph, buffer the main message
- Begin with a buffer. A *buffer* is a positive or neutral statement. The buffer may help readers to better receive your message, especially if the message is negative.

In the middle paragraph(s), explain and then state the main message
- Explain the main message. For instance, state the reason for a refusal or rejection. By first explaining the message, you prepare readers for the negative news.
- State the message.
- Suggest an alternative or remedy, if possible, when the message is negative. By suggesting an alternative or remedy, you may be able to keep the goodwill of your readers and show that you want to meet their needs.

In the final paragraph, close the correspondence
- End the correspondence with a gracious statement.

Most writers rarely use this approach, but it can be appropriate when the news is not urgent or doesn't require readers to respond or act. This approach may also be appropriate in correspondence with international readers who may be accustomed to a less-direct approach than is common in American business (Sims and Guice). The letter shown in Figure 12.10, page 371, illustrates the indirect approach.

ETHICS NOTE

Ethics and the Indirect Approach

Before you use the indirect approach, consider whether it is an ethically appropriate choice for the situation. The indirect approach can inappropriately obscure information. It also can mislead some readers into thinking that the message is good because the gracious, usually positive, opening delays the bad news. Readers who read only the opening may misinterpret the purpose of the correspondence. Thus, before using the indirect approach, carefully consider your readers and how they are likely to read the message.

FIGURE 12.9

Sample Letter Using the Direct Approach

402 Summer Court
Carrollton, TX 75007
June 11, 2009

Mrs. Norma Rowland
Denton County Appraisal District
3911 Morse Street
Denton, TX 76202-3816

Subject: Appraisal of block 5, lot 15, in Villages of Indian Creek phase 1

Dear Mrs. Rowland:

The writer tells the reader why he is writing.

I am requesting that you reconsider the 2008 appraisal of my home, 402 Summer Court in Carrollton. I have included information from the 2008 Dallas County appraisal and a market analysis by a local realtor. Based on this information, I request that you consider appraising the home between $101,300 and $105,352.

The Dallas County Appraisal

The writer explains the main message presented in the first paragraph. He also presents details about the message.

Dallas County appraised my home as follows in 2007 and 2008.

	Total Value	Improvements	Land	Sq. ft.	$ per sq. ft.
2007	$103,220	$86,220	$17,000	2,030	51
2008	$ 99,820	$84,820	$15,000	2,030	49

Denton County appraised my home as follows in 2007 and 2008.

	Total Value	Improvements	Land	Sq. ft.	$ per sq. ft.
2007	$106,896	$72,896	$34,000	2,170	49
2008	$122,931	$83,831	$39,100	2,170	56

As the above tables show, Denton County increased the appraisal by $16,035 while Dallas decreased the appraisal by $3,400. This increase in the appraised value is especially puzzling because similar homes in our neighborhood have not sold for more than $98,000.

The Market Analysis

A local realtor with Providence Realty, Ms. Ellen Babcock, reports that the price per square foot should be between $48 and $52 for our home. Ms. Babcock reports that similar homes in this neighborhood have not sold for more than $52 per square foot. As the tables in the above section show, the 2008 Denton County appraisal is $4 more per square foot than the upper end of the range reported by Ms. Babcock.

The writer closes by offering to supply documents or to answer questions. The writer tells the reader how to contact him.

I will be happy to supply documents from the Dallas appraisal or the realtor. If you would like these documents or have questions, please contact me at (972)555-5555 or at the above address.

Sincerely,

William W. Sims

William W. Sims

FIGURE 12.10

Sample Letter Using the Indirect Approach

Independent's Research, Incorporated

1010 West Main • Los Alamos, New Mexico 87890 • (505)565-3000

May 16, 2009

Ms. Lisa Jackson
402A Summer Court
Edwardsville, IL 67843

Dear Ms. Jackson:

The first paragraph includes a positive statement about the reader.

Last week, we told you that we were recommending you for a summer intern position with our Research and Development Department. Your excellent background and education would allow you and us to benefit from your interning.

The writer explains the main message in the first sentence and then states the message in the second and third sentences. The final sentence suggests an alternative.

Last week, the board of directors announced a hiring freeze for all positions until the end of the year. We hoped this freeze would not include the internship positions, but sadly it does. Therefore, we will not be able to offer you an internship this summer. The board feels certain that these intern positions will once again be available next summer. Because you are currently a sophomore, please reapply next year.

The letter concludes with a gracious statement.

We appreciate your interest in our company and look forward to your application next year.

Sincerely,

Peggy Fagner

Peggy Fagner
Manager, Recruitment

TIPS FOR SELECTING THE DIRECT OR INDIRECT APPROACH

- **If your readers expect the bad news, use the direct approach**. Because your readers expect the bad news, put it in the first paragraph. Readers may assume that the news is good if you move it to the second paragraph.
- **If your readers may read only the first paragraph or skim the correspondence, use the direct approach** (Locker). The indirect approach may mislead readers; they may not see the main message and may misinterpret the purpose of the correspondence.
- **If you know that your readers will resist the news or "won't take no for an answer," use the direct approach** (Locker). In these situations, the indirect approach may mislead readers into thinking that they can persuade you or your organization into changing the news or the "answer."
- **If the news is urgent or if you have sent the message repeatedly, use the direct approach**. With the indirect approach, readers may again ignore your message or may not respond as you intend.

CREATING A PROFESSIONAL IMAGE THROUGH YOUR CORRESPONDENCE

You create a professional image through your e-mail when you do the following:
- Put yourself in the readers' shoes.
- Use a tactful, professional tone.
- Avoid overused phrases.
- Use specific language.
- Follow grammar and punctuation rules.

Put Yourself in the Readers' Shoes

When you open an e-mail from your instructor or your university, you want to know how the information will affect you or what you will have to do. Similarly, in the workplace, your readers want to know how your letter, memo, or e-mail will affect them. Particularly with e-mail, readers may ask these questions:
- Why should I read this correspondence?
- What, if anything, do I have to do after reading this correspondence?
- How does this correspondence affect me?

To help readers answer these questions and correspondence, you want to "put yourself in the readers' shoes"; that is, you want to consider how your message affects the reader not how it affects you.

To put yourself in the readers' shoes, consider how readers will respond to your message. Compare the letters shown in Figures 12.11 and 12.12, pages 373-374. The writer of the letter in Figure 12.11 didn't carefully consider the tone of his letter. He uses language focused on himself and on his company. This writer-focused, "we," tone is evident in the pronouns

FIGURE 12.11

Sample Writer-Focused Letter

1212 canyon drive
boulder, co 67899
(303)555-4986
drr@co.com

May 18, 2009

Mrs. Annie Shepard
1244 Fork Road
Socorro, NM 54233

Dear Mrs. Shepard:

We here at Colorado Outfitters are always pleased to hear from our customers. We try to please our customers with quality recreational gear and equipment. Our newest feature for customers is our Colorado Outfitters Catalog, a way to shop by telephone, e-mail, or the Web. However, this feature does have one drawback—our mailing list for the catalog is incomplete.

Recently, we received your letter about a problem with our service. Your neighbor purchased a Flashmagic 2-person tent for $250 during our spring catalog sale, but you bought the same tent at the full price of $350 during January.

It is a shame that you weren't on our mailing list, so we could have offered you the Flashmagic 2-person tent for $250. We will put you on our mailing list today, so you won't miss any more of our sales. If we can serve you in any way, please call, write, or e-mail us—Colorado Outfitters is here to make your recreational activities fun and easy.

Happy camping,

David R. Rowland

David. R. Rowland
Manager

FIGURE 12.12

Sample Reader-Focused Letter

1212 canyon drive
boulder, co 67899
(303)555-4986
drr@co.com

May 18, 2009

Mrs. Annie Shepard
1244 Fork Road
Socorro, NM 54233

Dear Mrs. Shepard:

Your recent letter about your purchase of a Flashmagic 2-person tent in January concerned us. You explained that a neighbor had purchased the same tent during the spring catalog sale; however, you paid $100 more than your neighbor. We understand your concern, so we have enclosed a 50% discount coupon good on your next purchase from Colorado Outfitters.

You must wonder why you didn't receive a catalog. The spring catalog is the first one that we sent to our customers as part of our new shop-at-home service. Because this service is new, we are still adding long-standing customers on the mailing list. We have now entered your name and address on the mailing list. You will receive all future catalogs and sales notices. Soon, you—like your neighbor—can shop at home and take advantage of sales available exclusively to customers on our mailing list.

Mrs. Shepard, please let us know if we can serve you further. You are a valued customer.

Happy camping,

David R. Rowland

David. R. Rowland
Manager

Enclosure: discount coupon

that refer to him and his company ("we" and "our") and the way he focuses on company actions and policy instead of on the reader and her problem. In the final paragraph of his letter, he seems to be saying that the reader was negligent.

In the letter shown in Figure 12.12, the writer achieves a reader-focused, "you," tone. He focuses on the reader and her interests instead of on the company. He uses a positive tone and refers to the reader frequently by name and with second-person pronouns ("you" and "your").

Evaluate the tone and format of letters to and from a business in the Interactive Student Analysis online at www.kendallhunt.com/ technicalcommunication.

Use a Tactful, Professional Tone

Readers resist messages that carry bad news or point out their mistakes. Just as you prefer to receive positive news, so do your readers. They will respond more favorably if a message concentrates on the positive, deemphasizes their mistakes, and, when possible, focuses on ways of doing better. Even when you can't focus on the positive or deemphasize mistakes, use a tone that creates goodwill for you and your organization.

Focus on the Information, Not on the Person
Compare the impression made by the following two messages:

Focuses on the reader's actions

You failed to read the instructions at the top of the form. If you had read them, you would have signed the back of the form on the appropriate line. Without this signature, your application for a patent cannot be processed.

Focuses on the information

We will gladly process your patent application. Please sign the back of the enclosed form on line 28 and return it to us.

Focuses on the reader's actions

If you paid attention to details, you would have noticed that I sent the e-mail at 3:46 p.m. today.

Focuses on the information   I am forwarding you the e-mail
                                        that I sent earlier today.

In the passages that focus on the reader's actions, the writer does not create goodwill. These statements focus on what the reader has done rather
than on the information. These passages point out the reader's mistakes
instead of offering a way to correct those mistakes. Follow the tips on the
next page for choosing words and phrases that will help readers perceive
your messages as you intend them.

Avoid Overused Phrases

Over the years, many phrases have become associated with correspondence;
you have probably read and possibly used some of these phrases. While these
phrases may be easy to use, they are overused, insincere, and often inflated.
Figure 12.13, page 379, lists some overused phrases in correspondence.

Use Specific Language

To help create a professional image through your correspondence, use specific language. When you use specific language, you do the following:
- eliminate questions that readers may ask when the language is ambiguous or vague
- lessen the chance of miscommunication

When the language is not specific, your readers have to guess what you
intend, and they may guess incorrectly. Let's look at an example:

Not specific   Please send your revised report ASAP.

The writer does not give the reader a specific date for sending the revised
report. By using "ASAP," the writer intends for the reader to send the revised
report very quickly. However, "ASAP" (short for "as soon as possible") is vague.
The writer means "send the report immediately." However, the reader may read
"ASAP" as "as soon as possible for me." The reader will send the revised report
when convenient. Instead, the writer should have written something like this:

Specific   Please send your revised report by April 11.

In this version, the reader knows specifically when to send the report.

TIPS FOR CREATING A TACTFUL, PROFESSIONAL TONE[1]

- **Avoid words and phrases that point out readers' mistakes in a negative tone or a tone that makes readers feel inferior or ignorant.**
  > You neglected to . . .
  > You failed to . . .
  > You ignored . . .
  > We fail to see how you could possibly . . .
  > We are at a loss to know how you . . .
  > We cannot understand how you . . .
- **Avoid phrases that *inappropriately or unnecessarily* demand or insist that readers act.** These phrases may cause readers to resist or ignore your perceived demand.
  > You should . . .
  > You ought to . . .
  > You must . . .
  > It is imperative that you . . .
  > We must insist that you . . .
  > We must request that you . . .
- **Avoid ambiguous words and phrases that may sound fine to you but may make readers feel inferior.**
  > No doubt . . .
  > Obviously . . .
  > Of course, you understand . . .
- **Avoid implying that your readers are lying.**
  > You claim that . . .
  > Your letter (memo, e-mail) implies that . . .
  > You insist that . . .
- **When possible, avoid negative words when referring to readers, their actions, or their requests.**

impossible	deny	unable
will not	inferior	fail
neglect	inconvenient	difficulty
incorrigible	complaint	wrong

- **Don't click "send" when you are angry or upset. Instead, click "save as draft" and wait until you have cooled down.** When you are angry or upset, you may write an e-mail that includes information or language that is not tactful or professional. If you are angry or upset, wait to send the e-mail or ask a coworker to read your e-mail before you click "send."

[1] Some tips adapted from the work of Elizabeth Tebeaux.

TAKING IT INTO THE WORKPLACE
How's Your E-Mail Netiquette?

Like face-to-face conversations, e-mail allows for spontaneous responses and feedback (Lakoff; Ong). E-mail writers can misuse this spontaneity when they misspell words and incorrectly use lowercase or uppercase letters. When you begin working for an organization, read the e-mail of others before you send your own e-mail. Determine the level of formality—or informality—expected. You should also determine what guidelines the organization may have for e-mail. For example, does the organization allow you to send and receive personal e-mail?

To ensure that you are following proper e-mail *netiquette* ("etiquette on a network"), follow these guidelines:

- **Make your messages easy to read and the paragraphs short.** Follow the standard rules for capitalizing letters, and add a line between paragraphs. Use a readable type.
- **Get to the point.** Make your message brief and put the main message in the first paragraph.
- **Cut the clutter.** Eliminate any words or information that the readers do not need. Don't make the reader scroll through irrelevant information to find the main message.
- **Use a polite tone—don't flame.** *Flaming* is sending rude or angry e-mail messages.
- **Send messages only when you have something to say—don't send "junk mail."** Unnecessary or uninformative e-mail wastes readers' time.
- **Remember that e-mail is permanent and that it is not private.** Most organizations archive all e-mail written by their employees, so don't write anything in an e-mail that you wouldn't put in print or want others to read.
- **Don't copy or forward e-mail unless the readers need to see the e-mail.** When you copy or forward e-mail to people who do not need it, you are "junking" their inboxes and, therefore, creating work for them.
- **Proofread!** Even when e-mail is informal, it should not be sloppy.

Assignment

Interview representatives from at least two organizations and ask them the following questions. Report your findings in an e-mail to your instructor.

- Does your organization have a policy governing e-mail? If so, would you share a copy with me?
- Does the policy allow employees to send and receive personal e-mail? Why or why not?

FIGURE 12.13

Overused Phrases Common in Correspondence

Your cooperation in this matter is greatly appreciated
Any assistance would be appreciated
Attached please find . . .
Enclosed please find . . .
Enclosed you will find the information you requested
Pursuant to our agreement
To Whom It May Concern
Thanks in advance
Thank you for your assistance in this matter
Please do not hesitate to call
Feel free to call me any time

Follow Grammar and Punctuation Rules

Your correspondence is an extension of you and your organization. Therefore, your correspondence should present a positive, professional picture of you and your organization. When your correspondence contains grammar, punctuation, and spelling errors, you and your organization look sloppy. These errors are especially common in e-mail.

Your correspondence says a lot about you: it tells readers how you pay attention to details—or how you *overlook* details. Take the time to proofread your correspondence.

CASE STUDY ANALYSIS
When Private E-Mail Becomes Public:
How Microsoft Learned an Embarrassing Lesson

Background

To protect itself against potential legal scrutiny several decades ago, Microsoft implemented a system for documenting internal communications between key employees and project managers in the company. Approximately every 30 days, IT experts would download and make copies of key employees' computer hard drives and e-mail archives. Then the files would be sent to outside law offices, where the files would be reviewed for lawsuit potential. Any material deemed pertinent to ongoing cases was immediately turned over to the U.S. Department of Justice or to companies that had filed suit against Microsoft.

Because these files became part of legal actions, many of the private e-mail correspondence became public evidence in trial. In an Iowa antitrust lawsuit, one company executive's revealed e-mail said, "If I didn't work here, I'd buy a Mac." In another e-mail a Microsoft senior executive harshly criticized Bill Gates's leadership of the company.

In addition to the scrutiny Microsoft was under due to the 20 or so antitrust suits it faced, the company also suffered public humiliation when these supposedly internal, "private" communications became public.

Assignment

1. Pretend that you are outside legal counsel to Microsoft. Draft a memo to company executives advising them of the rules they should follow when drafting internal memos and e-mail. Refer to the antitrust cases and the exposure of previous internal memos. Turn your memo in to your instructor.

2. Imagine that you are Bill Gates. Draft a letter to investors addressing the controversial e-mail that was released in the antitrust suits. Reassure investors that the corporate atmosphere at Microsoft is still strong and that rifts between key executives have been mended. Turn your letter in to your instructor. Add a page of notes that discusses which approach you used in your letter—direct or indirect—and why you selected that approach.

EXERCISES

Download a Worksheet for Writing Reader-Focused Letters, Memos, and E-Mail online at www.kendallhunt.com/technicalcommunication.

1. Revise these sentences to improve their tone. If the sentence contains vague or ambiguous language, make the language specific.

   a. We are sorry that we cannot fill your order for our product. We get so many orders that we find it utterly impossible to fill them all.

   b. You understand, of course, that we cannot credit your account for the price of the unused cell phone.

   c. We are searching for the revised report that you claim to have mailed to us on September 17.

   d. Your cooperation in this matter is greatly appreciated.

   e. We cannot understand how you could have left out your quarterly check when you mailed your copy of the statement.

   f. Your frivolous responses to our questions show why we are behind schedule. Your responses demonstrate a profound lack of attention to simple details.

   g. In filling out your warranty information, you failed to fill in the serial number of the printer.

   h. On the mechanical drawings, you do not indicate the size of the windows—as the drawings should. Please specify a size for the windows.

   i. Understand me. The research proposal is VERY important—we don't want to just "stick in" some thoughtless, unsubstantiated budget estimates because the company may have to LIVE WITH these estimates for the next three years.

   j. After reviewing the dimensions for the elevator shaft, it seems you made an error.

2. Write an e-mail or a memo to your instructor explaining how the direct and indirect approaches differ. Then suggest three specific writing situations where you would use the direct approach and three where you would use the indirect approach. Explain your suggestions.

3. You manage several project teams for an environmental engineering firm. Recently, many members of the project teams have been charging personal expenses to their corporate credit cards. Several team members have been unable to pay for these expenses when the statements are due. When the members pick up their credit cards each year,

they receive a copy of the company policy for using corporate charge cards for personal expenses. The policy reads as follows:

> "You may use your corporate credit card for travel and other business-related expenses during the year. You may not use your card for personal expenses such as meals, gifts, and personal travel."

Write a memo to your project teams restating the policy and explaining that the firm will take cards away from any employees who misuse them and that these employees may lose their jobs if they violate the policy.

4. Think of a service, procedure, or policy that your college or university should change. Write a letter or e-mail message to the appropriate official at your college or university suggesting the change.

5. Think of a service or product with which you have recently experienced problems. Write a letter to the appropriate person about the problems, and ask for an appropriate solution or remedy. If you cannot find the name of the appropriate person, use the AMS simplified style (see Figure 12.6, page 363).

6. **Web exercise**. Find the e-mail address of a company, non-profit organization, or government agency that has sample materials or information that you need for a class project. Write an e-mail to this company, organization, or agency requesting the information or materials. Send a copy of the e-mail to your instructor.

7. You are the network manager of the computer network for your company. Over the past few weeks, many users have tied up the network by downloading graphics for personal use during peak business hours, by sending large files to their home computers for personal use, and by surfing the Web for personal business during regular business hours. The downloaded graphics and surfing consume enormous amounts of memory on the network server, slowing the speed of the network responses. You have sent several e-mails to all network users about this problem; but the number of users who continue to download graphics and to surf the Web for personal use has not lessened. Yesterday, you confiscated a downloaded and printed graphic of Mickey and Minnie

---

FIGURE 12.14

E-mail Message for Exercise 7

Today, two senior managers could not take care of company business between 11:30 and 1:00 because the network was saturated with users surfing the Web and downloading and printing graphics for personal use. During the past week, during regular business hours, I have confiscated printouts of the President's dog and Mickey and Minnie Mouse. I have also watched groups of employees watching videos on YouTube. I will say it again: IT IS AGAINST COMPANY POLICY TO SURF THE WEB OR TO DOWNLOAD GRAPHICS FOR PERSONAL USE DURING BUSINESS HOURS. IT IS ALSO AGAINST COMPANY POLICY TO WATCH VIDEOS ON THE WEB. In the future, IT will report all employees who use their computers for personal use to their supervisors and to senior managers. IT will also turn off your access to the Web.

Mouse and of the President's dog from the White House. Today, two senior managers could not log on to the Web between 11:30 a.m. and 1:00 p.m. because employees were downloading graphics and surfing the Web for personal use. You are unusually frustrated and angry. You write the e-mail in Figure 12.14 to all network users. The message is direct, abrupt, and writer-focused. You mean the threat humorously; however, your message offends several employees.

a. Write an e-mail to all employees retracting your "humorous" threat and explaining the company's Web policies. Make the message reader-focused and, when possible, positive.

b. Write a memo to your manager explaining the problems the network users are creating and the "threatening" e-mail message that you sent to all network users. Explain to your manager how you are smoothing out the situation, and ask your manager for suggestions for solving these network problems.

EXERCISES

REAL WORLD EXPERIENCE
Case of the Broken Blender

You work for the customer service manager for Blast Blenders, Inc. Your company manufacturers high-end blenders for homes and commercial restaurants. The blenders are especially popular with restaurants that sell smoothies. Yesterday, you received a letter, a broken blender, and an original oil painting from Bob Garcia. The text of the letter appears in Figure 12.15. Mr. Garcia evidently has not read the warranty or the instruction manual that he received with the blender. The manual clearly states that users should use blenders only for blending food and drinks. Because Mr. Garcia used the blender for mixing paint, he not only damaged the motor, but also introduced toxic substances into the blender. Even if Mr. Garcia paid for your company to repair the blender, the toxic substances could still get into any food or drinks mixed in the blender.

You want to keep Mr. Garcia's business; after all, he sent you an original oil painting of his dog, Houdini. However, his warranty is no longer valid because he used the blender to mix paint. You must also relay to Mr. Garcia that he can no longer use the blender to safely blend food or drinks because of the toxic substances left behind by the paints. If you recommend that Mr. Garcia or your company repair the blender, you and your company could be liable if someone became sick from the toxic substances that would remain.

Assignment

Using what you have learned about writing reader-focused correspondence and writing ethical documents (Chapter 2), write a letter to Mr. Garcia explaining that you will not repair or replace the blender because, even if repaired, it can't be safely used. However, let him know that you have decided to give him a coupon for 20 percent off a new blender.

EXERCISES

FIGURE 12.15

Real World Experience: Broken Blender Letter

September 30, 2008

Customer Service Manager
Blast Blenders, Inc.
4525 Benders Blvd.
Phoenix, Arizona 45043

Dear Customer Service Manager:

I have sent you my Blast Blender, model 240T. I bought the blender six months ago. I bought it from a great salesperson, Joe, at my local appliance store in my hometown. The store is called Rodenbeckers Home Appliances. I have enclosed the registration card and warranty (along with the receipt) for the blender. Last week, the blender stopped working. I took the blender apart to see if the paint that I had been mixing had clogged the motor. (When I'm not at work, I paint oil paintings that I sell in a local store. I've not yet sold a painting, but I keep trying; one of these days, someone will buy my paintings. My Blast Blender works great at mixing my paints.) Paint had not clogged the motor. However, I did find paint on the inside of the housing—you know, around the switch that you use to turn the blender on and off. I think that paint got in the housing because one day I forgot to put the lid on the blender pitcher. I was mixing a beautiful shade of purple—unfortunately, I had purple paint on the blender and on the ceiling and cabinets in the kitchen. (It made a huge mess—my wife wasn't too happy about the purple, especially since I decided that the best way to deal with the paint spills was to paint the cabinets purple. Personally, I think they look great. If you would like, I can send you a picture of the cabinets.)

I cleaned the housing by soaking it in hot soapy water for several hours and then dried the housing with my hair dryer. I didn't clean the motor because I knew that putting the motor in water would violate the warranty. The blender will still not work. Please repair my blender since it is still under warranty and return it to me in the same box. If you can't repair it, I will happily accept a new blender.

Happy blending,

Bob Garcia

Bob Garcia
Soon to be famous painter

Enclosures
P. S. Enjoy the oil painting of Houdini, my dog. He passed away about two months ago. I really miss him.

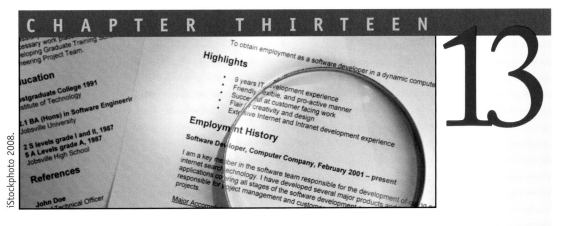

Writing Reader-Focused Job Correspondence and Resumés

Nicole will graduate from college in three months. She is excited about finding a job in her chosen field and beginning her career, but first she has to find that job. Nicole realizes that all the jobs she applies for will require a resumé. Further, a letter of application, or cover letter, may be required. Even if the letter of application is not required, Nicole is putting her best foot forward by providing one.

Like Nicole, you soon will graduate and look for a job. You might also want to apply for an internship and will need a cover letter and resumé. This chapter presents guidelines and techniques to help you do the following:

- plan your job search
- write resumés and letters of application
- prepare for interviews
- write follow-up letters
- prepare electronic portfolios

PLAN YOUR JOB SEARCH

Many students believe that a job search is simply posting your resumé on a Web site and waiting for employers to contact you. However, an effective job search takes planning and time. Figure 13.1, page 388, presents the stages of an effective job search.

FIGURE 13.1

The Stages of an Effective Job Search

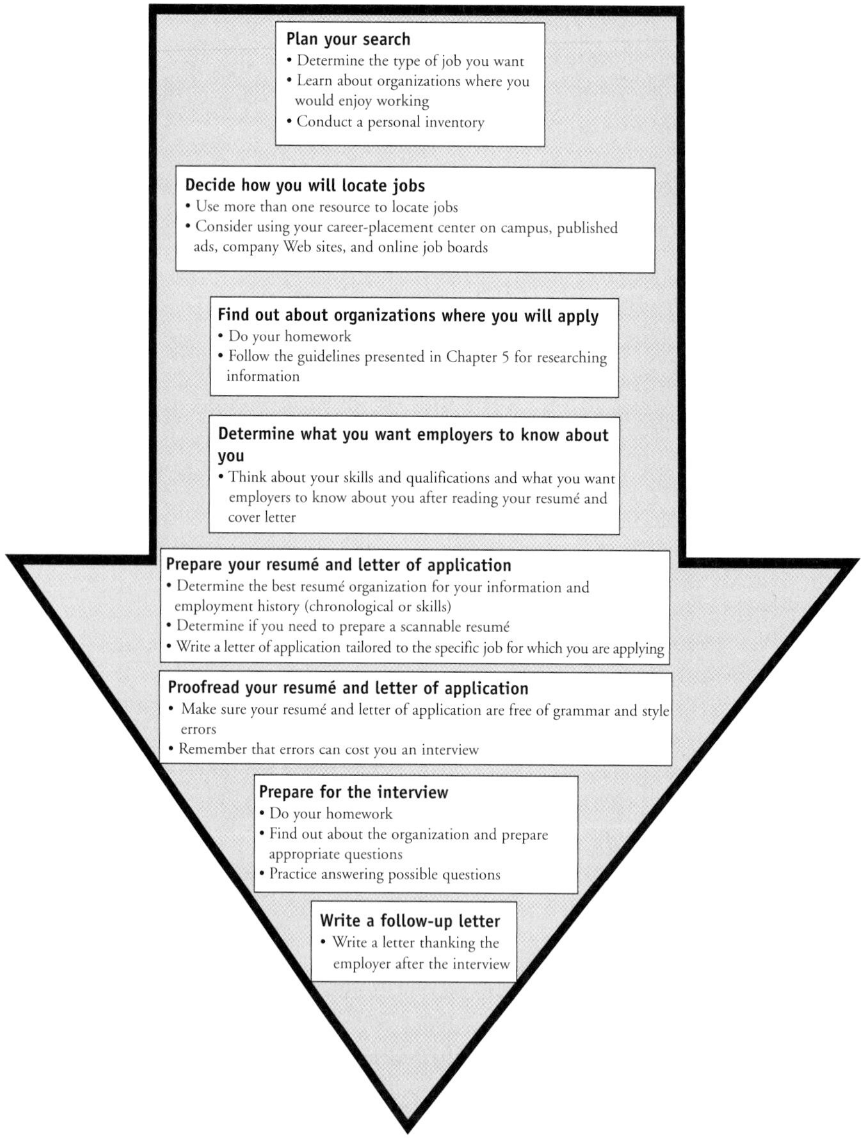

Plan your search
- Determine the type of job you want
- Learn about organizations where you would enjoy working
- Conduct a personal inventory

Decide how you will locate jobs
- Use more than one resource to locate jobs
- Consider using your career-placement center on campus, published ads, company Web sites, and online job boards

Find out about organizations where you will apply
- Do your homework
- Follow the guidelines presented in Chapter 5 for researching information

Determine what you want employers to know about you
- Think about your skills and qualifications and what you want employers to know about you after reading your resumé and cover letter

Prepare your resumé and letter of application
- Determine the best resumé organization for your information and employment history (chronological or skills)
- Determine if you need to prepare a scannable resumé
- Write a letter of application tailored to the specific job for which you are applying

Proofread your resumé and letter of application
- Make sure your resumé and letter of application are free of grammar and style errors
- Remember that errors can cost you an interview

Prepare for the interview
- Do your homework
- Find out about the organization and prepare appropriate questions
- Practice answering possible questions

Write a follow-up letter
- Write a letter thanking the employer after the interview

What Type of Job Do You Want?

Eli Davidson, business coach, reports that many job searchers don't do their homework and send applications "just to apply." Instead, you should apply only for jobs and to organizations that interest you. Don't waste your time or that of an employer by sending out resumés for jobs you don't really want. Davidson adds, "The best way to get a great job is to have a laser beam focus. The more targeted and specific you are, the more powerful your job search will be."

To determine the type of job you want, you can conduct a personal inventory by asking yourself the following questions:
- **What kind of organization appeals to you?**
  - Do you want to work for a large or a small organization?
  - Do you want a government position or an industry position?
  - Do you want to work for a non-profit organization?
- **Where do you want to live?**
  - Do you want to stay in your current location?
  - Can you or do you want to commute?
  - Do you want an international position?
- **What are your strengths and weaknesses, and how do they affect the type of job you are seeking?**
- **Do your skills and education match the jobs you want?**

To learn more about the kind of job you want, you can attend job fairs and find out about trends in your field. You might also visit with professors in your field and with the people in the career-placement center at your college or university. These people will help you learn about jobs available in your field.

How Will You Locate Job Opportunities?

After you have determined the type of job you want, you can look for that job using the following methods. To ensure that you find the right job for you, use more than one method.
- **Contact your college or university placement center**. Most colleges and universities have career-planning and placement centers that help graduating seniors and recent graduates find jobs. These placement centers link organizations and their recruitment officers with qualified

prospective employees. Most placement centers require you to register with them before you can interview. As part of the registration procedure, you probably will need to create a file (sometimes called a dossier or electronic portfolio) that includes an information sheet about you and your job interests, your resumé, and your college or university transcripts. In your file, you also may include samples of your work and other information that may interest potential employers. After receiving your dossier, the placement center will give copies of it to recruiters from businesses, government agencies, and industry; these recruiters will use the placement center to set up interviews on campus. Visit the placement center on your campus to find out how to set up a file and schedule an interview. The placement center is an excellent way to begin your job search: it is free, simple, and convenient.

- **Respond to job ads in print publications**. Businesses, industries, and government agencies sometimes place ads in print publications, such as newspapers (especially the Sunday classified sections in large city newspapers), journals in your field, and public-relations catalogs (such as *College Placement Annual*). When responding, carefully follow the instructions in the ad; you don't want your application for a job thrown out simply because you did not follow instructions.
- **Respond to ads published on an organization's Web site**. Most organizations post job ads on their Web sites. Most of these sites will ask you to complete an online application and upload your resumé.
- **Use an online job site**. You can search for jobs online through Web sites sponsored by federal agencies, professional organizations, and private organizations. These sites are often called *job boards*. You will find two types of job boards:
  - Some job boards only list positions. You then can apply for those positions by mailing or e-mailing your resumé and letter of application.
  - Some job boards allow you to submit your resumé electronically so employers can search for qualified candidates and contact them directly.

  If you decide to post your resumé to a site, you have no control over that resumé—who sees it, who uses it, or how it is used. Before you post your resumé, consider the tips on page 392 for using an online job board.
- **Network with others in your field, with people who know you personally, and with your professors**. When you begin your job search, tell

people in your field, your professors, and your personal and family friends that you are looking for a job. Consider sending them a copy of your resumé. These people may have contacts to help you locate job opportunities.

- **Send out unsolicited letters of application**. If you are interested in working for a specific organization, send an unsolicited letter of application to that organization. Many organizations do not advertise job opportunities, so unsolicited letters can be effective. Unsolicited letters do have a disadvantage: the organization may not have any openings. However, if you are truly interested in working for a particular organization, an unsolicited letter of application may be worth your time.
- **Use professional placement agencies**. Professional placement agencies present your resumé to potential employers. These agencies work much like a college placement center but often charge a fee paid either by potential employers or by the job seeker. Most of these agencies cater to more-experienced job seekers.

> *To ensure that you find the right job for you, use more than one search method for discovering employment opportunities.*

What Do You Want Employers to Know About You?

Before you put together a resumé or send out any letters of application, think about what information you want employers to know about you and what information employers want to know about potential employees like you. The information you provide should give potential employers a positive, accurate picture of you and what you can offer their organization.

You might begin by determining what information is likely to interest prospective employers. You might concentrate on these categories: education, work experience, activities, goals, and skills. After selecting your categories, brainstorm to create lists of information about yourself in each category. Write down any information that you think will help an employer understand you and your qualifications—information that will impress an employer. For instance, under education, list the degree you will receive when you graduate, the date that you will receive the degree, your major, and significant projects that you completed in your major field of study. Figure 13.2, page 393, shows the brainstorming list that Nicole created. Although she may not use all the information on her list, it gives her information to work with when she begins preparing her resumé and letter of application.

TIPS FOR USING AN ONLINE JOB BOARD

- **Determine if the online job board meets your needs**. Before you post your resumé or any information on a job board, ask yourself these questions:
    - **Who has access to your resumé?** If the site is not secure and anyone can view your resumé, consider whether you should post your resumé. If you decide to post your resumé on an unsecured Web site, remove your phone number and address.
    - **Will the job board charge you a fee to post or to update your resumé?** Some job boards charge a fee for posting your resumé, and some charge a fee each time you update your resumé. Make sure you know about the charges upfront. If a job board charges for updating your resumé, you might consider using a free site.
    - **Can you update your resumé?** If not, consider using other job boards.
- **Find out how you will be notified when an employer requests your resumé**. Some job boards will notify you while others will not. If you know an employer has requested your resumé, follow up to find out about potential job opportunities.
- **Find out whether your current employer or manager will see your resumé**. Depending on your employer or manager, you may jeopardize your current job if he or she finds out you are searching for another position. If your manager knows you're looking for another position, use the job board. If not, consider using other methods for finding job opportunities.
- **Use more than one online job board**. A potential employer may not see your resumé or personal information if you use only one job board. Be sure to post your resumé on boards specifically for your field. Some of the more popular job boards are
    - Monster
    - American Job Bank (the U.S. Department of Labor sponsors this board)
    - CareerBuilder
    - AfterCollege
    - Yahoo! HotJobs

You can also consider the information that specific employers may want to know about you. If you are applying for several jobs at the same time, you may not be able to pull together this information for each potential employer. Therefore, you may want to prepare a resumé first, concentrating

---

FIGURE 13.2

Sample Brainstorming List

Education
B.S. in mechanical engineering from University of Washington
Expect to graduate in December 2009
Dean's list three semesters—fall 2007, spring 2007, spring 2009
GPA 3.45

Work Experience
Internship at General Dynamics
Helped design robotics machinery for automated assembly lines; used CAD in refining designs
Learned to work as a team member

Trinity Pharmacy
Pharmacy technician since January 2007
Began as cashier and did general cleanup of store
Operate the cash register
Enter prescription information into the computer system
Help customers needing information about over-the-counter drugs and other items in the pharmacy area
Deal with confidential customer information

Lifegard and Swimming Instructor
Summers since high school
Know CPR
Certified Red Cross lifeguard at YMCA pool at home
Received Lifeguard of the Month award four times (get years and months)
Certified Red Cross swimming instructor
Taught private, semiprivate, and group swimming to children through the YMCA

Skills
Java
Have designed Web pages for Trinity Pharmacy—however, not experienced here
Know statistics packages, spreadsheets, CAD, and *Dreamweaver*
People skills—have learned to interact with customers and to be a team player
Work well with children

Activities
Mortar Board, senior year
American Society of Mechanical Engineers
Did volunteer work for Habitat for Humanity

Career Goal
To find an engineering position in robotics

---

on information that will demonstrate what you offer to employers. Later, you can customize your letter of application and your resumé for each employer, including information that will particularly interest each employer or that relates directly to a specific job opportunity.

PREPARE AN EFFECTIVE PAPER RESUMÉ

Robert Greenly of Lockheed Martin writes that "your resumé is the first impression you make. It should be eye-catching, clearly written, and easy to read." Your resumé and letter of application generally are the first information that an employer sees about you, so you want these documents to persuade employers to interview you. Follow these guidelines for writing an effective resumé:

- Organize your resumé to highlight your qualifications.
- Use dynamic, persuasive language that demonstrates what you can do.
- Proofread and remove errors.
- Create an eye-catching, accessible design for paper resumés.

One of the most commonly asked questions about resumés is "How long should my resumé be?" Your resumé should be long enough to provide employers with the information they need to understand what you can offer their organizations, but not so long as to irritate potential employers. Davidson writes that "sending a 10-page resumé is a mammoth error"; instead, highlight your abilities on one page. If you're having trouble, invite someone to help you (Davidson). Remember that "the person reviewing resumés has 15 seconds to decide to bring you in" (Davidson). If you have less than 10 years of experience in your field, try to keep your resumé to one page.

Organize Your Resumé to Highlight Your Qualifications

You can organize your resumé using one of two styles:
- chronological
- skills (sometimes called a functional resumé)

If you organize your resumé *chronologically*, you will present the information in both the work experience and the education categories in reverse chronological order: you will begin with your most recent or current job or degree and end with the least recent. The resumé shown in Figure 13.3 illustrates chronological style. In the education category, the writer begins with her most recent college work and ends with her least recent work. In the work experience category, she begins with her current work at the University of North Texas and ends with her least recent work as a server at Red Lobster restaurant. Figure 13.4, page 396, shows a chronological resumé for a non-traditional student.

Most job-seekers "prefer the logical progression of a chronological resumé" (Greenly); and for recent college graduates or for job-seekers looking for their first career job, skills resumés generally are less effective

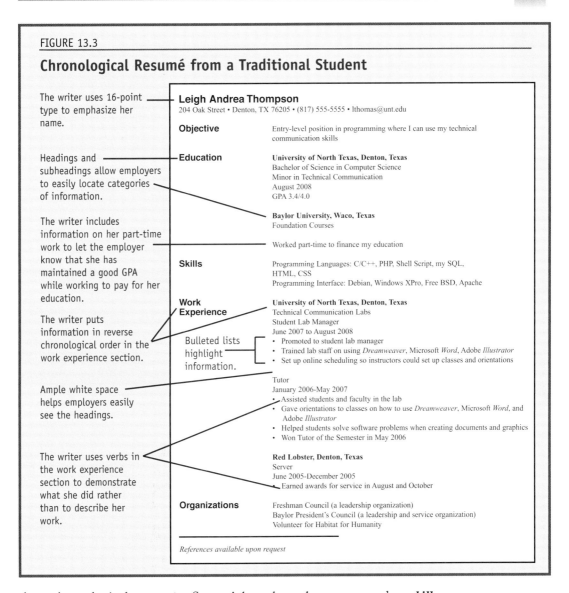

FIGURE 13.3

Chronological Resumé from a Traditional Student

The writer uses 16-point type to emphasize her name.

Leigh Andrea Thompson
204 Oak Street • Denton, TX 76205 • (817) 555-5555 • lthomas@unt.edu

Objective
Entry-level position in programming where I can use my technical communication skills

Headings and subheadings allow employers to easily locate categories of information.

Education
University of North Texas, Denton, Texas
Bachelor of Science in Computer Science
Minor in Technical Communication
August 2008
GPA 3.4/4.0

Baylor University, Waco, Texas
Foundation Courses

The writer includes information on her part-time work to let the employer know that she has maintained a good GPA while working to pay for her education.

Worked part-time to finance my education

Skills
Programming Languages: C/C++, PHP, Shell Script, my SQL, HTML, CSS
Programming Interface: Debian, Windows XPro, Free BSD, Apache

Work Experience
University of North Texas, Denton, Texas
Technical Communication Labs
Student Lab Manager
June 2007 to August 2008
• Promoted to student lab manager
• Trained lab staff on using *Dreamweaver*, Microsoft *Word*, Adobe *Illustrator*
• Set up online scheduling so instructors could set up classes and orientations

The writer puts information in reverse chronological order in the work experience section.

Bulleted lists highlight information.

Tutor
January 2006-May 2007
• Assisted students and faculty in the lab
• Gave orientations to classes on how to use *Dreamweaver*, Microsoft *Word*, and Adobe *Illustrator*
• Helped students solve software problems when creating documents and graphics
• Won Tutor of the Semester in May 2006

Ample white space helps employers easily see the headings.

The writer uses verbs in the work experience section to demonstrate what she did rather than to describe her work.

Red Lobster, Denton, Texas
Server
June 2005-December 2005
• Earned awards for service in August and October

Organizations
Freshman Council (a leadership organization)
Baylor President's Council (a leadership and service organization)
Volunteer for Habitat for Humanity

References available upon request

than chronological resumés. Some job-seekers, however, need a ***skills resumé***—one that focuses the reader's attention on the writer's marketable job skills and accomplishments rather than on a chronological listing of work experience. A skills resumé is especially effective in two situations:

• When you want to present your most important accomplishments or skills early in the resumé or at least in a lead-off position within categories (Greenly).

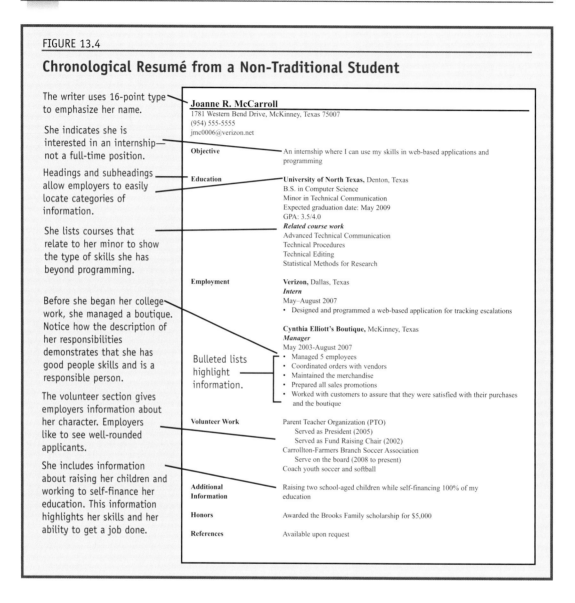

FIGURE 13.4

Chronological Resumé from a Non-Traditional Student

The writer uses 16-point type to emphasize her name.

She indicates she is interested in an internship—not a full-time position.

Headings and subheadings allow employers to easily locate categories of information.

She lists courses that relate to her minor to show the type of skills she has beyond programming.

Before she began her college work, she managed a boutique. Notice how the description of her responsibilities demonstrates that she has good people skills and is a responsible person.

The volunteer section gives employers information about her character. Employers like to see well-rounded applicants.

She includes information about raising her children and working to self-finance her education. This information highlights her skills and her ability to get a job done.

Bulleted lists highlight information.

Joanne R. McCarroll
1781 Western Bend Drive, McKinney, Texas 75007
(954) 555-5555
jmc0006@verizon.net

Objective An internship where I can use my skills in web-based applications and programming

Education **University of North Texas,** Denton, Texas
B.S. in Computer Science
Minor in Technical Communication
Expected graduation date: May 2009
GPA: 3.5/4.0
Related course work
Advanced Technical Communication
Technical Procedures
Technical Editing
Statistical Methods for Research

Employment **Verizon,** Dallas, Texas
Intern
May–August 2007
• Designed and programmed a web-based application for tracking escalations

Cynthia Elliott's Boutique, McKinney, Texas
Manager
May 2003-August 2007
• Managed 5 employees
• Coordinated orders with vendors
• Maintained the merchandise
• Prepared all sales promotions
• Worked with customers to assure that they were satisfied with their purchases and the boutique

Volunteer Work Parent Teacher Organization (PTO)
    Served as President (2005)
    Served as Fund Raising Chair (2002)
Carrollton-Farmers Branch Soccer Association
    Serve on the board (2008 to present)
Coach youth soccer and softball

Additional Information Raising two school-aged children while self-financing 100% of my education

Honors Awarded the Brooks Family scholarship for $5,000

References Available upon request

- If you want to change careers and a chronological organization might undermine your search (Greenly).

The resumé in Figure 13.5 illustrates a skills resumé. The major accomplishments category focuses on the job-seeker's skills and accomplishments in two areas: financial planning and financial analysis. The resumé

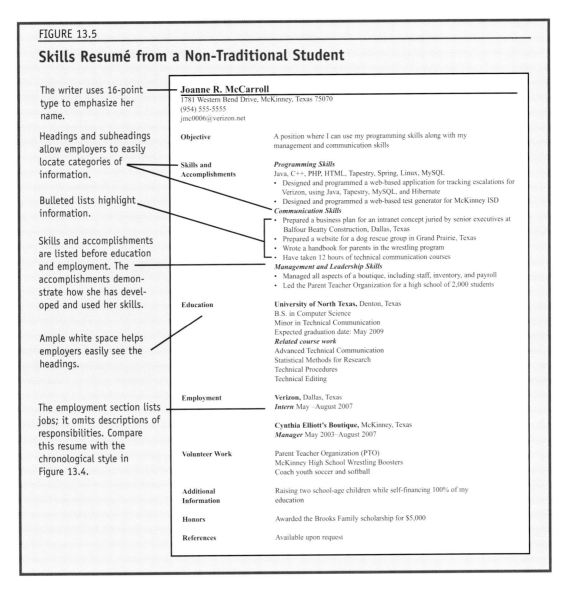

FIGURE 13.5

Skills Resumé from a Non-Traditional Student

The writer uses 16-point type to emphasize her name.

Headings and subheadings allow employers to easily locate categories of information.

Bulleted lists highlight information.

Skills and accomplishments are listed before education and employment. The accomplishments demonstrate how she has developed and used her skills.

Ample white space helps employers easily see the headings.

The employment section lists jobs; it omits descriptions of responsibilities. Compare this resume with the chronological style in Figure 13.4.

Joanne R. McCarroll
1781 Western Bend Drive, McKinney, Texas 75070
(954) 555-5555
jmc0006@verizon.net

Objective

A position where I can use my programming skills along with my management and communication skills

Skills and Accomplishments

Programming Skills
Java, C++, PHP, HTML, Tapestry, Spring, Linux, MySQL
• Designed and programmed a web-based application for tracking escalations for Verizon, using Java, Tapestry, MySQL, and Hibernate
• Designed and programmed a web-based test generator for McKinney ISD

Communication Skills
• Prepared a business plan for an intranet concept juried by senior executives at Balfour Beatty Construction, Dallas, Texas
• Prepared a website for a dog rescue group in Grand Prairie, Texas
• Wrote a handbook for parents in the wrestling program
• Have taken 12 hours of technical communication courses

Management and Leadership Skills
• Managed all aspects of a boutique, including staff, inventory, and payroll
• Led the Parent Teacher Organization for a high school of 2,000 students

Education

University of North Texas, Denton, Texas
B.S. in Computer Science
Minor in Technical Communication
Expected graduation date: May 2009
Related course work
Advanced Technical Communication
Statistical Methods for Research
Technical Procedures
Technical Editing

Employment

Verizon, Dallas, Texas
Intern May –August 2007

Cynthia Elliott's Boutique, McKinney, Texas
Manager May 2003–August 2007

Volunteer Work

Parent Teacher Organization (PTO)
McKinney High School Wrestling Boosters
Coach youth soccer and softball

Additional Information

Raising two school-age children while self-financing 100% of my education

Honors

Awarded the Brooks Family scholarship for $5,000

References

Available upon request

includes work experience, but simply lists the jobs; it does not list the responsibilities and duties relevant to each job.

Whether you use the chronological or the skills style, select information that highlights your qualifications and prompts employers to interview you. Include appropriate and effective information in these sections:

- career objective
- education
- work experience
- skills and specialized training
- personal information
- references
- other relevant information

You may choose not to include all of these sections in your resumé.

Career Objective

A **career objective** states the kind of work you are seeking in the form of a brief phrase. For example, you might write the following for your objective: "Entry-level position in commercial building construction." Many hiring managers consider these statements important because they indicate that the writer has goals. However, other managers find these statements vague, especially if the statements are general or don't relate directly to the advertised job. The following career objective could cause a manager to pass over a qualified job applicant: "An entry-level position in computer programming with the opportunity to advance into management." A broad statement like that can have unintended consequences. After reading such a statement, a potential employer might decide not to consider the applicant for a job that will not quickly lead to a management position. Resumés that omit an objective can give employers "greater flexibility in considering you for any number of peripheral positions that your experience and training qualify you for, perhaps even future openings that do not yet exist" (Greenly). If you decide to include a career objective, follow these guidelines:

- Use brief, specific statements directly related to the job for which you are applying.
- Include only the position, goals, or tasks specifically stated in the job advertisement.
- Avoid general, broad statements such as "a position where I can use my programming skills."

Education

In the education section, you give employers information about your degrees. In this section, you identify the following:

- your college or university degrees
- the institution awarding the degrees

- the location of the institution
- the date you received or will receive the degree

If you haven't yet graduated, list the colleges or universities that you have attended beyond high school, their location, and the expected date of your degree. For example, a senior at San Diego State University might write the following:

> San Diego State University, San Diego, California
> B. S. in Mechanical Engineering
> Expected graduation, May 2008

In addition to listing your degree (or degrees), you can include information such as the following:
- courses that qualify you for the type of job you are seeking
- minors or double-majors
- academic scholarships or fellowships that you received
- a high grade-point average (above 3.0 on a 4-point scale)
- academic (merit-based) honors or awards that you received as a college or university student
- any outstanding accomplishments, such as special projects or research that you did as a student

If you list courses related to the work you are seeking, include primarily upper-level courses in your major. List courses by title, not number. For example, if you want to mention a course in advanced automated systems, write "Advanced Automated Systems," not "MECH 4302." Figure 13.4, page 396, shows how you might list courses in the education section of your resumé.

Include appropriate and effective information in each section of your resumé to highlight qualifications and prompt employers to interview you.

The education section shown in Figure 13.6, page 400, includes several of the optional items, along with the writer's degree information. The writer lists his grade-point average (3.7) along with a reference point (4.0). He also mentions his academic honors. Notice that he states in this section that he worked part-time to help pay for his education. With this format, he is highlighting his impressive achievement of earning a high grade-point average and receiving honors while holding down a job.

FIGURE 13.6

Sample Education Section

Education	**University of New Mexico**, Albuquerque, New Mexico
	B.S. in Mechanical Engineering
	Expected graduation in August 2009
	GPA: 3.7/4.0
	Honors
	Dean's List (Fall 2006, Spring 2007, Fall 2007, Spring 2008)
	Alpha Lambda Delta (Freshman Honor Society)
	Tau Beta Pi (General Engineering Society)
	The National Society of Collegiate Scholars
	Worked part-time to finance my education

Work Experience

In the work experience section, you include information on your jobs. You might also title this section "Employment." List your jobs in ***reverse chronological order***, beginning with your most recent experience. For each job, include the following information:

- name and location of the organization for which you worked
- the years (or months, if less than a year) of your work with that organization
- your job title
- verb phrases describing the work you did (your job responsibilities)

As you describe your responsibilities, show how well you did your job and demonstrate that you can produce results. "The most qualified people don't always get the job. It goes to the person who presents himself [or herself] most persuasively in person and on paper. So don't just list where you were and what you did ... tell how well you did. Were you the best salesperson? Did you cut operating costs? Give numbers, statistics, percentages, increases in sales or profits" (Simon).

Figures 13.7 and 13.8 illustrate how two writers describe their work experience. In Figure 13.7, instead of using vague, unimpressive language, the writer, an experienced job-seeker, uses specific information to demonstrate how well he did his job:

Vague and unimpressive	Facilitated work teams Improved production
Specific and impressive	Facilitated two self-directed work teams of 26 team members Reduced cycle time from 5 days to 2 days Saved $250,000 annually

The writer doesn't just tell prospective employers that he saved the company money or reduced the cycle time. He states the specific reduction in days and the amount saved in dollars.

FIGURE 13.7

Work Experience Section for a Job Seeker with Experience

Employment **Texas Instruments, Inc.,** Dallas, Texas
Manufacturing Facilitator (2005-present)
- Facilitated two self-directed work teams of 26 team members performing screen printing and painting operations
- Led screen printing team to win Gold Teaming for Excellence Award
- Converted coating system to low VOC formulations that comply with existing air-quality standards

Reengineering Team Leader (2002-2005)
- Reengineered screen printing work flow to eliminate non-value added effort and reduce task handoffs from one person to another. Reduced cycle time from 5 days to 2 days, increased productivity by 25%, and saved $250,000 annually
- Received Site Quality Improvement Award two years for reducing cycle time and improving quality (2003, 2004)

Process Improvement Engineer of Finish and Assembly Areas (2000-2002)
- Designed and installed custom equipment and machine upgrades that reduced manual labor required by $100,000 annually
- Improved part racking on plating line, reducing scrap by $20,000 annually

---

FIGURE 13.8

Work Experience Section for a Job Seeker with No Experience

Work Experience **University of North Texas,** Denton, Texas
Lab Manager, Technical Communication Labs (2006-present)
- Received Tutor of the Semester Award (Fall and Spring 2006)
- Promoted to lab manager
- Trained new tutors on how to use the software
- Answered students' questions about Microsoft *Word, Dreamweaver, InDesign,* and Adobe *Illustrator*
- Helped students and faculty with computer and software problems

Chili's, Lewisville, Texas
Server (2004-2006)
- Received the Outstanding Service Award (2005)

The experience section in Figure 13.8 is excerpted from a college student's resumé. This student is looking for her first job in her field. She doesn't have the extensive work experience of the writer in Figure 13.7. However, she demonstrates how well she did the jobs she held as a student at the University of North Texas. She lists her award for service at Chili's restaurant, along with the year she received the award. She also lists her Tutor of the Semester award. These awards and her promotion demonstrate that she did her jobs well.

Like the writer in Figure 13.8, you may not have work experience in your field. However, you can list your summer or part-time jobs. If you were promoted or received any awards while in these jobs, include this information to help potential employers see that you are reliable and hardworking.

Skills and Specialized Training

Some writers include a section listing their skills or any specialized training or education they have received. These sections are most common in skills resumés. In a skills resumé, this section usually appears prominently near the beginning and can have various headings such as "Major Accomplishments" or "Skills." Figure 13.9 shows the skills section from the resumé of a job-seeker wanting to change careers. Many job-seekers in computer science list their programming languages in a skills section. Figure 13.10 shows a specialized training section. If you decide to use these sections in your resumé, include only information relevant to the type of job you are seeking.

Personal Information

You may want to include personal information that gives readers "a glimpse of the personal you" and furthers "the image you've worked to project in the preceding sections" of the resumé (Simon). You can list any of the following information if it will enhance a prospective employer's picture of you:
- volunteer work (such as work for charitable and other non-profit organizations, membership in community service organizations, or leadership or work with community youth organizations)
- college activities (such as membership on teams and in organizations, offices held, and awards won)
- professional memberships (such as in organizations in your field, including any leadership positions you've held) You can also put professional memberships information in a separate section.

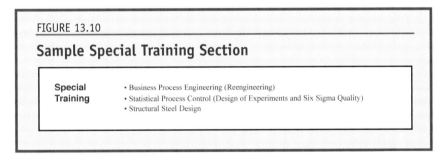

FIGURE 13.9

Sample Skills Section

Skills	
	Management Skills
	• Hired and supervised 14 employees in a $1.3 million catering business
	Communication Skills
	• Trained new employees in a catering business
	• Designed marketing brochures and the Web site for a catering business
	• Mentored new employees to improve their performance
	People Skills
	• Coached youth soccer, baseball, and basketball
	• Tutored math for the Boys' and Girls' Club

FIGURE 13.10

Sample Special Training Section

Special Training	
	• Business Process Engineering (Reengineering)
	• Statistical Process Control (Design of Experiments and Six Sigma Quality)
	• Structural Steel Design

- personal interests and hobbies, especially if they relate to your qualifications.
- sports or recreational activities that you enjoy.

Perhaps you've been a Girl Scout or Boy Scout leader in your community, or you've organized a blood drive for your college or university campus, or you've won awards for your leadership abilities. These activities and awards show that you are a team player, that you care about your community, and that you are disciplined.

As you prepare the personal section, do not include information that might invite employers to discriminate—for example, information about your religion, ethnicity, marital status, age, or health.

Evaluate a paper
resumé in the
Interactive Student
Analysis online at
www.benttreepress.
com/technical
communication.htm.

Use Dynamic, Persuasive Language that Demonstrates What You Can Do

The guideline for word choice in your resumé is simple: keep your writing style clear and uncluttered. Exclude extraneous information; and use dynamic, persuasive language (see "Tips for Using Dynamic, Persuasive Language in a Resumé" below). The language—along with the design—of your resumé gives employers their first impression of you. You want that impression to be positive.

The following examples illustrate phrases that use dynamic, persuasive language (the dynamic action verbs appear in bold type):

Not Dynamic/Persuasive   Created a computer program for students logging into the lab

Dynamic/Persuasive   **Designed** and **programmed** software that **reduced** the number of employees needed in the student

 TIPS FOR USING DYNAMIC, PERSUASIVE LANGUAGE IN A RESUMÉ

- **Keep the information and language simple and direct**. The employer reading your resumé may be reading hundreds of other resumés for the same job.
- **State your information or qualifications directly; omit any unnecessary information.** Be brief. Give employers the information they need to know about your abilities and background—and then stop.
- **Use dynamic action verbs**. Use verbs such as "saved," "designed," "supervised," "directed," and "designed." Avoid words and phrases that don't describe action or demonstrate what you have achieved and can achieve. Avoid phrases such as "my responsibilities included" or "my tasks and duties were."
- **Use specific language that emphasizes your accomplishment and what you can do**. Use specific language that demonstrates what you have accomplished and what you can do. When possible, use statistics and numbers to show rather than to describe. Statistics and numbers can effectively demonstrate your abilities.
- **Use verb phrases, not sentences**. Because you want to focus on what you have accomplished, use verb phrases, not full sentences.

	computer labs and **saved** the university $16,640 annually
Not Dynamic/Persuasive	In charge of charity gala for my sorority
Dynamic/Persuasive	**Coordinated** the Red Dress charity gala that **raised** $8,600 for the American Heart Association
Not Dynamic/Persuasive	Was responsible for designing and installing custom equipment and upgrading machines
Dynamic/Persuasive	**Designed** and **installed** custom equipment and machine upgrades that **reduced** the manual labor required by $100,000 annually

Create an Eye-Catching, Accessible Design for Paper Resumés

Design your resumé so an employer can quickly get a good idea of your qualifications (Greenly). You want to design your resumé so that "employers don't have to hunt for your qualifications"; employers should be able to quickly locate your qualifications without "playing detective" (Parker and Berry). If they have to play detective, they may overlook or ignore your resumé. To help employers, create visual categories with white space, headings, type sizes, and bulleted lists.

In the resumés shown in Figures, 13.3, 13.4, and 13.5, pages 395-397, employers can easily spot the categories of information because the headings are surrounded by white space, making them stand out. These headings allow employers to quickly locate information about the writer's qualifications. The writers use boldface type and different type sizes within the categories to highlight and prioritize information. The writers also use bulleted lists to help employers easily read about their work.

ETHICS NOTE

Honesty in Your Resumé

Good jobs are competitive. Only 1 of 1,470 resumés put into circulation ever actually results in a job offer (Bolles). Given such odds, young professionals, like you, are often tempted to tamper with the information on their resumés. According to one survey, fully one-third of people between the ages of 15 and 30 were willing to lie on their resumés; and experts predict that the percentage of job hunters who actually do so may make up one-third of applicants or more.

Such deception occurs at the most elite levels: Jean Houston, the infamous psychologist who counseled Hillary Rodham Clinton to imagine herself in dialogue with deceased luminaries like Eleanor Roosevelt, reported on her resumé having received a doctorate in the philosophy of religion from Columbia University. However, Joseph Berger, a writer for the *New York Times* reported that Houston had never completed her dissertation, a requirement for the doctorate. She had lied on her resumé.

Houston claimed that an aide had selected a resumé from the "bottom of the barrel" of resumés she kept on file. Houston's ethical lapse on her resumé—falsifying her credentials—is one that some resumé writers will make. Other common resumé deceits include inflating one's title or responsibilities. Employers have become more alert to the probable areas of deception and are double-checking advanced degrees, unexplained periods in employment history, and job titles.

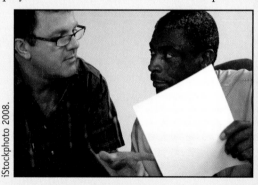

iStockphoto 2008.

So the bottom line: *Your resumé must be honest and accurate.*

TIPS FOR DESIGNING EFFECTIVE PAPER RESUMÉS

- **Use headings and subheadings to create visual categories**. Use type size to differentiate among the headings, subheadings, and text of the resumé. If you want to further differentiate headings and subheadings from the text, use bold type.
- **Surround the headings with enough white space for employers to easily see them**. Highlight headings with white space (see Chapter 11 for information on white space.)
- **Use bulleted lists instead of paragraphs**. You can help employers locate information by using bulleted lists instead of paragraphs, especially in the work experience section. A prospective employer glancing at a paragraph might miss important information, such as an award received for excellence. A bulleted list highlights the award.
- **Select a typeface that is professional and easy to read**. Many writers use a serif typeface, such as Times Roman, but traditional sans-serif typefaces, such as Helvetica or Tahoma, are also appropriate for resumés. Whether you choose a serif or sans-serif typeface, select one that is easy to read and looks professional.
- **Use 8 ½ x 11-inch white bond paper**. Use good-quality paper. Because some employers will scan or copy your resumé, use white paper; even off-white paper will darken on a scanned image.
- **Proofread—then proofread again!** Make sure your resumé is free of grammatical, spelling, and punctuation errors. Even the smallest punctuation error can cost you an interview.

PREPARE EFFECTIVE ELECTRONIC RESUMÉS

Most employers use an applicant-tracking system to post job openings, screen resumés, and generate interview requests to potential job candidates. These employers will expect you to send your resumé electronically; these resumés may be referred to as e-resumés. An *e-resumé* can be located and used by search engines, searched by keywords, converted into other electronic file types, such as a database, and printed (Ireland). You can send your resumé electronically in three ways:
- an e-mail attachment (as a pdf or word-processing file)
- a text-based, scannable resumé
- a Web resumé

The e-mail and the Web resumé have the same formatting as the traditional paper resumé. The Web-based resumé contains links to items such as documents in your electronic portfolio and your e-mail.

E-Mail Attachment

For many employers, you can simply attach your resumé as a word-processing or pdf file. If possible, use a pdf format because pdf files are becoming more commonly used and are an excellent way to send your formatted resumé. With a pdf file, your resumé will appear the same, regardless of the employer's Internet browser; this file type ensures that your formatting does not change when an employer opens your resumé.

If you use a word-processing file other than one created by Microsoft *Word*®, save your document as a pdf file. If the employer requests a plain-text document sent in the body of the e-mail message, do not attach the file. Follow the guidelines in the next section for creating a text-based, scannable resumé.

Text-Based, Scannable Resumé

A text-based, scannable resumé has the same information and major categories as a traditional paper resumé. However, it has no indentations or graphic elements, such as lines, bullets, or symbols. In your text-based, scannable resumé, you will use a simpler design and focus on keywords (usually nouns) to accommodate the applicant-tracking system and scanning. For example, if you are applying for a job requiring good oral and written communication skills, you might include "writing," "public speaking," "editing," and "presentations." If you are seeking a job in health management, you might include keywords such as "health science," "management," "interpersonal communication skills," "public health," and "health care management."

Figure 13.11 is a text-based, scannable version of the traditional, paper resumé in Figure 13.3, page 395. The writer has used no formatting (only letters, numbers, and basic punctuation marks) and has eliminated all boldface, bullets, italics, and horizontal lines. The writer has also included a list of keywords.

FIGURE 13.11

Text-Based, Scannable Resumé

The writer uses a sans-serif type for easy scanning.

The writer aligns all text on the left margin.

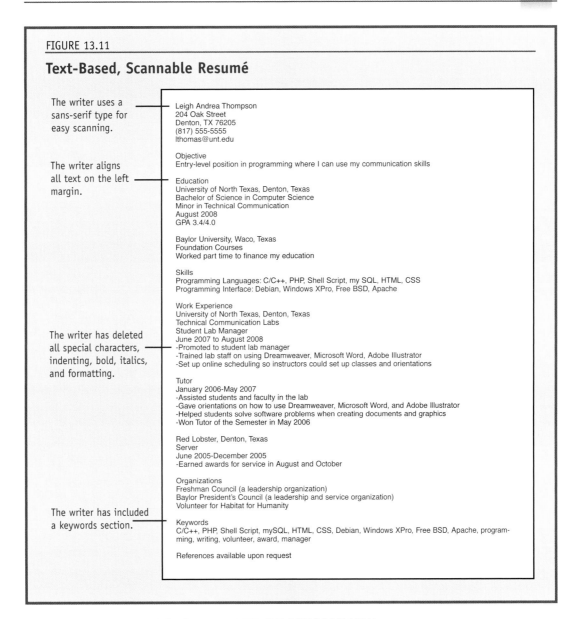

Leigh Andrea Thompson
204 Oak Street
Denton, TX 76205
(817) 555-5555
lthomas@unt.edu

Objective
Entry-level position in programming where I can use my communication skills

Education
University of North Texas, Denton, Texas
Bachelor of Science in Computer Science
Minor in Technical Communication
August 2008
GPA 3.4/4.0

Baylor University, Waco, Texas
Foundation Courses
Worked part time to finance my education

Skills
Programming Languages: C/C++, PHP, Shell Script, my SQL, HTML, CSS
Programming Interface: Debian, Windows XPro, Free BSD, Apache

The writer has deleted all special characters, indenting, bold, italics, and formatting.

Work Experience
University of North Texas, Denton, Texas
Technical Communication Labs
Student Lab Manager
June 2007 to August 2008
-Promoted to student lab manager
-Trained lab staff on using Dreamweaver, Microsoft Word, Adobe Illustrator
-Set up online scheduling so instructors could set up classes and orientations

Tutor
January 2006-May 2007
-Assisted students and faculty in the lab
-Gave orientations on how to use Dreamweaver, Microsoft Word, and Adobe Illustrator
-Helped students solve software problems when creating documents and graphics
-Won Tutor of the Semester in May 2006

Red Lobster, Denton, Texas
Server
June 2005-December 2005
-Earned awards for service in August and October

Organizations
Freshman Council (a leadership organization)
Baylor President's Council (a leadership and service organization)
Volunteer for Habitat for Humanity

The writer has included a keywords section.

Keywords
C/C++, PHP, Shell Script, mySQL, HTML, CSS, Debian, Windows XPro, Free BSD, Apache, programming, writing, volunteer, award, manager

References available upon request

WRITE A READER-FOCUSED LETTER OF APPLICATION

You will need a letter of application, or cover letter, to send with your resumé. If you send your resumé electronically, you might also include a letter of application as an e-mail. With this letter (and your resumé), employers want to know as quickly as possible "what do you want to do and what are you good at" (Hansen, K.). The employer does not want to "wade through lots of text to find out" (Hansen, K.). Put yourself in the employer's position: ask yourself what you can do for the employer—not what the employer can do for you.

A letter of application introduces your resumé and gives the employer additional information about you and your experience. Address the letter

---

TIPS FOR PREPARING TEXT-BASED, SCANNABLE RESUMÉS

- **Include keywords an employer might use to search for qualified job candidates**. For example, if an employer is looking for someone who has experience designing Web sites, you might include keywords such as "Web site," "Web pages," "Java," "Web-based applications," or "HTML." You might also include industry-specific language that an employer might search for in a resumé database. Include keywords only if they refer to skills you actually have: never lie or mislead on a resumé. You might consider adding a keywords section, as did the writer in Figure 13.11, page 409.
- **Include nouns as keywords**. "While action verbs are still important, you need to add key phrases and nouns that could be used as search terms by your potential employer" (Hansen, R.).
- **Use a 10 or 12-point sans-serif type**. You might select Helvetica, Arial, or Tahoma. Omit all bold, italics, underlining, special characters, and formatting, such as horizontal or vertical lines and graphics.
- **Align all text to the left margin**. Don't indent, center, or use double columns.
- **Use a line length of no more than 65 characters**. If you use lines longer than 65 characters, the lines may not appear as you intend.
- **Use spaces instead of tabs**. Readers may have their default tabs set differently than yours.
- **When you save your resumé file to post online, save it as plain text or ASCII.**
- **Open your resumé in a text editor, such as Notepad®, or e-mail your resumé to yourself before sending it to an employer.** Make sure your resumé appears as you intend. For example, if you use Hotmail as your e-mail provider, send your resumé to someone who uses Yahoo! Mail to see how your resumé transmits. You may have to adjust the design of your resumé.
- **When sending the scannable resumé on paper, use a high-quality laser or inkjet printer and use only white, 8$\frac{1}{2}$ x 11 inch paper.**

personally to the executive or manager most likely to make the hiring decision (Greenly; Simon). Before writing your letter, do your homework; find out where to send the letter and who will read it.

- Address your letter to a specific person, and spell the addressee's name correctly.
- If you don't know who should receive your letter, don't address it to "Dear Sir or Madam," "To Whom It May Concern," a department, or a person's job title unless the advertisement says to address the letter in that manner.
- Call the organization to find out who will receive your resumé. If you can't find out by phone, address the letter to an executive, such as the president or chief executive officer, and use the person's name.

Once you have determined who will read your letter of application, customize your letter for that reader and appeal directly to his or her needs (Greenly; Hansen, K.). For example, if you have classroom, internship, or employer experience relevant to the job for which you are applying, discuss that experience or classwork in the body of your letter. Remember, employers want to know the following:

- what you can do for the organization (Hansen, K.)
- how you and your skills and qualifications will benefit the organization (Hansen, K.)
- how you will fit into the team

Your customized, generally one-page letter, will have three sections:

- **your purpose for writing:** the introductory paragraph
- **your qualifications:** the education and experience paragraphs
- **your goal** (what you want from the employer): the concluding paragraph

Your Purpose for Writing: The Introductory Paragraph

In the introductory paragraph, tell the employer why you're writing. As you write, follow these guidelines:

- **Identify the position for which you are applying**. Employers often receive many letters of application for several jobs at the same time, so identify the specific job you are seeking.

- **Tell the employer where you found out about the job**. Because employers may be soliciting resumés in more than one place, they often want to know where you found out about the job. This information is especially important if you learned of the job from an employee, coworker, or acquaintance of the employer. This information may lead the employer to show more interest in your resumé. If you are writing an unsolicited letter, "quickly explain why you are approaching the company," and then ask whether a job is available (Simon).

Figure 13.12 presents three sample introductory paragraphs. Each specifically identifies the job and the writer's purpose for writing. The first writer uses a personal contact (Dr. Kathryn Raign) to open the paragraph and get the employer's attention. The second writer mentions a specific job advertisement. The third writer is sending an unsolicited letter. That writer is not responding to a specific job advertisement and doesn't know whether the company currently has job openings.

As these introductions illustrate, the tone of a letter of application must be positive and confident—not tentative or boastful. State your qualifications in a positive manner without focusing on your weaknesses, but be careful not to sound arrogant. You want to appear confident about your education, experience, and abilities while indicating that you know you have much to learn and are eager to work in your profession.

FIGURE 13.12

Sample Introductory Paragraphs

Using a personal contact

Dr. Kathryn Raign suggested that I contact you about the project engineer position you currently have open in the Orlando office. My experience as an intern for Balfour Beatty provides me with the qualifications you are seeking. Please consider me for the project engineer position.

Using a specific job advertisement

My course work in computer science and my experience as an intern for Microsoft qualify me for the Web-applications designer position that you posted on CareerBuilders.com on May 16. Please consider me for this position.

Sending an unsolicited letter

My experience as an intern at BlueCross/BlueShield, my work as a health technician at Parkland Hospital, and my double major in biology and health management give me a solid foundation in health management. Please consider me for a position in your management training program.

Your Qualifications: The Education and Experience Paragraphs

After you have told the employer why you are writing, present information about your education and experience. As you write the education and experience paragraphs, follow these guidelines:

- **Follow the order of your resumé when discussing your education and work experience**. If your resumé gives information about your education first, then discuss your education first. If your resumé gives information about your work experience or skills first, then discuss your experience or skills first. When you have many years of work experience, you can eliminate the education paragraph and include two or more experience paragraphs.

- **Highlight, add to, or expand on the information in your resumé**. Don't simply repeat the information in your resumé or give the details of your education and experience in chronological order. Instead, highlight or add to information that may especially interest the employer or that is particularly relevant to the job for which you are applying. For example, in your resumé, you may have stated that you had an internship; in the letter, expand on that internship information by including a project on which you worked or by detailing some of your responsibilities.

- **Create a unified theme in these paragraphs**. Avoid the temptation to simply list (often unrelated) information about your education or work experience. Instead, begin each paragraph with a topic sentence and then develop that topic in the sentences that follow. As you discuss your education, consider how it uniquely qualifies you for the job that you seek. For instance, if the job advertisement says that applicants should write well, you might discuss projects where writing was a significant component.

- **Consider how your experience uniquely qualifies you for the job**. This task is especially difficult if your experience does not directly relate to the job you are seeking. For example, Rodney is a new college graduate looking for an engineering job. He has never worked in the field of engineering, but he worked as a tutor in a computer lab for three years and was promoted to student manager of the lab. He has several skills that will impress employers. He was promoted because of his ability to work well with others and his ability to supervise his peers. In addition, his university implemented several of his ideas, such as putting a flat-screen monitor with the weekly schedule of classes outside the lab so that students would know when their classes met in the lab and when the lab was open for general access. This idea saved the lab $400 a year in paper costs. Although Rodney's experience is not directly related to engineering, he can write a paragraph focusing on his abilities to work with others and to be a team player by suggesting money-saving ideas.

As Katherine Hansen, author of *New Grad Resumés and Cover Letters*, explains, "Experience is experience. It doesn't have to be paid. Anything you've done that has enabled you to develop skills that are relevant to the kind of job you seek is worth consideration for resumé and cover letter mention. That's especially true if you don't have much paid experience. The key . . . is relevance. Consider the following in evaluating what experience and skills you've gained that are relevant to what you want to do when you graduate: internships; summer jobs; campus jobs; sports; entrepreneurial/self-employed jobs; temporary work; volunteer work." Hansen also suggests that you consider your research papers or projects, your campus activity positions, fraternity/sorority/social club positions, and any extracurricular or sports leadership positions.

Figures 13.13 and 13.14 illustrate how two writers approached the experience and education paragraphs for a letter of application. The writer of the paragraphs shown in Figure 13.13 has no work experience in her field. The writer of the paragraphs shown in Figure 13.14 has work experience. The writer with work experience begins with his experience and moves to education because his resumé follows that order. This writer also includes his major, degrees received, and school in the education category. The tone of both writers is confident as they mention facts about their education and experience and state qualities and experiences that are relevant to potential employers.

Your Goal: The Concluding Paragraph

In the concluding paragraph, directly state what you want from the reader: an opportunity to meet the employer and discuss your qualifications. In the paragraphs preceding the conclusion, you provided specific, detailed information about yourself—information to convince the employer to invite you for an interview. In the concluding paragraph, do the following:

• **Refer the employer to your resumé**.

- **Request an interview**.
- **Tell the employer how to contact you by phone and e-mail**. Give the employer your phone number, and mention the best time to call. You can encourage the employer to act by including this specific information in the concluding paragraph. As Katherine Hansen suggests, powerful letters of application "include every possible way to reach you." This information should appear both in the resumé and the letter.
- **Make sure that your e-mail address is professional**. Your e-mail address says something about you. Bill Behn, a national director of staffing for SolomonEdwardsGroup, reports that "I actually had an interviewee tell me to contact her via e-mail at likes2party@aol.com. Needless to say, that person was not offered the job."

FIGURE 13.13

Experience/Education Paragraphs for a Job Seeker with No Experience

At Chambers University, I have taken many courses requiring writing. In an advanced technical communication course, I used *InDesign* to produce a 40-page user's manual for inventory software used by Minyards, Inc. (a regional grocery-store chain). Currently, all Minyards stores use the manual to train new employees on the inventory system and as a reference guide for employees after initial training.

For the past three years, I have worked in the Technical Communication Computer Lab at Chambers University. I began as a lab tutor, assisting students with software questions, especially related to Microsoft *Word, InDesign,* and Microsoft *PowerPoint*. After eighteen months, I was promoted to student lab manager. As manager, I work with the faculty to schedule classes in the lab, work with the lab tutors to set up their schedules, and conduct meetings each week with the lab tutors. Most recently, I set up a scheduling system that uses e-mail instead of paper. This system saved $400 in paper costs annually. As manager and tutor in the lab, I have developed interpersonal skills that would benefit Writers, Inc.

FIGURE 13.14

Experience/Education Paragraphs for a Job Seeker with Experience

While at Texas Instruments, I worked as an innovative design engineer. I have more than 15 years of research, production, and manufacturing experience, especially in the areas of machine design, power transmission, and structural analysis. I began as a process improvement engineer and was promoted to reengineering team leader and finally to manufacturing facilitator. As manufacturing facilitator, I supervised two self-directed work teams of 26 total team members. I led one of these teams, the screen printing team, to win the Gold Teaming for Excellence Award. I also received the Site Quality Improvement Award in 2003 and 2004 for increasing productivity by 25% annually.

Along with my experience as a design engineer for Texas Instruments, I have a Bachelor of Science and a Master of Science degree in agricultural engineering from Texas A&M University. As part of my academic experience, I worked as a research assistant in the agricultural engineering department. I designed and constructed custom equipment and instrumentation used in energy conservation research.

Use specific language in the concluding paragraph. Avoid vague language, as in the following examples. The writers don't confidently state the goal of meeting the employer or encourage the employer to contact them:

Vague   I look forward to hearing from you soon. Thank you for considering my resumé.

Vague   If possible, may I meet with you or someone in your company to discuss my resumé and my qualifications?

The writers of the following paragraphs refer the employer to their resumé and directly ask the employer to contact them. (For more information on tone, see Chapter 12.) These writers also use a polite, respectful, confident tone:

Specific   My resumé provides additional information about my education and work experience. I would enjoy discussing my application with you. Please write me at raign@yahoo.com, or call me anytime at (307) 555-9061.

Specific   You can find more information about my education and experience in the enclosed resumé. I would appreciate the opportunity to discuss my resumé with you at your convenience. Please e-mail me at jscott@verizon.net or call me at (505) 555-9033 weekdays or at (505) 555-0034 evenings and weekends.

Figure 13.15 illustrates an effective letter of application. The writer uses a respectful yet confident tone and includes specific information to persuade the employer to invite her for an interview.

PREPARE FOR A SUCCESSFUL INTERVIEW

You've written a successful resumé and letter of application, and now you have an interview. As when writing your resumé and letter of application, you need to prepare for the interview. Successful job seekers spend time

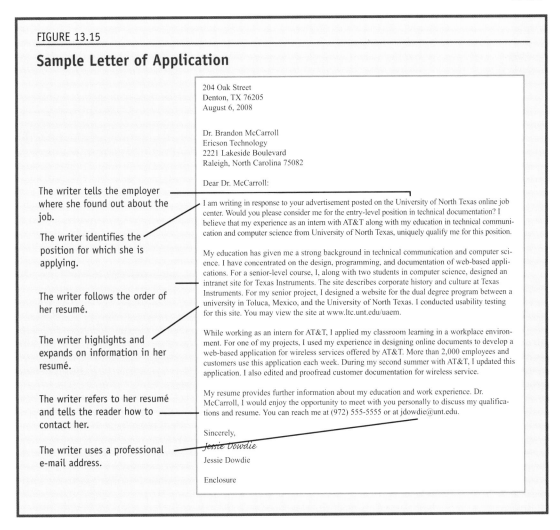

FIGURE 13.15

Sample Letter of Application

204 Oak Street
Denton, TX 76205
August 6, 2008

Dr. Brandon McCarroll
Ericson Technology
2221 Lakeside Boulevard
Raleigh, North Carolina 75082

Dear Dr. McCarroll:

The writer tells the employer where she found out about the job.

I am writing in response to your advertisement posted on the University of North Texas online job center. Would you please consider me for the entry-level position in technical documentation? I believe that my experience as an intern with AT&T along with my education in technical communication and computer science from University of North Texas, uniquely qualify me for this position.

The writer identifies the position for which she is applying.

My education has given me a strong background in technical communication and computer science. I have concentrated on the design, programming, and documentation of web-based applications. For a senior-level course, I, along with two students in computer science, designed an intranet site for Texas Instruments. The site describes corporate history and culture at Texas Instruments. For my senior project, I designed a website for the dual degree program between a university in Toluca, Mexico, and the University of North Texas. I conducted usability testing for this site. You may view the site at www.ltc.unt.edu/uaem.

The writer follows the order of her resumé.

While working as an intern for AT&T, I applied my classroom learning in a workplace environment. For one of my projects, I used my experience in designing online documents to develop a web-based application for wireless services offered by AT&T. More than 2,000 employees and customers use this application each week. During my second summer with AT&T, I updated this application. I also edited and proofread customer documentation for wireless service.

The writer highlights and expands on information in her resumé.

My resume provides further information about my education and work experience. Dr. McCarroll, I would enjoy the opportunity to meet with you personally to discuss my qualifications and resume. You can reach me at (972) 555-5555 or at jdowdie@unt.edu.

The writer refers to her resumé and tells the reader how to contact her.

Sincerely,

Jessie Dowdie

Jessie Dowdie

The writer uses a professional e-mail address.

Enclosure

doing their homework before the interview. Follow the tips on page 420 to ensure your job interview is successful.

USE LETTERS TO FOLLOW UP

Follow-up letters are important to your job search. Make sure your letter is error free. You can write a follow-up letter in these situations:

- **when you have sent a letter of application and resume and have not received a response within three or four weeks**—If you have not received a response, write a brief, polite letter. Mention your previous letter and its date, and include another copy of your resumé. To know when to write such letters, keep copies of all the letters of application that you send, and keep a file of the responses you receive from employers. Without these copies and a detailed file, you may not know when to send follow-up letters.

- **after an interview**—Within two days following an interview, write a brief thank-you letter addressed to the manager who will decide whether to hire you. If you had extensive interviews with more than one person, write to all the people who interviewed you. In your letters, state your interest in the job and the organization. Use these letters to reinforce what you offer the organization—what you can bring to the job. Mention the organization by name, and mention the names of people in the organization with whom you met.

- **when you accept a job**—When you accept a job, write a brief letter confirming your acceptance. In this letter, you can confirm details such as when you will begin work.

- **when you reject a job offer or no longer want an employer to consider you for a job**—When you accept a job, don't forget to write the other organizations that seriously considered you for a job. You may want to work for or with one of those organizations in the future, so do them the courtesy of writing a brief letter. Thank the organization and the people who interviewed you for their interest in you. State that you have taken a job with another organization. You don't have to identify the specific job offer that you accepted; instead, you can simply write, "I have decided to accept another offer." Include only positive comments about the organization and your experiences with the interviewer. End your letter with a brief statement of goodwill, such as, "Thank you for the interest you showed in my application."

Follow-up letters are most effective after an interview. Post-interview letters offer an excellent opportunity to restate your qualifications and to add any information about your application that you didn't have the opportunity to discuss during the interview (Simon). Figure 13.16, on the next page illustrates a follow-up letter that you would send after an interview.

FIGURE 13.16

Sample Follow-Up Letter

---

402 Spring Avenue, Apt. 6C
Alexandria, VA 23097

May 4, 2009

Mr. Dwight Wilson
Senior Production Engineer
I-2 Technology, Inc.
San Diego, CA 92093

Dear Mr. Wilson:

Thank you for taking time from your busy schedule yesterday to show me I-2 Technology's facilities and to discuss the quality control job. I especially enjoyed meeting many of your coworkers. Please thank Ms. Johnson in the quality control division.

As a result of our visit, I have a good understanding of I-2 Technology and appreciate its progressive approach to maximizing production without sacrificing quality control. I feel confident that my work as an intern at Boeing provides the experience you are looking for in your team.

I-2 Technology's place in the semiconductor industry and your colleagues in the quality control division confirm my impression that I-2 Technology would be an exciting place to work. If I can answer further questions, please call me at (703) 555-0922.

Best regards,

Cynthia Demsey

Cynthia Demsey

TIPS FOR A SUCCESSFUL JOB INTERVIEW

Before the Interview
- **Do the research necessary to understand the position and the company to which you are applying.** Kip Hollister, founder and CEO of Hollister, Inc. staffing, explains that "one of the biggest turn-offs for a hiring manager is when an interviewing candidate has not done the research necessary to understand both the position and the company" (Zupek). Find out what products the company produces or what services it provides. You can begin by visiting the company's Web site: read their mission statement, find out about their locations, etc.
- **Create a list of good questions to ask.** Near the end of the interview, the interviewer will probably ask you if you have any questions. If you do not ask questions or if you do not ask good questions, the interviewer may assume that you are not interested in the position or that you are unprepared. Rachel Zupek of CareerBuilder.com recommends asking open-ended questions—questions that require more than a one or two-word response. For example, you might ask, "How do you see me fitting in at your company?" or "What will make the person who takes this position successful?" (Zupek).
- **Study lists of common interview questions.** You can visit job boards and your college or university placement center for lists of common interview questions. These questions will help you prepare for what an interviewer may ask you.
- **Hold a mock interview.** Rehearse the interview by asking a friend, family member, or professor to hold a mock interview with you. Your college or university placement center may also hold such interviews.
- **Decide what you will wear.** Don't neglect your appearance. Bill Behn, a national director of staffing for SolomonEdwardsGroup, recommends that you "dress for the position you want to have." You should dress conservatively; avoid clothing that is too revealing, too casual, or too outrageous. Also, don't wear too much jewelry or cologne, and don't chew gum.
- **Make sure you know where you are going for the interview.** If you will be driving, make sure you know how to get there. If you are unfamiliar with the location, consider driving to the location a day or two before the interview.

The Day of the Interview
- **Look over your list of questions and your questions for the interviewer.**
- **Arrive early.** If you arrive late, the interviewer may assume that you will have sloppy work habits.

At the Interview
- **Shake the interviewer's hand and look him or her in the eyes**. Give the interviewer a firm handshake.
- **Use the interviewer's title, such as Mr., Dr., or Ms**. unless the interview says something like, "Please call me Brenda."
- **Give more than "yes" or "no" responses to questions when appropriate**. Hollister says that interviewers want you to answer directly, but "it is OK to support your point with specific examples that are relevant to your work experience." For example, if an interviewer asks if you had courses in technical communication, you might respond, "Yes, and in my technical communication class, I completed a volunteer's handbook for the local Boys and Girls Club."
- **Avoid speaking negatively about past employers (Zupeck)**. If the interviewer asks you about a previous job, be prepared to show how you valued the experience. For example you might say, "I learned about how to solve problems" or "The job taught me how to work with people with different work styles."

PREPARING AN ELECTRONIC CAREER PORTFOLIO

An *electronic career portfolio* is a Web-based collection of materials related to your job search and your career. As a first-time job seeker, you would use the portfolio to provide employers with a traditional resumé along with other materials that might enhance your job search. After you have a job, you might use a career portfolio to show your skills, experiences, and accomplishments to your coworkers, your supervisors, and potential employers.

A portfolio usually includes the following sections:
- introductory page
- Web-based resumé
- references (letters of recommendation)
- transcripts
- skills
- samples of your work (sometimes called artifacts)

Figure 13.17, page 422, shows a sample electronic portfolio. Your university may require standard templates for students who use the career placement center. If your university does not require or have such templates, you can create your own design. As you plan your electronic portfolio, follow the design guidelines for building effective Web sites in Chapter 19.

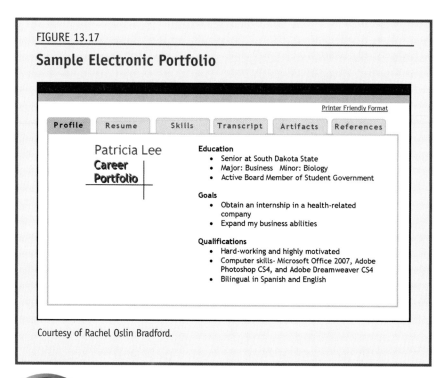

FIGURE 13.17

Sample Electronic Portfolio

Printer Friendly Format

| Profile | Resume | Skills | Transcript | Artifacts | References |

Patricia Lee
Career Portfolio

Education
- Senior at South Dakota State
- Major: Business   Minor: Biology
- Active Board Member of Student Government

Goals
- Obtain an internship in a health-related company
- Expand my business abilities

Qualifications
- Hard-working and highly motivated
- Computer skills- Microsoft Office 2007, Adobe Photoshop CS4, and Adobe Dreamweaver CS4
- Bilingual in Spanish and English

Courtesy of Rachel Oslin Bradford.

CASE STUDY ANALYSIS
Notre Dame Embarrassed Over Resumé Padding Scandal

Background

Notre Dame has built a reputation on football success and has leveraged that success to recruit top students, making it one of the best in the nation. When a losing streak forced the athletic department to replace its football coach in 2001, the nationwide search for a replacement began with the best of the best.

Hiring directors thought they had found the best coach in George O'Leary, a former Georgia Tech coach who had big goals for the Notre Dame program. His resumé featured a master's degree in education from New York University, as well as several years of football playing experience at New Hampshire, where he had earned three letters. In addition, he had a long career of coaching at prominent universities around the country.

O'Leary was quickly signed to a six-year contract and introduced to the media as the school's new football coach. However, only 5 days later, he resigned after a reporter discovered discrepancies in O'Leary's resumé. O'Leary had never received a master's degree from New York University, as he had claimed. In addition, he had not earned even one letter from New Hampshire because he had not played in a game in his two years on the team.

When the story broke, O'Leary was forced to resign in shame. Even though O'Leary apologized for the discrepancies, Notre Dame athletic department officials were understandably embarrassed.

Assignment

1. Pretend you are the fundraising director for Notre Dame's athletic department. Your job is to meet with donors and convince them to give to the school's football program. In light of the O'Leary scandal, however, donors are feeling hesitant about giving to the program. Write a letter to your biggest donors, explaining how the athletic department is working to restore the faith of donors. Be careful about your tone and your denunciation of O'Leary. Try to keep the letter as positive as possible. Turn your letter in to your instructor. (For information on writing letters, see Chapter 12.)

2. Imagine you are the public relations director for Notre Dame. Write a letter to the editor responding to the reporter's story exposing George O'Leary. Focus on mending the broken trust the deception created. Highlight the improvements this incident has prompted in hiring staff and recruiting students to Notre Dame. E-mail your letter to your instructor.

Compiled from information downloaded from the World Wide Web, May 6, 2008: http://www2.ljworld.com/news/2001/dec/15/scandal_stuns_irish/

TAKING IT INTO THE WORKPLACE
Designing Your Resumé to Land an Internship

You can enhance your experience and learn about your field by landing an internship. An internship is an excellent way to take what you are learning in the classroom and apply it in the workplace. To get that internship, you will need an up-to-date resumé. As Marianne Green of JobWeb.com explains, "The resumé you design to land an internship will have much in common with the employment resumé." For your internship resumé, you will use the chronological style with these modifications:

- **In the objective category, include the word "internship" and your field of interest**. For example, you might write "Internship in Web-based programming" or "Internship in health management."
- **Include details about your academic background**, including relevant courses, GPA, honors, scholarships, and projects.
- **Include high school information, if helpful**. Most job seekers leave out high school activities. However, if you are a sophomore or junior in college, you may want to include high school information if it will give employers relevant information about you.
- **Include your expected date of graduation**.
- **Include detailed information about activities**: volunteer work, research, and leadership
- **List all your jobs, even though they probably do not relate to your career goals**. These jobs will communicate a strong work ethic and demonstrate some of your skills and accomplishments.
- **Identify your skills**. Put yourself in the hiring manager's place. What skills will he or she expect in an intern?
- **"Limit your resumé to one page"** (Green)

Assignment

1. Visit your college or university placement center and find an internship for which you are qualified. If your placement center does not post internships or you can't find one, visit an online job board and search for internships in your field.
2. Prepare a resumé for applying for the internship.

EXERCISES

1. Find a job opportunity in your field that you are qualified for or will be qualified for when you graduate. You can look for these opportunities on online job boards, at your college or university placement center, or on the Web. You can also locate job opportunities by talking to family and friends in business and industry. After you have located a job opportunity, complete one of these steps:

   a. If you located the job through a print publication, copy or cut out the advertisement.

   b. If you located the job on an online job board or Web site, print a copy of the advertisement.

   c. If you talked to someone about the job, ask for a business card from that person or get a copy of the job announcement.

   Download a copy of the Worksheet for Writing Reader-Focused Job Correspondence online at www.benttreepress. com/technical communication.htm.

2. Decide on the categories of information that you might include in your resumé for the job you located in Exercise 1. Then create a list of the information that you could include in each of these categories. Your list might look like the brainstorming list in Figure 13.2, page 393.

3. Using some or all of the information from the list that you created in Exercise 2, prepare a paper resumé. Use the questions in the online "Worksheet for Writing Reader-Focused Job Correspondence" as you write your resumé.

4. Create a text-based, scannable version of the resumé that you created for Exercise 3. Remember to follow the tips for creating text-based, scannable resumés on page 410. Send your resumé to your instructor as an attachment.

5. Write a letter of application for the job you located in Exercise 1. Use the online "Worksheet for Writing Reader-Focused Job Correspondence" as you write your letter.

6. Using the questions in the online "Worksheet for Writing Reader-Focused Job Correspondence," evaluate your paper resumé; text-based, scannable resumé; and letter of application. Your instructor

may also ask you to use these questions to evaluate the resumés and letters of application of two of your classmates.

7. Write a paragraph evaluating the follow-up letter, below. Use the questions for follow-up letters in the online "Worksheet for Writing Reader-Focused Job Correspondence" to guide you as you evaluate the letter.

> Dear Penny:
>
> Meeting you and all your coworkers was great fun. The company seems to be a wonderful place to work. Thanks for showing me the facilities and for taking me to lunch. I would love to become one of your coworkers.
>
> Again, I would enjoy working with you and your coworkers. And I believe that I have a lot to offer your company. Have a great week.
>
> Best,
>
> Gretchen Delpero

8. Rewrite the follow-up letter from Exercise 7.

CHAPTER FOURTEEN

14

Writing Reader-Focused Definitions and Descriptions

You see definitions almost daily. You might read a definition of a technical term used in one of your textbooks, or you might read a more extended definition of a complex term or concept such as bioengineering or nanotechnology. A definition tells readers what something is; it can be as simple as a word or sentence or as complex as several paragraphs.

You also use descriptions daily. You might describe how something looks or how something works. Descriptions create a picture using words and graphics. Descriptions in technical communication provide information on a product or a process for readers who need to know about it or to use, produce, or possibly purchase it. For example, if you work for a company that sells solar panels, you might write a description about how the solar panels work to generate electricity.

Definitions and descriptions often work together. For example, if you work for a pharmaceutical company, you may write a fact sheet to accompany a drug. The fact sheet would include a definition, but also a description of how the drug works. Reader-focused definitions and descriptions use

- reader-focused, precise language
- a logical organization
- visual information when necessary

In this chapter, you will learn techniques for writing reader-focused definitions and descriptions. We will also focus on a common type of description: technical marketing materials.

PLANNING DEFINITIONS AND DESCRIPTIONS

To plan your definitions and descriptions, you will do the following:
- Find out about your readers.
- Anticipate and answer your readers' questions.
- Plan for precise language.
- Design the visual information.

Find out about Your Readers

As with any technical document, you will begin planning your document by learning about your readers' needs and expectations. You want to determine your readers' knowledge about what you are defining and what you are describing. You also want to know why your readers need the definition or description. For example, do they simply need to understand a technical term while they are reading a report? Or do they need to understand a technical concept so they can complete a task or use a product? Use the following questions to help you gather specific information about your readers. Once you have answered these questions, you can determine the format and level of technical terminology appropriate for your readers.
- What do your readers know about what you are defining or describing?
- Why do they need the definition or description? To gather information? To understand information or technical terminology? To complete a task? To make a decision? To find out about your product?
- Are they internal or external to your organization?
- Will more than one group of readers read your definition or description?
- In what context will they read the description or definition? As consumers? As users? As decision-makers?
- What level of technical terminology will they understand or expect?

Anticipate and Answer Your Readers' Questions

To help readers understand your definitions and descriptions, you must gather the information necessary to answer their questions. This task may be as simple as determining whether they understand specific terminology or as complex as employing primary and secondary research techniques (see Chapter 5). Although the specific questions vary with your purpose for writing, your readers will expect your definitions and descriptions to answer the questions in Figure 14.1.

Plan for Precise Language

Readers will expect precise

- measurements, weights, and dimensions
- locations and spatial relationships
- positions

Consider these examples:

Not Precise	**Precise**
the hybrid car	a 2008 VUE Green Line Hybrid Saturn
near the power button	above the red power button

When you use precise language, you eliminate language that focuses on individuals or opinions as in the examples on page 430.

FIGURE 14.1

Questions Readers May Ask when Reading Definitions and Descriptions

- Why should I read this definition or description?
- What is the purpose of this definition or description?
- How do I know that the information is honest, accurate, and complete?
- Do the information and the document look credible?

| **Focuses on Opinions** | **Precise** |
| The car has **great** gas mileage. | The Toyota Prius gets 48 MPG on the highway and 45 in the city. |

Design the Visual Information

As you plan your definitions and descriptions, decide what visual information, if any, your readers need. For example, if you are simply defining a technical term in a parenthetical definition, you probably do not need visual information. If you are describing for general readers what causes acid rain, a drawing will help them understand the process. For expert readers, you probably would not need a drawing.

The most commonly used graphics in definitions and descriptions are
- line drawings
- diagrams
- flowcharts
- photographs

Regardless of the graphics that you decide to use, plan for them early in the writing process. They may take time to prepare; for example, if you plan on using line drawings or diagrams, you may need to hire someone to prepare these graphics.

WRITING READER-FOCUSED DEFINITIONS

You can use three types of definitions:
- parenthetical definitions
- sentence definitions
- extended definitions

In this section, we will discuss each of these types of definitions and focus on writing effective extended definitions.

Parenthetical Definitions

Parenthetical definitions briefly define a term by rephrasing it with words that readers can understand. These definitions usually occur in parentheses immediately following the word(s) being defined. In the examples below, the parenthetical definition appears in bold type.

> Store the camera in an area where the ambient **(surrounding)** temperature reaches no higher than 86°F **(30°C)**.

> The Mini Cooper is equipped with a six-speed manual transmission or a continuously variable **(automatic)** transmission.

Sentence Definitions

If your definition or your readers require more than a word or phrase, you may want to use a ***sentence definition*** with the elements of a formal definition: term, class or *genus*, and *differentia*. ***Differentia*** refers to what makes the term different from all other similar terms. Figure 14.2 shows examples of sentence definitions divided by the elements of a formal definition. Each of these examples use words that are familiar to general readers.

Particularly for definitions, you want to determine your readers' knowledge about what you are defining and what you are describing.

FIGURE 14.2

Sample Sentence Definitions

Term	Class or Genus	Differentia
Ozone is	a gas	that occurs both in the Earth's upper atmosphere and at ground level.
Radon is	a natural radioactive gas	that you can't see, smell, or taste and that causes cancer.
Influenza (the flu) is	a contagious respiratory illness	that is caused by an influenza virus.

You might use sentence definitions when your readers require only a basic understanding of a term or concept. For example, the definition of radon in Figure 14.2 is appropriate for an informal report to a homeowner whose house you are inspecting. However, if you are writing a formal report for the Environmental Protection Agency (EPA) on the effects of radon on public health, you might use an extended definition.

Focus on Extended Definitions

For some readers and purposes, you may need an extended (expanded) definition. **Extended definitions** are explanations of a term, concept, or process. They are usually one or more paragraphs in length depending on your purpose and your readers. You have probably read extended definitions in textbooks or perhaps in users' manuals. Extended definitions often begin with a sentence definition, as in the following example:

> Bacteria are invisible one-celled organisms that are self-sufficient and multiply by subdivision.

This definition tells you the classification of the organism and how it reproduces. However, it doesn't tell you how bacteria cause illness or which bacteria are beneficial and which are harmful. You would extend the definition to provide that information.

You can extend a definition using any of the following techniques:
- examples
- partition
- principle of operation
- comparison and contrast
- negation
- history
- etymology
- stipulation
- graphics

Extending with Examples

Examples are one of the most commonly used methods for extending definitions. Examples are especially helpful when you want to make an abstract term, a concept, or an unfamiliar process familiar to your readers. In the following extended definition of "pesticide," the writer includes examples of common household pesticides. The examples help readers understand what pesticides are by associating a less-familiar term (pesticide) with familiar examples (household products):

> A pesticide is any substance or mixture of substances intended to prevent, destroy, repel, or mitigate any pest; a pest is any living organism that occurs where it is not wanted or may damage crops, or injure humans or animals. Some common examples of pesticides are these household products:
> - cockroach sprays and baits
> - insect repellents
> - rat and other rodent poisons
> - flea and tick sprays, powders, and pet collars
> - kitchen, laundry, and bath disinfectants and sanitizers
> - products that kill mold and mildew
> - some lawn and garden products, such as weed killers
> - some swimming pool chemicals[1]

Extending with Partition

Partition is the division of an item into its individual parts, so readers can better understand the item as a whole (see Chapter 6). In the following example from the United States Department of Agriculture, the writers divide "grains" into two types: whole grains and refined grains (mypyramid.gov). With this partitioning, readers can better understand the definition of all grains.

> Grains are divided into 2 subgroups, **whole grains** and **refined grains**. Whole grains contain the entire grain kernel: bran, germ, and endosperm. Some examples of whole grains are
> - whole-wheat flour
> - bulgur (cracked wheat)
> - oatmeal
> - whole cornmeal
> - brown rice

[1] Adapted from "What is a Pesticide?" http://www.epa.gov/pesticides/about/index.htm#what_pesticide

Evaluate a descriptive document in the Interactive Student Analysis online at www.kendallhunt.com/technicalcommunication.

Refined grains have been milled, a process that removes the bran and germ. Although this process gives grains a finer texture and improves their shelf life, it also removes dietary fiber, iron, and many B vitamins. Some examples of refined grain products are

- white flour
- degermed cornmeal
- white bread
- white rice

Extending through Principle of Operation

You can often effectively extend a definition by explaining how something works, especially if you are defining a process or an object. To define "gas turbine," the writers of the definition below use the principle of operation. They first define the gas turbine as an internal combustion engine; however, some readers may not understand how this engine works, so the writers explain how the engine—in this case, a gas turbine—works.

A gas turbine engine is a type of internal combustion engine. Essentially, the engine can be viewed as an energy conversion device that converts energy stored in the fuel to useful mechanical energy in the form of rotational power. The term "gas" refers to the ambient air that is taken into the engine and used as the working medium in the energy conversion process.

This air is first drawn into the engine where it is compressed, mixed with fuel and ignited. The resulting hot gas expands at high velocity through a series of airfoil-shaped blades transferring energy created from combustion to turn an output shaft. The residual thermal energy in the hot exhaust gas can be harnessed for a variety of industrial processes (Solar Turbines).

Extending through Comparison and Contrast

You can extend a definition by comparing or contrasting it to similar concepts, processes, or terms that are familiar to the reader. In Figure 14.3 the writer extends the definition of "bonds" to stocks and money market instruments. Notice how the writer has organized the definition by the type of investment tool and then used the same criteria (definition and risk) to define each tool.

FIGURE 14.3

Using Comparison and Contrast to Extend a Definition

A bond is a debt security, issued by a government, municipality, corporation, or other entity. The issuer promises to repay the buyer the amount of the bond (the principal) plus a specified rate of interest during the life of the bond. In the chart below, you can see how bonds compare with two other common investment tools: stocks and money-market instruments.

Investment Tool	Definition	Level of Risk	Examples
Bond	A debt security, similar to an I.O.U., issued by a government, municipality, corporation, or other entity	Moderate	• U.S. government securities • Municipal bonds • Corporate bonds • Mortgage and asset-backed securities • Foreign government bonds
Stock	Ownership shares in a company	High	• Shares in McDonalds • Shares in Apple • Shares in Dow Chemical
Money Market	Short-term, interest-bearing investments	Low	• Money market accounts • Certificates of Deposit (CDs)

Extending through Negation

You can extend a definition by telling readers what the term, concept, or process is *not*. You must use negation with other techniques for extending a definition because once you have told a reader what a term (concept or process) is not, you should tell the reader what it is, as in the following example:

> In the U.S., football is not soccer. Football is played with a spherical ball whereas soccer is played with a round ball and has different rules.

TAKING IT INTO THE WORKPLACE
Understanding the Technical Professional's Role in Writing Marketing Materials

Even when companies have marketing divisions responsible for preparing technical marketing materials, these divisions need experts to draft the technical information, such as specifications and process descriptions. In some smaller companies, the technical experts may be responsible not only for drafting the technical information, but also for writing and perhaps designing the marketing materials. You may work for such a company when you graduate.

Assignment

1. Interview a technical professional who works for a company that manufacturers a product. You may interview by phone or by e-mail. (For information on conducting interviews, see Chapter 5.) The purpose of your interview is to determine the professional's role in preparing the technical marketing materials.
2. Write a memo to your instructor summarizing what you learned from your interview. Attach the list of questions you used for the interview. (For more information on writing memos, see Chapter 12.)
3. Write a follow-up thank-you letter or e-mail to the professional whom you interviewed. (For information on writing follow-up letters, see Chapter 13.)

Extending through History

You can extend a definition by giving historical or background information to help readers understand a term, concept, or process. In the example below, the writer extends the definition of mouse by giving its history:

> Invented by Douglas Engelbart of Stanford Research Center in 1963 and pioneered by Xerox in the 1970s, the mouse is one of the great breakthroughs in computer ergonomics because it frees the user from using the keyboard. The mouse is important for graphical user interfaces because you can simply point to options and objects and click a button on the mouse. The mouse is also useful for graphics programs that allow you to draw pictures by using the mouse like a pen, pencil, or paintbrush. Users can select from three basic types of mice:

- **Mechanical.** On its underside, this mouse has a rubber or metal ball that can roll in all directions. Mechanical sensors within the mouse detect the direction the ball is rolling and move the screen pointer accordingly.
- **Optomechanical.** This mouse operates similarly to the mechanical mouse, but uses optical sensors to detect the motion of the ball.
- **Optical.** This mouse uses a laser to detect the mouse's movement. The user moves the mouse along a special mat with a grid so that the optical mechanism has a frame of reference. These mice do not have mechanical moving parts. [2]

Extending through Etymology

For some terms and concepts, you can extend through etymology. *Etymology* is defining a word by tracing its derivation. You will generally combine etymology with other techniques for extending a definition. In the following definition, etymology is used to define "phishing."

> Phishing is the act of sending an e-mail to a user and falsely claiming to be a legitimate organization. The sender is attempting to scam the user to share private information. The e-mail redirects the user to a bogus Web site where he or she is asked to update personal information. The word phishing "comes from the analogy that Internet scammers are using e-mail lures to fish for passwords and financial data from the sea of Internet users. The term was coined in 1996 by hackers who were stealing AOL Internet accounts by scamming passwords from unsuspecting AOL users. Since hackers have a tendency to replace "f" with "ph" the term phishing was derived" (Beal).

You can also use etymology to define an acronym, as in the following example.

> LASIK stands for laser-assisted *in situ* keratomileusis and is a procedure that permanently changes the shape of the cornea.

---

[2] Adapted from the Webopedia. http://www.webopedia.com/TERM/m/mouse.html

Extending through Stipulation

You use stipulation when you restrict (stipulate) the meaning of a term for a particular situation. In the following example, the writers stipulate the meaning of "pesticide." The readers then know that any time the writers use "pesticide," they are referring to insecticides, herbicides, and fungicides.

> In this report, we use the term "pesticide" to refer to insecticides, herbicides, and fungicides.

Extending through Graphics

You can use graphics to extend a definition because they help readers visualize a concept or a process. If you use graphics, consider using callouts to help readers focus on what you want them to see. In the extended definition in Figure 14.4, the writers use a diagram to show how ozone develops.

A Sample Extended Definition

In the sample extended definition in Figure 14.5, page 440, the writers define a cochlear implant. The definition begins with a sentence definition, and the writers then use the following techniques to extend the definition: graphics, partition, examples, comparison/contrast, and principle of operation.

WRITING READER-FOCUSED DESCRIPTIONS

As a technical professional, you may need to describe any of the following:
- **Processes**. A *process* is an action that brings about a result. Usually, the action takes place over time. You are familiar with many processes, such as digesting food, or logging onto a network. A process description differs from instructions; when you describe a process, you tell what happens, whereas in instructions, you tell the user how to perform the steps of the process.
- **Mechanisms**. A *mechanism* is a machine—an object with parts that work together. For example, a helicopter, a car, and a coffeemaker are mechanisms. Each of these mechanisms has identifiable parts that work together.

FIGURE 14.4

Using Graphics to Extend a Definition

What is Ozone?

Ozone: Good Up High - Bad Nearby

Ozone (O3) is a highly reactive gas composed of three oxygen atoms. Depending on where it is in the atmosphere, ozone affects life on Earth in either good or bad ways.

Stratospheric ozone is formed naturally through the interaction of solar ultraviolet (UV) radiation with molecular oxygen (O2). The stratospheric "ozone layer" extends from approximately six to thirty miles above the Earth's surface and reduces the amount of harmful UV radiation reaching the Earth's surface.

Tropospheric, or ground-level, ozone forms primarily from reactions between two major classes of air pollutants: volatile organic compounds (VOCs) and nitrogen oxides (NOx). These reactions depend on the presence of heat and sunlight, meaning more ozone forms in the summer months.

NOx is emitted by cars, power plants, industrial plants, and other sources. Significant sources of VOC emissions include gasoline pumps, chemical plants, oil-based paints, auto body shops, print shops, consumer products and some trees. Significant human-made sources of VOC emissions include gasoline pumps, chemical plants, oil-based paints, auto body shops, print shops, and some consumer products.

NOx + VOC + Heat & Sunlight = Ozone
Ground-level or "bad" ozone is not emitted directly into the air, but is created by chemical reactions between NOx and VOCs in the presence of heat & sunlight.

Emissions from industrial facilities and electric utilities, motor vehicle exhaust, gasoline vapors, and chemical solvents are some of the major sources of oxides of nitrogen (NOx) and volatile organic compounds (VOC).

Source: Downloaded from the World Wide Web, November 2008: www.epa.gov/airomsmo/airaware/day1-ozone.html.

- **Objects**. An *object* is a single item. That item might be part of a helicopter, a car, or a coffeemaker. For example, if you describe the propeller of a helicopter, you are describing an object.

Descriptions of these items usually appear in many documents used in the workplace: technical marketing brochures, instructions, manuals, propos-

FIGURE 14.5

Extended Definition

The writer begins with a sentence definition.

The writer uses partition.

The writer uses graphics to extend the definition.

The writer uses the principle of operation.

What is a cochlear implant?

A cochlear implant is a small, complex electronic device that can help to provide a sense of sound to a person who is profoundly deaf or severely hard-of-hearing. The implant consists of an external portion that sits behind the ear and a second portion that is surgically placed under the skin (see figure). An implant has the following parts:

- A microphone, which picks up sound from the environment.
- A speech processor, which selects and arranges sounds picked up by the microphone.
- A transmitter and receiver/stimulator, which receive signals from the speech processor and convert them into electric impulses.
- An electrode array, which is a group of electrodes that collects the impulses from the stimulator and sends them to different regions of the auditory nerve.

Credit: NIH Medical Arts
Ear with Cochlear implant.
View larger image.

An implant does not restore normal hearing. Instead, it can give a deaf person a useful representation of sounds in the environment and help him or her to understand speech.

How does a cochlear implant work?

A cochlear implant is very different from a hearing aid. Hearing aids amplify sounds so they may be detected by damaged ears. Cochlear implants bypass damaged portions of the ear and directly stimulate the auditory nerve. Signals generated by the implant are sent by way of the auditory nerve to the brain, which recognizes the signals as sound. Hearing through a cochlear implant is different from normal hearing and takes time to learn or relearn. However, it allows many people to recognize warning signals, understand other sounds in the environment, and enjoy a conversation in person or by telephone.

Who gets cochlear implants?

Children and adults who are deaf or severely hard-of-hearing can be fitted for cochlear implants. According to the Food and Drug Administration (FDA), at the end of 2006, more than 112,000 people worldwide had received implants. In the United States, roughly 23,000 adults and 15,500 children have received them.

Adults who have lost all or most of their hearing later in life often can benefit from cochlear implants. They learn to associate the signal provided by an implant with sounds they remember. This often provides recipients with the ability to understand speech solely by listening through the implant, without requiring any visual cues such as those provided by lipreading or sign language.

Cochlear implants, coupled with intensive postimplantation therapy, can help young children to acquire speech, language, and social skills. Most children who receive implants are between two and six years old. Early implantation provides exposure to sounds that can be helpful during the critical period when children learn speech and language skills. In 2000, the FDA lowered the age of eligibility to 12 months for one type of cochlear implant.

Credit: Centers for Disease Control and Prevention (CDC)

How does someone receive a cochlear implant?

Use of a cochlear implant requires both a surgical procedure and significant therapy to learn or relearn the sense of hearing. Not everyone performs at the same level with this device. The decision to receive an implant should involve discussions with medical specialists, including an experienced cochlear-implant surgeon. The process can be expensive. For example, a person's health insurance may cover the expense, but not always. Some individuals may choose not to have a cochlear implant for a variety of personal reasons. Surgical implantations are almost always safe, although complications are a risk factor, just as with any kind of surgery. An additional consideration is learning to interpret the sounds created by an implant. This process takes time and practice. Speech-language pathologists and audiologists are frequently involved in this learning process. Prior to implantation, all of these factors need to be considered.

What does the future hold for cochlear implants?

With advancements in technology and continued follow-up studies with people who already have received implants, researchers are evaluating how cochlear implants might be used for other types of hearing loss.

NIDCD is supporting research to improve upon the benefits provided by cochlear implants. It may be possible to use a shortened electrode array, inserted into a portion of the cochlea, for individuals whose hearing loss is limited to the higher frequencies. Other studies are exploring ways to make a cochlear implant convey the sounds of speech more clearly. Researchers also are looking at the potential benefits of pairing a cochlear implant in one ear with either another cochlear implant or a hearing aid in the other ear.

Source: Downloaded from the World Wide Web, November 2008: www.nidcd.nih.gov/health/hearing/coch.asp.
"Cochlear Implants." National Institute on Deafness and Other Communication Disorders.

TIPS FOR WRITING READER-FOCUSED EXTENDED DEFINITIONS

- **Use language familiar to your readers.** If you define terms, concepts, or processes using unfamiliar language, your definition will not help your readers. Make sure that you know enough about your readers to write a definition that is appropriate for their level of knowledge and experience.
- **Use precise language.** For example, if you are defining "ozone," don't write, "Gases combine with the air to create ozone." This language is not precise. Instead, write, "Ozone is created when NOx (nitrogen oxide) and VOC (volatile organic compounds) combine with heat and sunlight."
- **Use language accessible to readers of other cultures and languages.** Use Simplified Technical English to ensure that readers of other cultures and languages can understand your definitions. (See Chapter 8 for information on Simplified Technical English.) Avoid analogies that may not work for your readers.
- **Use the standard patterns of organization to structure your extended definition.** Some of the more commonly used patterns include spatial order, partition, and comparison and contrast. (See Chapter 6 for information on these patterns.)
- **Use graphics when possible.** Graphics help readers to quickly visualize a process or concept. (See Chapter 11 for information on creating graphics.)
- **Place the definition where the readers need it.** If your extended definition is part of a longer document, place the definition so readers have the information they need to understand the rest of the document.

als, Web sites, or even reports. For example, if your company is writing a brochure on a new product, you may be called on to write a description of how the product works. If you are redesigning equipment in a manufacturing plant, you might describe how the redesigned equipment will work or how the parts of the equipment will work together.

Descriptions do not have a conventional structure like a proposal or a report. However, you can use the tips on the following page to ensure that you achieve your purpose and answer your readers' questions.

TIPS FOR WRITING DESCRIPTIONS OF PROCESSES, MECHANISMS, AND OBJECTS

- **Introduce the description**. If the description will appear in a separate document, give it a title. If it will appear as part of another document, include an informative heading identifying what you are describing. (See Figure 14.6.)
- **Use the appropriate level of detail**. To determine how much detail is appropriate, you must know your readers and their expectations. For example, if you are describing the effects of nature on barrier islands for general readers, you would use less detail than if you were writing a report for the U.S. Geological Service.
- **Use the standard patterns of organization**. *For process descriptions*, use the chronological pattern. *For mechanisms and objects*, use a spatial organization or general-to-specific description. If you use the general-to-specific description, begin with a general definition of the object or mechanism and move to the specifics of how it functions or how it is used. For more complex objects or mechanisms, you may combine the organizational patterns. For example, if you are describing the human body, you could describe it spatially from the head to the toes. You could also describe it using a general-to-specific pattern, beginning with a definition of the human body and then describing the specific systems of the body: the nervous system, the digestive system, the musculoskeletal system, and the respiratory and circulatory systems.
- **Use language that is familiar to your readers**. Make sure that you know about your readers' knowledge of your subject and use language they will understand. If you fill your description with unfamiliar words, it will be useless to your readers.
- **Use present tense**. Unless you are describing a process, mechanism, or object that occurred in the historical past, use the present tense. For example, if you are describing how the Grand Canyon was created, you would use the present tense because the Grand Canyon still exists. If you were describing how the Twin Towers, destroyed on 9/11, were built, you would use past tense because the towers no longer exist.
- **Use graphics**. To effectively describe most objects and mechanisms, use a graphic so readers get a picture of what you are describing. Graphics are especially important if you are describing a complex object or mechanism that is new or unfamiliar to most readers. The most commonly used graphics for descriptions are photographs, drawings, diagrams, and flowcharts. Figure 14.6 uses a drawing with numbers to orient the description and to help readers visualize the oven.
- **Place the description where readers will need it**. For example, if you are describing a product in a manual, place the description early in the manual before you tell the user how to use the product.

FIGURE 14.6

Description Using a Graphic

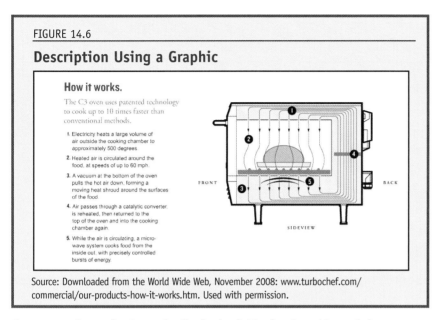

How it works.

The C3 oven uses patented technology to cook up to 10 times faster than conventional methods.

1. Electricity heats a large volume of air outside the cooking chamber to approximately 500 degrees.

2. Heated air is circulated around the food, at speeds of up to 60 mph.

3. A vacuum at the bottom of the oven pulls the hot air down, forming a moving heat shroud around the surfaces of the food.

4. Air passes through a catalytic converter, is reheated, then returned to the top of the oven and into the cooking chamber again.

5. While the air is circulating, a microwave system cooks food from the inside out, with precisely controlled bursts of energy.

FRONT   BACK

SIDEVIEW

Source: Downloaded from the World Wide Web, November 2008: www.turbochef.com/commercial/our-products-how-it-works.htm. Used with permission.

Focus on Descriptions in Technical Marketing Materials

Technical marketing materials include descriptions of products. As a technical professional, you may write descriptions to be included in these materials. You will have to describe the product accurately while using language to persuade your readers to purchase your product. The readers may have any level of knowledge of your product and field: from general readers to expert readers. Technical descriptions help these readers visualize a product and determine how its features and specifications can meet their needs.

Technical marketing documents come in many forms. Some of the more common forms include the following:

- **Web pages**. Companies often put descriptions of products on their Web sites. These sites provide a flexible, inexpensive medium for presenting extensive descriptions. The Web page in Figure 14.7, page 444, is the home page for Cadillac. This Web page uses positive language, provides both general and precise language, and includes an engaging photo. Compare this page with Figure 14.9, page 449, another page from the same Web site. In Figure 14.9, the writers use precise language to back up some of the information on the home page (Figure 14.7).
- **Brochures**. A brochure might be a popular tri-fold or an entire booklet. An effectively designed and written brochure can show readers the

quality of your product. Brochures can be expensive to produce, and you must reprint to update them. To save money, many organizations put pdf's of their brochures and fact sheets on their Web sites so readers can download them.

- **Specification sheets, sometimes called fact sheets**. Specification or fact sheets focus on basic information on the product. They present the information in a more straightforward manner than do most Web pages and brochures. They are usually printed double sided on $8\frac{1}{2}''$ x $11''$ paper. These pages can be more difficult to design because they are information-rich, yet they must be attractive, inviting, and easy to access. See the sample document in Figure 14.10, page 450, for an example of a fact sheet.

FIGURE 14.7

Technical Marketing Web Page

Source: Downloaded from the World Wide Web, November 2008: www.cadillac.com/cadillacjsp/model/landing.jsp?model=cts&year=2008. Used with permission.

Using Persuasive, Accurate Language in Technical Marketing Materials

While your primary purpose in writing technical marketing material is to persuade, you also must give readers accurate, honest information about the product. You want to attract readers with positive, upbeat language to describe your product; however, for readers to trust that your claims are honest and accurate, you also must include precise information about the features and specifications. Readers may be suspicious of persuasive language that is not accompanied with specific information about the features of the product. For example, if you have a testimonial from a client who states that the software was great and helped to improve efficiency, you must balance that testimonial with how the software can improve efficiency. In the following example from the Cadillac Web site, the writers state the following:

> As is the case with its design, the CTS Coupe Concept extends the acclaimed capabilities of the sedan in terms of performance technology. This includes the capability to support a broad engine range of gasoline and diesel engines. The CTS Coupe Concept of course supports the sedan's 3.6L V-6 engines, including the 304-horsepower (227 kW) Direct Injection power plant. The CTS Coupe Concept also is designed for a new 2.9L turbo diesel being developed for international markets. This new engine, tailored for use in the CTS, will deliver an estimated 250 horsepower (184 kW) and 406 lb.-ft. of torque (550 Nm). A six-speed manual transmission backs the engine, sending torque to an independently sprung rear axle. The CTS Coupe Concept's sport-tuned suspension gives it a slightly lower ride height than a production CTS – a look enhanced by the car's rakish shape and large, 20-inch front and 21-inch rear wheels. (adapted from Cadillac)

Although the writers begin with positive, upbeat promotional language in the first sentence, they back up that language with specific information in the sentences that follow.

Sample Descriptions

Figure 14.8, page 446, is an excerpt from a process description of cataract surgery. The description helps general readers who want to understand what to expect when having cataract surgery.

FIGURE 14.8

Process Description

The writer uses
language familiar
and appropriate for
the users.

Cataract Surgery

What is a cataract?

A cataract is a clouding of the eye's naturally clear lens. The lens focuses light rays on the retina — the layer of light-sensing cells lining the back of the eye — to produce a sharp image of what we see. When the lens becomes cloudy, light rays cannot pass through it easily, and vision is blurred.

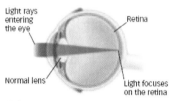

Light rays entering an eye with a normal lens.

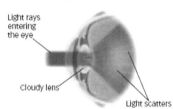

Light rays entering an eye with a cataract. When a cataract forms, the lens of your eye is cloudy.light cannot pass through it easily, and your vision

What causes cataracts?

Cataract development is a normal process of aging, but cataracts also develop from eye injuries, certain diseases or medications. Your genes may also play a role in cataract development.

How can a cataract be treated?

A cataract may not need to be treated if your vision is only slightly blurry. Simply changing your eyeglass prescription may help to improve your vision for a while. There are no medications, eyedrops, exercises or glasses that will cause cataracts to disappear once they have formed. Surgery is the only way to remove a cataract. When you are no longer able to see well enough to do the things you like to do, cataract surgery should be considered.

In cataract surgery, the cloudy lens is removed from the eye through a surgical incision. In most cases, the natural lens is replaced with a permanent intraocular lens (IOL) implant.

Source: Downloaded from the World Wide Web, November 2008:
www.medem.com/MedLB/article_detaillb_for_printer.cfm?article_ID=ZZZY9VMAC8C&sub_cat=119.
Courtesy of American Academy of Ophthalmology

FIGURE 14.8 CONTINUED

What can I expect if I decide to have cataract surgery?

Before Surgery
To determine if your cataract should be removed, your ophthalmologist (Eye M.D.) will perform a thorough eye examination. Before surgery, your eye will be measured to determine the proper power of the intraocular lens that will be placed in your eye. Ask your ophthalmologist if you should continue taking your usual medications before surgery.

You should make arrangements to have someone drive you home after surgery.

The Day of Surgery
Surgery is usually done on an outpatient basis, either in a hospital, an outpatient surgical center, or an ambulatory surgery center. You may be asked to skip breakfast, depending on the time of your surgery.

When you arrive for surgery, you will be given eyedrops and perhaps a mild sedative to help you relax. A local anesthetic will numb your eye. The skin around your eye will be thoroughly cleansed, and sterile coverings will be placed around your head. Your eye will be kept open by an eyelid speculum. You may see light and movement, but you will not be able to see the surgery while it is happening.

Under an operating microscope, a small incision is made in the eye. In most cataract surgeries, tiny surgical instruments are used to break apart and remove the cloudy lens from the eye. The back membrane of the lens (called the posterior capsule) is left in place.

The writers organize the process using a chronological organization.

The writers use simple graphics to help readers visualize the process.

During cataract surgery, tiny instruments are used to break apart and remove the cloudy lens from the eye.

An intraocular lens (iol) implant.

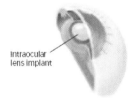

Intraocular lens implant

In cataract surgery, the intraocular lens replaces the eye's natural lens.

After surgery is completed, your doctor may place a shield over your eye. After a short stay in the outpatient recovery area, you will be ready to go home.

Figure 14.9, page 449, illustrates a frequently used description: the specification. A specification usually includes a graphic and a set of statistics and characteristics about the object or mechanism. Specifications often appear in product descriptions and technical marketing materials. You may also see them on consumer Web sites, such as *Consumer Guide*. This specification from the Cadillac Web site includes a photograph, the exterior dimensions, and precise information about the engine, transmission, and fuel system. The writers use precise language and an appropriate amount of detail.

Figure 14.10, page 450, is a fact sheet for a car. Notice how the writers have used headings and a table to help readers locate the features of the car.

ETHICS NOTE

Earning the Trust of Your Readers

When writing technical marking materials, you may be tempted to exaggerate what your product can do for the readers in an attempt to persuade them. While your purpose is to persuade readers of the merits of your product, your goal is also to inform.

You have an ethical obligation to make claims that you can back up. While you can appropriately use language to persuade your readers, you must be able to back up that language with data. When you use accurate information and are honest with your readers, you will earn their trust and, hopefully, they will purchase your product. When you are honest and back up your claims, you are simply doing good business.

FIGURE 14.9

Excerpt from a Specification

The writers include photos from different viewpoints to give the exterior dimensions of the car.

The writers use precise language.

The writers use a chart to help readers easily locate the specifications.

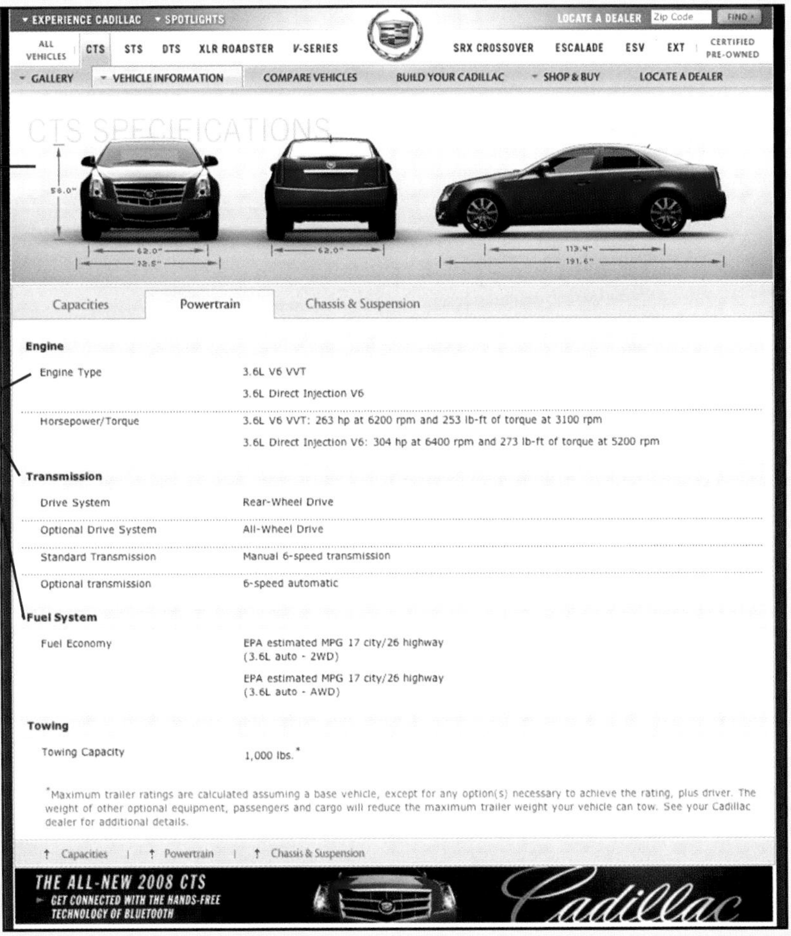

Source: Downloaded from the World Wide Web, November 2008:
www.cadillac.com/cadillacjsp/model/po_specification.jsp?model=cts&year=2008&section=Powertrain.
Used with permission.

FIGURE 14.10

Fact Sheet

Source: Downloaded from the World Wide Web, November 2008: www.cadillac.com/_res/pdf/08Cadillac_CTS.pdf. Used with permission.

CASE STUDY ANALYSIS
How Expanding the Definition of "Obesity" Would Impact Children

Background

The old children's rhyme goes, "Sticks and stones may break my bones, but words will never hurt me." However, in 2006, a proposed change by a committee of the American Medical Association (AMA) and some federal government agencies could have hurt an untold number of children with one word, "obesity."

The AMA committee wanted to expand the definition of childhood obesity, effectively labeling nearly 25 percent of American toddlers and 40 percent of children between the ages of 6 and 11 with a medical condition known as "overweight and obese."

Critics of the proposed extended definition said the change would cause many children undue stress and could lead to teasing, to eating problems, and to avoiding physical activity. The critics were also concerned that labeling children with a medical condition would impact their future health care even though childhood obesity is not a good predictor of future weight or health problems as children grow into adults. The critics further cited that "overweight and obese" are arbitrary labels determined by a set cutoff point on the Body Mass Index, rather than a verifiable medical condition.

The AMA wanted to extend the definition of childhood obesity to align it with its adult definition of obesity. After receiving criticism for its proposed change, the AMA decided to reconsider the label.

Assignment

1. Pretend you work for the AMA and are asked to write the extended definition of childhood obesity. Research the criteria for "obese," and write a definition that informs readers of the revised obesity definition encompassing children. Be careful to avoid inflammatory language.

Download a copy of the Worksheet for Writing Reader-Focused Definitions and Descriptions at www.benttreepress. com/technical communication.htm.

EXERCISES

1. Write a sentence definition for the following terms:
   a. USB flash drive
   b. solar panel
   c. catalytic converter
   d. computer virus
   e. e-mail spam

2. Write an extended definition for a term in your field or for one of the terms below. Your definition should be 750 to 1,000 words. Be sure to cite any sources that you use.
   a. bacteria
   b. cloning
   c. vaccine
   d. electricity
   e. telecommuting
   f. stem cells
   g. recycling

3. Write a memo addressed to your instructor to accompany the extended definition you wrote in Exercise 2. (See Chapter 12 for information on memos.) The memo should include the following:
   a. the techniques you used to extend your definition
   b. why you selected the techniques
   c. the standard pattern(s) of organization that you used

4. **Web exercise**: Visit the Web sites of three similar products. For example, you might visit the Web sites for three hybrid vehicles or three digital cameras. At the Web sites, look for descriptions of the products. Print a copy of the descriptions and answer the following questions:
   a. Do the Web sites appropriately balance persuasive language with informative language? Explain your answer using examples.
   b. Do the descriptions look credible and trustworthy? Explain your answer using examples.
   c. How appropriate is the language for the intended readers and for readers of other cultures and languages? Explain your answer using examples.

5. Write a 750- to 1,000-word process description of one of the following processes or a similar process that you understand and can describe. Along with your description, write a memo to your instructor stating the intended readers. In the description, remember to cite your sources.
   a. how electricity is generated
   b. how digestion works
   c. how cheese is made
   d. how cell phones work
   e. how beaches erode
   f. how wind is used to generate electricity

6. Write a 500- to 750-word description (for general readers) of one of the following objects or mechanisms, or select one yourself. Cite any sources that you use.
   a. garage door opener
   b. Geiger counter
   c. turbine
   d. toaster
   e. helicopter

7. **Collaborative exercise**: Divide into teams of four. Your team works for a company that produces various types of mechanisms. Your team has been assigned the task of preparing technical marketing materials for one of the products. To complete your task, follow these steps:
   a. Decide on a mechanism that you will market. The mechanism should be simple enough that all members of the team could understand or use it. You do not have to invent the mechanism; you may use an existing one.
   b. Create a name and logo for your company.
   c. Prepare a brochure and a fact sheet for the product.
   d. Turn in your materials to your instructor.

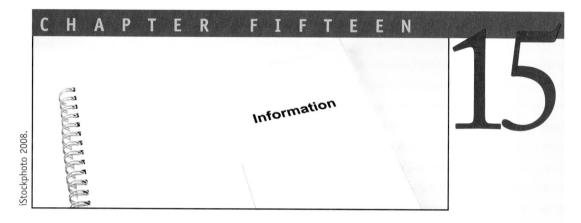

iStockphoto 2008.

CHAPTER FIFTEEN

15

Information

Writing User-Focused Instructions and Manuals

We use instructions at home and in the workplace. Whether we are cooking microwave popcorn or installing new software, we use instructions. Some instructions are short, simple, and informal—perhaps only a few steps. For example, Figure 15.1, page 456, presents brief instructions that tells users how to remove a battery. Other instructions may appear in the form of a manual and may be hundreds of pages.

The instructions that you write may be as simple as the list shown in Figure 15.1 or as complex as an entire manual. You might write instructions for tasks that you want your coworkers to complete or for tasks that consumers will follow when using your company's products. Regardless of the number and complexity of the tasks, the same techniques apply. In this chapter, you will learn techniques to help you write effective, user-focused instructions.

PLAN THE INSTRUCTIONS OR MANUAL

To plan your instructions or manual, you will
- learn about your users
- design the visual information
- design ethical safety information

FIGURE 15.1

Simple Instructions

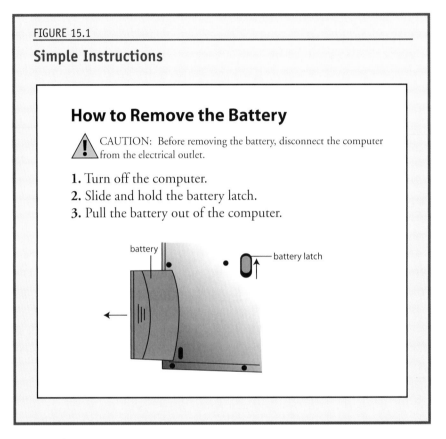

How to Remove the Battery

⚠ CAUTION: Before removing the battery, disconnect the computer from the electrical outlet.

1. Turn off the computer.
2. Slide and hold the battery latch.
3. Pull the battery out of the computer.

Learn about Your Users

To write effective instructions, you need to know about your users. Who will use the instructions? How much do they know about the task? Will more than one group of users use the manual? This information will help you decide how much detail users will need to perform the task correctly and easily. You might also find out about users' backgrounds and training by answering these questions:

- Have users performed the task before? Have they performed similar tasks?
- Do all users have the same background or knowledge of the task?
- Will users have different purposes for using the instructions? Will their purposes for using the instructions change?
- How much detail do users need to complete the task? Do they want only minimal instructions?

- What are users' attitudes toward the task and toward the equipment used for the task (Brockmann)?

If you know how familiar users are with the task, you can select the appropriate amount of detail. For example, if you know users have used a microscope, you won't have to tell them how to operate a microscope; however, if you know they have never used a microscope, you will need to include basic operating instructions. You can also help your users if you know whether they have performed similar tasks or used similar equipment. If you know users are familiar with a similar task, you can compare the familiar task with the new task to help them feel comfortable. You also can help users feel comfortable with a task if you know their attitudes toward the task and the equipment (Brockmann). For instance, if you know users may resist a change in routine or habits caused by the instructions, frequently reassure them and, whenever possible, link the new task to procedures they already know.

Depending on the needs of your users, you may want to include detailed instructions and explanations or perhaps even the theory behind a procedure. For example, novice users may need detailed information to perform a task correctly, whereas expert users may need—and want—only minimal instructions. As you decide how much detail and explanatory information to provide, consider users' reasons for reading your instructions or manual. Users "with specific tasks in mind need little or no elaboration," but users "without specific goals benefit from explanations of how to apply procedures," not from elaborations that describe general concepts (Charney, Reder, and Wells). Users who have specific tasks to complete need little, if any, explanation of the procedure or the concept behind the procedure, and users without specific tasks will benefit from "how-to" statements, rather than "why" and "when" statements.

> *You can help users feel comfortable with a new task by building on knowledge they have of similar instructions or equipment.*

Consider the elaboration in a document on controlling thatch (Koop and Duble).

General elaboration   Mowers tend to scalp lawns that have excess thatch.

| How-to elaboration | Use vertical mowers specifically designed to remove thatch. When using these mowers, make sure the blades penetrate through the thatch to the soil surface. |

The general elaboration will not help users control thatch. However, the how-to elaboration gives users specific information about the type of mower to use and how to use that mower to remove thatch.

Find out if users will have varying backgrounds and knowledge of the task or if users will read the instructions for varying purposes. This information will help you determine how to organize the instructions to best meet the needs of all users. For instance, new users of a software application may first consult the manual to learn how to get started; so they will want basic information. As these users learn more about the software, they may have questions about more-advanced commands; they will then want to use the manual as a reference guide, not as a tutorial. The manual thus needs a tutorial-type section for new users and a reference section for users who have become more familiar with the software.

Design the Visual Information

As you plan your instructions and manuals, decide how you will design the visual information. You should plan for an appropriate typeface, layout, color, and page size for your purpose and your users (Brockmann). As you plan these elements, ask yourself how and where users will use the instructions. For example, will they use the instructions at a computer workstation? If so, you might use a spiral binding so the pages will lie flat, and a smaller-than-traditional page size so the instructions will fit easily at a workstation. Perhaps users will use the instructions outside a typical office environment. If so, what is the environment, and how will it affect how users reference the manual?

Design Instructions Suitable for the Users' Environment

If users will use the instructions or manual outside a typical office environment, select a paper, a typeface, a type size, and a page size that will make the instructions easy to use. For example, if users will use the instructions in an environment such as a manufacturing or maintenance

TAKING IT INTO THE WORKPLACE
Liability and Safety Information—Can You or Your Company Be Liable?

Are companies liable for damages when instructions for their products are imprecise or inaccurate? In Martin v. Hacker, the New York Court of Appeals unanimously decided that companies indeed are liable. This decision is especially interesting to those who write instructions because the court carefully analyzed the language of instructions in a lawsuit over a drug-induced suicide. Eugene Martin was taking hydrochlorothazide and resperpine for high blood pressure; although he "had no history of mental illness or depression, [he] shot and killed himself in a drug-induced despondency" (Caher). His widow alleged that the warning supplied with the drugs was insufficient. The court stated that the case centered on the drug manufacturer's obligation to fully reveal the potential hazards of its products. Therefore, the court specifically examined the accuracy, clarity, and consistency of the safety warnings. The court scrutinized specific language that the writers used. The court dismissed the lawsuit, stating that the warnings "contained language which, on its face, adequately warned against the precise risk" (Martin v. Hacker).

According to this case and a growing trend, courts will carefully analyze the specific language of technical documents and will hold companies liable for that language (Parson). Companies and their writers, then, must be diligent in writing instructions—especially in terms of accuracy, clarity, and consistency because the "stakes are substantial" (Caher). When a writer's work is unclear and a user "inadvertently reformats a hard drive, that's unfortunate"; but if a writer's inaccurate or unclear language claims a life, that's another matter altogether" (Caher).

Assignment

1. Using the Web, locate a similar case where a consumer sued an organization for damages based on presumably faulty instructions and safety information.
2. Summarize the case and be prepared to discuss it with your classmates.

facility, use a laminated paper or card stock that users can easily clean and a hard cover or notebook that protects the pages.

If users must read the instructions from a distance, the type size and page size may need to be larger than usual. For example, the instructions for clearing the air passages of a choking victim in a restaurant might be on a poster in large type, so users can see the instructions while working with a victim. If users will use the instructions in a small area, the pages must be small enough to fit easily in the work space. For example, quick-reference information for some computer software programs is printed on the front and back of a card that users can place next to their computer.

Design Easy-to-Use Instructions

Instructions must be easy to use, so design the information to help users easily search for and locate specific instructions. When reading instructions, users have a problem to solve or a task to complete; and they want to quickly locate the information needed to solve the problem or complete the task. For example, they may need to learn how to install a software program, how to maximize the storage space on a hard drive, or how to install a new filter in a refrigerator.

 TIPS FOR DESIGNING EASY-TO-USE INSTRUCTIONS

- **Use ample white space.** Users prefer open, uncluttered pages where they can easily scan the page and locate the instructions.
- **Place the graphics near the related information.** Locate a graphic near its related instruction.
- **Use callouts to highlight information in the graphic.** As in Figure 15.2, use callouts to highlight visual information that you want readers to notice.
- **Use a modified hanging or left-hanging format for headings** (See Chapter 10.) The hanging format helps users locate instructions and scan for specific tasks.
- **When possible, use color consistently to highlight important elements such as first-level headings and lists**. Use the same color throughout the document so that users won't wonder what different colors mean (Brockmann). For example, if you use blue for one first-level heading, use blue for all first-level headings.
- **Use typefaces, type sizes, and other design elements consistently**. When you use typefaces, type sizes, and other design elements consistently, you guide users through your instructions. For example, if you use a left-hanging format and a sans-serif type consistently for first-level headings, you help users quickly locate those headings.

FIGURE 15.2

Effectively Designed Instructions

Change the style, symbol, color, or size of a bullet or number

1. To make changes to the bullets or numbers in your slides, on the **Home** tab, in the **Paragraph** group, click the arrow on either the **Bullets** or **Numbering** button, and then click **Bullets and Numbering**.

The writer uses task-oriented headings with verbs to focus the action.

2. In the **Bullets and Numbering** dialog box, do one or more of the following:

A copy of the screen helps users follow the directions.

- To change the style of the bullets or numbering, on the **Bulleted** tab or the **Numbered** tab, click the style that you want.
- To use a picture as a bullet, on the **Bulleted** tab, click **Picture**, and then scroll to find a picture icon that you want to use.
- To add a character from the symbol list to the **Bulleted** or **Numbered** tabs, on the **Bulleted** tab, click **Customize**, click a symbol, and then click **OK**. You can apply the symbol to your slides from the style lists.
- To change the color of the bullets or numbers, on the **Bulleted** tab or the **Numbered** tab, click **Color**, and then select a color.
- To change the size of a bullet or number so that it is a specific size in relation to your text, on the **Bulleted** tab or the **Numbered** tab, click **Size**, and then enter a percentage.
- To convert the existing bulleted or numbered list to a SmartArt graphic, on the **Home** tab, in the **Paragraph** group, click **Convert to SmartArt Graphic**.

Change list levels (indent), spacing between text and points, and more

1. To create an indented (subordinate) list within a list, place the cursor at the start of the line that you want to indent, and then on the **Home** tab, in the **Paragraph** group, click **Increase List Level**.

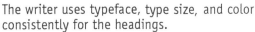

The writer uses typeface, type size, and color consistently for the headings.

The writer helps the user to focus on the pertinent information in the graphics by using redlines and callout numbers.

1 Decrease List Level (indent)

2 Increase List Level (indent)

2. To move text back to a less indented level in the list, place the cursor at the start of the line, and then on the **Home** tab, in the **Paragraph** group, click **Decrease List Level**.

3. To increase or decrease the space between a bullet or number and the text in a line, place the cursor at the start of the line of text. To view the ruler, on the **View** tab, in the **Show/Hide** group, click the **Ruler** check box. On the ruler, click the hanging indent (as shown in diagram below) and drag to space the text from the bullet or number.

NOTE There are three different markers that appear on the ruler to indicate the indentation defined for a text box.

A logical, consistent layout like the one in Figure 15.2 helps users find the information they need. The layout of this page from a user's guide is effective because of the headings, lists, color, and graphics. The headings suggest tasks that users will understand, and the flush-left format makes the headings easy to see. The numbered lists visually separate each step, so users can easily identify and perform each instruction. To design easy-to-use instructions, follow the tips on page 460 and also review the guidelines for effective design in Chapter 10.

Design the Safety Information

The American National Standards Institute suggests specific terms for alerting users to hazards. These terms appear in Figure 15.3.

- Use the word "Danger" to indicate that death or serious injury *will* occur.
- Use "Warning" to indicate that death or serious injury *could* occur.
- Use "Caution" to indicate that a minor injury *might* occur.

Notice these terms in the warning shown in Figure 15.4, page 464.

While preparing your instructions, consider these questions:

- Will the procedure or equipment endanger the users, their surroundings, or their equipment?
- Is the safety alert adequate for the circumstances and severity of the hazard (Brockmann)?
- Is the safety alert located where users will see it before they perform the task that will endanger them?

DRAFT USER-FOCUSED LANGUAGE

To create user-focused language, use a "talker" style: write as if you were speaking directly to your users (Brockmann; Haramundanis). The following examples contrast talker style with writer style:

Talker Style	**Writer Style**
Press the <enter> key for help.	The user should press the <enter> key for assistance.
Press the shutter-release button.	The user should press the shutter-release button.

FIGURE 15.3

Sample Safety Information

Signal Word	Meaning	Example Graphic	Example Language
Danger	Alerts users to an immediate and serious hazardous situation, which will cause death or serious injury.	⚠ DANGER	Danger: Moving parts can crush and cut.
Warning	Alerts users to a potentially hazardous situation, which may cause death or serious injury.	⚠ WARNING	Warning: To reduce the risk of electrical shock, do not use this equipment near water.
Caution	Alerts users to the potential of a hazardous situation, which may cause minor or moderate injury.	⚠ CAUTION	Caution: Do not use to exhaust hazardous materials and vapors
Note	Gives users a tip or suggestion to help **them** complete a task or use equipment successfully.	*NOTICE*	Note: Turn off the mixer before scraping the bowl.

Source: Downloaded from the World Wide Web: Danger/Warning/Caution/Notice signs:
www.compliancesigns.com/index.shtml?kw=osha%20labels&gclid=CL6A4Y63upMCFQoRswodGAmUCA-1.
Courtesy of www.compliancesigns.com, supplier of high quality safety signs.

Talker Style

Tighten the knobs on each side of the handle.

Writer Style

The knobs on each side of the handle should be tightened.

You create *talker style* by using
- action verbs
- imperative sentences
- task-oriented headings
- simple, specific language
- language that users will understand

TIPS FOR WRITING AND DESIGNING SAFETY INFORMATION

- **Place the safety alert before or next to the directions for the hazardous task—not after them.** If the hazard is severe, put the safety alert at the beginning of the instructions or manual and repeat it before–or next to–the task where the hazard occurs.
- **Use a symbol or an icon to indicate a warning, danger, or caution**. The warnings shown in Figure 15.4 use an exclamation point (!) inside a triangle to indicate a hazard. This symbol is the international symbol for safety alert. These warnings also use simple drawings to illustrate the potential hazard. Graphics are especially important for international users or for users who aren't native or expert speakers of technical English.
- **Use a distinct color and/or icon consistently for notes.**
- **Separate the safety alert from the text with white space or a border**.

FIGURE 15.4

Sample Safety Signs

Source: Downloaded from the World Wide Web: www.safetylabel.com/productsafetylabels/customization.php.

Use Action Verbs

Action verbs tell users what to do. When writing step-by-step directions, use verbs such as *run, adjust, press, type*, and *loosen*—not verbs such as *is* or *have*. Notice the action verbs (in boldface type) in these instructions:

Replace the battery with a new one.
Turn off all surface units.
Remove the spill with a paper towel.
Press the OK button.

Each action verb clearly indicates the action that the user should perform. Each action verb is also in the active voice (to review the difference between the active and passive voice, see Chapter 7).

Use Imperative Sentences

Effective step-by-step directions consist of talker-style sentences with the understood "you" as the subject, as in "Replace the battery"—meaning "[You] replace the battery." When the subject is understood to be you, the sentence is imperative. An *imperative sentence* is a command. Its main verb tells the user to carry out some action. Imperative sentences focus attention on what the user is to do, not on the user. Compare these two sentences:

Writer-style   You should locate a wall stud in the area where you want to install your flat-screen television.

Imperative   Locate a wall stud in the area where you want to install your flat-screen television.

The writer-style sentence focuses the user's attention on the subject *you*. In the imperative sentence, the user understands the subject to be *you* and focuses on the main verb, *locate*, which expresses the action.

Let's look at two more examples:

Writer-style   Testing of emergency numbers should be performed during off-peak hours, such as in the early morning or late evening.

Evaluate instructions in the Interactive Student Analysis online at www.kendallhunt.com/technicalcommunication.

Imperative   Test emergency numbers during off-peak hours, such as in the early morning or late evening.

In the writer-style sentence, the verb is in the passive voice ("be performed"), and the action that the user is to perform is in the noun *testing*. The imperative sentence expresses the action in the verb *test*. By using imperative sentences, the writer eliminates passive-voice constructions and focuses the user's attention on the action.

Use Task-Oriented Headings

Task-oriented headings focus on action by using verbs or the words "how to." Consider these headings:

Not task-oriented   Slide Shows

Task-oriented   Running the Slide Show
How to Run the Slide Show

The not task-oriented heading uses nouns instead of verbs to identify the tasks described in the section that it introduces. The task-oriented headings use verbs to identify the tasks. Task-oriented headings
• suggest activities that users may already understand and can find immediately applicable (Brockmann)
• help users locate the instructions for specific tasks

Use Simple and Specific Language

Try to resist the temptation to use "fancy" or less-familiar words. Instead, use words that users can quickly and easily understand. If you are unsure whether users will understand a word, choose another word or provide a definition that they will understand. See Chapter 8 for more information on using simple language.

When writing instructions and manuals, use specific language. Otherwise, users might not gather the proper equipment and materials, or they might misunderstand the instructions and injure themselves or damage equipment. Consider these instructions for changing the oil in an automobile:

Before changing the oil, run the engine until it reaches normal operating temperature. The engine has reached this temperature when the exhaust pipe is warm to the touch. You should run the engine to mix the dirt and sludge with the oil in the crankcase, so the dirt and sludge drain along with the oil.

Experienced automobile mechanics would know how long to let the engine run before the exhaust pipe became so hot that it would burn their hands. Other users might not know how long to let the engine run. Thus, the writer should specifically state how long to let the engine run, so that novice users won't risk burning themselves on a hot exhaust pipe. Specific language is crucial if users are to correctly and safely follow instructions.

Use Language Users Will Understand

Use terminology that users will understand. To determine how "technical" your manual or instructions can be, consider these questions:
- What sort of background and training do your users have? Will they understand technical terminology related to the task?
- Are the users native or expert speakers of technical English?

ETHICS NOTE

Protecting Your Users

You have a legal and ethical responsibility to protect users from possibly injuring themselves or others and from possibly damaging equipment and materials. You should include whatever information is necessary to fully inform readers of possible dangers when following your instructions or when using equipment involved in those instructions. To ensure that you ethically provide users with safety information, do the following:
- Make the safety information easy to see and to read.
- Put the safety information where users need to see it. You cannot simply include the safety information in the introduction. You should place the information where users could possibly injure themselves or damage equipment.
- Use graphics for warnings and danger so users of all languages will be protected.
- Include all information necessary to protect users.

TIPS FOR USING USER-CENTERED LANGUAGE

- **Use action verbs**. When you use action verbs, you focus the sentence on the task.
- **Use imperative sentences**. Imperative sentences have an understood "you" subject; for example, "Turn the speed control to off," rather than "You should turn the speed control to off." Imperative sentences focus the users' attention on the action.
- **Use simple language**. Select words that users can quickly and easily understand.
- **Use specific language**. Without specific language, users may misunderstand the instructions and possibly damage equipment or injure themselves or others.
- **Use language that users will understand**. Use technical terminology only if your users are familiar with this terminology. If you must use technical terms that your users won't understand, define those terms.
- **Use technical terminology when your users expect it**. If your users are familiar with the technical terminology, they will expect you to use it.
- **If your users are not native speakers of English, avoid connotative and ambiguous language and terms and follow the guidelines for Simplified Technical English**. Such language and terms may have different meanings in other languages. See Chapter 8 for information on Simplified Technical English.

If you know that users have background or training in a field related to the task, use technical terms that they will know and expect. If your users have little relevant background or training, avoid technical terms and instead use language that they will understand. If your users are not native or expert speakers of technical English, avoid connotative and ambiguous language and terms that may have different meanings in other languages.

When writing for users from other cultures, be sensitive to the tone of your writing. For instance, imperative sentences will make users in some cultures uncomfortable. For other users, you might use labeled drawings and graphics, especially when the instructions are relatively simple and require few words (Brockmann).

DRAFT THE ELEMENTS OF INSTRUCTIONS AND MANUALS

The structure, length, and formality of the instructions or manual varies depending on the procedure and users. If you want to explain to your coworkers how to use a new copier, you might first briefly introduce the procedure and then present the step-by-step instructions. You could send these instructions to the users in a memo or by way of e-mail, or you might write the instructions on a card placed on a wall near the copier. Your coworkers just want to know how to use the copier; they don't need to know how a copier works or a formal list of needed materials. However, suppose you are writing instructions that come in the carton with the copier. These instructions will have to be more formal than the ones for your coworkers, and the instructions must meet the needs of a diverse group of users. These instructions would appear in a manual that has a table of contents and an index. Such a manual would also list the materials, equipment, and instructions for installing and using the copier. The manual would also likely have a troubleshooting section for solving common problems.

For informal instructions, you will include these sections:
- Introduction
- Step-by-step directions
- Troubleshooting (not included with all instructions)
- Reference aids (not included with all instructions)

For manuals, you will include these sections:
- Front matter (cover, title page, table of contents)
- Introduction
- Step-by-step directions
- Troubleshooting (not included with all instructions)
- Reference aids (not included with all instructions)
- Index

For information on front matter and indexes, see Chapter 18.

Introduction

The *introduction* gives users the basic information they need to understand how to use the instructions. Introductions for manuals often have titles other than "Introduction." They might be titled "Overview," "Getting Started," or "How to Use This Manual." In a manual, the introduction may have some additional elements; it may explain how the manual is organized, who should use the manual, conventions used in the manual, and where to get additional information. For example, Figure 15.5, page 470, illustrates the introduction from a software manual.

FIGURE 15.5

Introduction from a Software Manual

The introduction explains what the software does and what the user needs to know before using the software.

The introduction explains who should use the software.

Welcome to OptQuest®

Welcome to OptQuest® for Crystal Ball® 2000!

OptQuest enhances Crystal Ball by automatically searching for and finding optimal solutions to simulation models. Simulation models by themselves can only give you a range of possible outcomes for any situation. They don't tell you how to control the situation to achieve the best outcome.

OptQuest, through a new optimization technique, finds the right combination of variables that produces the best results possible. If you use simulation models to answer questions such as, "What are likely sales for next month?" now you can find the price points that maximize monthly sales. If you asked, "What will production rates be for this new oil field?" now you can additionally determine the number of wells to drill to maximize net present value. And if you wonder, "Which stock portfolio should I pick?" with OptQuest, you can choose the one that yields the greatest profit with the least risk.

Like Crystal Ball, OptQuest is easy to learn and easy to use. With its wizard-based design, you can start optimizing your own models in under an hour. All you need to know is how to create a Crystal Ball spreadsheet model. From there, this manual guides you step by step, explaining OptQuest terms, procedures, and results.

Who this program is for

OptQuest is for the decision-maker—from the businessperson analyzing the risk of new markets to the scientist evaluating experiments and hypotheses. With OptQuest, you can make decisions that maximize the use of your resources, time, and money.

OptQuest has been developed with a wide range of spreadsheet uses and users in mind. You don't need highly advanced statistical or computer knowledge to use OptQuest to its full potential. All you need is a basic working knowledge of your personal computer and the ability to create a Crystal Ball spreadsheet model.

- **Glossary**

  A compilation of terms specific to OptQuest as well as statistical terms used in this manual.

- **Index**

  An alphabetical list of subjects and corresponding page numbers.

Additional resources

Decisioneering, Inc. offers these additional resources to increase the effectiveness with which you can use our products.

Technical support

If you have a technical support question or would like to comment on OptQuest, there are a number of ways to reach Technical Support. See the accompanying Crystal Ball README file for more information.

Consulting referral service

Decisioneering, Inc. provides referrals to individuals and companies alike. The primary focus of this service is to provide a clearinghouse for consultants in specific industries who can provide specialized services to the Crystal Ball and OptQuest user community.

If you wish to learn more about this referral service, call 800-289-2550 Monday through Friday, between 9:00 A.M. and 5:00 P.M. Mountain Standard Time.

The introduction directs users to additional resources.

FIGURE 15.5 CONTINUED...

"How this manual is organized" gives an overview of the major sections of the manual.

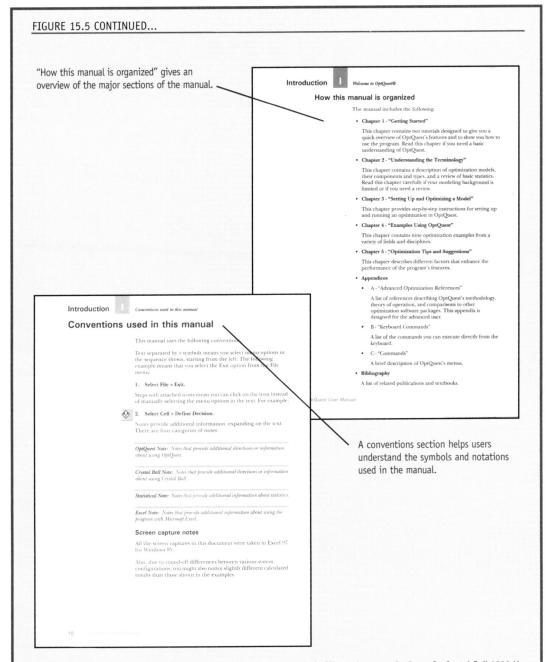

Introduction | **I** | *Welcome to OptQuest®*

How this manual is organized

The manual includes the following:

- **Chapter 1 - "Getting Started"**

  This chapter contains two tutorials designed to give you a quick overview of OptQuest's features and to show you how to use the program. Read this chapter if you need a basic understanding of OptQuest.

- **Chapter 2 - "Understanding the Terminology"**

  This chapter contains a description of optimization models, their components and types, and a review of basic statistics. Read this chapter carefully if your modeling background is limited or if you need a review.

- **Chapter 3 - "Setting Up and Optimizing a Model"**

  This chapter provides step-by-step instructions for setting up and running an optimization in OptQuest.

- **Chapter 4 - "Examples Using OptQuest"**

  This chapter contains nine optimization examples from a variety of fields and disciplines.

- **Chapter 5 - "Optimization Tips and Suggestions"**

  This chapter describes different factors that enhance the performance of the program's features.

- **Appendices**

  - A - "Advanced Optimization References"

    A list of references describing OptQuest's methodology, theory of operation, and comparisons to other optimization software packages. This appendix is designed for the advanced user.

  - B - "Keyboard Commands"

    A list of the commands you can execute directly from the keyboard.

  - C - "Commands"

    A brief description of OptQuest's menus.

- **Bibliography**

  A list of related publications and textbooks.

OptQuest User Manual

Introduction | *Conventions used in this manual*

Conventions used in this manual

This manual uses the following conventions.

Text separated by > symbols means you select menu options in the sequence shown, starting from the left. The following example means that you select the Exit option from the File menu:

1. **Select File > Exit.**

Steps with attached icons mean you can click on the icon instead of manually selecting the menu options in the text. For example:

2. **Select Cell > Define Decision.**

Notes provide additional information, expanding on the text. There are four categories of notes:

OptQuest Note: *Notes that provide additional directions or information about using OptQuest.*

Crystal Ball Note: *Notes that provide additional directions or information about using Crystal Ball.*

Statistical Note: *Notes that provide additional information about statistics.*

Excel Note: *Notes that provide additional information about using the program with Microsoft Excel.*

Screen capture notes

All the screen captures in this document were taken in Excel 97 for Windows 95.

Also, due to round-off differences between various system configurations, you might also notice slightly different calculated results than those shown in the examples.

10

A conventions section helps users understand the symbols and notations used in the manual.

When reading the introduction to instructions, and especially to a manual, readers may ask the following questions:

- What is the purpose of the instructions?
- What, if anything, should I know before beginning the task or using the equipment?
- What materials and equipment do I need?
- Is the equipment or the task safe? What do I need to do to protect the equipment or my surroundings from damage? What do I need to do to protect myself from injury?
- Where will I find the information that I need in the manual?

The introduction for instructions and manuals may include any or all of the information listed in the tips on page 473.

Step-by-Step Directions

Step-by-step directions tell users exactly how to carry out a procedure. To determine how to organize the directions, consider these questions (Brockmann):

- What action begins each task?
- What are the specific steps for performing the task? Can you group these steps into subtasks?
- What action ends each task?
- Can you perform the task in more than one way? If so, do users need to know both ways?

Once you have considered readers' questions, categorize the task into major steps and then divide the steps into appropriate substeps. For example, if you are explaining to a novice how to change the oil in a car, you might divide the task into four major steps: (1) drain the old oil, (2) remove the old oil filter, (3) install the new filter, and (4) refill the crankcase with oil. Then, you might subdivide each step into substeps. When you divide the steps into substeps, you help users understand the task and follow your directions. One long list of uncategorized steps can intimidate users. The simple task of changing the oil could have as many as thirty steps. Most users would prefer four major steps with substeps to a list of thirty steps.

TIPS FOR WRITING AN INTRODUCTION

- **State the purpose of the instructions**. Tell the users the goal of the instructions.
- **State clearly who should use the instructions and what they should know about the task**. For example, the installation guide for a garbage disposal states that the instructions are "for electricians with a Level 3 or higher certification according to Section 3 of the state electrical code." The writer, therefore, assumes that users know how to install similar equipment and that they understand the electrical code and related ordinances.
- **List all the materials and equipment that users need**. List all the materials and equipment in one place, so users can gather them before beginning the task.
- **Use drawings of some materials and equipment to ensure that users know what tools they need.** For example, the screw identification page shown in Figure 15.9, page 481, uses drawings and words to make sure users correctly identify screws when removing components from a laptop. The drawings are the actual size, so users can keep track of and identify the screws. Some writers identify the materials and equipment in a separate section after the introduction.
- **Explain typographical conventions and terminology used in the instructions**. If you use any typographical conventions, icons, or terminology that your users may not understand or recognize, explain them in the introduction. If you use many terms that users may not understand, define them in a glossary and introduce the glossary in the introduction.
- **Explain the parts of the equipment**—if equipment is involved. For example, Figure 15.6, page 475, shows a drawing of a cell phone with callouts of the phone's features. These graphics help orient the user before reading the manual or using the phone.
- **Include any other information necessary to help users use and understand the instructions**. For instance, if you've divided the manual into sections, tell users what information each section includes and perhaps when and how to use each section.
- **Explain all safety information and how you will note that information.** When users see the warning graphics and signal words, they will better understand the nature of the hazard.

As you write the directions, follow the tips on page 474. The sample instructions shown in Figures 15.9 and 15.10 on pages 481-483 follow these tips.

TIPS FOR WRITING STEP-BY-STEP DIRECTIONS

- **Categorize the task into major steps and then divide those steps into substeps**. Categories help users understand and follow your directions. (See Chapter 6 for information on using classification and division.)
- **Describe only one action in each step or substep**. If you put more than one action in a step, users may not see the second action. The users' eyes will leave the page as they perform the required action. When users return to the page, they might go to the next numbered step and miss the second action.
- **Number each step**. Without a number, a step may not stand out, and users may miss it.
- **Use a list format, so users can follow directions easily**. If you bury directions in paragraphs or extremely long lines of text, users may skip a step or miss directions.
- **Use graphics when appropriate**. Graphics can help users follow directions and can clarify the required actions.
- **Give step-by-step directions in chronological order**. Include any necessary safety alerts and materials or equipment (if the equipment or materials vary.)

Troubleshooting Guide

A *troubleshooting guide* helps users solve commonly encountered problems. It often appears in a table format with the problem in one column and the solution in another column (see Figure 15.7, page 476). Alternatively, you can simply use left-hanging headings for the problem and bulleted lists for the solutions. Brief, simple instructions may not need troubleshooting guides.

Troubleshooting guides can save you, your organization, and your users time and money by helping users solve problems on their own. When users can solve their problems without having to call your organization, you can reduce the amount of time and money your organization spends for online and phone support. By anticipating users' problems, troubleshooting guides allow users to solve common problems in less time than a trial-and-error approach would take and without enlisting technical support from your company.

Reference Aids

To help users find the information they need in your instructions and manuals, you can provide a variety of reference aids, depending on the needs of your users and the length and complexity of the instructions. Common reference aids include the following:

- **Table of contents**. Include a table of contents for all manuals (see Chapter 18 for information on preparing a table of contents.)

FIGURE 15.6

Introduction from a Cell Phone Manual

The writer uses a small page size. The small size allows the manual to easily fit in the cell-phone box and the user to easily carry the manual. Even though the pages are small, they are uncluttered and easy to read.

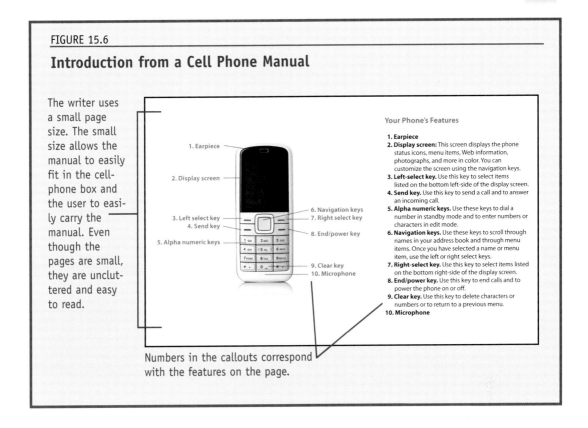

Your Phone's Features

1. **Earpiece**
2. **Display screen:** This screen displays the phone status icons, menu items, Web information, photographs, and more in color. You can customize the screen using the navigation keys.
3. **Left-select key.** Use this key to select items listed on the bottom left-side of the display screen.
4. **Send key.** Use this key to send a call and to answer an incoming call.
5. **Alpha numeric keys.** Use these keys to dial a number in standby mode and to enter numbers or characters in edit mode.
6. **Navigation keys.** Use these keys to scroll through names in your address book and through menu items. Once you have selected a name or menu item, use the left or right select keys.
7. **Right-select key.** Use this key to select items listed on the bottom right-side of the display screen.
8. **End/power key.** Use this key to end calls and to power the phone on or off.
9. **Clear key.** Use this key to delete characters or numbers or to return to a previous menu.
10. **Microphone**

1. Earpiece
2. Display screen
3. Left select key
4. Send key
5. Alpha numeric keys
6. Navigation keys
7. Right select key
8. End/power key
9. Clear key
10. Microphone

Numbers in the callouts correspond with the features on the page.

- **Index**. Include an index for all manuals. Indexes help users find specific information and solve problems quickly. (See Chapter 18 for information on preparing an index.)
- **Headings**. Headings help users locate specific information and provide a road map to the instructions. (See Chapter 10 for information on designing headings.)
- **Quick-reference cards**. Quick-reference cards provide an overview of capabilities or commands. These cards help users who are familiar with the task and who don't need the elaboration of more-extensive instructions. Quick-reference cards are usually a separate document from the manual; they may be in the form of a brochure, a small card, a template, or a flyer. You can also place a quick-reference "card" on the back cover, or inside the front or back covers of a book. You might place the "card" on the walls or on doors as with Figure 15.8, page 477. This quick-reference card provides instructions for hand washing. The card includes

FIGURE 15.7

Troubleshooting Guide

Troubleshooting Guidelines

In the event that something is not working correctly on the oven, the display will show an error message and suggest that you call for service. Before calling for service, reference the following table for problems that you may be able to fix yourself.

PROBLEM	POSSIBLE SOLUTION(S)
Displays and indicator lights are not working	Check that oven is receiving power.
Cook Navigator Screen is too dark or light	Adjust the brightness of the display – see Oven Setup, page 20.
Sounds are not working	Check that the volume is turned on – see Oven Setup, page 20.
Oven sounds are too loud or soft	Adjust the volume – see Oven Setup, page 20.
Menus are in the wrong language	Make sure desired language is selected – see Oven Setup, page 20.
Units and Measurements are displayed in metric and I want standard or vice versa	Change the Units and Measurements – see Oven Setup, page 20.
I forgot to save my changes to a recipe recently cooked	See Cooking a Recently Cooked Dish, page 7. Select "Save as Favorites."
Clock is set to the wrong time	Use the Set Timer Knob to reset. See page 6.
Oven light bulb is burned out	Call Customer Service at 866.44SERVE to order a replacement bulb. Instructions and all necessary components included with each bulb.
Oven Timer does not count down	Make sure the Set Timer Knob is pressed back into its original position.
I experienced interference with my wireless phone	900MHz cordless phones are recommended to limit interference also try operating the wireless network on channel 1 if possible.

Source: Downloaded from the World Wide Web, www.turbochef.com. *TurboChef 30" Double Wall Speedcook Oven Use and Care Guide*, page 37. Used with permission of TurboChef.

graphics to instruct the users and provides instructions in both Spanish and English.

TEST YOUR INSTRUCTIONS

When time and money allow, test your product and your instructions before you release them to your primary users to determine whether your product or documents are usable. A usable document contains accurate, complete information that is easy to use. ***Usability testing*** is a process of conducting experiments with people who represent the users of a document. The goal of usability testing is to determine how well users will

FIGURE 15.8

Quick-Reference Card

The card includes numbers to show users the
sequence for washing their hands.

The instructions appear in both English and
Spanish. Users can easily distinguish the
English and Spanish versions because the
English instructions appear in blue and the
Spanish in purple.

The simple graphics allow users to wash
their hands properly without reading the
text.

Source: Compiled from information downloaded from the World Wide Web, July 16, 2009:
https://fortress.wa.gov/doh/here/materials/PDFs/12_GermBust_B06L.pdf. DOH Pub 130-012, 8/2006.
Courtesy of Washington State Department of Health.

understand the document and whether they can safely and easily use it. You can conduct usability testing on any technical document; however, you will primarily test instructions and manuals. Usability testing can uncover places where users cannot understand what you have written, where you have given too much or too little information, and where users need graphics. Usability testing tells you whether or not users can

- locate the information they need to perform the task
- use the information to perform the task successfully and easily

To perform an effective usability test, you must put users in a realistic setting that simulates the actual situations where they will use the document (Zimmerman, Muraski, and Slater; Redish and Schell). These users must be people who will actually use the document. Most organizations use a setting that simulates the actual situations in which users will perform the tasks and use the documents; however, some technical communicators conduct usability testing in the field (Zimmerman, Muraski, and Slater). For example, a group of researchers wanted to test the effectiveness of pesticide-warning labels, so they asked a group of farmers to participate in usability tests (Zimmerman, Muraski, and Slater).

You can test instructions several times during the writing process by doing the following:

- Test the prototype, or first draft, of a chapter or section. Prototype testing occurs early in the writing process before you draft the entire document. Prototype testing helps you see whether the layout, design, and style you have selected will work for your users.
- Test a complete, but preliminary, draft of the instructions.
- Test a revised, but not yet final, draft of the instructions.

Prototype testing provides feedback before you write the entire document. If you wait to test until you are completely satisfied with the instructions, you probably will be near your final deadline and may not have time to revise the document after it is tested.

To conduct an effective usability test, you will

- prepare for the test
- conduct the test
- interpret the test results to revise the document

Prepare for the Test

To conduct an effective usability test, you and your team will need to plan the test. To plan your test, complete the following steps:

- **Determine the needs of the users**. Many organizations conduct focus groups to determine what users need and want. These groups may be made up of users who use similar products and documents. Other organizations conduct on-site interviews and observations of users in the workplace. These focus groups, on-site interviews, and observations help testers find out what users already know about the task and in what situations users will use the product or instructions.

- **Determine the purpose of the test**. What do you want to learn from the test? For example, you might want to test a prototype of the instructions. You can also test an advanced draft of your instructions.

- **Design the test**. For most usability tests, you will work with a team. Each member of the team will be responsible for carrying out one aspect of the test. The team as a whole or one member of the team will develop the test. Make sure that the test asks participants to focus on specific problems such as the following (Daugherty):

  - *Content*. Are the instructions complete?
  - *Ease*. Can users easily complete tasks when following the instructions?
  - *Design*. Can users easily locate and read the information?
  - *Sentence structure and language*. Is any of the information confusing?
  - *Graphics*. Would using more or fewer graphics help users follow the instructions?

- **Select the test participants**. Select real users to participate in the test, not your friends, family, or coworkers (unless you're writing a manual for your coworkers). Test participants should mirror your intended users (Brockmann).

- **Reserve the testing facilities**. If you're simulating the situation in which users will perform the task, you will need two rooms: the participants' room with a camera to record the testing and the observers' room. The observers' room should, if possible, have a one-way mirror; so observers can see participants, but participants can't see observers. If you do not have access to this two-room setup, you can set the test up in one room and video the users during the test.

- **Prepare a schedule of the testing-day events**. Make sure that all participants and observers know when testing begins.

Conduct the Test

To ensure that the test is reliable, consider giving the test twice (Daugherty), and follow these guidelines:

- **Explain the purpose of the test, either orally or in writing**.
- **Tell participants how they should note problems or make suggestions**. For example, you might give them a questionnaire to complete as they work through the test, or you can give them a notepad for jotting down ideas.
- **Observe participants while they use the instructions**. Watch for any problems they encounter.
- **Conduct an exit interview or debriefing with each participant**. Prepare a brief questionnaire to give each participant at the end of the test. You may ask participants to complete this questionnaire in writing, or you or another member of your team can simply ask the questions orally, using the written questionnaire as a guide. As part of the debriefing, be flexible. If you or your team members have questions about the participants' actions or answers, ask—even if the question isn't on the questionnaire. If possible, put one team member in charge of the exit interview to keep the interviews consistent and organized.

Interpret the Results

After you have completed the test, you will have much data to tabulate and analyze. After your analysis, write a clear, detailed report (usually an informal report) that includes the results and draws conclusions. (See Chapter 17 for information on informal reports.) For example, you may determine that you need to revise the instructions and conduct more usability testing.

SAMPLE INSTRUCTIONS

The sample instructions presented in Figure 15.9, pages 481-482, are an excerpt from an online user's guide. The instructions illustrate an effectively designed list of materials and method for identifying parts. Figure 15.10, page 483, is an excerpt from a laptop manual. This excerpt includes a clear graphic that shows users how to connect a network cable box. Many users will only use the diagram to connect the cable; however, the writers have included text to support the graphic.

FIGURE 15.9

Online Instructions from a Laptop User's Guide

These links direct the reader to a drawing of the laptop and a table that identifies the screws that the user will be removing from the laptop.

This section of the manual includes an introduction explaining what the user should know before beginning the task.

"Recommended Tools" tells users the tools they need before beginning.

The introduction includes important safety information.

The writer uses imperative sentences and action verbs.

Getting Started

⚔ Understanding the laptop parts
⚔ Identifying the screws

This guide gives you procedures for removing and installing your laptop components. Before beginning these procedures,
- follow the steps in Turn Off Your Laptop and in Before Working Inside Your Laptop
- read the safety information in your Owner's Manual

Recommended Tools

For these procedures, you will need the following tools:
- Phillips screwdriver
- Flathead screwdriver
- Small plastic scribe
- Hex nut driver

Turn Off Your Laptop

→*Notice:* If you do not properly exit programs and close files, you could lose data.
1. To shut down the operating system, follow these steps:
   a. Save and close open files.
   b. Exit open programs.
   c. Click on Turn Off Your Computer.
   d. In the Turn Off window, click Turn off. The laptop will automatically turn off after the operating system completes the shutdown process.
2. Make sure that the laptop is turned off. If your laptop does not automatically turn off when you shut down the operating system, press and hold the power button for 3 seconds.

FIGURE 15.9 CONTINUED

The graphic helps users to complete the task.

Identifying the Screws

Print this screw identification chart. As you work, keep the screws organized by placing them on their respective squares.

M2 x 3mm   M2 x 4mm   M2 x 5mm   M2 x 6mm   M2 x 8mm

This section has drawings of the actual size of the screws. The writer recommends using the chart as a "placemat" to keep track of the screws.

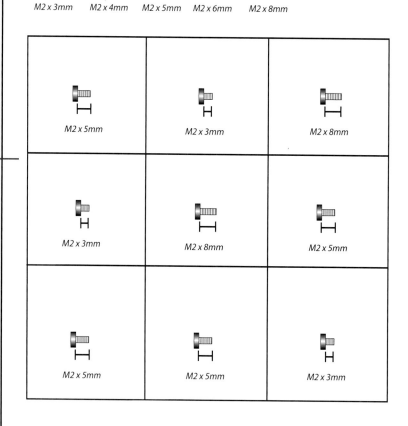

FIGURE 15.10

Using Graphics to Instruct

Setting Up a Home and Office Network

Connecting Your Computer to a Network Adapter

To connect your computer to a network, you will need a network cable.
If you did not receive a network cable with your computer, contact customer service at 1-866-543-1256.

1. Insert the network cable into the network adapter on your computer.
  → **TIP:** Insert the cable until it clicks into place.

2. Connect the other end of the cable to a network connection such as a network wall jack.
  → **TIP:** Do not use a network cable with a telephone jack. You could damage your computer.

Setting Up a Home and Office Network • 25

CASE STUDY ANALYSIS
The Importance of the "Human Factor" in Product Design and Usability Testing

Background

When a medical device's design leads to patient death, more often than not, it can be attributed to human error. Human error has been attributed annually to 98,000 deaths annually in U.S. hospitals.[1] Often, the cause of human error can be traced to inadequate usability testing. The "human factor" has been recognized as a common problem in devices failing for some time. In 1996, the U.S. Food and Drug Administration published a report entitled "An Introduction to Human Factors in Medical Devices,"[2] in which it stated, "...device design and related use errors are often implicated in adverse events." The purpose of the report was to "encourage manufacturers to improve the safety of medical devices and equipment by reducing the likelihood of user error." Usability testing was one method recommended by the Food and Drug Administration to address the problem.

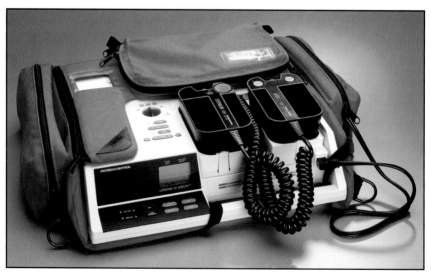

The Lifepack 10 defibrilator featured improvements designed to make it easier to use. However, the company did not test the changes before the product was released, and unforeseen usability problems resulted in delayed treatment in real-world situations.

Photo courtesy of Physio-Control.

Usability testing is important even when a change will benefit a device's design. Consider the design of the Lifepak 10 defibrillator. An earlier version of the defibrillator featured paddles that had to be lifted and tilted to be removed from the base. The new model, however, featured paddles that slid out toward the user, making them easier to disengage from the base. This change was promoted as an enhanced feature of the Lifepak 10, but it resulted in medical delays when staff had difficulty disengaging the paddles. The problem was compounded because the machine did not include any visual clues, instructions, or warnings about how to remove the paddles. Usability testing would have identified the problem. Instead, delays resulted in patient distress and death.[3]

Assignment

1. Examine the photo of the Lifepak 10 defibrillator on the previous page. Create a graphic, instruction, or warning label for the device, explaining how to disengage the paddles. Turn your design in to your instructor with an explanation about where the label will be located on the defibrillator.

2. Design a poster explaining how to use the defibrillator for placement on a wall. For this exercise, because you may not know all the steps for using a real defibrillator, use these five steps: 1) turn the defibrillator on; 2) set the voltage; 3) remove the paddles; 4) administer the shock; 5) examine the patient for signs of revival. Use graphics and text to design your poster. Turn your poster in to your instructor with a short memo describing your design selections.

[1] Downloaded from the World Wide Web: www.humanfactorsmd.com:hfandmedicine_reducerror_magnitude.html. Human Factors MD Inc., "The Human Factors MD Index."
[2] Downloaded from the World Wide Web: http://www.fda.gov/cdrh/humfac/doitpdf. U.S. Food and Drug Administration, "An Introduction to Human Factors in Medical Devices," Do It By Design.
[3] Compiled from information downloaded from the World Wide Web:
http://www.health.qld.gov.au/patientsafety/documents/psmseptoct2006.pdf. May 22, 2008

EXERCISES

1. Locate instructions that are hard to follow. Print or copy the instructions. (Your instructor may specify a length for these instructions.) Then complete these tasks:

   a Identify the intended users of the instructions. Analyze the content, organization, conventional elements, language, and design to see whether they are appropriate for the users. Using this analysis, determine why the instructions are hard to follow.

   b. Write a memo to your instructor describing the intended users of the instructions and summarizing your analysis. Attach a copy of the instructions to your memo. (For more information on memos, see Chapter 12.)

   c. Rewrite the instructions (or a portion of the instructions, as assigned by your instructor) so users can follow them easily. When you give the revised instructions to your instructor, attach a copy of the original version.

2. Write instructions for one of the following tasks or for a task related to your major. When you have written the instructions, attach a brief memo to your instructor stating the users and the purpose of the instructions. Be sure to include graphics where appropriate and to follow the guidelines presented in this chapter. (For more information on memos, see Chapter 12.)

   a. how to change the filters for the heating and air conditioning system in your house or apartment

   b. how to locate, download, and import a graphic from a Web site

   c. how to change a tire

   d. how to grill a steak

3. Write a troubleshooting guide for the instructions that you wrote in Exercise 2.

4. Analyze the language and design of the instructions in Figure 15.11 on page 488. Write a memo presenting your analysis to your instructor.

5. Rewrite the instructions in Figure 15.11, page 488, eliminating the problems in language and design that you identified in Exercise 4.

6. **Web exercise**: Locate online instructions for a product or service related to your major. Evaluate the online instructions in an e-mail to your instructor. Tell your instructor where to locate the instructions. In your e-mail, answer these questions:
   a. Could you easily navigate through the instructions?
   b. Did the screen design help you locate information?
   c. How effectively did the writer design the instructions?
   d. How would you improve the instructions?

REAL WORLD EXPERIENCE
Working with a Team to Write a Manual

You and your team will write a manual for a task for which written instructions are not available or are inadequate for users to complete the task. To write the manual, you and your team will complete these steps:

1. Select a task. Make sure that all members of your team are familiar with the task. If the task requires equipment, make sure that your team has access to the equipment to prepare and to test your instructions. Your instructor may want to approve your topic before you move to the next step.
2. Prepare a detailed outline of the manual. Submit this outline to your instructor for approval.
3. Decide on the reference aids that you will include with the manual.
4. Prepare prototype pages showing the language, headings, and design that you plan to use in the manual.
   a. Find sample users. Ask these users to follow the step-by-step directions on the prototype pages. Watch these users as they follow the instructions, noting problems. After these users have completed the instructions, ask them whether the language, headings, and layout helped them follow the instructions and how you can improve any of these elements.
   b. Revise the language, headings, and layout as necessary.
5. Write the manual, dividing the writing tasks among your team members. Test the manual again. Be sure to follow the guidelines for usability testing in this chapter.
6. Revise the manual based on the test.
7. Turn in the final version of the manual to your instructor.

FIGURE 15.11

Sample Instructions for Exercises 4 and 5

Operating Instructions: Humidity Sensor

Covered

Vented

Dry off dishes so they don't mislead the sensor.

When using the Humidity Sensor...

This microwave automatically adjusts cooking times based on the amount of humidity released during cooking. Various types and amounts of food release varying levels of humidity, guiding the sensor in detection of doneness.

Using the Humidity Sensor feature twice within a short period on the same food will likely result in severely overcooked or burnt food. If additional heating/cooking is desired after the first cooking period (using the Humidity Sensor), use the Time Cook feature.

The type of container used for heating food affects the Humidity Sensor. Tight sealing plastic containers do not allow steam to escape. Therefore, the Humidity Sensor is not able to detect cooking progress, resulting in over-cooked foods. The proper containers to use are those designed "microwave-safe." Cover them with their matching lids or use vented plastic wrap.

If wet or damp containers are used for heating food, the added steam will mislead the Humidity Sensor and food will likely be undercooked. This applies to the inside of the microwave as well–it must be dry in order for the Humidity Sensor to work accurately.

Operating Instructions

21

iStockphoto 2008.

CHAPTER SIXTEEN

16

Writing Persuasive Proposals

Proposals are documents designed to persuade someone to follow or accept a specific course of action. Proposals offer to solve a problem, to provide a service or product, or to perform research. They suggest a specific plan for delivering that solution, service, or product, or for performing that research. This chapter discusses how to write a persuasive proposal. Before learning how to write a proposal, we'll begin by exploring the types of proposals.

TYPES OF PROPOSALS

You might write a proposal, sometimes called a bid, in response to several scenarios. Let's look at five scenarios and how the writers responded:

Scenario 1   Bill Martinez works for a large construction company. His company is building a new manufacturing plant. The manufacturing process will create large amounts of ash. According to EPA guidelines, the company must dispose of the ash properly. Bill is responsible for hiring a company to develop recommendations for disposing of the ash. Therefore, he sends out requests to several environmental service companies, inviting them to submit their proposals to develop the recommendations. (The sample proposal in the Interactive Student Analysis, online, is a proposal from one of the environmental service companies.)

View the Interactive Document Analysis online at www.kendallhunt.com/ technicalcommunication.

Scenario 2   A government agency in Wisconsin decides that it should renovate its office building. The building is fifty years old and has several problems: it doesn't meet current fire code regulations; the windows and outside doors provide little insulation from the weather; the bathrooms have plumbing problems; and the offices are not appropriately wired for current telecommunications and fiber optic technology. The agency wants to find a qualified construction company to renovate the building at a reasonable price. The agency advertises the proposed work in newspapers across Wisconsin. Any companies interested in submitting proposals can then request the specifications of the project.

Scenario 3   The National Interagency Fire Center requests bids for providing non-perishable food for smokejumpers. The center put an IFB (information for bid) in the *Commerce Business Daily* (see Figure 16.1). The IFB describes the specifications for the food, gives the due date for the bid, and provides information on obtaining specifics about the request.

Scenario 4   Susan Rowland writes software documentation. In recent months, her workload—as well as that of other documentation writers in her office—has dramatically increased. Susan knows that her company currently has a hiring freeze, so she can't hire another writer. She believes the only way to get help with the increased workload is to update the computer software and hardware used to write and produce the documentation. She writes a proposal to her regional manager, requesting updated software and hardware for her department.

Scenario 5   Nancy Griffin wants to research whether or not the concentrations of mercury and lead allowed by the government are safe. She wants to determine whether people living in areas with allowable concentrations of mercury or lead have a history of health problems. To conduct this

---

FIGURE 16.1

Information for Bid (IFB) from the *Commerce Business Daily*

COMMERCE BUSINESS DAILY

A daily list of U.S. Government procurement invitations, contract awards,
subcontracting leads, sales of surplus property and foreign business opportunities.

Non-perishable food items for smokejumpers food boxes

General Information

Document Type:	PRESOL
Posted Date:	Mar 19, 2008
Category:	Subsistence
Set Aside:	N/A

Contracting Office Address

BLM-FA NATIONAL INTERAGENCY FIRE CENTER3833 DEVELOPMENT AVE BOISE ID 83705

Description

The Bureau of Land Management, National Interagency Fire Center, will post Request for Quotation solicitation #RAQ083004 for non-perishable food items for smokejumpers fire food boxes, which will include, but not limited to canned soups, canned meats, candy bars, energy bars, and dried fruit. This solicitation falls under NAICS code 424410, and Product Service code 8970. This will be set a side totally for small businesses. Small business classification per SBA table is 100 employees. The solicitation will be made available electronically on March 12, 2008 from the Electronic Commerce site "http://ideasec.nbc.gov." Contractors must be registered and active at the Central Contractor Registration website at: www.ccr.gov to participate, receive notifications of amendment, award notification, and to faciitate payment. No hard copies will be sent. The anticipated solicitation closing is March 31, 2008. The Government reserves the right to make multiple awards, as a result of the solicitation responses. Questions regarding this announcement may be directed to Dawn Graham at (208)387-5544.

Original Point of Contact

POC Dawn Graham Purchasing Agent 208387554 Dawn_Graham@nifc.bim.gov

Source: Downloaded from the World Wide Web, November 2008:
www.cbdweb.com/index.php/search/show/21773566.

---

research, Nancy needs funds from a public or private agency, but she cannot find a specific request related to her research topic. She prepares a proposal to send to various private and public agencies, hoping these agencies will fund her research.

These five scenarios illustrate the primary types of proposal situations: solicited and unsolicited, and internal and external.

Solicited Proposals

Solicited proposals originate when a person, an organization, or a government agency requests qualified organizations and individuals to deliver a

Both solicited and unsolicited proposals offer to solve a problem, to provide a service or product, or to perform research.

product and/or services or to perform research. In scenarios 1 through 3, companies or government agencies are soliciting proposals or bids. In scenario 1, Bill Martinez is requesting proposals for a *service*: he wants companies to propose plans for disposing of the ash. He is not requesting that the companies actually dispose of the ash, only to develop a comprehensive, feasible plan for disposing of the ash. Similarly, in scenario 2, the agency is requesting proposals for providing a service, in this case, renovating a building. In Scenario 3, the National Interagency Fire Center is requesting bids for a *product*, in this case, non-perishable food. In each of these scenarios, the organizations and individuals write proposals (or bids) in response to a specific request: an RFP (request for proposal) or IFB (information for bid).

Unsolicited Proposals

Scenario 4 illustrates a situation requiring ***unsolicited*** proposals—proposals not requested by the organization, individual, or government agency that receives them. Unlike solicited proposals, unsolicited proposals must convince readers that a specific need or problem exists before explaining the plan, the cost, or the qualifications. You can write unsolicited proposals to people in your own organization, as in Scenario 4, or you can write them to people outside your organization.

Scenario 5, Nancy Griffin's proposal, is unsolicited because she is not responding to a request for proposals. This scenario is also different from the other proposals because Nancy is not proposing a service or product. Instead, she is proposing that an organization fund her research. Some research proposals are solicited in that an organization may have posted a request for research proposals.

Susan Rowland's and Nancy Griffin's task—writing an unsolicited proposal or a research proposal—can be more difficult than writing a proposal in response to a specific request. Susan and Nancy not only have to convince readers that they have good projects and can carry out those projects at a reasonable cost, but they also have to persuade readers that their projects are worthwhile and will fulfill a need. Susan wants to convince her readers that her department has an increased workload and that the computer software and hardware used by the documentation writers is preventing the department from completing its work in a reasonable time and with the quality the company expects. Nancy wants to persuade her readers that the allowable levels of mercury and lead may be too high and that a scientific study can determine whether or not the levels actually are too high. Even if she is responding to a request for research proposals, she must convince her readers that her research proposal is reasonable and worth funding.

Internal and External Proposals

The previous five scenarios illustrate how proposals originate and where readers are located. Readers of proposals can work within or outside your organization. If the readers work within your organization, as in Susan's scenario (Scenario 4), the proposal is *internal*—written to someone within the writer's organization. If the readers work outside your organization, as in Scenarios 1, 2, 3, and 5, the proposal is *external*. This chapter will help you write effective proposals, whether they are solicited or unsolicited, internal or external.

PREPARING TO WRITE A PERSUASIVE PROPOSAL

An effective proposal is persuasive; it convinces readers to accept and possibly to pay for the work that it proposes. If you propose a project for a fee, you will want to persuade your readers that you can carry out the project within a reasonable time and at a reasonable cost. A *proposal*, then, is a "sales" document: it sells a proposed action and possibly your services or the services of your organization to carry out that action. If you are proposing to perform research, your proposal should convince the readers that your research is valid and that you have a sound plan for conducting that research. To write a persuasive proposal, you need to

- find out about your readers
- anticipate and answer their questions
- present a professional, ethical picture of you and/or your organization

Find Out about the Readers

Before writing any proposal, find out about the people who are most likely to read it. You can write an effective proposal only if you understand your readers and have some idea about how they will respond to the

TAKING IT INTO THE WORKPLACE
Examining the Varying Forms of Bids and Proposals

Organizations of all sizes use proposals and bids to evaluate projects. Some organizations require a formal procedure for submitting proposals and bids and specific formats for the information. Others use a less formal approach, expecting fewer details and using an unspecified format. However, most use a written proposal and bid to compare, select, and award contracts. By examining how proposals and bids are gathered in the workplace, you can better understand the variety of forms that bids and proposals take.

Assignment

Visit a department of your choosing on your campus. Consider a department associated with your degree, an administrative department, or a maintenance or food services department at your college or university.
- Talk to the person in charge of accepting bids or writing proposals.
- Ask that person to show you an example of a recent bid or proposal received by his or her department. Review it together.
- Discuss what aspects of the proposal or bid are the most important when considering the item, and which are the least helpful.
- Ask the person considering the proposal or bid what he or she liked or disliked about the item.
- Be prepared to make an oral presentation to your class about how real-world proposals and bids compare to what you learned in class. (See Chapter 20 for information on oral presentations.)

work that you propose. To find out about your readers, ask yourself these questions:
- **What positions do your readers hold in the organization? If the readers work in the organization that employs you, where are their positions in relation to yours in the organizational hierarchy?** When you know readers' positions in the hierarchy, you can more accurately determine who will approve or disapprove of your proposal, who will understand the topic and the background of your proposal, and who will be your primary and secondary readers. Are your readers above you, below

you, or at the same level as you in your organization? If they are at a higher level than you, you may want to use a more formal approach or have your manager read a draft of your proposal before you send it to your primary readers. If you and your readers are at the same level in the organization, or if you are at a higher level, they may expect a less-formal approach.

- **Will more than one group read the proposal? If so, what sections of the proposal will each group read?** Several groups of readers may read a proposal. For example, the managers or executives may read the summary to determine whether the proposal has merit. If they decide that it has merit, they may send the proposal to the accountants to look at the budget and to the technical experts to look at the solution and the plan of work.
- **What do your readers know about the problem or need that prompted your proposal?** If your proposal is unsolicited, readers probably will know little about the problem or need addressed in your proposal. If your proposal is solicited, readers will understand the problem.
- **What do your readers know about you or your organization? Have their previous experiences with your organization or with you been positive? If not, why not?** Find out whether your readers have had previous experiences with your organization. Was the experience extended or brief, positive or negative? What impression are the readers likely to have of you or your organization? When you know the answers to these questions, you can better write the qualifications section of the proposal and lessen any negative concerns your readers may have about you or your organization.

If you are aware of the answers to these questions, you can more effectively write a proposal that meets your readers' needs and expectations. Once you have found out about your readers, you can determine what they may ask about your proposal.

Anticipate and Answer Readers' Questions

The success of your proposal depends on how
- persuasively and logically you argue for your proposed solution, service, product, or research
- convincingly you argue that you or your organization is best qualified to carry out the plan

- persuasively you argue that you or your organization can complete the work within a reasonable time and at a reasonable cost

A series of questions relating to three areas will help you to anticipate what readers expect from your proposal and what they may ask as they read it: the problem or need addressed in the proposal; the proposed solution, product, or service; and the plan of work. You will find these questions in Figure 16.2. This figure relates the questions to the conventional sections of proposals discussed in this chapter.

To answer readers' questions, you need to research the readers and their organization. You might find out about the organization's history, its financial standing and goals, and its organizational hierarchy. You also can research the organization's corporate culture to see how it might affect your readers' perspectives and ideas. With this information, you can better anticipate your readers' questions and provide persuasive answers.

In addition to anticipating readers' questions, consider what you and your organization can realistically propose and look at the strengths and weaknesses of your proposal. Figure 16.2 lists questions that you can consider as you evaluate what you are proposing. You want to make sure that your proposal will appeal to readers, without compromising what you and your organization can actually do.

USE THE CONVENTIONAL STRUCTURE FOR PROPOSALS

As the five scenarios presented earlier illustrate, you may write proposals in response to different situations. The formats for the proposals may vary with the situations. If your proposal is formal, you might choose to include front matter, such as a letter of transmittal, title page, table of contents, and executive summary (see Chapter 18). If your proposal is informal, you might select a letter or memo format. Whether you use a format with front matter or a letter or memo format, your proposal will have some or all of the conventional sections discussed in this chapter. From one organization to another, these sections may have different titles; but their purpose remains the same. For example, instead of "problem definition," some organizations may use the phrase "scope of work"; and instead of "budget," some organizations may use "cost estimates." If your organization

FIGURE 16.2

Questions to Consider when Writing the Conventional Sections

Section of the Proposal	Writer's Questions	Readers' Questions
Introduction and Problem Definition	• How can the proposal demonstrate that you understand the problem or need? • Should you restate the problem or need to show readers that you understand it?	*If the proposal is solicited:* • What do readers expect from your proposal? *If the proposal is unsolicited:* • Why should readers be interested in your proposal? • What problem or need does your proposal address? • Why is the problem or need important to readers?

Proposed Solution, Product, Service, or Research	• How can you reasonably solve the problem or meet readers' needs?
	 • What are the strengths of your solution, product, service, or research?
	    – How can you emphasize those strengths?
	 • What are the weaknesses of your solution, product, service, or research?
	    – How can you ethically counter those weaknesses and readers' objections to them?
	 • How does your solution, product, service, or research meet readers' needs or those of the "community" or of your field?
	 • Can you and/or your organization reasonably carry out what you propose?
	 • How can you make the solution, product, service, or research attractive to readers without compromising what you and/or your organization can actually do?
		• How will you solve the problem or satisfy the need?
		 • What do you specifically propose to provide or to do?
		 • Are other solutions, products, services, or research methods possible? If so, why have you chosen the solution, product, service, or method presented in your proposal?
		 • How does your choice compare with other possibilities?
		 • How will readers view the solution, product, service, or research that you propose?

Plan of Work	• What are the strengths of your plan? • What are the weaknesses of your plan? • How can you counter those weaknesses and the readers' possible objections to them? • Can your organization reasonably carry out the plan? • How can you ethically make your plan attractive to readers without overstating what you and/or your organization can do?	• What do you propose to do, to make, or to provide? • How long will the plan take? • Is the plan ethical and reasonable?
Qualifications	• How can you demonstrate that you and/or your organization are qualified and that the readers can depend on you to do what you propose?	• Why should readers believe that you and/or your organization have the expertise to do what you propose? • Why should readers believe that they can depend on you and/or your organization to deliver what you propose?
Budget	• Does the budget reflect the actual cost of the plan? • Have you justified your budget and anticipated readers' possible objections to it?	• How much will the proposed solution, product, service, or research cost? • Is the cost reasonable?

ETHICS NOTE

Is Your Proposal Honest?

How can you provide readers with the information they need, without overstating what you or your organization can provide?
- Determine what your readers expect and what questions they need answered as they read your proposal.
- Decide what you or your organization can reasonably propose.

You have an ethical responsibility to be honest in your proposal. You must explain specifically and accurately the work that you or your organization can provide, so readers will not expect more or less than you intend to deliver and will not hold you or your organization responsible for more or less work than you propose.

When you prepare an honest proposal, you are
- protecting yourself and your organization from possible legal trouble
- doing the right thing

In your eagerness to get a proposal accepted, don't exaggerate or overestimate the work that you can perform or the qualifications of those who will perform the work.

prefers certain titles for the sections or requires particular sections to appear in its proposals, use those titles and include those sections. If you are responding to an RFP (request for proposal) or IFB (information for bid), the request may specify the sections and the format that you should follow. Follow the specifications of the requesters; otherwise, your proposal or bid may be rejected.

Most proposals contain the following sections:
- Executive Summary
- Introduction
- Problem Definition
- Proposed Solution
  - Plan of Work

- Qualifications
  - Personnel
  - Resources
- Budget
- Conclusion

For some proposals, you may not need all of these sections—especially the qualifications and budget. These two sections may not be necessary based on your relationship with the readers and the purpose of your proposal. For example, if you are writing an informal proposal to your manager, you wouldn't include a qualifications section because your manager knows your qualifications. However, if you are writing a more formal proposal to external readers who are not familiar with you, you would include a qualifications section.

The following sections describe these conventional sections. They correspond to the questions that readers may ask and questions that you should ask (see Figure 16.2, page 497).

Executive Summary

The *executive summary* is a powerful tool; it is a condensed version of your proposal. A well-written summary persuades readers to read more about your proposed project, while a poorly written proposal may persuade readers to ignore your proposal or to reject it.

Because readers look at the summary first, it is often crucial to the success of a proposal for these reasons:
- Many readers will use the executive summary to decide whether or not to consider a proposal.
- Many readers may read only the summary.
- Readers who have received many proposals in response to one RFP may use the summaries to determine which proposals to consider seriously.

The executive summary contains essential information about the proposed solution, product, service, or research; the plan of work; and the cost. Figure 16.3, page 500, presents the summary from an unsolicited propos-

FIGURE 16.3

Summary from an Unsolicited Proposal

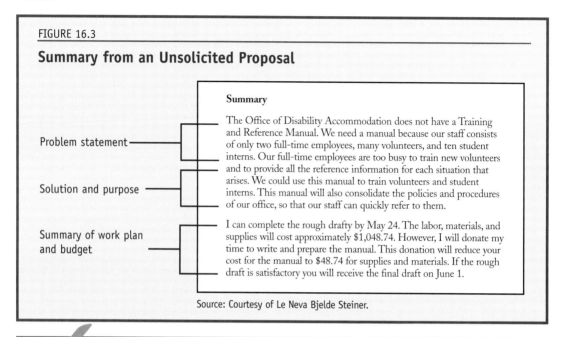

Problem statement

Solution and purpose

Summary of work plan
and budget

Summary

The Office of Disability Accommodation does not have a Training
and Reference Manual. We need a manual because our staff consists
of only two full-time employees, many volunteers, and ten student
interns. Our full-time employees are too busy to train new volunteers
and to provide all the reference information for each situation that
arises. We could use this manual to train volunteers and student
interns. This manual will also consolidate the policies and procedures
of our office, so that our staff can quickly refer to them.

I can complete the rough drafty by May 24. The labor, materials, and
supplies will cost approximately $1,048.74. However, I will donate my
time to write and prepare the manual. This donation will reduce your
cost for the manual to $48.74 for supplies and materials. If the rough
draft is satisfactory you will receive the final draft on June 1.

Source: Courtesy of Le Neva Bjelde Steiner.

 TIPS FOR WRITING EXECUTIVE SUMMARY

- **Concisely state the problem or need addressed in the proposal**. Include only the
  information necessary for readers to understand the problem or need.
- **Summarize the proposed solution, product, service, or research**. Show how it
  meets the readers' needs and requirements and why your proposal is valuable to the
  readers, the community, or your field.
- **Describe your plan for carrying out the proposed solution, product, or service**.
  Exclude details about the plan of work. Instead, present only the basics of the plan.
- **Summarize your qualifications or those of your team and/or your organization** (if
  included). Some proposal writers include only the qualifications of the organization
  in the summary.
- **Summarize the budget** (if included). Summarize by simply stating the total cost of
  the proposed solution, product, service, or research. Some sales proposals purposeful-
  ly leave the budget information out of the summary so the proposed product or service can be "sold" before giving the financial data. As Neil Cobb, executive director for
  AT&T, explains, if your product isn't the lowest priced on the market, tell readers
  about the product and the company before you give them the "bottom line," so they
  will see its value in relation to its cost (2008).

al for a service-learning project; for this project, the student is proposing to write a training and reference manual for a university Office of Disability Accommodation.

Introduction

The *introduction* briefly describes the problem, product, service, or research that you are proposing and tells why you are proposing it. For example, you might state that you are writing in response to a request from readers or to a specific request for proposals. In the introduction, you also can tell readers what follows in the proposal. The writer of the example shown in Figure 16.4, page 502, combines the introduction and the summary; this combination is common in short, informal proposals. The writer identifies the problem that is being addressed: "to develop a conceptual closure plan for the Caney Branch Ash Disposal Area." The writer then briefly explains how the plan will be developed and describes what follows in the proposal. As you write your introduction, follow the tips below.

Problem Definition

Once you have told readers what you are proposing, convince them that you understand the work that you are proposing and that you designed your solution and work plan after studying their needs. If the proposal is unsolicited, convince readers that your proposal addresses a significant

---

 TIPS FOR WRITING THE INTRODUCTION

- **Identify the problem or need that the proposal addresses**. Briefly describe that problem or need.
- **State the purpose of your proposal**. Although your readers will likely know the purpose, state it to ensure that you and your readers clearly understand the purpose of your document.
- **Describe what follows in the proposal**. Tell readers the organizational pattern that you will follow in the proposal. For example, the writer of Figure 16.4 writes, "The following proposal describes our understanding of the required scope of work and our work plan."

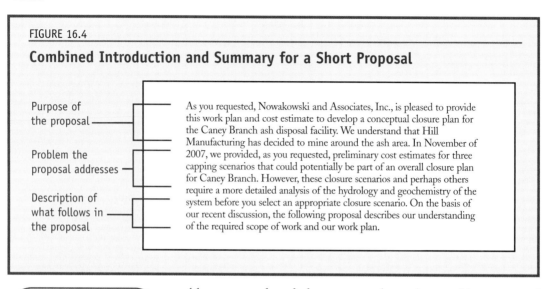

FIGURE 16.4

Combined Introduction and Summary for a Short Proposal

Purpose of the proposal

Problem the proposal addresses

Description of what follows in the proposal

As you requested, Nowakowski and Associates, Inc., is pleased to provide this work plan and cost estimate to develop a conceptual closure plan for the Caney Branch ash disposal facility. We understand that Hill Manufacturing has decided to mine around the ash area. In November of 2007, we provided, as you requested, preliminary cost estimates for three capping scenarios that could potentially be part of an overall closure plan for Caney Branch. However, these closure scenarios and perhaps others require a more detailed analysis of the hydrology and geochemistry of the system before you select an appropriate closure scenario. On the basis of our recent discussion, the following proposal describes our understanding of the required scope of work and our work plan.

> *When you anticipate and answer readers' questions in your proposals, they know that you clearly understand their situation.*

problem or need and demonstrate how that problem or need affects them. If you are writing an unsolicited research proposal, convince the readers that your proposal is significant to their field or their interests. Persuade them first that a problem or need exits and then that it is important to them.

Anticipate and answer any questions that readers may have, so they know that you clearly understand their situation. (Figure 16.2 on page 497 lists questions that readers might ask as they read a problem definition. Your readers may have other questions about the specific subject addressed in your proposal.) Depending on whether the proposal is solicited or unsolicited, the problem definition may do any of the following:

- **Define the problem**—in detail for an unsolicited proposal and with less detail for a solicited proposal.
- **Give the background of the problem or situation or explain how it developed** (primarily for unsolicited proposals). The background may help readers understand that a significant problem exists or that you understand their needs. For a research proposal, the background demonstrates that you understand prior research related to your proposed project.
- **Explain why the proposed solution, product, service, or research is necessary** (for unsolicited proposals). For instance, if you are proposing a research project, explain why the research is important.

✓ TIPS FOR WRITING THE PROBLEM DEFINITION

- **Define the problem or need.** If you are writing an unsolicited proposal, give the readers the details they require to understand the problem or need and explain why they should then consider your proposed solution, product, research, or plan. If you are writing a solicited proposal, include fewer details. Instead, give the readers only enough details to know that you understand the problem. This information is important because proposals result in a deliverable, and you want a written record of your understanding of the problem or need. A *deliverable* is what you will "deliver" at the end of a project; for example, if you are proposing to design and build a Web site, the deliverable is the Web site.

- **Present the necessary background for the readers to understand the problem or need and the proposed solution or plan.** Show the readers that you have carefully studied the problem and/or their request—that you have done your homework. For example, if you conducted library research, briefly summarize your research. If you examined company reports or interviewed people affected by the problem or need, use that information to give readers what they require to understand the context of your proposal.

- **Give the readers the information they require to understand why they should accept your proposal.** Explain why the proposed solution, product, service, or research is necessary to the readers, their organization, the community, or their field.

Figure 16.5, page 504, presents the problem definition (called "Scope of Work" in this case) of a solicited proposal. The readers know what their needs are, so the primary purpose of this problem definition is to show that the writer understands the readers' needs. Thus, the writer restates the work that the readers expect of his organization as well as what the readers do not expect.

Figure 16.6, page 505, shows the problem definition of an unsolicited proposal for a university Office of Disability Accommodation. Using examples and terms the readers will understand, the writer specifically explains the problems resulting from the turnover of volunteers and students. This problem definition (called "Current Problems") points out why the problems are significant. The writer explains, for example, that these problems

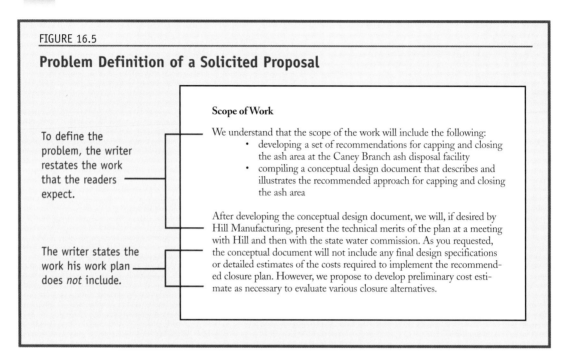

FIGURE 16.5

Problem Definition of a Solicited Proposal

To define the problem, the writer restates the work that the readers expect.

Scope of Work

We understand that the scope of the work will include the following:
- developing a set of recommendations for capping and closing the ash area at the Caney Branch ash disposal facility
- compiling a conceptual design document that describes and illustrates the recommended approach for capping and closing the ash area

After developing the conceptual design document, we will, if desired by Hill Manufacturing, present the technical merits of the plan at a meeting with Hill and then with the state water commission. As you requested, the conceptual document will not include any final design specifications or detailed estimates of the costs required to implement the recommended closure plan. However, we propose to develop preliminary cost estimate as necessary to evaluate various closure alternatives.

The writer states the work his work plan does *not* include.

prevent the office from creating "the positive image that the students and the faculty need," and "slow the work of the office."

Proposed Solution and Plan of Work

After you describe the problem or need, your readers will want to know how you plan to solve the problem or meet their needs—what you will deliver. They will especially want to see how you clearly link your proposed solution to the problem or need. Readers will also expect you to present a detailed plan for carrying out the work. As you present this plan, consider the readers' questions. For example, they might ask: How does your plan compare with other possible solutions? Why should we adopt your plan instead of another plan? Why is this plan better than the others? (For more questions, see Figure 16.2 on page 497). You must persuade your readers that you have not only crafted a detailed plan, but also that your plan is the best solution to the problem.

In some work plans, you may want to explain not only what you and your organization will do, but also what you will not do. Your readers have a

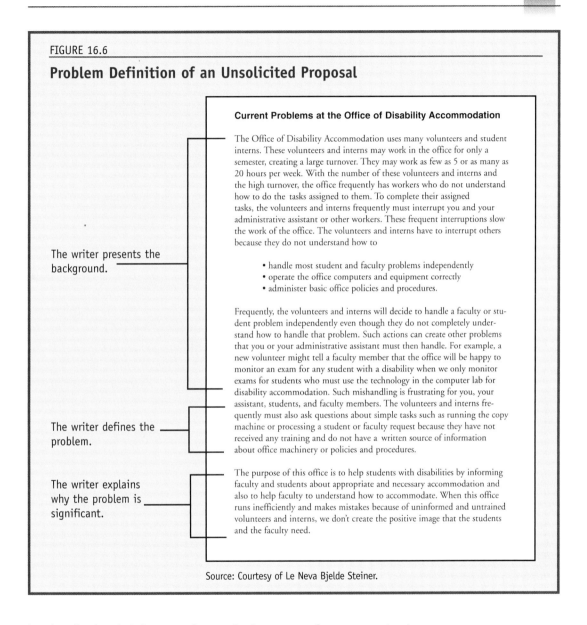

FIGURE 16.6

Problem Definition of an Unsolicited Proposal

The writer presents the background.

The writer defines the problem.

The writer explains why the problem is significant.

Current Problems at the Office of Disability Accommodation

The Office of Disability Accommodation uses many volunteers and student interns. These volunteers and interns may work in the office for only a semester, creating a large turnover. They may work as few as 5 or as many as 20 hours per week. With the number of these volunteers and interns and the high turnover, the office frequently has workers who do not understand how to do the tasks assigned to them. To complete their assigned tasks, the volunteers and interns frequently must interrupt you and your administrative assistant or other workers. These frequent interruptions slow the work of the office. The volunteers and interns have to interrupt others because they do not understand how to

- handle most student and faculty problems independently
- operate the office computers and equipment correctly
- administer basic office policies and procedures.

Frequently, the volunteers and interns will decide to handle a faculty or student problem independently even though they do not completely understand how to handle that problem. Such actions can create other problems that you or your administrative assistant must then handle. For example, a new volunteer might tell a faculty member that the office will be happy to monitor an exam for any student with a disability when we only monitor exams for students who must use the technology in the computer lab for disability accommodation. Such mishandling is frustrating for you, your assistant, students, and faculty members. The volunteers and interns frequently must also ask questions about simple tasks such as running the copy machine or processing a student or faculty request because they have not received any training and do not have a written source of information about office machinery or policies and procedures.

The purpose of this office is to help students with disabilities by informing faculty and students about appropriate and necessary accommodation and also to help faculty to understand how to accommodate. When this office runs inefficiently and makes mistakes because of uninformed and untrained volunteers and interns, we don't create the positive image that the students and the faculty need.

Source: Courtesy of Le Neva Bjelde Steiner.

legal and ethical right to understand what you and your organization are proposing and also what you are not proposing. If you do not tell readers what you will and will not do, they may expect more than you intend. You can prevent such miscommunication by spelling out exactly what you are proposing and not proposing.

The proposed solution and plan of work in the proposal for Hill Manufacturing appear in the sections titled "Work Plan" and "Schedule" in Figure 16.8 (see page 510). (Proposal writers and companies use different titles to refer to the proposed solution and plan of work. Use the titles that the RFP or IFB requires or those that your organization prefers. If your organization doesn't have requirements, select the most appropriate titles for your readers.) In the proposal for Hill Manufacturing, the writer links the proposed solution and plan of work to the problem. Figure 16.7 shows how the writer (Nowakowski and Associates, Inc.) links the solution and plan of work directly to the readers' (Hill's) problems and needs. In the problem definition (called "Scope of Work" in Figure 16.8 on page 510), the writer mentions three areas of work: "developing a set of recommendations for capping and closing the ash area," "compiling a conceptual design document," and "a meeting with Hill and then with the state water commission." In the "Work Plan," the writer mentions these three areas in the list of tasks and explains how the tasks will fulfill the work requested by Hill. The proposal lists and describes the tasks necessary to carry out the work and then presents a schedule.

FIGURE 16.7

The Link Between Hill Manufacturing's Problem and the Work Proposed by Nowakowski and Associates, Inc.

Hill Manufacturing's Problem/Needs	How Nowakowski and Associates Proposes to Solve the Problem and Meet the Needs
Set of recommendations for capping and closing the ash area at the Caney Branch Ash Disposal Area	**Task 1:** Review data and develop preliminary closure options **Task 2:** Collect and analyze additional data **Task 3** (partial): Evaluate the closure alternatives
Conceptual design document that describes and illustrates the recommended approach for capping and closing the ash area	**Task 3** (partial): Develop a conceptual closure plan
Presentation of the technical merits of the plan at meetings with Hill Manufacturing and the state water commission	**Task 4:** Attend strategic planning meetings

As you prepare the proposed solution and work plan, consider the questions in Figure 16.2 (page 497) and the following tips:

TIPS FOR WRITING THE PROPOSED SOLUTION AND WORK PLAN

- **Link the proposed solution (deliverable) to the problem by explaining how the solution will solve the problem and/or meet the needs of the readers.** Tell your readers specifically how your proposed solution will solve the problem or address the needs that you described in the problem-definition section. Don't assume that readers will see the link; instead, directly state the link even in a solicited proposal.
- **Present a detailed, step-by-step plan for carrying out the work.** You want the readers to understand the scope of the work you are proposing. Be sure to justify your plan and to anticipate readers' possible questions.
- **If necessary, explain specifically what you are and are not proposing to do.** You want readers to clearly understand not only what you and/or your organization will do, but also what you will not do.
- **Consider including a graphic to illustrate the schedule.** Many proposals include a graphic, such as a timeline or chart, illustrating the schedule (for example, see the Gantt chart in Figure 16.8 on page 510.)
- **Create a realistic schedule.** Be careful not to create an overly optimistic schedule in your eagerness to get your proposal accepted. Allow ample time to complete each phase of the work plan. Your readers and your organization want a realistic schedule, not one that is impossible to meet.

Qualifications

The qualifications section—sometimes called "Project Team," "Facilities," or "Personnel"—is important for readers who want to know whether you and/or your organization are capable of carrying out the work that you propose. This section includes some or all of the following information:

- **Qualifications of the people (including yourself, if necessary) who will carry out the plan of work.** You can include a paragraph summarizing each person's qualifications for the project, or you can attach resumés in an appendix. If you attach resumés, use the qualifications section to introduce the project personnel or summarize each person's qualifications and then refer readers to the appendix.

• **Qualifications of the organization.** For some proposals, you may need to "sell" your organization's qualifications. You can demonstrate to readers that the organization has carried out similar work and has the facilities and knowledge to successfully and efficiently complete the work. In this section, you might give readers a brief background of your organization and projects that it has completed successfully. You also can include information on specific facilities that the organization will use to complete the work, especially if those facilities compare favorably with industry standards or with your competitors.

The sample proposal in Figure 16.8 includes a qualifications section. The writers summarize the qualifications of the personnel who will carry out the proposed work. The writers also include specific information about each person's responsibilities on the project and explain why each person is uniquely qualified to work on the project. The writers don't present resumés in an appendix because their organization has previously worked with Hill Manufacturing.

TIPS FOR PREPARING THE BUDGET

• **Carefully estimate the cost of labor, equipment, and any materials needed to carry out the work plan.** To ensure that you have included all costs, look at the budgets for similar projects or proposals or ask an experienced coworker to look at your budget.
• **Estimate accurately.** If you underestimate your costs in trying to get a proposal accepted, you may be bound by your estimate. If your estimate is too low, you or your organization could lose the goodwill of your client if you have to charge more money for the services. If you have contracted for a specific amount, you may not be able to ask the client for additional money; you or your organization may have to fund the additional expenses out of pocket.

Budget

The *budget*, sometimes called a "Cost Estimate" or "Cost Proposal," is an itemized list of the estimated costs of the plan of work. For some proposals, readers will expect a justification along with the budget. The budget justification explains each budgeted item and its purpose for the proposed solution. As you prepare your budget, follow the tips at left.

The budget for the Hill Manufacturing proposal (see Figure 16.8) is in a section titled "Cost Estimate." There, the writers present a total estimate for the project and separate estimates for each task.

Conclusion

In most proposals, the conclusion briefly restates the problem or need and the proposed solution. The conclusion also restates

- what the proposal offers readers
- how the proposal will benefit readers
- why readers should accept the proposal
- why readers should accept you and/or your organization to carry out the proposed solution

For shorter, less-formal proposals in memo or letter format, the conclusion is likely to be a brief statement of whom readers should call if they have questions.

TWO SAMPLE PROPOSALS

Figures 16.8, pages 510-514, and 16.9, pages 515-517, illustrate the conventional sections for proposals. In Figure 16.8, an environmental services and engineering firm is responding to an RFP. The writer proposes to prepare a plan for capping and closing an ash area at a lignite mine owned by Hill Manufacturing. In Figure 16.9, as part of a service-learning project, a student proposes a manual for the Office of Disability Accommodation at her university. This office helps students with disabilities receive appropriate accommodation from the university and helps faculty members who have these students in their classes. The readers of her proposal are the director and assistant director of that office.

FIGURE 16.8

Sample Solicited Proposal

Nowakowski & Associates

Nowakowski & Associates, Inc.
243 26th Street, Suite 808
Montrose, Colorado 80303-2317
303/555-1823 Fax: 303/555-1836

March 24, 2009

The writer uses the letter format.

Ms. Ginny Thompson
Hill Manufacturing
400 North Vintage Street
St. Louis, Missouri 75201

The writer identifies the document as a proposal.

Re: Work Plan and Cost Estimate to Develop a Conceptual Closure Plan for the Caney Branch Ash Disposal Facility

Dear Ms. Thompson:

The writer combines the summary and introduction. Here, the writer describes the problem and the organization of the proposal.

As you requested, Nowakowski and Associates, Inc., is pleased to provide this work plan and cost estimate to develop a conceptual closure plan for the Caney Branch ash disposal facility. We understand that Hill Manufacturing has decided to mine around the ash area. In November of 2008, we provided, as you requested, preliminary cost estimates for three capping scenarios that could potentially be part of an overall closure plan for Caney Branch. However, these closure scenarios and perhaps others require a more detailed analysis of the hydrology and geochemistry of the system before you select an appropriate closure scenario. On the basis of our recent discussion, the following proposal describes our understanding of the required scope of work and our work plan.

Because the proposal is solicited, the writer uses "Scope of Work" to identify the problem definition and only briefly states the scope of work. The writer also states the work that the project will not include.

Scope of Work

We understand that our scope of work will include the following:

- developing a set of recommendations for capping and closing the ash area at the Caney Branch ash disposal facility
- compiling a conceptual design document that describes and illustrates the recommended approach for capping and closing the ash area

After developing the conceptual design document, we will, if desired by Hill Manufacturing, present the technical merits of the plan at a meeting with Hill and then with the state water commission. As you requested, the conceptual document will not include any final design specifications or detailed estimates of the costs required to implement the recommended closure plan. However, we propose to develop preliminary cost estimates as necessary to evaluate various closure alternatives.

The writer provides an overview of the work plan.

Work Plan

To develop the conceptual closure plan, we will balance the level of effort and associated costs for preparing the plan with the requirements to provide sufficient documentation and justification for addressing possible questions from the state water commission. We will work closely with Hill to ensure that the level of effort and work are consistent with Hill's objectives. We will complete four tasks to complete the work that Hill requires.

FIGURE 16.8 CONTINUED

G. Thompson, March 24, 2009, p. 2

The writer describes in detail each task in the work plan.

Task 1. Reviewing Data and Developing Preliminary Closure Options
We request that Hill Manufacturing provide any water level data, sump discharge volume data, and sump water quality data obtained at Caney Branch since the Phase II Investigation (2008-2009). We will review these data, along with the historical data provided in the Phase II Report, to further develop possible closure options. In developing a preliminary list of closure options, we will review correspondence from the state water commission and registration papers regarding Caney Branch. We will also examine the water commission's precedents for closure of ash disposal facilities at other lignite mines in the state. We will also use these reviews to further evaluate possible data gaps in the various closure options.

Task 2. Collecting and Analyzing Additional Data
During Task 2, we will evaluate the need for collecting additional data; we currently envision two specific needs. The latest water-level readings, taken in selected ash and overburden wells at Caney Branch, were measured in early May of 2008. The last complete set of readings for all wells occurred in October of 2008. Historically, water levels in the ash have generally exhibited an upward trend but may be approaching a quasi-static condition. The quasi-static water level in the ash under present stratigraphic conditions is important because it serves as a "baseline" when estimating the long-term water-level conditions in a post-mining scenario. Therefore, we propose to obtain another set of water-level measurements in wells within and near Caney Branch to evaluate the baseline condition.

We also foresee the need for new data on the ground-water chemistry for ash and overburden wells. Because attenuation of ash leachate constituents will likely be a critical basis for limiting the scope of closure activities, we will collect samples from selected wells to confirm that ash leachate constituents are continuing to attenuate. We presently anticipate that collecting and analyzing the samples from six monitoring wells will be sufficient to document that attenuation.

We will collect and analyze the samples in the same manner as in the Phase II investigation. Field analyses would include Ph, Eh, specific conductance, and temperature. However, we will analyze the samples in the lab for a shorter list of constituents than those evaluated during the Phase II investigation. The proposed list of constituents includes calcium (dissolved), magnesium (dissolved), sodium (dissolved), potassium (dissolved), chloride, sulfate, alkalinity, boron (dissolved), and selenium (dissolved). We will be most interested in the key parameters of sulfate, boron, and selenium. To estimate cost, we have assumed that Core Laboratories, which analyzed the water samples for Phase II, will analyze the samples. However, if desired, the Hill Manufacturing lab could analyze the samples.

Task 3. Evaluating Closure Alternatives and Developing a Conceptual Closure Plan
The current mine plan calls for mining within 400 feet of Caney Branch. This plan will have substantial effects on both the short- and long-term hydrogeologic conditions near the ash area. Therefore, we will consider these effects when evaluating the closure options:
- effects of mining on the rate of ash water leaching
- extent of ash area dewatering caused by adjacent mining and the possible effects on induration of the ash
- time required to resaturate the spoil and ash material after mining
- post-mining hydrologic conditions (effects of spoil characteristics, ponds, etc., on post-mining hydrology)
- historical and future attenuation of ash water constituents
- hydrologic effects of constructing a flow barrier between the ash area and Barrier Lake (A slurry wall is currently not a preferred option, but we will review it to address possible

FIGURE 16.8 CONTINUED

G. Thompson, March 24, 2009, p. 3

questions from the water commission or to provide an alternative.)
- cost-feasibility of construction options
- applicability of the water commission's Draft Risk Reduction Rules to Caney Branch
- potential post-closure care requirements and costs

Although we don't anticipate substantial modeling efforts associated with developing the conceptual plan, the existing model developed during the Phase II investigation will help us to address some of the hydrologic issues. In some cases, we can use previous model simulations to assess an issue; and in other cases, we anticipate performing additional simulations. We anticipate using the HELP (Hyrologic Evaluation of Landfill Performance) model to evaluate the hydrology of various capping scenarios. In other cases, we will draw on experience at the Barrier Lake Mine and perhaps simple hydraulic calculations of water imbalance to develop estimates. The actual level of effort and list of critical issues will depend upon which closure options we evaluate.

We will develop a narrative to justify and describe the conceptual closure plan and simple conceptual design drawings to illustrate the closure concept. We will present the draft closure plan to Hill Manufacturing at a meeting (Task 4). The report to Hill will not include complete documentation of model results and analytical analyses as part of the conceptual planning.

Task 4. Attend Strategic Planning Meetings
We will estimate the time and material required to prepare for and present the technical merits of the proposed conceptual closure plan in meetings with Hill Manufacturing and the water commission. Our cost estimate assumes two meetings, one with Hill Manufacturing and one with the water commission.

Schedule
We estimate eight weeks for drafting the conceptual closure plan. We assume that Hill Manufacturing can provide during the first week of the project any additional data that they have collected since 2008 (see Task 1). We anticipate performing the field work associated with Task 2 within the first two weeks of the project. We will proceed with Tasks 1 and 3 while the lab analyzes the additional water samples (two- to three-week turnaround). After we receive the sample results, we estimate that we can finalize the conceptual closure plan in approximately three weeks. Table I summarizes the tentative schedule.

> The writer describes the schedule for completing the work.

Cost Estimate
We propose to execute the work plan on a time and materials basis according to our 2009 Schedule of Charges. We will provide periodic reports to the Hill Manufacturing Project Manager detailing the progress of the project and will work closely with Hill Manufacturing personnel to streamline the work and ensure that the project deliverables are consistent with Hill Manufacturing expectations. We will not exceed the following total cost estimate without authorization from Hill Manufacturing. If the project requires less work, the invoiced amount will be less than the estimated budget. The itemized cost estimate appears in Table II (see page 5).

> The writer introduces the budget.

FIGURE 16.8 CONTINUED

G. Thompson, March 24, 2009, p. 4

Table I: Schedule of Tasks

Task	In weeks, beginning April 1, 2009							
	1	2	3	4	5	6	7	8
Review data and develop preliminary closing options	4/1		4/21					
Collect data in the field	4/1	4/15						
Analyze collected data			4/15		5/7			
Evaluate closure alternatives and develop a closure plan					5/1		5/21	
Present closure plan to Hill and to the water commission								5/21–5/30

Qualifications

Bob Congrove, P.E., Senior Engineer, will serve as project manager and will provide much of the technical analysis and input for developing the conceptual closure plan. Mr. Congrove served as project manager for the Caney Branch Phase II Investigation and knows the conditions and water commission permitting processes and requirements.

Alejandro Martinez, P.E., Senior Civil Engineer, will provide conceptual design input on engineered components of the closure plan, such as capping specifications, and will assess the technical and financial feasibility of various closure options. Mr. Martinez has over 22 years of applied engineering experience, including designing landfill caps, liner systems and slurry walls, and developing landfill closure plans.

Brandon McCarroll, Principal Hydrogeologist, will provide input about geochemical processes affecting the mobility of ash leachate constituents. In addition, Donna Camp, Ben Armstrong, and Terry Huey (support staff) have recently worked on ash disposal projects and associated water commission permitting. As necessary, they will provide technical input and review. They will also participate in meetings with Hill Manufacturing and the state water commission.

If you have any questions about our recommended approach for developing the conceptual plan or about any other aspect of this proposal, please call me at (303) 555-1823.

Sincerely,

Tom J. Nowakowski

Tom J. Nowakowski
Senior Engineer

[Margin note:] The writer describes the qualifications of all the people who will complete the tasks.

[Margin note:] The writer concludes by offering to answer questions.

FIGURE 16.8 CONTINUED

The writer presents a detailed budget for each task.

Table II: Itemized Cost Estimates

G. Thompson, March 24, 2009, p. 5

Task	Cost per unit in dollars	Total units	Total cost in dollars
Task 1: Labor			
Principal	100/hr	4	400
Senior	90/hr	24	2,160
Staff	55/hr	4	220
Support Staff	30/hr	3	90
Task 1: Expenses			
Communications/Shipping			75
Photocopies			25
Task 1 Total			**2970**
Task 2: Labor			
Principal	100/hr	4	200
Senior	90/hr	24	1,440
Staff	55/hr	4	1,650
Support Staff	30/hr	3	60
Task 2: Expenses			
Vehicle			200
Per Diem	70/day	2.5	175
Sampling Equipment			100
Communication/Shipping			200
Photocopies			25
Task 2: Outside Services			
Lab Analysis	108/sample	6	648
10% Handling			65
Task 2 Total			**4,813**
Task 3: Labor			
Principal	100/hr	32	3,200
Senior	90/hr	140	12,600
Project	75/hr	16	1,200
Staff	55/hr	40	2,200
Drafting	40/hr	24	960
Support	30/hr	12	360
Task 3: Expenses			
Communication/Shipping			200
Photocopies/Reproducing			200
Task 3 Total			**20,920**
Task 4: Labor			
Principal	20/hr	100/hr	2,000
Senior	20/hr	90/hr	1,800
Support Staff	4/hr	30/hr	120
Task 4: Expenses			
Travel			200
Communication/Shipping			150
Photocopies/Reproducing			25
Task 4 Total			**4,295**
Estimated Project Total			**32,998**

FIGURE 16.9

Sample Unsolicited Proposal

The writer uses a formal format.

Proposal to Create a Training and Reference Manual for the Office of Disability Accommodation

Summary

The writer summarizes the problem, solution, schedule, and budget.

The Office of Disability Accommodation does not have a Training and Reference Manual. We need a manual because our staff consists of only two full-time employees, many volunteers, and ten student interns. Our full-time employees are too busy to train new volunteers and to provide all the reference information for each situation that arises. We could use this manual to train volunteers and student interns. This manual will also consolidate the policies and procedures of our office, so that our staff can quickly refer to them.

I can complete the rough draft by May 24. The labor, materials, and supplies will cost approximately $1,048.74. However, I will donate my time to write and prepare the manual. This donation will reduce your cost for the manual to $48.74 for supplies and materials. If the rough draft is satisfactory, you will receive the final draft on June 1.

The writer uses informative headings.

Current Problems at the Office of Disability Accommodation

The Office of Disability Accommodation uses many volunteers and student interns. These volunteers and interns may work in the office for only a semester, creating a large turnover. They may work as few as 5 or as many as 20 hours per week. With the number of these volunteers and interns and the high turnover, the office frequently has workers who do not understand how to do the various tasks assigned to them. To complete their assigned tasks, the volunteers and interns frequently must interrupt you and your administrative assistant or other workers. These interruptions slow the work of the office. The volunteers and interns have to interrupt others because they do not understand how to

The writer describes the nature of the problems.

- handle most student and faculty problems independently
- operate the office computers and equipment correctly
- administer basic office policies and procedures

The writer cites specific examples of the problems.

Frequently, the volunteers and interns will handle a faculty or student problem independently even though they do not completely understand how to handle the problem. Such actions can create other problems that you or your administrative assistant must then handle. For example, a new volunteer might tell a faculty member that the office will be happy to monitor an exam for any student with a disability, when we only monitor exams for students who must use the technology in the computer lab for disability accommodation. Such mishandling is frustrating for you, your assistant, students, and faculty members. The volunteers and interns frequently must also ask questions about simple tasks such as running the copy machine or processing a student or faculty request because they have not received any training and do not have any written source of information about office machinery or policies and procedures.

The purpose of this office is to help students with disabilities by informing faculty and students about appropriate and necessary accommodation and also to help faculty to understand how to accommodate. When this office runs inefficiently and makes mistakes because of uninformed and untrained volunteers and interns, we don't create the positive image that the students and the faculty need.

FIGURE 16.9 CONTINUED

2

Proposed Solution: A Training and Reference Manual

The proposed Training and Reference Manual will provide new volunteers and student interns with the training and information to
- handle most student and faculty problems independently
- operate the office computers and equipment correctly
- quickly learn the basic policies and procedures.

I will write the Training and Reference Manual from my experience and will consult with your administrative assistant for technical information. The Training and Reference Manual will include the following sections.

Answering Student Inquiries
- Telephone Inquiries
- In-House Visits

Working with New Students
- Necessary Forms and Procedures
- Campus Offices of Assistance

Working with Faculty Members
- Pre-Semester Notification Packets
- Faculty Guide

Using Office Equipment and Computers
- Equipment and Furniture
- Computers

Understanding Procedures
- Registering Students
  Early Registration
  Regular Registration
- Testing
  Regular Semester Exams
  Final Exam Procedures
- Filing
  Confidentiality
  Location of Different Files
- Issuing Elevator Keys
  Assigning Keys
  Returning Keys

Working with Various Groups with Disabilities
- Hearing Disabilities
- Visual Disabilities
- Motor/Mobility Disabilities
- Learning Disabilities
- Head-Injury Disabilities
- Hidden Disabilities
- Speech Disabilities

The writer links the solution to the problems listed in "Current Problems."

The writer presents a detailed outline to show how the proposed manual will solve the problem.

FIGURE 16.9 CONTINUED

3

Working with Volunteers
- Applications
- On-Campus Service Groups

I can complete the Training and Reference Manual during the next seven weeks. During the week of May 17, the office volunteers and students will use the rough draft and give me their comments. I will then submit the draft to you on May 24. If the draft is satisfactory, you will receive the completed manual on June 1.

My Qualifications for Writing the Proposed Manual

I have worked in the Office of Disability Accommodation and in the computer lab for disability accommodation for three years as a student volunteer. I have seen and experienced first-hand the problems and frustrations of not understanding how to handle a student or faculty problem. I have worked in all phases of the office and am able to write the proposed manual. In addition, I am a senior majoring in rehabilitation therapy, so I understand the legal and ethical issues involved in working with students with disabilities.

Budget

The following table reflects the estimated cost of writing and printing the manual:

Items	Time and Supplies	Cost (dollars)
Writing and Editing the Manual	100 hours @ $10.00 per hour	$1,000.00
Binding Costs	Vinyl Front and Back Spiral Binding	$5.18
Colored Illustrations	10 pages @ $2.50 per page	$25.00
Tab Inserts	$.89 for 5 pages x 4 packets	$3.56
Copying Costs	300 pages @ $.05 a page	$15.00
Total Cost		**$1,048.74**

I will donate my time in return for a receipt for my donation. Your cost for materials and supplies will by $48.74.

Conclusion

I am excited about the possibility of preparing this much-needed manual for our office. This manual will resolve our ongoing problem with training volunteers and student interns. I look forward to the possibility of working with you on this manual.

The writer includes a schedule.

The writer describes her experience.

The writer uses a table to present the budget.

Source: Courtesy of Le Neva Bjelde Steiner.

CASE STUDY ANALYSIS
With an "Eye" on Quality, Price Becomes Secondary in Awarding Bid[1]

Background

A contract does not always go to the lowest bidder. When the city of Georgetown, Texas, decided to accept bids for residential and commercial garbage collection, it left the guidelines open on its request for bids. Garbage disposal companies were encouraged to be creative in their bids to win the 4-year contract, which offered a renewal option for 10 years. The process was weighted toward quality, rather than price. The city council wanted to ensure it contracted a premium service for its nearly 11,000 combined residential and business accounts.

As city officials expected, they received a range of bids; but one, in particular, caught their attention. It was from Texas Disposal Systems (TDS) in Austin, Texas. Even though TDS didn't come in with the lowest price, it did offer an innovation to ensure quality that the city couldn't resist. The disposal company offered to install video cameras on its garbage trucks to record its service.

TDS was awarded the contract, and the disposal company made an upfront investment of approximately $2,500 to install the cameras on each of its eight trucks. Still, some city officials were unconvinced that the added cost of awarding TDS the contract would pay off for either the city or the disposal company.

Soon after installation, however, the city and the disposal company began to realize rewards. Cameras on the front and back of the trucks captured customers' canister numbers, workers emptying containers, and the date and time of service. As a result, if a customer complained that his or her garbage was not picked up, TDS could go back and review videotapes to verify service to that customer's serial number on a particular day. In addition, TDS was able to review tapes to see which customers were illegally disposing of hazardous wastes and to improve worker procedures to ensure quality. In one instance, the tapes were even used to avoid a costly lawsuit when a customer claimed that a garbage truck had damaged her vehicle. The tapes showed no liability on TDS's part, and the driver dropped the claim. For its part, the city of Georgetown has received exemplary service as a result of TDS's cameras. Because the cameras catch every move, workers are more likely to pick up spills and errant papers, which improves quality and service in trash collection throughout the city.

Assignment

1. Assume you are submitting a bid for a service that you know will be higher priced than your competitor's. You may be tempted to lower the price up front and then ask the customer to pay more later. Write a paragraph exploring the ethics of lowering the price simply to win a bid. Turn your paragraph in to your instructor.
2. Make a two-column list. On the left side, list some types of work where price is more important than quality. On the right side, list other types of work where quality is more important than price. Be willing to defend your list to your classmates.

¹Compiled from information downloaded from the World Wide Web:
http://wasteage.com/mag/waste_eye_collection_wins/. "An eye on collection wins contract,"
Wasteage. Saturday, May 24.

EXERCISES

1. For your technical communication class, you will prepare a report or manual. Follow these steps to select a topic and write a proposal for that report or manual.

    a. Make a list of possible topics, and gather information about each one. You might select topics related to your major, to a need in your community, or to your current job. As you gather information, use some of the research techniques and tools that you learned in Chapter 5.

    b. Brainstorm the feasibility of each topic. As you brainstorm, list the pros and cons of each topic and consider the amount of time you will have to complete the project and the importance of the topic to you, your field, the readers, your community, or your workplace.

    c. After you've considered the pros and cons of each possible topic, select a topic and write a memo asking your instructor to approve your choice. In your memo, give your instructor enough information to understand and evaluate the topic.

    d. Include a list of possible sources of information in your memo to your instructor. Remember to follow either MLA, CBE, or APA style as you list your sources.

    e. After your instructor approves your topic, write a proposal addressed either to your instructor or to a person who would make a decision about your proposed topic. Include the conventional sections relevant to your proposal.

2. Evaluate your proposal using the questions in the online "Worksheet for Writing Persuasive Proposals." After you have evaluated your proposal, revise it based on your answers to the questions.

3. Bring three copies of the proposal that you revised for Exercise 2 to class. Using the copies of your proposal, three of your classmates will read and evaluate your proposal based on questions in the online "Worksheet for Writing Persuasive Proposals." Your instructor may ask you to post your proposal on the class Web site or bulletin board so classmates can use the comment function of a word-processing program to comment on your proposal (see Chapter 3).

4. Based on your classmates' answers to the questions in Exercise 3, revise your proposal as needed. Then prepare a final draft for your instructor.

REAL WORLD EXPERIENCE
Writing a Proposal to Solve a Local Problem

You probably are aware of many problems and needs on your college or university campus or in the surrounding community. For instance, your campus may not have sufficient lighting for students walking across campus after dark; or in your community, some children may not receive new toys at Christmas or Hanukkah, or some senior citizens may need their homes painted and work done in their yards. These are only a few of the possible problems and needs that you might find on your campus or in your community.

Assignment

Your team will write a proposal to solve a problem on your campus or in the surrounding community. To write this proposal, complete these steps:

1. Select a problem on campus or in the surrounding community. Make sure that you and all members of your team are familiar with the problem and that team members can gather the needed information about the problem to propose a persuasive solution in the allotted time.
2. Gather information about the problem and about readers of the proposal. Develop possible solutions.
3. Discuss the advantages, disadvantages, and feasibility of each solution. Then determine the most effective solution for the readers.
4. Write the proposal and send it to campus or community leaders who would want to see such proposals. Be sure to give a copy to your instructor.

iStockphoto 2008.

CHAPTER SEVENTEEN

17

Writing Reader-Focused Informal Reports

You will give many reports during the course of your career. These reports will be both oral and written. You've given reports during much of your school life. For example, in grade school, you probably wrote or presented a report on a book that you read. In middle school, you might have reported on a historical figure; and in high school, you may have written a lab report for your physics class. If you have a job, your manager may have asked you to complete a project and you might have reported on the status of the project. Even though each of these reports may have a different format and a different subject matter, they are all reports. **Reports** are oral or written communications that help people to understand, to analyze, to act, and/or to make decisions. Reports can be formal or informal. In this chapter, we will discuss informal reports. In Chapter 18, we will discuss formal reports.

An **informal report** communicates information about routine, everyday business. Informal reports can cover any number of topics, from a memo clarifying the dress code to a memo requesting that your manager approve $3,500 to buy a laptop. You might present an informal report using a memo, an e-mail, a letter, or a preprinted form. Even though we use the word "informal," it doesn't mean that the information or purpose of the report is insignificant. For example, an informal report informing the architect of a problem in a bridge design could save lives.

Evaluate an informal
report in the Interactive
Student Analysis online
at www.kendallhunt.com/
technicalcommunication.

How do you know if a report is formal or informal? Your readers and your organization may give you some clues. Your readers may require that the report be formal or informal. In some organizations, a report might be formal while in another organization, the same report might be informal. For instance, a field report in most companies is an informal report; yet at Luminant, an energy company, these reports generally are formal reports. To determine if a report is formal or informal, find out what your organization and your readers expect. You might also look at similar reports written by your coworkers. After you analyze these expectations and look at similar reports, you may still be uncertain about whether to treat your report as formal or informal. However, generally, you will use informal reports to communicate information about everyday business. Even though your "everyday business" may be informal, it may still be essential to you, your career, your organization, or the public.

In this chapter, you will first learn the process for drafting an informal report. Then, we will discuss the following types of informal reports:
- Directives
- Progress reports
- Meeting minutes
- Field and lab reports
- Trip reports

DRAFTING INFORMAL REPORTS

To draft an informal report, you will
- find out about your readers
- anticipate and answer your readers' questions
- select the appropriate format

Find Out about Your Readers

Before you can decide on the format (or perhaps even whether the report is formal or informal), find out about your readers. This task may be easy for some reports. For example, if you are writing a progress report on your monthly activities for your manager, you will know what your reader expects. You will probably even know the report format that he or she prefers. However, if you are reporting progress on a project for external readers, you may be writing to more than one person and you may know little about your readers' needs and expectations. Because you don't routinely write for these external readers, your task may be a little more difficult.

As you analyze the readers of your report, consider these questions:
- What do your readers know about the topic of your report?
- Why are they reading your report. To gather information? To complete a task? To make a decision?
- What questions will they ask as they read the report?
- Are the readers internal or external?
- What positions do your readers hold in the organization? If they are internal readers, where are their positions in relation to yours in the organizational hierarchy?
- Will more than one group read the report?
- What do your readers know about you or your organization? Have their previous experiences with you or your organization been positive? If not, why not?

Once you know the purpose of your writing and the needs of your readers, you can better prepare an effective report. Whether you're writing a report that presents the progress of your work or the minutes from a meeting, begin your task by finding out about your readers because your report will succeed only if you meet their needs and expectations.

Anticipate and Answer Your Readers' Questions

To help your readers understand your report, you must gather the information necessary to answer their questions. This task may be as simple as printing a spreadsheet or as complex as using primary and secondary research techniques (see Chapter 5 for information on researching information). For some reports, once you have gathered the information, you will need to analyze that information and then make appropriate recommendations or draw conclusions.

Your readers want to know the purpose of your report, why they are receiving your report, and how your report affects them and/or their organization.

Although the specific questions will vary with the topic and purpose, your readers need you to answer these basic questions:
- What is the purpose of your report?

- Why are they receiving the report?
- How does the report affect them and/or their organization?

For each of the informal reports described in this chapter, you will find a figure that lists specific questions that readers may ask as they read that type of report. These questions will help you prepare your informal reports.

Select the Appropriate Format

Once you find out about your readers, you can determine the most appropriate format. You can use any of the following formats for informal reports:

- **Memos**. Organizations use memos for informal, internal documents. You send memos to people within your organization. You might, for example, send a memo about the status of a project to your manager; however, you would send a letter to report that status to a client, an external reader. (See Chapter 12.)

- **Letters**. Most organizations expect you to use the letter format for informal reports with people outside the organization. However, if you are sending an informal report to someone several levels above you in your organization, you might consider using the letter format. Check with your coworkers to determine what your organization prefers in this situation. (See Chapter 12.)

- **E-mail**. Many organizations send informal reports via e-mail because of its speed, convenience, and flexibility. When you send an informal report by e-mail, readers and writers can easily adapt the reports. For example, if you are writing meeting minutes for your team, you can send a draft of the minutes to all the meeting participants. They can examine the draft, insert comments, and e-mail the revised draft back to you. You, then, can revise the minutes and save time at the next meeting because participants have already seen a draft.

- **Forms and templates**. Some organizations have forms and templates for some informal reports. The form or template may simply be a cover sheet for the report, or it may be a format for the entire report. For example, Balfour Beatty, a construction company, has templates for meeting minutes. Employees use this template to create and archive the minutes. The template prompts the writer to insert the project name, project number, date, and so on. Luminant uses an "Examination Action Item List" template to accompany field reports. This template includes sections for the observation, the recommendation, and the engineer's comments. Other organizations have preprinted forms for progress reports. Before you select the format for your informal reports, find out if your organization has preprinted forms and templates for these reports. These preprinted forms and templates may save you time.

- **Formal report format**. Some organizations expect all reports, whether informal or formal, to include the front matter typical in a formal report (see Chapter 18). In other words, an informal report communicating information about routine, everyday business may have to be presented in a formal report format.

When you have determined the appropriate format for your readers and your organization, you are ready to draft your report. If you are writing an informal report that doesn't fit into one of the categories discussed above, use one of the standard patterns of organization presented in Chapter 6 or the direct approach for correspondence presented in Chapter 12.

WRITING DIRECTIVES

Directives explain a policy or procedure that readers are expected to follow. For example, you might write a directive about required procedures for submitting receipts for business expenses. You might receive a directive on procedures for purchasing safety glasses and shoes. When writing a directive, you have two purposes:

- to inform or remind readers of the policy or procedure
- to explain why the policy or procedure is important to the readers and/or the organization

You want to pay particular attention to the tone of a directive. Although you can require readers to follow a directive, you want to convince them that the policy or procedure is desirable, or at least necessary. The questions in Figure 17.1, page 528, will help you to include the information needed to convince readers of the importance of the policy or procedure. The directive in Figure 17.2, page 529, explores the reasons for a company dress code.

FIGURE 17.1

Readers' Questions Prompted by a Directive

- Why am I receiving this directive?
- Why is this directive important to me and/or the organization?
- How will the directive impact me or my job?

WRITING PROGRESS REPORTS

Progress reports describe the current status of an ongoing project or the activities of a department or division. Progress reports may also be called *status reports* or *activities reports*. Progress reports are often the intermediate document between a proposal and the end of a project. Readers of progress reports may be managers, clients, coworkers, or sponsors of a project. Progress reports have three primary purposes:

- to describe progress on one or more projects—or to describe the activities of a department or division—so readers can monitor the work over a period of time
- to provide a written record of progress
- to document problems with, and changes to, a project

Progress reports are a way for you to let your managers, your customers, your clients, and/or your coworkers know how a project is progressing; it is a way to check in with readers. These reports not only give you the opportunity to help readers monitor your progress, the documents also provide a written record of that progress. This record gives readers evidence of your work. Suppose your manager decides to postpone work on your project for one year. The monthly progress reports that you wrote before the delay will provide a record of completed work; so when the project starts up again, you or your coworkers can refer to the reports to determine where to resume work. The reports could prevent you or your coworkers from duplicating work completed before the company postponed the project.

FIGURE 17.2

Directive on Company Policy

SKATEBOARD-ON

Memo

May 19, 2009

To: All office employees
From: Kimberlee May, Human Resources
Subject: Business casual Fridays

The writer explains why the policy is important.

Recently, I have noticed that some employees are dressing too casually on Fridays. This dress creates problems for managers and presents an unprofessional image to our clients. Because the company has not clearly defined "casual" dress, we have developed a dress code that will help all employees maintain a professional personal appearance on business casual Fridays. Many of you are already following the guidelines of the new dress code for business casual Fridays. For others, the policy will clearly define what is and is not acceptable for business casual dress.

The writer uses a positive tone to recognize those employees who are following the guidelines.

The Dress Code for Business Casual Fridays
Business casual dress will be attractive, comfortable, and professional. You may not wear the following clothing on Fridays:

The writer clearly states the policy that the employees are expected to follow.

- Shorts
- Athletic shoes
- Spandex
- Clothing of any kind that promotes the products and services of competitors
- Hats
- Clothing with cartoons or drawings
- Ultra low-rise pants or shorts
- Bare feet
- Noticeably missing or exposed undergarments
- Tee shirts

The writer explains the consequences of not following the policy.

If employees do not follow the dress code, managers may send those employees home to change clothes. The time spent away from work for this reason will be treated as vacation or leave without pay. When employees' clothing is questionable, the manager will decide whether or not the employee must go home and change. If you wish to report any employee not following the dress code, please contact your manager.

As you write your progress reports, consider your readers and your purpose for writing. Ask yourself:
- Why will readers read your report?
- How will they use it?
- What questions will they want answered?

Readers will ask questions about the project or work that the report covers, your progress (and how that progress may affect the deliverable), future work, and the overall status of the project. Readers are interested in your specific accomplishments during the reporting period to see whether your work is progressing as planned. Readers will want to know your schedule and, possibly, your budget for the next reporting period or for the remainder of the project, especially if you will not be writing any other progress reports. They also may want to know what kinds of results they can expect during the next reporting period.

ETHICS NOTE

Honestly Reporting Your Progress

If the project is not progressing as planned, you may be tempted to withhold bad news—perhaps hoping that things will improve before the next progress report. However, when you withhold information from your readers, you are misleading them. You have an ethical responsibility to report the progress as it is—not as you want it to be or expect it to be. Your readers have a right to know how your project is progressing. Your job is to report the progress on the project clearly and completely and to accurately predict the future of the project.

If you are faced with a problem that has affected the progress of a project, explain why the project is not progressing as planned and how the project will be affected. For example, if your construction project is behind schedule because of excessive rain, specifically state the number of days that rain affected the work and how you expect the overall schedule to be impacted. As you report problems in the project, do the following:
- **Use an objective tone.** Don't make excuses.
- **Be specific.** Provide specific details that led to the problems.
- **Give a realistic estimate of how the problem will affect the schedule and the budget.**

Using the Conventional Structure for Progress Reports

Most progress reports contain these conventional sections:
- Introduction
- Discussion of the progress
- Conclusion

Figure 17.3 presents questions that readers may ask and relates those questions to the conventional sections of a progress report.

Introduction

The introduction identifies your project and the purpose of your report. In the introduction, you will
- identify the project or projects that the report covers
- state the time period that the report covers
- state the objectives of the project or projects (if readers need this information)

You can usually identify a project and its time period in one sentence, such as "This report covers progress on the Midwest Relocation project from January 1 through March 31." You would explain the objectives of your

FIGURE 17.3

Readers' Questions about the Conventional Sections of a Progress Report

Conventional Section	Readers' Questions
Introduction	• What project does the report cover? • What time period does the report cover? • What are the objectives of the project?
Discussion of the Progess	• What work have you accomplished since the beginning of the project or since the last progress report? • Is the project progressing as planned? • What work have you planned for the next reporting period or for the remainder of the project? • What are the results of the work? • What problems, if any, have you encountered?
Conclusion	• What changes, if any, do you recommend? • How will these changes affect the project? • What is the overall status of the project?

FIGURE 17.4

Sample Introduction to a Progress Report

This report describes the current status of, and progress on, the lignite handling system for the Old Eagles plant. The purpose of this project is to renovate the lignite-handling system at the plant. This report covers our work on the obsolescence study, the mechanical engineering, and the electrical engineering from July 1, 2008, to August 30, 2008.

project to help readers put your report in context or to remind them of its purpose. You also can state the objectives as you defined them in the project proposal. Figure 17.4 presents a paragraph in which the writer states the objectives of a project and also identifies the project and the time period that the progress report covers.

Discussion of the Progress

The discussion of the progress answers readers' questions about how the project is proceeding and about what work you've planned for the next reporting period or for the remainder of the project (see Figure 17.3, page 531).

Readers will be especially interested in the following items:
- how your progress compares with what you planned to accomplish during the reporting period
- any problems that you have encountered
- the results of your work

You can organize the discussion of the progress in two ways:
- **By the progress made during the reporting period and the progress expected during the next reporting period** (see Figure 17.5). This pattern emphasizes the total amount of progress you have made as well as the work you expect to do during the next reporting period.
- **By the tasks or projects to be completed** (see Figure 17.6). This pattern emphasizes the task or the project, not the amount of progress made or not made.

Using the progress-made/progress-expected pattern, shown in Figure 17.5, you can organize the discussion around the progress made on one or several tasks or projects during the reporting period and the progress expected during the next reporting period. Using the task/project pattern shown in Figure 17.6, you can organize the discussion around one or several tasks or projects, discuss the progress made and expected on task or project 1, and then move to task or project 2, and so on. (See Chapter 6 for more information on using a comparison/contrast organization.) As you write the discussion of your progress, follow the tips on the next page.

FIGURE 17.5

Progress-Made/Progress-Expected Pattern

Progress made during the current reporting period (the work that you accomplished during the time period covered by the progress report)
a.   Task 1
b.   Task 2
c.   Task 3
Progress expected during the next reporting period (the work that you expect to complete during the next reporting period)
a.   Task 1
b.   Task 2
c.   Task 3

FIGURE 17.6

Task/Project Pattern

Task 1 (or Project 1)
a. Progress made during the current reporting period (the work that you accomplished during the time period covered by the progress report)
b. Progress expected during the next reporting period (the work that you expect to complete during the next reporting period)
Task 2 (or Project 2)
a. Progress made during the current reporting period
b. Progress expected during the next reporting period
Task 3 (or Project 3)
a. Progress made during the current reporting period
b. Progress expected during the next reporting period

TIPS FOR WRITING THE DISCUSSION OF THE PROGRESS

- **Explain how your progress compares with what you planned to accomplish during the reporting period**. Readers want to know if your project is on schedule.
- **Report any problems that you have encountered**. Explain how these problems affected your progress and the status of your project. Will these problems affect the budget, the schedule, or the personnel? If so, explain how.
- **Include the major results of your work, if applicable**. Readers need information about results and problems, so they can approve or change the project, budget, schedule, or personnel. By including information about problems, you document those problems and possibly lay the foundation for changes later in the project.
- **If your progress report is long, consider putting the results and problems in a separate section**. You can title that section "Results and Problems" or "Evaluation of the Progress."

Conclusion

The conclusion summarizes the overall progress made on the project and, if necessary, recommends changes. In this section, you can recommend ways to overcome problems that you are experiencing, or you might suggest ways to alter the project to get better results. You also can briefly mention these problems and recommended changes in the discussion of the progress.

In the conclusion, report the overall status of your project:
- summarize the progress made on the project
- summarize any problems experienced during the reporting period
- evaluate the overall progress
- recommend ways to improve or change the project or future work, if necessary

The two sample progress reports presented in this chapter illustrate the conventional sections. In Figure 17.7, pages 535-536, a student reports on her progress on a manual for the Double Oaks Golf Shop; she uses the progress-made/progress-expected pattern. In Figure 17.8, page 537-538, an engineering company reports on its progress on a lignite-handling system; it uses the task/project pattern.

FIGURE 17.7

Progress Report Using the Progress Made/Progress-Expected Pattern

The writer uses the memo format for her internal readers.

MEMO

March 5, 2009

To: Professor Patricia McCullough
From: Nicole Sanders
Subject: Progress report on the employee manual for Double Oaks Golf Shop

The introduction lists the project and time period.

This report covers my progress on an employees' manual for the Double Oaks Golf Shop since February 15. This manual will include up-to-date procedures for opening, maintaining, and closing the shop, and for maintaining the inventory; it will also include the current policies for the employees. Although my work is progressing satisfactorily, I have experienced some problems in gathering information from the management of the shop.

The writer uses the progress made/progress expected pattern.

My Progress During the Past Three Weeks

During the past three weeks, I had planned to interview the sales staff and the manager of the shop and to update and revise the procedures and policies for the shop. I have completed the interviews; and based on the information gathered in these interviews, I have updated the procedures currently in writing. However, I have not updated and revised all procedures and policies because my interviews revealed that many of the procedures are not in writing. To write the remaining procedures, I need information from the manager and from the managers of other small golf shops.

The writer describes the project, including the progress made.

Interviews with the Sales Staff and the Manager

I interviewed five members of the sales staff: Bart Thompson, Stella Smith, Melissa Connors, Dan Sandstedt, and Bradley Davis. Bart and Melissa explained that few of the current procedures and policies appear in writing. Before the interview, I assumed that the current employees' manual contained most of the procedures and policies but that the procedures and policies were out of date. However, Bart and Melissa explained that the manual did not contain any inventory procedures—a major responsibility for the sales staff.

I also interviewed the manager of the club, Donna Shoopman. She explained how the manual differed from the current policies of the shop. She also confirmed that the procedures for maintaining the inventory and closing the shop were not in writing.

Updating and Revising the Policies and Procedures

I have updated and revised the policies on employee dress, customer satisfaction, substance abuse, disciplinary issues, and work schedules. I updated and revised these poli-

---

FIGURE 17.7 CONTINUED

---

cies based on information from Donna, the shop manager. I have also updated and revised the opening procedures for the shop. These were the only procedures in writing.

Problems with Updating and Revising the Procedures

I cannot completely update the procedures for maintaining the inventory and for maintaining and closing the shop because they are not in writing. Donna plans to give me the procedures for maintaining the inventory, closing the shop, and maintaining the shop by March 12. These procedures will serve as a foundation for the procedures section of the manual. I cannot finish the procedures section of the manual without the information from Donna.

Progress Expected During the Next Three Weeks

During the next three weeks, I will
- interview managers of other small golf shops
- finish updating and revising the procedures
- prepare a completed draft of the manual

Interviewing Managers at Other Golf Shops
I will interview Don Price, manager of Briarwood Golf Shop, and Bonnie Barger, manager of Brentwood Golf Shop. These interviews will help me appropriately revise the procedures for maintaining the inventory and the shop.

Completing the Updating and Revising of the Procedures
I will update and revise the procedures for maintaining inventory and for maintaining and closing the shop after receiving the procedures from Donna. I will also use the information from the interviews with managers of Brentwood and Briarwood Golf Shops to help me appropriately revise these procedures.

Conclusions

My interviews with the sales staff and with the managers gave me valuable information and guidelines for writing the manual. Even though I am behind schedule, I can complete the draft of the manual by March 30 if Donna gives me the procedures by March 12.

The writer includes the problems she has encountered.

The writer includes specific information on work she has planned.

In the conclusion, she reports on the overall status of the project.

FIGURE 17.8

Progress Report Using the Task/Project Report Pattern

E-E-C Corporation

Material Handling Systems Division
4000 Highline Court
Chalfant, Pennsylvania 18914
(412) 555-4200

This letter format is appropriate for the external reader.

October 15, 2009

Mr. William Holtz
Morris and Associates, Inc.
400 East Fifth Street
Midwest City, Iowa 65091

Subject: Progress on the lignite handling system at the Old Eagles plant

Dear Mr. Holtz:

The introduction lists the project, purpose, and time period.

This letter describes the current status of, and progress on, the lignite-handling system for the Old Eagles plant. The purpose of this project is to renovate the current lignite-handling system at the plant. This letter covers our work on the obsolescense study, the mechanical engineering, the structural engineering, and the electrical engineering from July 1, 2009, to August 30, 2009.

Obsolescense Study

We identified all obsolete equipment. This equipment includes the Merrick scales, heating and air conditioning vents, Westman level detectors, and rotary plow relay logic.

The writer uses the task/project pattern.

Mechanical Engineering

Progress made

We have submitted 90 mechanical drawings to Britt Engineers. They have approved or approved-with-comments all the drawings. These drawings show the following items:
- changes to conveyors C-3A, C-3B, BF-16, BF-17, BF-18, and BF-19 because of obsolete speed reducers
- revised conveyor C-2 because of obsolete fluid couplings
- small scale drawings with 2009 changes and additions
- drawings with miscellaneous 2009 changes and additions to the Active Storage Building, Rotary Plow Area, Reclaim Hoppers, Crusher House, and Transfer Vault
- miscellaneous minor pre-2009 changes to the Boiler Surge Bin Tower and Transfer Towers 1A and 1B

Future work

In the final reporting period, the mechanical engineering division will
- provide a quotation for the fire protection system based on the June 14, 2009, letter from Britt Engineers
- complete the mechanical portion of the remaining obsolescence study items listed in our letter of October 6, 2009

FIGURE 17.8 CONTINUED

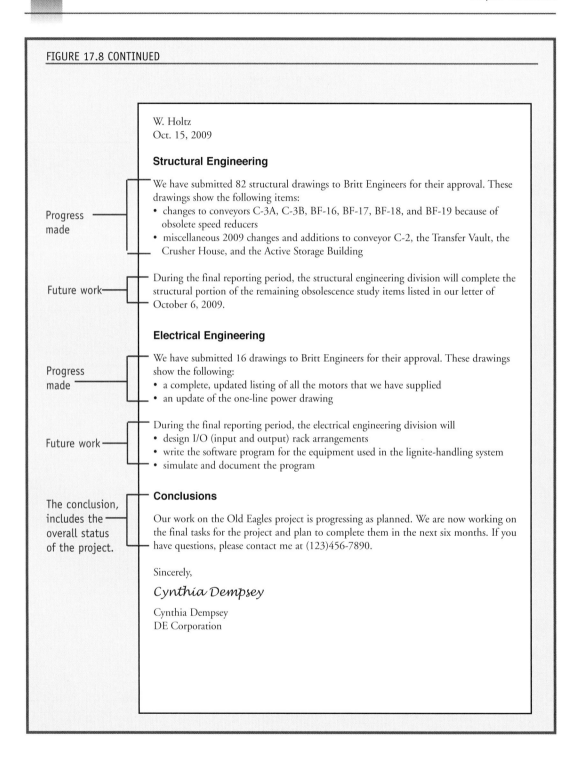

W. Holtz
Oct. 15, 2009

Structural Engineering

Progress made

We have submitted 82 structural drawings to Britt Engineers for their approval. These drawings show the following items:
- changes to conveyors C-3A, C-3B, BF-16, BF-17, BF-18, and BF-19 because of obsolete speed reducers
- miscellaneous 2009 changes and additions to conveyor C-2, the Transfer Vault, the Crusher House, and the Active Storage Building

Future work

During the final reporting period, the structural engineering division will complete the structural portion of the remaining obsolescence study items listed in our letter of October 6, 2009.

Electrical Engineering

Progress made

We have submitted 16 drawings to Britt Engineers for their approval. These drawings show the following:
- a complete, updated listing of all the motors that we have supplied
- an update of the one-line power drawing

Future work

During the final reporting period, the electrical engineering division will
- design I/O (input and output) rack arrangements
- write the software program for the equipment used in the lignite-handling system
- simulate and document the program

Conclusions

The conclusion, includes the overall status of the project.

Our work on the Old Eagles project is progressing as planned. We are now working on the final tasks for the project and plan to complete them in the next six months. If you have questions, please contact me at (123)456-7890.

Sincerely,

Cynthia Dempsey

Cynthia Dempsey
DE Corporation

WRITING MEETING MINUTES

Minutes are the official record of a meeting. The meeting could be as informal as a meeting of your product-development team or as formal as a meeting of a city council. You may read or write minutes from meetings within your organization or meetings between representatives of your organization and representatives of external organizations. For example, construction company representatives may meet with the owners of a project they are handling; the company representatives will keep minutes of meetings with the owners to provide a record of what occurred at the meetings. Meeting minutes are sent to those who belong to the group or organization represented at the meeting; these minutes also may be made public as in the minutes of a city council meeting. For these meetings, the minutes may be published in a newspaper or made available online for anyone to read.

Informal Meeting Minutes

The type of meeting will dictate what you put in the minutes. For example, if the meeting was informal, you might simply send an e-mail to those who attended the meeting. The e-mail would
- summarize what was discussed
- summarize agreements
- identify action items

Action items identify what the group decided to do, who is responsible, and what deadlines were set.

Formal Meeting Minutes

As discussed in "Taking It Into the Workplace," page 550, the minutes from a formal meeting actually begin with the agenda for the meeting. (An *agenda* is an outline for the meeting.) An effectively written agenda gives you an outline for the minutes. An agenda includes the following information:
- Time and date of the meeting
- Location of the meeting
- Items to be discussed

As you prepare the items that you will discuss, you should do the following:

- Consider including desired outcomes for each item.
- Start with the most important items. If you run out of time, the items not discussed will be the less-important ones.
- Consider including times for each item.
- Identify the person(s) responsible for each item.

The following section provides guidelines for the conventional structure of formal minutes. In general, you need to include the following:

- information about the meeting and the attendees
- items discussed
- action items
- your information

Figure 17.9 presents the questions that readers may ask as they read meeting minutes.

Information about the Meeting and Attendees

Because the minutes serve as the official record of a meeting, they should include the following information that identifies the meeting:

- **The name(s) of the group(s) involved in the meeting**. Identify the name of the division, committee, companies, etc.
- **The location, date, and time of the meeting**. Only include the beginning time, not the ending time, here.
- **The type of meeting**. Identify the type of meeting. Was it a regularly scheduled meeting or a one-time or special meeting?
- **The attendees**. List the first and last names of those people who attended, those who were absent, and any guests who attended. At a large public meeting, such as a city council meeting, you may have an auditorium filled with guests. For these situations, you only need to list the guests who spoke or participated directly in the meeting.
- **The time the meeting adjourned**.
- **Name and title of the person who wrote the minutes**.

FIGURE 17.9

Readers' Questions on the Conventional Sections of Meeting Minutes

Conventional Section	Readers' Questions
Information about the meeting and the attendees	• What is the name of the group(s) involved in the meeting? • Where did you meet? • What was the date and time of the meeting? • Was the meeting a regularly scheduled or a one-time (special) meeting?

Items discussed and action items	• Did you act on business discussed at a previous meeting?
	  – If so, what did you do or discuss?
	• What major topics did the group discuss?
	• What motions, if any, were introduced?
	  – Did the motions pass or fail?
	• What actions did the groups agree to do?
	  – Who is responsible for the action item?
	  – What is the deadline for completing the action?

Information on the recorder of the minutes	• Who prepared the minutes?

The first three pieces of information appear in a heading for the minutes. The list of attendees generally appears in a sentence or paragraph immediately after the heading, and the time the meeting adjourned appears as the last item of business in the minutes.

Items Discussed and Action Items

If the meeting is a recurring meeting, you will organize this section by old business and new business. If it is a one-time meeting, you will simply organize the discussion items. In either case, follow the order of the agenda.

If the meeting is not the first one of your group or committee, you will have **old business**. For many meetings, the old business is simply approving the minutes of the previous meeting. In some meetings, however, the attendees may discuss and update or amend items introduced at a previous meeting. For example, at a previous meeting, a member of the group may have introduced installing and getting approval from the Department of Transportation for traffic lights near a commercial development. At the next meeting, the group may want to discuss whether they received approval or when the lights will be installed. This business would be "old business." The group would put these items on the agenda before *new business*. For old business, you simply record what action is taken on the minutes or on a previously introduced item of business. For example, you might write, "Johnson moved to approve the minutes of the last meeting. Smith seconded the motion. The motion passed by a vote of 6 to 0." (Or you could write, "The motion carried by a vote of 6 to 0.")

To write effective minutes, your job is to record the major topics discussed and any action taken at the meeting. In a perfect meeting, the group will discuss the items as they are listed on the agenda. However, because meetings often deviate from the agenda, you may find it difficult to take accurate minutes. Your goal is to create a "snapshot" of items discussed in the meeting, not to record every word. For any action items, you should identify

- the specific action
- who will complete the action
- the deadline for completing the action

The "Tips for Writing Accurate Meeting Minutes," page 543, will help you write effective meeting minutes.

Your Information

As the writer of the minutes, you will end the minutes with information about you. As part of your contact information, you may also give readers the opportunity to amend the minutes. You might end the minutes as follows:

> Respectfully submitted,
> John Smith, Senior Project Engineer

Figure 17.10, page 544, illustrates effective meeting minutes. Notice how the writer has included all of the conventional information and used a format to help readers locate information.

WRITING FIELD AND LAB REPORTS

You may write a field or lab report after you complete an experiment or after you inspect some machinery or other equipment. You might also write a field or lab report after you examine a site. For example, your company may be considering

TIPS FOR WRITING ACCURATE MEETING MINUTES

- **Include each major topic discussed**. For most meetings, especially formal ones, you will have a list of these topics on the agenda. If the meeting is less formal, include the major topics or items discussed.
- **Record any motions made, who make the motion, and the outcome of the motion**. Include whether the motion was passed, defeated, or tabled. Include the vote count—"the motion passed by a vote of 7 to 2." If the group amended the motion, record specifically how the motion was amended.
- **Record the names of people who read reports, introduced action items, and so on**. As part of the minutes, include the names of people who, for example, make motions and read reports. Exclude the names of people who simply participated in the discussion; you will include the names of those people when you list the attendees in the first part of the minutes.
- **Summarize the discussions**. Effective meeting minutes summarize the discussions. Minutes aren't transcripts, so include enough information to create an accurate record, but omit information that readers will not need.
- **Focus on the actions, not on the emotional exchanges**. Because minutes aren't a transcript, separate the actions from the emotional exchanges. For example, if the group argues, don't write, "The motion to paint the new office space taupe with hunter-green accents passed 6 to 1 after Garcia and McCarroll argued about the color. At one point, McCarroll said that 'the organization had no taste when it comes to interior decorating.'" Instead, simply write, "After discussion, the motion to paint the new office space taupe with hunter-green accents passed 6 to 1."
- **Include enough detail on an issue or action item so readers will understand the item** especially those readers who may not have attended the meeting.
- **If the discussion moves too quickly or you didn't hear and understand something, interrupt the discussion and ask the group to clarify**. Your job is to create an accurate record; so if you don't understand something or need someone to repeat a motion or action item, ask the group to clarify or repeat the information.
- **Ask someone to write the minutes and/or to take notes if you are in charge of the meeting**. You cannot take notes for accurate minutes while you are running the meeting.
- **Proofread the minutes before sending them.**

FIGURE 17.10

Meeting Minutes

The writer identifies the group, date, time, and attendees.

The writer uses headings to help readers locate information.

The writer numbers each major topic discussed in the order of the agenda.

The writer identifies the persons introducing and seconding motions.

The writer identifies the persons introducing new business.

The writer specifically identifies action items—including the person responsible and the deadline.

The writer identifies the recorder of the minutes.

Monthly Regional Directors' Meeting

Wednesday, October 14, 2008

Attendees: Jo Bernhardt; Anne Constanides; Marj Smith; Caleb Robinson; James Kelly; and Patrick Sims. Stuart Smith was not present because he is on leave.

Jo Bernhardt called the meeting to order at 3:30 p.m. in the conference room.

Old Business
1. The minutes of the February 13 meeting were approved by a unanimous vote.
2. Jo Bernhardt distributed information on the progress of the United Way campaign. She reported that we have two weeks left in the campaign and that the southwest region had reached its goal.

New Business
1. Jo Bernhardt reported on the plans for the implementation of the company's new intranet portal. The group discussed how the new portal would affect the work of the divisions.
   Action item: Patrick Sims will work with IT to ensure that information for all current projects will be located in one area of the portal. He will report on his progress by November 15.
2. After a discussion of the new engineer training, Patrick Sims moved that we extend the training to a whole day instead of a half-day session and that we ask some engineers to join Human Resources in planning the training. Anne Constanides seconded the motion. The motion carried by a vote of 5 to 1.
   Action item: Jo Bernhardt will contact Human Resources and ask the project managers to suggest engineers to participate in the planning. She will ask the managers to send their suggestions by October 31.
3. James Kelly introduced information on the new hotel and convention center project. He requested that the engineers receive a briefing on the project. After discussion, James changed his request into a motion. Caleb Robinson seconded the motion. The motion passed unanimously.
   Action item: James Kelly will set up a date, time, and location for the briefing.
4. Jo Bernhardt adjourned the meeting at 4:45 p.m.

Respectfully submitted by Anne Constanides, project engineer.

buying a manufacturing plant. Your managers send you to the plant to inspect the equipment. When you return, you might write an informal report recording what you found during your inspections and listing what you recommend.

Using the Conventional Structure of Field and Lab Reports

A field or lab report has the following conventional sections:

- **Introduction**. State the purpose of your report and the problem addressed in the report.
- **Methods**. Describe the methods you used. Because this report is informal, you only need to briefly describe the methods. In field and trip reports, most writers deemphasize the methods.
- **Results**. State the results of your experiment, inspection, and so on. These results may be preliminary. If so, be sure to state that they are only preliminary.
- **Conclusions**. Tell readers what you learned from the results.
- **Recommendations**. If your readers expect recommendations, include them in the report.

In many field and lab reports, you may not have a heading for all of the conventional sections. Figure 17.11, on page 546, presents questions that readers might ask as they read the conventional sections of a field or lab report. The field report in Figure 17.12, on page 547, illustrates how a writer used the conventional structure to report on his inspection of a chimney at a power plant.

WRITING TRIP REPORTS

You may write a trip report after you return from a business trip. You might, for example, attend a seminar on new software for project management. When you return to the office, your manager may want you to report on the seminar. Your manager will want to know what you learned and what you recommend. Your manager will not, however, be interested in a minute-by-minute itinerary of what you did and what occurred at the seminar. You may also write a trip report after you visit with customers. Your report might include information on how your organization can follow up with these customers and better serve them. You might also write a trip report after you visit with employees who work away from the home

FIGURE 17.11

Readers' Questions about the Conventional Sections of Lab/Field Reports

Conventional Section	Readers' Questions
Introduction	• What is the purpose of your report? • Why did you conduct the inspection, experiment, etc.? • What led to the inspection, experiment, etc.?
Methods	• What method did you use for your inspection, experiment, etc.?
Results	• What were the results? • What did you learn? • What problems, if any, did you observe?
Conclusions	• What do the results mean? • Are the results conclusive? • If you observed problems, are these problems major or minor?
Recommendations (optional)	• What do you recommend? • If you observed problems, how do you recommend solving the problem?

or regional office. For instance, if you supervise field representatives who work in regional offices, you may visit them to find out about their work. After such a trip, you would write a report about how their work is proceeding or how the company can improve their work environment.

Using the Conventional Structure of Trip Reports

When you write trip reports, consider using these conventional sections:
- **Introduction**. Include the place, the date, and the purpose of your trip.
- **Summary**. Summarize what you observed, learned, and recommend.
- **Discussion**. Present the details of the important information that you gathered during the trip. Your readers aren't interested in reading, "First, I did this; then, I did that." Instead, they want to know what you learned and why that information is important to them and the organization.
- **Recommendations**. State and explain your recommendations.

FIGURE 17.12

Field Report

The writer uses a letter format because he is writing to an external reader.

The writer identifies the purpose of the report.

The writer identifies the methods used.

The writer uses lists to highlight the results.

The writer includes recommendations.

Structural Safety
Engineering
1545 Steam Road
Brazelton, KY 75609
(663)555-9001

May 20, 2009

Ms. Joyce Carr, Plant Manger
Big Brown Steam Electric Plant
New Power
Santa Fe, New Mexico 86090

Re: Examination of Chimney in unit 3

Dear Joyce,

We performed a level 5 examination and structural assessment of the concrete chimney, brick liner, and associated components at Big Brown Electric Plant, unit 3. Our examination follows up the baseline assessment performed in 2006. We will use the findings of this follow up to update the database for future examinations and to plan maintenance and possible repairs.

Methods
We conducted the examination from April 20 through May 8, 2009. To examine the chimney and brick liner, we used five vertical drops (one in the interior of the brick liner, two in the annular space, and two along the exterior surface of the concrete shell). To examine the chimney components, we visually inspected the breeching duct, seals, platforms, ladders, lightning protection system, and other related appurtenances.

Results of the Examination and Assessment
The overall condition of the chimney is good. However, our examination and assessment revealed the following moderate structural deficiencies:
- Part of the brick liner that passes through the breeching duct is missing.
- The three aviation obstruction lights at the upper gas monitory platform are not working.
- The handrail around the perimeter of the top platforms does not continue between grating sections. This defect causes loose handrails and excessive lateral deflections.

We also found these safety issues:
- nonconforming safety chains across the openings
- inadequate safety rail clamps on the ladder

Conclusions and Recommendations
Except for the safety items and the missing liner bands in the breeching duct, the chimney is in good condition. However, we recommend repairing the following items to minimize the potential for structural problems:
- Replace the missing brick liner bands in the breeching duct. The severe environment requires that the replacement rods be coated with coal tar epoxy and encased in 316L stainless steel pipe.
- Replace the aviation obstruction lights.
- Add a handrail section to minimize excessive lateral deflections and to meet OSHA directives.
- Correct all safety items noted in this report.

If you have questions, please call me at (305) 555-5555.

Regards,

Mathew McCarroll

Mathew McCarroll
Structural Engineer

Because you will usually write a trip report to an internal reader, most trip reports are written in memo format. Figure 17.13 presents the questions that your readers might ask as they read a trip report. Figure 17.14 presents a trip report written after an employee attended demonstrations of hands-free cell phone systems.

SAMPLE INFORMAL REPORTS

Throughout this chapter, you have read samples of the following types of informal reports:

- Directives (Figure 17.2, page 529)
- Progress reports (Figure 17.7, page 535, and Figure 17.8, page 537)
- Meeting minutes (Figure 17.10, page 544)
- Field and lab reports (Figure 17.12, page 547)
- Trip reports (Figure 17.14, page 549)

Your organization or your manager may have a specific format or template for these informal reports. Ask your coworkers or your manager if your organization has a preferred or an expected format for these reports. If so, use that format or template. If not, use these sample reports to guide you.

FIGURE 17.13

Readers' Questions about the Conventional Sections of Trip Reports

Conventional Section	Readers' Questions
Introduction	• Where did you go? • When did you go? • What was the purpose of your trip?
Summary	• What did you observe or do? • What, briefly, did you learn? • What, briefly, do you recommend?
Discussion	• What did you learn? • Why is the information important to the organization?
Recommendations	• What do you recommend? • Why?

FIGURE 17.14

Trip Report

The writer uses a memo format because he is writing to an internal reader whom he is equal to in the organizational hierarchy.

> **Memoradum**
>
> Date: September 21, 2009
> To: Robert Congrove, Director of Safety
> From: Eric Hacker
> Subject: Trip to IT Dynamics
>
> This memo summarizes the information I gathered from a trip to IT Dynamics in Madison, Wisconsin, on September 18, 2009. The purpose of my trip was to attend demonstrations of several hands-free cell phones.

The writer summarizes what he observed and what he recommends.

> **Summary**
> IT Dynamics demonstrated three hands-free cell phones. I recommend Nokia's CARK-91 for employees who have a company vehicle and Motorola's Timeport 270c for vehicles in the company fleet.

The writer presents detailed information about what he learned. He selects only the information that is important to his reader.

> **Discussion**
> IT Dynamics offers three hands-free cell phones:
> • Nokia's CARK-91 for $175 plus installation
> • Sharper Image's speaker/microphone cradle for $129
> • Motorola's Timeport 270c for $349
>
> The Nokia cell phone can be retrofitted into any vehicle; with this integrated setup, the cell phone uses the vehicle's antenna to boost reception. This cell phone had the best reception of the three models. However, the phone is not portable.
>
> The Sharper Image cell phone has a cradle to hold the phone; however, with this model, the reception was poor. I had trouble hearing the speaker and tended to lean into the phone when talking. As I leaned, I took my eyes off the road.
>
> The Motorola cell phone was easy to use and to transport. This cell phone has voice activation and speakerphone capabilities. Users simply clip the cell phone to the sun visor. The reception was excellent. However, the cost of this phone is double that of the Nokia model.

The writer presents and explains his recommendations.

> **Recommendations**
> I recommend that we purchase the Nokia CARK-91 and have them installed into 200 fleet vehicles. Because of the cost of the Motorola model, I recommend that we purchase the Motorola Timeport 270c only for the employees who have a company-issued vehicle. By purchasing both models, we will have a hands-free option in 60 percent of the company vehicles. I would like to talk with you and Gene about these recommendations. Call me at x7980 to set up a time to talk.

TAKING IT INTO THE WORKPLACE
Conducting Effective Meetings

Meetings are a regular part of the workplace; yet meetings are unpopular, in part, because people see them as a waste of time. As *GovLeaders.org* reports, many meetings are "time-wasting" and poorly run. *GovLeaders.org* (2008) suggests six "Golden Rules" for conducting an effective meeting:

- Run your meetings as you would have others run meetings that you attend.
- Be prepared and ensure that all participants are prepared as well.
- Stick to a schedule.
- Stay on topic.
- Don't hold unnecessary meetings.
- Wrap up meetings with a clear statement of the next steps and who is to take them.

To help follow these rules, use these guidelines before the meeting:

- **Ask yourself, "Is this meeting necessary?"** Some meetings should never be held. Instead, you may be able to accomplish your goals by sending an e-mail or memo or by simply making the decision on your own. If you can make the decision alone, don't call a meeting. If you want to explain your decision, a meeting might be a good way to do so.
- **Pick an appropriate time and location before announcing the meeting**. *MyMeeting.com* reports that the best meeting times are 9:00 a.m. and 3:00 p.m. Pick a location where you will not be distracted.
- **Prepare an agenda**. An agenda helps you organize the meeting, tell others why the meeting is important, and stay on task during the meeting. An agenda answers the question, "Why should I come to this meeting?" Without an agenda, people may not understand the purpose and/or importance of the meeting. An agenda also helps you organize the meeting and, thus, accomplish your objectives. During the meeting, if the conversation drifts away from the topic, an agenda provides a "road map" to guide the conversation back on track.
- **Announce the meeting**. Attach the agenda to your meeting notification.

Once you have announced the meeting, you are ready to hold the meeting.

- **Start on time**. When you do not start on time, you waste everyone's time and encourage people to assume that the meeting is not important.

- **Tell people why they are there**. Start with a brief, clear statement of the purpose of the meeting. If you don't know the purpose of your meeting, you won't know when you've achieved the purpose or met your objectives.
- **Follow the agenda and help people stay on track**. You can help people stay on track by recapping frequently. If the discussion gets off track, politely interrupt with a reference to a specific point under discussion and recap previous items.
- **Know when to move to the next agenda item and how to deal with side issues**. Side issues occur when some or all of the participants discuss issues not on the agenda. These issues waste time and distract everyone from the agenda. To politely deal with side issues, try these techniques:
    - Ask specific questions of the person(s) discussing the side issues.
    - Use the person's name(s) when asking the question. For example, you might say, "John, how would you suggest handling the training for the new e-mail system?"
    - Know when to move to the next item on the agenda. Side issues/conversations may occur because an item has been fully discussed.
- **End on time**. Respect your participants' time. If you announced that the meeting will end at 4:30; then end the meeting at 4:30. If you did not complete the agenda, set up another meeting.

Once your meeting is over, follow up:
- **Send minutes to all the participants**.
- **Ask for their comments or corrections**.
- **Remind participants of any action items they are responsible for completing**.

As a professional, you will conduct many meetings. Make them effective meetings.

Assignment

1. Attend a meeting at a local business, school board, city council, or non-profit organization. Take notes on the effectiveness of the meeting.
2. Write a memo trip report to your instructor on what you learned from the meeting and recommend ways that the leader of the meeting could have more effectively conducted the meeting. If the meeting was effective, recommend that the leader continue to conduct meetings in the same way. In your report, give specific examples of what was effective and what was ineffective.

CASE STUDY ANALYSIS
City Council Meeting Erupts in Fist Fight

Background

In 1906, a meeting of the Passaic, New Jersey, city council erupted in a fist fight between two council members when one accused the other of refusing to curb the use of lighted business signs hanging over the sidewalk because he had a vested interest in keeping the electric company happy.

Council member Osborne wanted the signs removed because he thought they were a public nuisance. He accused council member Hammond of refusing to curb the sign's use, saying that Hammond wanted to make money off the contract. Hammond countered by calling Osborne a liar and saying that Osborne, himself, had said he carried a heavy mortgage and that the Lighting Committee would have thousands of dollars in the Public Service Corporation contract.

The argument escalated, with each calling the other a liar, until Hammond suddenly punched Osborne between the eyes. Osborne reeled; then came back with a blow of his own. Finally, other council members stepped in to separate the two and end the fight.

Assignment

1. Break into groups of 4. Assume you are on the Passaic, New Jersey, city council. Discuss what methods you could use to keep future meetings orderly. Discuss your discoveries with the class.

2. Prepare a "meeting" for your class. Prepare an agenda on three topics that you'd like to discuss. Use the agenda format discussed on pages 539-540. Turn in your agenda to your instructor. Be prepared to lead the meeting of your class if you are selected to do so.

3. Take notes on another student's meeting presentation. Turn your notes in to your instructor. Use the informal meeting notes format discussed on page 539.

EXERCISES

1. As the manager of Gotta Get Clean, a home cleaning service, you have received complaints that members of your staff are parking in front of mailboxes. When they park in front of mailboxes, mail carriers will not deliver the mail. Homeowners don't like it when they do not get their mail, and the Post Office has a policy that mail carriers are not to deliver the mail if a vehicle is blocking the mailbox. Instead, the mail carrier is to place a note on the vehicle explaining the following: "The Postal Service will not deliver mail to any mailbox that is obstructed by a vehicle or other item. If you wish to receive mail, do not obstruct the mailbox." You are tired of the complaints. Write a directive in memo format for the cleaning staff, explaining the following new policy: The cleaning staff must park the company cars at least 20 feet from a mailbox.

Download a copy of the Worksheet for Writing Reader-Focused Informal Reports at www.kendallhunt.com/ technicalcommunication.

2. Revise the progress report you analyzed for the online interactive document. Be sure to incorporate your recommendations from the analysis.

View the Interactive Document Analysis online at www.kendallhunt.com/ technicalcommunication.

3. Write a progress report for the project that you proposed in Exercises 1, 2, and 3 in Chapter 16.

4. If you are working on a collaborative project in one of your classes, write the minutes of one of your meetings. E-mail the minutes to each member of the team and to your instructor. Be sure to follow the conventional structure and to answer the readers' questions in Figure 17.9, page 541.

5. **Collaborative exercise**: Like many college campuses and communities, your campus or community most likely has an inconvenient or dangerous situation. For example, the campus may have an intersection where many accidents occur. For this exercise, your team will complete these steps:
   - Identify an inconvenient or dangerous situation on your campus or in your community.
   - Observe the situation or inspect the area on two or three consecutive days. (Be sure to observe during peak time; for example, don't observe the intersection at 2:00 a.m.)

- Write a field report to the appropriate school administrators or city officials. Be sure to use the conventional sections and to answer the readers' questions in Figure 17.11, page 546.
- Include a specific recommendation in your report.

REAL WORLD EXPERIENCE
Writing a Trip Report

Imagine that you are a purchasing representative for your company. Your job is to visit local businesses in your community to learn about the products or services they offer. Set up an appointment to discuss, for example, new cell phones, new furniture, cleaning services, etc.

Assignment

After your visit, write a trip report for your instructor. Be sure to follow the conventional structure and to answer the readers' questions in Figure 17.13, page 548. Include your recommendations for purchase of services or products.

iStockphoto 2008.

CHAPTER EIGHTEEN

18

Writing Reader-Focused Formal Reports

Cassie and Jon are technical writers employed by an engineering firm that works with companies to renovate and rebuild manufacturing plants. Cassie's group is working with a company to find out why a tower collapsed at one of its plants. Cassie will write a report explaining possible reasons for the collapse. Jon is working with another group to prepare a report recommending ways to repair the tower. Cassie and Jon are working on the same project, but are preparing two different types of reports. Both Cassie and Jon will prepare a formal report, but their purposes for writing are different. Cassie will present the results of her research and draw conclusions based on those results. Jon will present results, draw conclusions, and make recommendations based on those conclusions.

Like Cassie and Jon, you may write formal reports for a class or in the workplace. A *formal report* communicates less-routine business than an informal report. Like an informal report, a formal report can cover any number of topics. Unlike an informal report, a formal report, in most cases, has both front and back matter. A formal report might be the final step in a series of documents. This series might begin with a proposal, continue with progress reports, and end with a formal report. You might write a formal report that isn't preceded by a proposal or a progress report. For example, your manager might ask you to evaluate telecommuting: you might examine whether telecommuting improves employee productivity. For this project, you would research telecommuting using both primary and secondary sources and write a report presenting the results of your research, possibly making recommendations.

In this chapter, you will learn guidelines to help you write reader-focused formal reports. This chapter also specifically focuses on a frequently used report format, the feasibility report.

THE TYPES OF FORMAL REPORTS

You've heard the word *report* since you were in grade school. When you enter the workplace, you will be asked to write reports. **Report** is an umbrella term for a group of documents that inform, analyze, or recommend. To meaningfully discuss reports, we need standard terminology. However, the workplace and many fields don't have standard terminology for referring to reports. For example, some terms refer to the report topic, such as *meeting minutes*, *lab reports*, or *field reports*. Other terms refer to the phase of the research or project, such as *progress reports* or *completion reports*. Still other terms refer to the purpose of the report, such as *recommendation reports*.

If you are new to your organization and your manager asks you to write a report, spend some time talking with your manager to determine what type of report and format he or she expects. You might ask: "What is the purpose of the report?" "Do you have a preferred format for the report?" You can also look at similar reports written by coworkers.

For this chapter, we will use the purpose of the report to categorize formal reports into the following categories:
* information reports
* analytical reports
* recommendation reports

Information Reports

Information reports present results so readers can understand a particular problem or situation (see Figure 18.1). For example, the manager of a city's Web site might prepare an information report for the city council; the report would provide statistics on the number of people who visit the

FIGURE 18.1

Types of Reports

Type of Report	Presents Results	Draws Conclusions	Makes Recommendations
Information	✓		
Analytical	✓	✓	
Recommendation	✓	✓	✓

site, the number of people who pay their city water and sewage bills online, the number of links to other city-related Web sites, and the number of city departments that use the site to provide information to residents. The purpose of this type of report is to present facts—often called "results"—not to analyze the facts or to draw conclusions.

As you learned in Chapter 17, a report can be formal or informal. When you determine the formality of the report, you know what format to use and whether to include front and back matter. *Formal* and *informal* refer to the format of the report, not to the purpose or significance of the report or to the organization of the report's body. To determine if an information report should be formal or informal, determine whether the report is routine. For example, consider the Web site manager we discussed earlier. He or she might write an information report to the city's public information officer providing statistics on the site for the month. This report is routine and would be informal. The report to the city council, however, is not routine because the Web manager doesn't routinely present reports to the city council. The council might request such a report because it wants to know if the city is using too many or too few city resources on the Web site, and if current practices are improving communications to city residents. Because it is not routine, this report would be formal.

Information reports might do any of the following:

- **Present information on the status of current research or of a project**. For example, you might report on the status of a project to construct a new runway at an airport.
- **Present an update of the operations in your division**. For example, the Web site report for the city council might present an update of work done by the city's public information division.
- **Explain how your organization or division does something**. For example, you might report on how your division tracks the work of subcontractors.
- **Present the results of a questionnaire or research**. For example, you might present the results of an employee questionnaire on commuting.

Analytical Reports

Analytical reports go a step beyond presenting results. *Analytical reports* present results, analyze those results, and draw conclusions based on those results (see Figure 18.1, page 557). Analytical reports analyze and interpret the information or data in the results. They answer the question, "What do the results mean?" These reports attempt to describe why or how something happened and then to explain what it means. For example, let's again consider the Web manager's report to the city council. Along with presenting the results, the manager might analyze those results and present conclusions. Based on the low percentage of city residents who use the site to pay their water and sewage bills, the manager might conclude that either the residents don't know the service is available or that the residents can't easily find the service on the Web site. The manager's purpose in the report is only to draw conclusions based on the results, not to make recommendations.

Like information reports, analytical reports can be formal or informal and can present and analyze a variety of results. They do the following:

- **Explain what caused a problem or situation**. For example, you might present the results of a traffic study at an intersection where many accidents have occurred. In the report, you would draw conclusions about why the accidents occurred based on the results of the traffic study.
- **Explain the potential results of a particular course of action**. For example, based on the results of your research, you might conclude

whether or not opening a new office branch will increase business for your organization. You might then present conclusions on whether the potential income will offset the cost of running the new office.

- **Suggest which option, action, or procedure is best**. For example, you might present options for treating people with diabetes and report which treatment has the best outcome. However, you would not recommend a treatment.

> *Report categories— information, analytical, and recommendation— can blur when you are writing in the workplace.*

Recommendation Reports

Recommendation reports suggest a particular course of action. These reports may seem to be the same as analytical reports. So how do you differentiate? Think of it this way: you might conclude that treatment X is more effective than treatments Y and Z. This statement is your conclusion, but it is not necessarily a recommendation to use treatment X. You might, for example, recommend using treatment Z because of cost; or you might recommend not using any treatment at all.

Like information and analytical reports, recommendation reports may be formal or informal. Recommendation reports might address any of the following situations and advocate a specific course of action. These, again, are only examples; you may encounter other situations for writing recommendation reports.

- **What should we do about a problem?** You might recommend a course of action for dealing with a problem. For example, you might answer the questions: "What should we do about the problem of excessive absences in the manufacturing division?" or, "What should we do about the growing number of plastic bags in our landfills?"
- **Should we or can we do something?** You might look at whether your organization can or should do something. For example, you might answer the question, "Even though we currently have two engineers working on runway projects for the airport, should we hire another project engineer to work on the new construction project that we will begin next year?"
- **Should we change the method or technology we use to do something?** You might examine whether a change would benefit your organization. For example, your organization currently sends all printing to

an outside vendor. You might answer the question, "Should we continue to use an outside vendor or move all printing in-house?"

Although the categories of reports—information, analytical, and recommendation—make report writing seem easily defined, you may find that the categories blur. As you write reports, be flexible and ask questions. Seek help from people senior to you in the organizational hierarchy.

DRAFT THE FORMAL REPORT

Much of the work of drafting a formal report comes before you actually begin writing. Before you can begin to write, you need to find out about the purpose of your report and your readers, formulate questions, and answer those questions through appropriate and thorough research and, finally, draw conclusions and make recommendations. Figure 18.2 outlines this process.

Identify the Purpose and Readers of Your Report

Before you begin writing your report, talk with the readers or with the person who asked you to write the report. With most reports, you will know the purpose; with others—especially reports that you're writing for the first time or that someone asked you to write—you will need to spend time defining the purpose. These questions may help you define the purpose:
- What do you want readers to know, do, or learn from the report?
- Do you only want to present results?
- Do you want to draw conclusions?
- Do you want to make recommendations based on those conclusions?
- Is the report routine?

As with informal reports, find out as much as possible about your readers. To help you analyze the readers of your reports, consider these questions:
- What do your readers know about your field or the topic of your report?
- Why are they reading your report? To gather information? To complete a task? To make a decision?
- How much detail will readers need or expect? Does this need or expectation differ among the various readers? If so, how?
- Do your readers expect an informal or formal report?
- Are your readers internal or external?

- What positions do your readers hold in the organization? If they are internal readers, where are their positions in relation to yours in the organizational hierarchy?
- Will more than one group read the report?
- What do your readers know about you or your organization? Have their previous experiences with you or your organization been positive? If not, why not?

FIGURE 18.2

Steps in Drafting a Formal Report

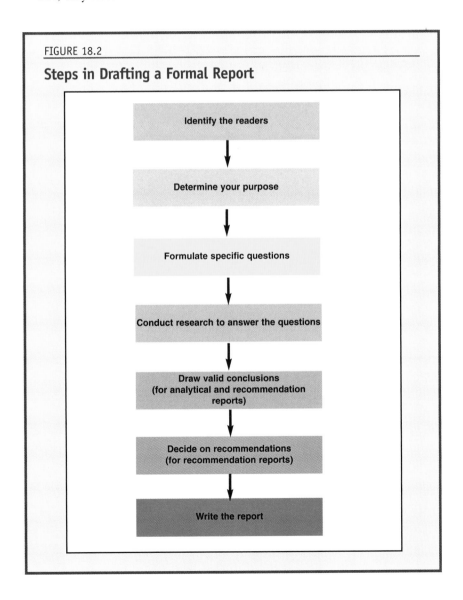

Identify the readers

Determine your purpose

Formulate specific questions

Conduct research to answer the questions

Draw valid conclusions
(for analytical and recommendation reports)

Decide on recommendations
(for recommendation reports)

Write the report

After you have a sense of what your readers know about the topic and what they expect, consider these basic questions that your readers may ask:

- What is the purpose of this report? To present results? To draw conclusions? Or to make recommendations?
- Why are we receiving the report?
- How does this report affect us and our organization?

Writing for Readers with Varied Knowledge and Purposes

Pay particular attention to two important factors that will help you discover what readers need or expect: 1) readers' familiarity with the topic, and 2) their purpose for reading (Holland, Charrow, and Wright). Readers who are familiar with the topic of your report will "find it easier to grasp new material about the topic than readers" who are not familiar with it (Holland, Charrow, and Wright). If readers are not familiar with, or do not understand your field, include adequate detail and explain technical terms and concepts.

If you know the knowledge level will vary among your readers, consider these three options:

- write separate reports
- direct the language and detail to readers with the lowest level of knowledge
- compartmentalize the report (Holland, Charrow, and Wright)

You will rarely write separate reports because it is time-consuming and expensive. If you direct the language and details to readers with the lowest level of knowledge, readers highly familiar with the topic may become impatient with what they see as simplistic and tedious explanations. Thus, compartmentalizing the report is the most efficient and effective method for writing for varied readers. When you compartmentalize, you create a separate section for each group of readers. You can compartmentalize by using headings, tables of contents, summaries, and indexes to help readers find the sections that interest them (Holland, Charrow, and Wright). You also can place definitions and explanations of technical terminology and concepts in footnotes, glossaries, appendixes, or other sections within the report (Holland, Charrow, and Wright).

Effective report writers tend to spend more time on macrowriting—"big picture"—issues (Baker). These writers spend time formulating specific questions and then determining the best way to answer those questions. Less-effective writers spend little time formulating specific questions and doing the necessary research to answer those questions. Consider these questions for an analytical report on the health risks of electric and magnetic fields:

Vague   Do electric and magnetic fields cause health problems?

Specific   What are the health risks of exposure to low-strength, low-frequency electric and magnetic fields produced by power lines and electric appliances?

The first question is vague because it does not specify the strength, frequency, or origin of the electric and magnetic fields. The second question gives the writer specific information to use when researching the topic.

When you have formulated specific questions, you're ready to determine what primary research techniques and secondary research strategies are appropriate for answering your questions. (For information on researching information, see Chapter 5.)

Make Conclusions and Recommendations Based on Sound Research

After you have formulated specific questions and researched your topic, you're ready to draw conclusions and make recommendations. For many reports, drawing valid conclusions is relatively simple if you have thoroughly researched your topic and gathered sufficient information or data. If you've not done adequate research, you may have a difficult time drawing sound conclusions, or you may draw invalid conclusions. Depending on the topic of your report, these invalid conclusions could put you, your organization, or your readers at risk. You have an ethical responsibility to conduct sufficient research to draw valid conclusions.

If you have thoroughly carried out the research plan up to this point, you will more easily see the recommendations that the conclusions suggest. For example, if your conclusions indicate that customers are bringing in their cars for service because of faulty ignition systems, the clear course of action would be to send a recall notice to each customer who purchased that model car. The company would then repair, at no charge to the customer, the ignition systems. Of course, this recommendation is probably not the one that you would have wanted to choose, but the results of your research—not your desire—should dictate your recommendation. Even if you don't like it, this recommendation is ethical and in the long run will protect you, the company, and most importantly the customers who drive the cars.

TIPS FOR DRAWING VALID CONCLUSIONS

- **When examining the results of your research, look for cause-and-effect relationships.** When you find these relationships, you may be able to draw valid conclusions. Or these relationships may indicate that you need to conduct further research. For example, if you are answering the question, "Does drinking eight ounces of skim milk increase weight loss when combined with exercise?" you may find that women who included six ounces of low-fat yogurt, but didn't drink skim milk, lost the same amount of weight as women who drank milk. You might assume that the relationship between dairy products—not just skim milk—and weight loss is an important cause-and-effect relationship.

- **Be wary of results that seem to point to the same conclusions**. In most projects, you want more than one or two results that point to the same conclusion. You typically need more evidence than just one or two results to draw a valid conclusion. Let's consider the question about skim milk and weight loss: your results indicate that women who drank milk and exercised lost an average of ten pounds or more over a six-month period and women who did not drink milk lost an average of four pounds. Both groups performed thirty minutes of aerobic exercise each day. With these results, you can reasonably conclude that drinking skim milk, when combined with exercise, may cause women to lose more weight.

- **Watch for areas where you have used illogical or unsupported arguments**. For example, you may find that ten percent of the study participants did indoor cycling for their thirty minutes of aerobic exercise and these participants lost a larger percentage of weight. You might, then, conclude that indoor cycling when combined with drinking eight ounces of skim milk causes women to lose more weight. This conclusion would not be valid because not all of the women included indoor cycling in their exercise routine. These women may have lost the same amount of weight if they had used other forms of aerobic exercise.

USE THE CONVENTIONAL STRUCTURE

When you have drawn your conclusions and determined the recommendations you will make (if required), you're ready to write the report. Unlike informal reports, most formal reports have common sections (see Figure 18.3). This discussion presents guidelines for using the conventional structure for reports. We will begin with the body of the report because even though your readers will see the front matter before the body, you will write the body of the report before creating most of the front matter.

The Body of the Report

The body of a formal report generally has these sections:
- Introduction
- Methods
- Results
- Conclusions (analytical and recommendation reports)
- Recommendations (recommendation reports)

FIGURE 18.3

Conventional Structure of Formal Reports

Structure of the Report	Conventional Sections
Front matter	• Letter of Transmittal • Cover • Title Page • Table of Contents • List of Illustrations (included only if the report includes graphics) • Abstract or Executive Summary
Body	• Introduction • Methods • Results • Conclusions • Recommendations
Back matter	• Works Cited or List of References • Glossary • List of Abbreviations or Symbols • Appendices • Index

The type of your report (information, analytical, recommendation) will determine what elements you will include. For example, an information report would not include a recommendations section. The body of your report should address some typical questions that readers may ask. See Figure 18.4 for "Questions Reader's May Ask When Reading the Body of a Formal Report."

Introduction

The introduction prepares the reader for the information presented in the report. The introduction
- identifies the purpose of the report
- identifies the topic of the report
- indicates how the report affects or relates to readers
- presents background information
- presents an overview of the report

FIGURE 18.4

Questions Readers May Ask When Reading the Body of a Formal Report

Conventional Sections	Questions Readers May Ask
Introduction	• What is the topic of the report? • What is the purpose of the report? • How does the report affect me and/or my organization? • What is the background of the report? • Have others researched similar topics? If so, how does that research relate to this report? • What follows in the report?
Methods	• How did you conduct the research? • How did you gather the information that led to your conclusions and/or recommendations? • If you conducted an experiment or study, how did you design it?
Results	• What did you find out?
Conclusions	• What do the results mean? What do they tell you?
Recommendations	• Based on the results and the conclusions, what do you recommend? What should be done?

The sample document in Figure 18.26, page 604, presents an effective introduction to a formal report from a telecommunications company. In the first paragraph, the writer introduces the general topic of the report: changing procedures in the proposal centers to improve productivity. The first paragraph also tells how the report affects the reader—the regional manager overseeing the proposal center. The first and second paragraphs give the reader background about the need to increase productivity without increasing costs. The third paragraph states the specific topic and purpose of the report: to evaluate three options for improving productivity. In paragraph four, the writer informs the reader that he will recommend one of these options later in the report. The introduction doesn't include the conclusions and recommendation. Those appear in an executive summary. The introduction also tells the reader what follows in the report. To help you write an effective introduction, follow the tips below.

TIPS FOR WRITING THE INTRODUCTION

- **State clearly the topic of the report**. If a proposal or progress report has preceded the report, you can most likely "cut and paste" the information on the topic directly from one of those documents.
- **State the purpose of your report**. Clearly state the purpose of the report, not the purpose of the project. For example, you might write, "In the report that follows, we recommend the most cost-effective option for increasing our productivity without sacrificing quality and client satisfaction."
- **Identify how the report affects or relates to the readers (optional)**. In the introduction, you may want to explain why readers should read the report.
- **Present the background that readers need to understand the report**. If other researchers have examined your research question, include a review of the current research to demonstrate that you have done your homework. Readers are more likely to accept your conclusions and recommendations if they know you have looked at other research on your topic. For some topics, little, if any, research is available; if that is the case, tell your readers.
- **Present an overview of the report**. Tell readers what follows in the report.

Methods

The methods section of a report answers the question, "How did I do the research or conduct the study?" Readers want to know exactly how you gathered your information. Some readers may be as interested in your methodology as in your results; so use specific, detailed language when describing your methodology. If readers might duplicate your methods, use language specific enough for others to reproduce your research methods.

You can view an excellent methods section in the sample report presented in Figure 18.27, page 613, at the end of this chapter. This report was prepared for the U.S. Department of Energy by scientists at the Oak Ridge National Laboratory in Tennessee. The report presents the results of a study of macroinvertebrates and fish in streams near two oil retention ponds. This methods section includes specific references to the procedures used to collect water samples, and to identify and quantify benthic macroinvertebrates. Notice how the writers use specific language to describe their procedures.

Results

Results are the data that you gathered from your research. The results section answers the question, "What did you find out?" When writing this section, only present the results; interpret them in the conclusions section. Readers are more likely to understand your logic and your conclusions if you present all the results before you present your interpretations. If you mix the results with the interpretations, you may confuse readers.

The arrangement of a results section will vary with the topic and the purpose of the report. For some reports, the results section may be a series of paragraphs and supporting graphics, such as tables and graphs. If you used a variety of methods in your research, you can organize the results section around those methods. You then can structure your discussion of the results in the order in which you present the methods. For example, in the report presented in Figure 18.27, page 613, the writers first discuss the methodology for sampling benthic macroinvertebrates and then for sampling fish. In the results section, they then discuss the results of the benthic macroinvertebrates sampling first. For other reports, you may use one of the standard patterns of organization for your results section (see Chapter 6). As you write the results section, review the tips on the next page and study the results sections in the sample reports at the end of this chapter.

TIPS FOR WRITING THE METHODS SECTION

- **Tell your readers how you gathered your information**. Explain how you did the research or conducted the study. By giving readers this information, you add credibility to your results, conclusions, and recommendations.
- **Use clear, specific language**. If you use vague language, readers may think you didn't use clearly defined methods and that your results or conclusions may not be valid. Without specific language, readers can't duplicate your methods.

TIPS FOR WRITING THE RESULTS SECTION

- **Include only the results**. In the results section, simply report the results—the data that you gathered. Interpret the results in the conclusions section of the report.
- **Use a standard pattern of organization to organize the results**. These patterns will help you present your results in an organized, logical manner.
- **Use graphics when appropriate**. If you have numerical data, use graphics such as tables, line graphics, or bar graphics to present the data. Be sure to introduce and explain any graphics that you use.

Conclusions

The conclusions section answers the question, "What do the results mean?" In some reports, this section is titled "Discussion of the Results." It interprets and explains the significance of the results. The conclusions and the recommendations sections are often the most important sections of many reports.

When you have conducted sound research and have adequately analyzed your results, you can state your conclusions clearly and confidently. To convey this confidence, avoid words and phrases that may undermine readers' confidence in your conclusions. For example, readers may think that the following conclusion indicates that the writers lack confidence in their conclusion or perhaps that they have not conducted sound research:

Lack of confidence   **We** believe that Option 1 will maintain current expense levels and may build on current expertise to reduce the time for writing the boilerplate. However, **we** think that Option 1 does not appreciably improve the quality of the text and could disrupt the work group.

The subject of both sentences in this conclusion is "we": "We believe" and "we think." When the sentences focus on the options, not on the writers, the conclusions are more direct and confident (subjects appear in bold):

Confident   **Option 1** maintains current expense levels and builds on current expertise to reduce the time for writing the boilerplate text. However, **Option 1** does not appreciably improve the quality of the text and could disrupt the work group.

When writing the conclusions section, you may discover that the results are inconclusive, that none of the options studied meets the criteria, or that the methodology was poor. The conclusions won't always fit into the neat categories that you expected. Nevertheless, you have an ethical responsibility to report clearly what the results mean, even when the conclusions are not what you or your readers expect or want. The sample report in Figure 18.26, page 604, includes a conclusions section that uses a clear, confident tone. Follow the tips on page 569 to help you write an effective results section.

Recommendations

The recommendations section answers the question, "Based on the results and the conclusions, what do you recommend?" In this section, you recommend a course of action (or perhaps, inaction) based on the results and the conclusions. The recommendations section may be shorter than the conclusions or results section. Some writers combine the recommendations and conclusions sections.

The sample report in Figure 18.26, page 604, presents a recommendations section. Here, the writer confidently states the recommendations in one

TIPS FOR WRITING THE RECOMMENDATIONS SECTION

- **State the recommendations in clear, direct language**. Tell your readers what specific course of action you recommend. Even if your readers may not expect or may resist your recommendations, state them directly.
- **Make sure your recommendations clearly follow the conclusions and results**. If your readers have closely read your conclusions, they will understand why you are recommending a particular course of action.
- **Eliminate unnecessary explanations of the recommendations**. If you have drawn well-supported conclusions, the conclusions section will support and explain your recommendations. You don't need to restate the conclusions in the recommendations section.

sentence, offering little explanation because the recommendations clearly follow from the results and conclusions. The sample report in Figure 18.27, page 613, presents the recommendations section from the research report prepared for the Department of Energy. The writers recommend more studies and explain the issues the studies should address. See the tips above for writing the recommendations section.

The Front Matter

Front matter consists of the reference aids that come before the body of a document. As you plan your report, you'll determine the type of front matter that best meets your purpose and your readers' needs. For example, you may find that you do not need a list of illustrations. You may also find that your organization has standard formats and style guidelines, or even printed forms, for some front matter. Before preparing the front matter, find out if your organization has guidelines or templates for the front matter. If it does, follow those guidelines or use those templates. If it doesn't, find copies of successful reports written by others in the organization. These reports will show you how other employees formatted the front matter of similar reports and will perhaps reveal some of your organization's unwritten preferences. If your organization doesn't have guidelines or examples of front matter, you can choose the format. Front matter consists of the following elements:

- Letter of Transmittal

- Cover
- Title Page
- Table of Contents
- List of Illustrations
- Abstract and Executive Summary

Letter of Transmittal

The letter of transmittal—sometimes called a cover letter—has the following objectives:
- to summarize the subject and purpose of the document
- to identify the occasion (the reason) for preparing the document
- to emphasize any information in the document that is of special interest to the readers, such as methods, conclusions, recommendations, or changes from the proposal or original plan for the document

The letter of transmittal is the first part of your document that readers will see. Therefore, it should create a good impression of you and your organization. You can attach the letter of transmittal to the cover of the document, place it inside the cover before the title page, or send it separately. If you decide to send the letter separately, be sure that it tells readers when you or your organization plan to send the document itself. Figure 18.5 shows a letter of transmittal written by a technical communication student. She prepared the report mentioned in the letter for the Louisiana Parks and Wildlife Department.

Cover

The cover of a document serves three basic purposes:
- to protect the pages
- to identify the document
- to create a positive impression of your document (and possibly your organization)

Before you design a cover, check to see whether your organization has a standard cover design for all documents. If it does, use that design. The cover may help to spark readers' interest or establish a certain tone. For example, if you're preparing a cover for a proposal to customize telecommunications functions for a company, you might customize the cover by using the company's colors or displaying a picture of the company's headquarters.

FIGURE 18.5

Sample Letter of Transmittal

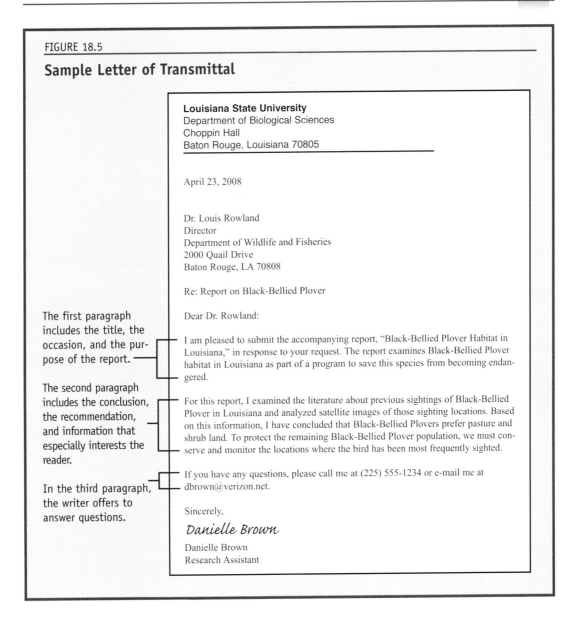

Louisiana State University
Department of Biological Sciences
Choppin Hall
Baton Rouge, Louisiana 70805

April 23, 2008

Dr. Louis Rowland
Director
Department of Wildlife and Fisheries
2000 Quail Drive
Baton Rouge, LA 70808

Re: Report on Black-Bellied Plover

Dear Dr. Rowland:

The first paragraph includes the title, the occasion, and the purpose of the report.

I am pleased to submit the accompanying report, "Black-Bellied Plover Habitat in Louisiana," in response to your request. The report examines Black-Bellied Plover habitat in Louisiana as part of a program to save this species from becoming endangered.

The second paragraph includes the conclusion, the recommendation, and information that especially interests the reader.

For this report, I examined the literature about previous sightings of Black-Bellied Plover in Louisiana and analyzed satellite images of those sighting locations. Based on this information, I have concluded that Black-Bellied Plovers prefer pasture and shrub land. To protect the remaining Black-Bellied Plover population, we must conserve and monitor the locations where the bird has been most frequently sighted.

In the third paragraph, the writer offers to answer questions.

If you have any questions, please call me at (225) 555-1234 or e-mail me at dbrown@verizon.net.

Sincerely,

Danielle Brown

Danielle Brown
Research Assistant

Use any of these methods to print the cover and attach it to the document:

- Print or copy the cover on card stock. Card stock can damage some printers, so you may want to print the cover on printer paper and then copy it onto card stock.
- Put a clear piece of plastic over the cover.

TIPS FOR WRITING THE LETTER OF TRANSMITTAL

In the first paragraph . . .
- State the title or subject of the document.
- State the occasion of the document—the reason for preparing it. For example, the occasion might be to complete a class assignment, to respond to a request from a manager or client, or to complete a research project.

In the middle paragraph(s) . . .
- State the purpose of the document. (Some writers include the purpose in the first paragraph.)
- Summarize your conclusions and recommendations.
- *Optional:* Include and possibly explain any changes to your work or the document since you last corresponded with the readers.

In the final paragraph . . .
- Offer to answer any questions that readers may have about the document or its contents.
- *Optional:* Mention and possibly thank any person, group, or organization who helped you prepare the document.

- Laminate the cover for durability and a more professional-looking appearance.
- Use a binder that has a clear pocket on the front to hold the cover that you have printed. For this type of binder, use printer paper.

These printing and binding options are inexpensive and give reports a professional appearance. Spiral or three-ring binders secure the pages of the document much more professionally than a paper clip or a file folder.

Most covers include some or all of this information:
- **Document title**
- **Your name and position** (If you are preparing a document for someone outside your organization, do not include your name or position on the cover. For some internal documents, you may also exclude this information. Look at other internal documents to determine whether the cover should include the writer's name and position.)
- **Your organization's name and/or logo**

ETHICS NOTE

Make Honest, Well-Supported Recommendations

You have a responsibility to present honest, well-supported recommendations even if your recommendations are not what readers want to hear or what they expect. You may know, for example, that your readers want you to recommend a particular option because it will make your organization look good, or they may want you to recommend the cheapest solution even if your research does not support that recommendation. If your research does not support the expected outcome, you should not recommend that outcome. Instead, you should make sure that you have conducted sound research, double-checked your results, and told the truth.

You may be tempted to set up your research in a way that will lead you to the outcome that your organization or your readers expect. This approach is dishonest; it could put you or your organization at risk for legal action. At worst, it could endanger a consumer.

As you set up your research and make your recommendations, ask yourself these questions:
- Is it legal?
- Is it consistent with company policy and my professional code of conduct?
- Am I doing the right thing?
- Am I acting in the best interests of all involved?
- How will it appear to others? Am I willing to take responsibility publicly and privately?
- Will it violate anyone's rights?

- **Name of the organization or client for whom you prepared the document** (For documents for the general public, you will not include this information.)
- **Date of submission**

For documents prepared for classes at your university, you will probably use a format like that in Figure 18.6, page 576. This cover is for Danielle Brown's report for her technical communication class. The cover includes the title of her report, the intended readers, the writer's information, and

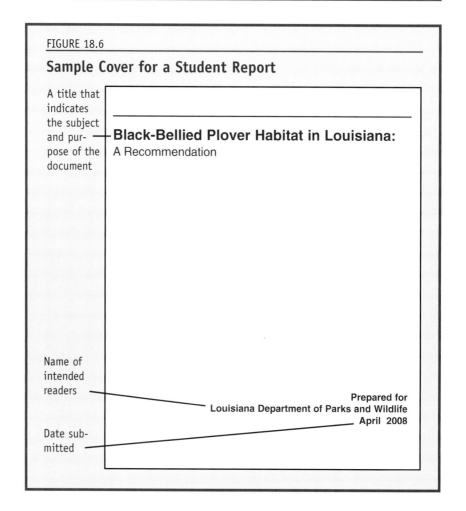

FIGURE 18.6

Sample Cover for a Student Report

A title that indicates the subject and purpose of the document

Black-Bellied Plover Habitat in Louisiana:

A Recommendation

Name of intended readers

Date submitted

**Prepared for
Louisiana Department of Parks and Wildlife
April 2008**

the date submitted. Figure 18.7 shows a cover for a document prepared for the general public. This cover identifies the issuing organization (U.S. Environmental Protection Agency), the title of the document, and the date. This type of cover is common for reports issued by organizations, while Figure 18.6 is less common outside of school.

Title Page

Some title pages look exactly like the cover; some repeat only certain parts of the cover; and some are completely different from the cover, having few, if any, graphics or color like the one in Figure 18.8, page 578. Title pages contain some or all of the following information:

FIGURE 18.7

Sample Cover for a Workplace Report

- document title
- name of the organization or client for whom you prepared the document
- your name and position
- your organization's name and/or logo
- date that you submitted the document to its intended readers

Many companies have a standardized format for title pages. These standardized formats may call for much more information than generally appears on a title page. Figure 18.9, page 579, illustrates a title page from the U.S. Environmental Protection Agency's 2008 report on the environment. This

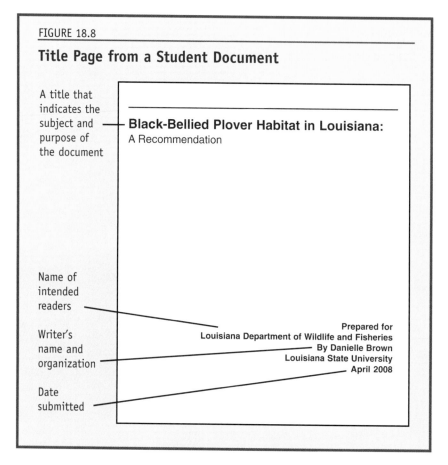

FIGURE 18.8

Title Page from a Student Document

A title that indicates the subject and purpose of the document

Black-Bellied Plover Habitat in Louisiana:
A Recommendation

Name of intended readers

Writer's name and organization

Date submitted

Prepared for
Louisiana Department of Wildlife and Fisheries
By Danielle Brown
Louisiana State University
April 2008

title page lists not only the title of the report, but also the document number, a disclaimer, and information about the organization. Like many title pages in the workplace, this title page does not identify the names of the writers or the names of the intended readers.

Table of Contents

The **table of contents** identifies what is included in a document and helps readers locate specific sections of that document. An effective table of contents lists more than just the first-level headings. (If your document has only one level of headings, consider subdividing some sections to help readers locate information. Be sure to follow the guidelines for outlining presented in Chapter 6.) Figure 18.10, page 580, shows a table of contents that contains only first-level headings that convey no specific information

FIGURE 18.9

Title Page from a Workplace Document

The writers include information about the organization and a document number.

U.S. Environmental Protection Agency
Washington, DC 20460

EPA/600/R-07/045F
May 2008

2008

EPA's Report on the Environment

The writers include a disclaimer.

DISCLAIMER
This document has been reviewed in accordance with U.S. Environmental Protection Agency policy and approved for publication. Mention of trade names or commercial products does not constitute endorsement or recommendation for use.

The writers do not include their names.

Preferred Citation:
U.S. Environmental Protection Agency (EPA). (2008) EPA's 2008 Report on the Environment. National Center for Environmental Assessment, Washington, DC; EPA/600/R-07/045F. Available from the National Technical Information Service, Springfield, VA, and online at http://www.epa.gov/roe.

Recycled/Recyclable—Printed with Vegetable Oil Based Inks on 100% Postconsumer, Process Chlorine Free Recycled Paper

Source: Downloaded from the World Wide Web: http://cfpub.epa.gov/ncea/cfm/recordisplay.cfm?deid=190806. Environmental Protection Agency, *EPA's Report on the Environment 2008*.

FIGURE 18.10

Uninformative Table of Contents

Contents

about the subject of the document. Readers curious about the "Discussion" section, for example, would have to search more than 15 pages to find a particular subsection or topic. To prepare an effective table of contents, follow the tips on page 583.

Some word-processing software has functions for generating a table of contents after you have typed your document. You can also use this same function to generate the list of illustrations, discussed in the next section. These functions can save you time and help you use the exact wording that appears in the body of the document.

The most commonly used style for the table of contents appears in Figure 18.11. Many writers and organizations, especially in the sciences, prefer a decimal system, which adds decimal numbers to the headings and subheadings in the table of contents (see Figure 18.12, page 582). Select the style that your readers or organization expect. If your readers and your organization have no expectations, look at similar documents to see how others in your organization or field have prepared the table of contents; then pattern your table of contents accordingly.

FIGURE 18.11

Effective Table of Contents

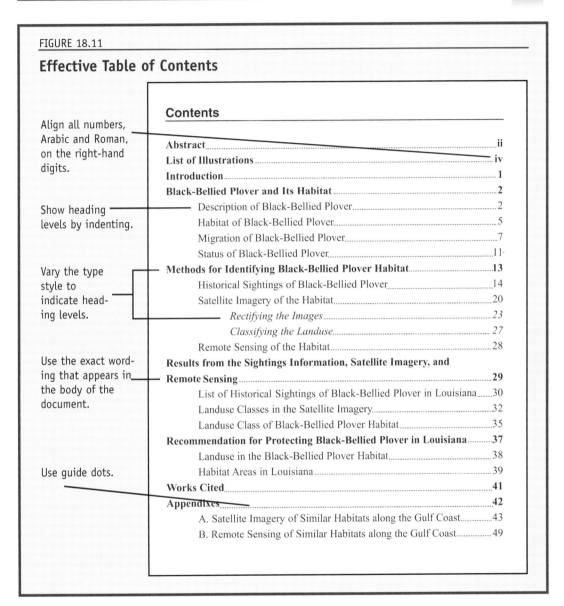

Align all numbers, Arabic and Roman, on the right-hand digits.

Show heading levels by indenting.

Vary the type style to indicate heading levels.

Use the exact wording that appears in the body of the document.

Use guide dots.

Contents

List of Illustrations

If you use graphics in your document, list the numbers and titles of those graphics in a list of illustrations (see Figure 18.13, page 583). The list of illustrations appears on a separate page after the table of contents. The list of illustrations is sometimes titled "List of Tables and Figures." To prepare a list of illustrations, follow the tips on page 584.

FIGURE 18.12

Decimal-Style Table of Contents

"1" indicates the chapter or section—in this case, Chapter 1.

"3" indicates the third subsection of Chapter 1.

This number indicates the first subsection of Subsection 2 in Chapter 2.

Contents

Abstract .... **ii**
List of Illustrations .... **iv**
Introduction **1**
1.0 Black-Bellied Plover and Its Habitat **2**
1.1 Description of Black-Bellied Plover 2
1.2 Habitat of Black-Bellied Plover 5
1.3 Migration of Black-Bellied Plover 7
1.4 Status of Black-Bellied Plover 11
2.0 Methods for Identifying Black-Bellied Plover Habitat **13**
2.1 Historical Sightings of Black-Bellied Plover 14
2.2 Satellite Imagery of the Habitat 20
2.2.1 *Rectifying the Images* *23*
2.2.2 *Classifying the Landuse* *27*
2.3 Remote Sensing of the Habitat 28
3.0 Results from the Sightings Information, Satellite Imagery, and Remote Sensing **29**
3.1 List of Historical Sightings of Black-Bellied Plover in Louisiana 30
3.2 Landuse Classes in the Satellite Imagery 32
3.3 Landuse Class of Black-Bellied Plover Habitat 35
4.0 Recommendation for Protecting Black-Bellied Plover in Louisiana **37**
4.1 Landuse in the Black-Bellied Plover Habitat 38
4.2 Habitat Areas in Louisiana 39
Works Cited **41**
Appendixes **42**
A. Satellite Imagery of Similar Habitats along the Gulf Coast 43
B. Remote Sensing of Similar Habitats along the Gulf Coast 49

Abstract and Executive Summary

The abstract and executive summary give an overview of the facts, results, conclusions, and recommendations of a document. Without having to read the entire document, readers can use the information in these sections

TIPS FOR PREPARING A TABLE OF CONTENTS

- **Use the exact wording that appears in the headings and subheadings in the body of the document**. If the heading in the body of the text is "Habitat of Black-Bellied Plover," the wording in the table of contents should be the same, not a shortened version such as "Habitat."
- **Show the heading levels by varying the style and indenting** (See Figure 18.11, page 581). By consistently varying the type style and indenting, you help readers locate first-level headings (or major sections). Then readers can locate related subheads.
- **List only the first three levels of headings if your document has more than three levels**. If you include four or more levels, the table of contents will be hard to read.
- **Use guide dots (. . . .)—sometimes called dot leaders—to connect the headings and the page numbers** (see Figure 18.11, page 581). Some organizations use other graphic elements to help readers easily find the related page number of a heading in the table of contents. For example, some organizations will place lines above and below each table of contents entry, so readers can easily see the entry and corresponding page number.
- **Include the list of illustrations, abstract or executive summary, and any other front matter**. Do not list the cover, title page, and letter of transmittal in the table of contents.
- **Use lowercase Roman numerals (i, ii, iii, etc.) for the page numbers for the list of illustrations, abstract, executive summary, and any other front matter**.

FIGURE 18.13

List of Illustrations

Illustrations

Guide dots

TIPS FOR PREPARING A LIST OF ILLUSTRATIONS

- **If you used separate numbering sequences for tables and figures, separate the tables from the figures in the list of illustrations**. Title the entire list "List of Illustrations" or "Illustrations."
- **If you used only one numbering sequence, title the list "Illustrations" or "List of Illustrations."** If your document contains only tables, you can use the title "List of Tables" or "Tables." If your document contains only figures, you can title the list "List of Figures" or "Figures."
- **Use the exact wording and number that appears with the illustration**. If the graphic number and title is "Figure 3.1. Migration Path of Black-Bellied Plover," the wording in the list of illustrations should be the same, not a shortened version, such as "Figure 3.1. Migration Path" or "Migration Path."
- **Use guide dots—sometimes called dot leaders—to connect the illustrations and the page number** (See Figure 18.13, page 583).

to decide whether they need to read the document. The abstract and executive summary appear before the body of the document. In this section, you will learn how to write an informative abstract, a descriptive abstract, and an executive summary.

Informative abstracts are primarily for readers knowledgeable about the topic of the document. These abstracts must stand independently from the document. When writing an informative abstract, follow these guidelines:
- **Identify the document**. Because an informative abstract must stand independently from the document, just writing "Abstract" above the text of the abstract will not give potential readers enough information. Instead, include the document title, your name, and perhaps the name of your organization.
- **State the topic and purpose of the document**. Don't assume that readers know what the document addresses. Instead, clearly state the topic and the purpose. With this information, readers can decide whether your document contains information that they need or want to read.
- **Conclude with the key results, conclusions, or recommendations**. Because an informative abstract must be able to stand alone, include the key results, conclusions, or recommendations of your research or document. Exclude all

examples and details. Include your methods only if they are new, unique, or vital to understanding your results, conclusions, or recommendations.

Generally, abstracts are one paragraph. Figure 18.14 presents an informative abstract of the student report on the habitat of the Black-Bellied Plover. The abstract identifies the report's title and writer, and states the objective of the report in the first two sentences. It concludes with the key results of the writer's research and a recommendation. You can also find examples of informative abstracts in some professional journals. Many journals limit abstracts to 200 words.

Descriptive abstracts are not a substitute for the document itself. Instead of reporting key results, conclusions, and recommendations, a descriptive abstract includes the major topics of the document. Its purpose is to help readers decide whether they want to read the document. Figure 18.15, page 586, presents a descriptive abstract for the student report on the habitat of the Black-Bellied Plover.

Executive summaries present the conclusions and recommendations of a document. An executive summary provides information on which readers need to act or to make a decision. Many readers, such as decision-makers, will only read the executive summary while other readers may bypass the executive

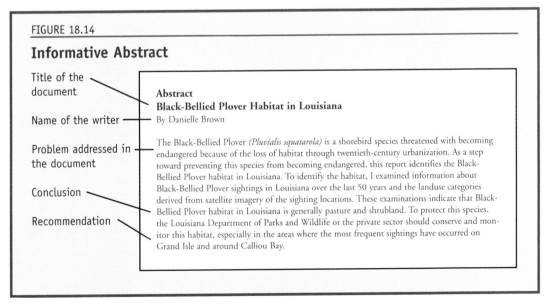

FIGURE 18.14

Informative Abstract

Title of the document

Name of the writer

Problem addressed in the document

Conclusion

Recommendation

Abstract
Black-Bellied Plover Habitat in Louisiana
By Danielle Brown

The Black-Bellied Plover *(Pluvialis squatarola)* is a shorebird species threatened with becoming endangered because of the loss of habitat through twentieth-century urbanization. As a step toward preventing this species from becoming endangered, this report identifies the Black-Bellied Plover habitat in Louisiana. To identify the habitat, I examined information about Black-Bellied Plover sightings in Louisiana over the last 50 years and the landuse categories derived from satellite imagery of the sighting locations. These examinations indicate that Black-Bellied Plover habitat in Louisiana is generally pasture and shrubland. To protect this species, the Louisiana Department of Parks and Wildlife or the private sector should conserve and monitor this habitat, especially in the areas where the most frequent sightings have occurred on Grand Isle and around Calliou Bay.

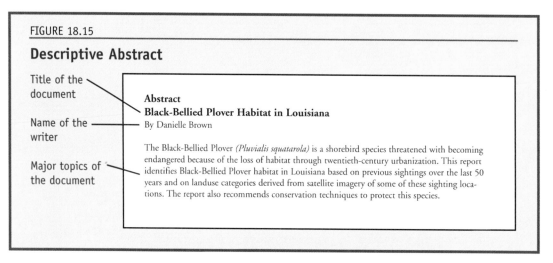

FIGURE 18.15

Descriptive Abstract

Title of the document

Name of the writer

Major topics of the document

Abstract
Black-Bellied Plover Habitat in Louisiana
By Danielle Brown

The Black-Bellied Plover (*Pluvialis squatarola*) is a shorebird species threatened with becoming endangered because of the loss of habitat through twentieth-century urbanization. This report identifies Black-Bellied Plover habitat in Louisiana based on previous sightings over the last 50 years and on landuse categories derived from satellite imagery of some of these sighting locations. The report also recommends conservation techniques to protect this species.

Evaluate an executive summary from a formal report in the Interactive Student Analysis online at www.kendallhunt.com/technicalcommunication.

summary and read the body of your document. The readers of an executive summary may not be the primary readers of the document. For example, engineers might write a report on pipe stress problems at a power plant. The primary readers of the report are other engineers and the plant operations manager—all experts in engineering and power plants. In the report, the writers discuss the methods they used to examine the problem, the specifications of the testing of the pipes, and the detailed results of those tests. They use technical language, knowing that the primary readers will understand it. However, many of the people who will decide whether to fund the proposed solution are not engineers. For these readers, the engineers write an executive summary using non-technical language and include only the information that these readers will likely need to knowledgeably judge the proposed solution.

Readers of executive summaries expect the following information:
- **A general overview of the topic and purpose of the document**. In this overview, succinctly state the topic of your document such as the problem addressed, the procedure or situation analyzed, and so on. For some situations and readers, you may need to provide background information to help readers understand the topic.
- **A concise statement of the key results, conclusions, and recommendations without excessive detail**.

Put yourself in the readers' shoes and try to anticipate the questions that they may ask as they read the executive summary:

- **What is the document about?** Specifically identify the topic and purpose of your document. Readers will be especially interested in how the information in the document directly affects them, their employees, their department, or their organization. They also will be interested in any costs related to conclusions and recommendations and how those costs will affect them.
- **How will the results, conclusions, and recommendations affect the department, employees, organization, and others?** Readers will be especially interested in costs and savings. Readers of executive summaries are less interested in details and evidence that supports the findings and more interested in information that will help them to make a decision or implement recommendations.
- **What are the key results, conclusions, or recommendations?** If you know that readers will understand the topic of your document, include your key results. Such readers may expect you to summarize the significant data concerning the results. These readers will use this data to make decisions, not to conduct further studies. Some readers, in contrast, may not be experts in your field or may not want to read about your results. For these readers, leave out the results and present only conclusions and recommendations. Such readers are interested in your recommendations based on your analysis or research—even if your recommendation is simply to study a situation or problem further or to "wait and see." The executive summary presented in Figure 18.16, page 588, focuses effectively on the conclusions and recommendations.

The Back Matter

Back matter appears after the body of a document. Back matter in technical documents generally includes the following:

- Works Cited or List of References
- Glossary
- List of Abbreviations or Symbols
- Appendices
- Index

Works Cited or List of References

If you cite the works of others in your document, include a works cited list or a list of references after the body of your document. You might use MLA, APA, CSE, or a company-approved style of documentation.

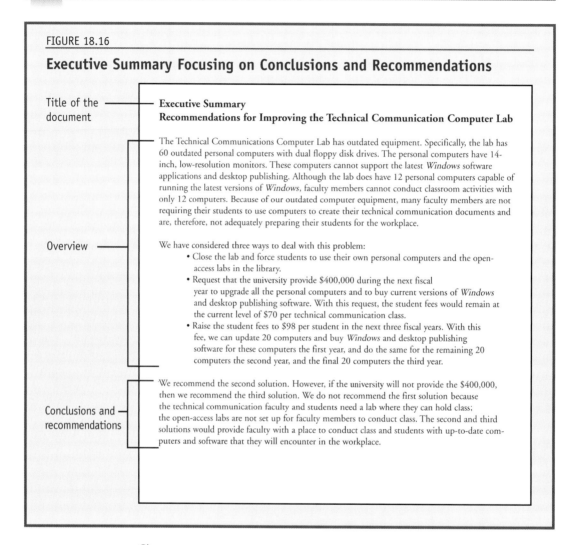

FIGURE 18.16

Executive Summary Focusing on Conclusions and Recommendations

Title of the
document

Executive Summary
Recommendations for Improving the Technical Communication Computer Lab

The Technical Communications Computer Lab has outdated equipment. Specifically, the lab has 60 outdated personal computers with dual floppy disk drives. The personal computers have 14-inch, low-resolution monitors. These computers cannot support the latest *Windows* software applications and desktop publishing. Although the lab does have 12 personal computers capable of running the latest versions of *Windows*, faculty members cannot conduct classroom activities with only 12 computers. Because of our outdated computer equipment, many faculty members are not requiring their students to use computers to create their technical communication documents and are, therefore, not adequately preparing their students for the workplace.

Overview

We have considered three ways to deal with this problem:
- Close the lab and force students to use their own personal computers and the open-access labs in the library.
- Request that the university provide $400,000 during the next fiscal year to upgrade all the personal computers and to buy current versions of *Windows* and desktop publishing software. With this request, the student fees would remain at the current level of $70 per technical communication class.
- Raise the student fees to $98 per student in the next three fiscal years. With this fee, we can update 20 computers and buy *Windows* and desktop publishing software for these computers the first year, and do the same for the remaining 20 computers the second year, and the final 20 computers the third year.

Conclusions and
recommendations

We recommend the second solution. However, if the university will not provide the $400,000, then we recommend the third solution. We do not recommend the first solution because the technical communication faculty and students need a lab where they can hold class; the open-access labs are not set up for faculty members to conduct class. The second and third solutions would provide faculty with a place to conduct class and students with up-to-date computers and software that they will encounter in the workplace.

Glossary

A ***glossary*** is an alphabetical list of specialized words and their definitions. A glossary provides definitions of terms unfamiliar to some readers. Effective glossaries allow you to meet the needs of readers with different levels of expertise. While glossaries generally appear after the body, they can appear immediately following the table of contents or list of illustrations. Figure 18.17, page 590, shows part of a glossary from a document on software used for forecasting. Follow the tips for writing an effective glossary on the next page.

TIPS FOR WRITING THE EXECUTIVE SUMMARY

- **Include the title of your document**. Some writers in the workplace also include the words "Executive Summary" with the title. You might consider identifying the summary in this way if your executive summary will circulate separately from the document.
- **Identify the topic and purpose of the document**. Include how the topic and purpose directly affects the readers, their employees, their department, or their organization.
- **Focus on the conclusions and recommendations, not on the details of your findings**. Avoid including details and evidence that support your conclusions. Include only information that helps readers make a decision or implement your recommendations.
- **Use non-technical language**. Many readers of executive summaries may not be experts in your field. Even if the document is for technical readers, write the summary for decision-makers who may not know the technical language of your field.
- **Give readers only the information they need to determine whether they should read your document or to make a decision**. Eliminate, for example, any information about your methods or about the theories behind your project.

TIPS FOR CREATING A GLOSSARY

- **In the body of the document, identify all words that appear in the glossary**. Put these words in italic or boldface type, or place an asterisk next to each one. Use the same system of marking throughout the document, and be sure to explain to readers what you are doing—perhaps in the introduction or in a footnote accompanying the first glossary word appearing in the text. This footnote will explain that terms defined in the glossary appear in the body of the text in, for example, boldface italic type: *"This and all other terms appearing in boldface italic type are defined in the Glossary, which begins on page 77."*
- **Define all terms that readers may not understand**. In the definition, use words that readers will understand and include cross-references to other closely related terms defined in the glossary (see Figure 18.17, page 590).
- **List the words in alphabetical order**. Alphabetical order helps readers locate words.
- **Use phrases or clauses, not sentences, for the primary definitions**. After the primary definition, you can use sentences for secondary definitions. If you use words defined in the glossary, cross-reference those words. For example, look at the definition of *continuous probability distribution* in Figure 18.17. In this definition, the writer uses *probability distribution*, which is also in the glossary; so the writer tells the reader to "See probability distribution."
- **Include the glossary and its first page number in the table of contents**.

FIGURE 18.17

Partial Glossary

Glossary

Assumption
An estimated value or input to a spreadsheet model.

CDF
Cumulative distribution function that represents the probability that a variable will fall at or below a given value.

Certainty level
The percentage of values in the certainty range compared to the number of values in the entire range.

Continuous probability distribution
A probability distribution that describes a set of uninterrupted values over a range. *(See probability distribution)*

Discrete probability distribution
A probability distribution that describes distinct values with no intermediate values. *(See probability distribution)*

List of Abbreviations or Symbols

If your document contains many abbreviations or symbols, include a list explaining what they mean (see Figure 18.18). This list usually appears at the end of a document before the appendix although some writers put it in

FIGURE 18.18

Partial List of Abbreviations

Abbreviations

ESS	Electronic Still Store. The graphics box over the anchor's shoulder.
NC	News Conference.
PREPS	Preparations. Used in story descriptions.
REAX	Reactions. Used in story descriptions.
SOT	Sound On Tape. Used in a script to tell the editor where to place the sound on tape.
VO	Video Only.

TAKING IT INTO THE WORKPLACE
Reports in the Information Age

In the 1990s, the personal computer, technology, and global communication revolutionized how companies conducted daily business and how they distributed information to their employees and to those outside the company. As the use of personal computers and the Web grew, so did the "user's expectation for information availability on or through that machine" and the Web (Foy).

Readers are increasingly demanding information and services that put information and services directly into their hands via the computer and the cell phone. For example, professionals are having to decide how to structure e-mail messages and other documents so that readers can read and navigate them on a hand-held device such as a Blackberry®. While this concept of providing universal access to information seems alluring to many readers and organizations, the real issue may be more work and mediocrity (Foy). For many organizations, the issue is twofold:

- deciding how to best create, share, protect, and distribute information for this universal access
- making the right information easy to access

Organizations want to "put their best foot forward." They want to put out information that adds value to their organization. They want to avoid a "data dump"—making all information available regardless of its value or its appropriateness for certain media such as a hand-held device. For example, can a hand-held device be an appropriate media for accessing a formal report? As organizations consider making information available in the Information Age, new formats for traditional reports will evolve. You may even help to create and shape these new formats.

Assignment

1. Using the Web, locate two articles that discuss how technology is changing the way organizations make information available both internally and externally. These articles should relate to information traditionally found in reports, not routine correspondence such as e-mail.
2. Write a brief summary of the articles and e-mail it to your instructor. Include the URL for the articles.

TIPS FOR CREATING A LIST OF ABBREVIATIONS OR SYMBOLS

- **List the abbreviations in alphabetical order**. Alphabetical order helps readers locate abbreviations.
- **Use phrases or clauses for explaining what the abbreviations or symbols mean**. You can simply write the words that an abbreviation or symbol represents; you can also include other information that readers may need to understand the abbreviation or symbol.

the front matter after the table of contents and list of illustrations. Follow the tips above as you prepare a list of abbreviations or symbols.

Appendices

Appendices appear after the works cited, glossary, or list of symbols. Material in appendices supports information in the body of the document. Appendices contain information that

- is not essential for readers to understand the document
- may interest only a few readers
- would interrupt the flow of the document

An appendix might include maps, large diagrams or other graphics, a sample questionnaire, a transcript of an interview, or other supporting documents. Any of these items should supplement the body of the document. To prepare an effective appendix, follow the tips on the next page.

Index

An *index* lists terms used in a document and the page number where the terms appear. An index appears after all other items in a document. The terms in an appendix appear in alphabetical order.

Number the Front and Back Matter Correctly

As you prepare the front and back matter, follow these standards for numbering the pages:

- Use lowercase Roman numerals (i, ii, iii, etc.) for the front matter.

TIPS FOR PREPARING AN APPENDIX

- **At the appropriate place in the body of the document, refer to each appendix**. For instance, if you summarize survey results in the body of the document and include the survey instrument and tabulated results in an appendix, you might write: "Sixty-five percent of the respondents reported that they used e-mail more frequently than voice mail (see Appendix B for the tabulated results)."
- **Put each major item into a separate appendix**. Identify each appendix with a letter or a title: Appendix A, Appendix B, and so on.
- **List each appendix in the table of contents**. If you've titled an appendix, include the complete title in the table of contents.
- **Put essential items in an appendix only when these items are so long or so large that putting them in the body of the document would severely interrupt the flow of information**. Otherwise, include only nonessential information in the appendix.

- Use Arabic numerals (1, 2, 3, etc.) for the body and the back matter.
- Put odd numbers on right-hand pages and even numbers on left-hand pages.
- Begin the body of the document on a right-hand page, even if doing so means that the facing left-hand page will be blank.
- Do not put a page number on the title page or any blank pages even though you will include them in your page count.

Figure 18.19, page 594, shows the page numbering of a formal document. In this document, the writer
- does not put a page number on the title page (even though the title page is actually page i)
- uses lowercase Roman numerals for all front matter
- begins the body on a right-hand page

Follow these guidelines for placing the page numbers:
- You can put page numbers in two effective places on a page: the top or bottom outside corners (see Figure 18.20, page 595). Page numbers on the outside corners are easier for readers to see than page numbers on the inside corners or centered on the top or bottom page margins.
- On left-hand pages, place the page number in the top or bottom left-hand corner.

FIGURE 18.19

Sample Page Numbering of a Formal Document

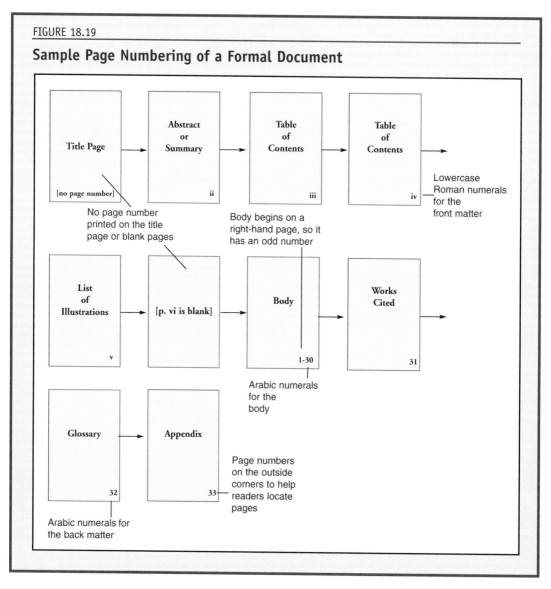

- On right-hand pages, place the page number in the top or bottom right-hand corner.

You can also include a header or footer with the page numbers. (For information on headers and footers, see Chapter 10.)

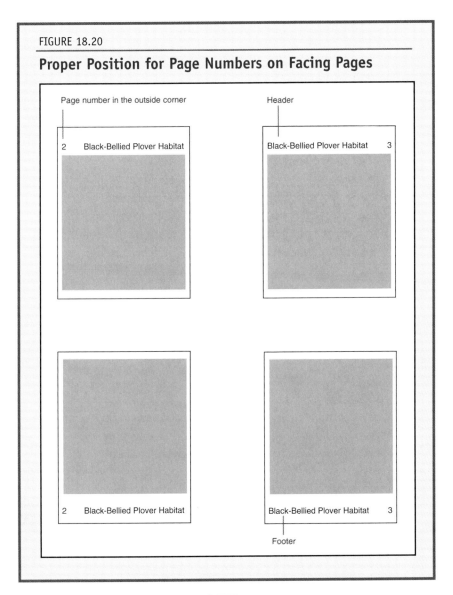

FIGURE 18.20

Proper Position for Page Numbers on Facing Pages

Page number in the outside corner

Header

2   Black-Bellied Plover Habitat

Black-Bellied Plover Habitat   3

2   Black-Bellied Plover Habitat

Black-Bellied Plover Habitat   3

Footer

FOCUS ON FEASIBILITY REPORTS

Feasibility reports are a type of recommendation report. *Feasibility reports* document a study of two or more courses of actions or options. These options are evaluated based on appropriate criteria. These reports answer questions such as, "Which method is best for repairing the tower?", "Should we buy a new computer or update the current one?", "Which product should we purchase?", or "Which product is most appropriate for our customers and our location?". To write a feasibility report, follow the basic plan presented in Figure 18.21.

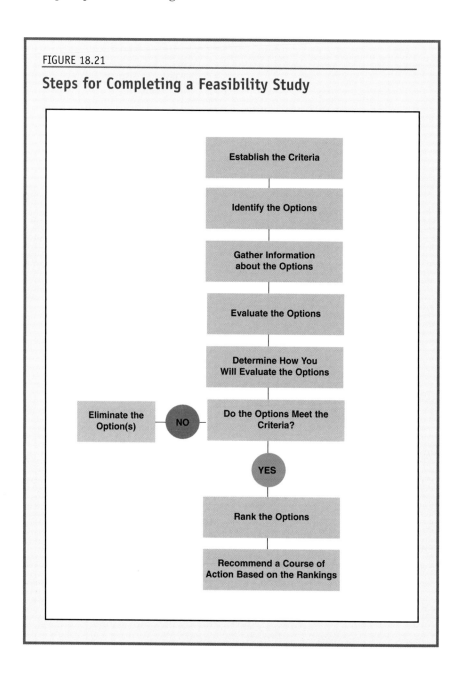

FIGURE 18.21

Steps for Completing a Feasibility Study

Establishing Criteria for Evaluating the Options

To ensure that you make sound recommendations, you must establish criteria. *Criteria* are requirements or benchmarks that you use to evaluate an option. Criteria may be quantitative or qualitative. For example, if you are deciding which laptop to buy for your office, you will have many choices as well as perhaps company guidelines that affect what you can purchase. You will probably have minimum quantitative criteria (sometimes called specifications) for cost, size, speed, and memory; criteria that you can quantify. Your quantitative criteria might include the following specifications:

- Cost: lowest cost not to exceed $3,500
- Processor speed: 2 GHz
- Display size: 15 inches
- Weight: not to exceed 16 pounds
- Hard drive size: 200 GB

Qualitative criteria might be the reputation of the company that produces the laptop or the comfort of the keyboard. You cannot quantify a reputation or comfort.

When you set up a feasibility study, you must specifically define the minimum criteria, so you can use those criteria to draw conclusions and recommend a course of action. If you are considering using a qualitative min-

TIPS FOR ESTABLISHING CRITERIA

- **Define the minimum criteria (requirements).** You must specifically define the minimum criteria, so you will know when an option has met or exceeded that criterion.
- **Determine how you will know when an option has met or exceeded a minimum criterion.** If you cannot define when an option has met or exceeded a minimum criterion, then you should revise or eliminate that criterion.
- **If a minimum criterion is cost, determine how you will factor in other criteria.** Can you include the cost of features in the total cost of the option? If options have about the same cost, how will you decide which option to recommend?

imum criterion, how will you know if an option meets or exceeds this criterion? If you cannot define how an option will meet a criterion, you should revise or eliminate that criterion.

In setting up any feasibility study that involves cost, consider how you will factor in the other criteria. Cost is generally the overriding criterion. If the lowest cost option meets the minimum criteria, you must select that option unless you have established other overriding criteria. You should consider this situation as you set up your study. For example, you may decide (or your organization may decide) that you can consider criteria other than cost if the cost of the options are within 5 percent of each other. Or you may decide to factor in features that affect cost. Let's consider laptops: Option 1 is the least expensive laptop; however, you will have to pay for shipping and a three-year maintenance package if you recommend purchasing this option. Option 2 is the most expensive option, but the shipping and the three-year maintenance package are free—making Option 2 the same cost as Option 1. You can then select the higher-priced option because you are factoring in other features that impact cost.

If someone has asked you to conduct a feasibility study, the criteria may be established for you. For example, if you are buying the laptop to use at your office, your manager may give you minimum criteria, such as a budget and a list of company-approved vendors. If the criteria aren't established for you, establish the criteria by researching your topic; you might interview subject matter experts, administer end-user questionnaires, and read journals and other technical documents related to your topic. For example, to determine the criteria for purchasing the laptop, you might interview people who work in information technology. Review the tips on page 597 for establishing criteria.

Identifying the Options

Once you have established the criteria, you can determine the options available. *Options* are possible solutions or courses of action. You should look for all options that meet your minimum requirements. If you find that you have too many options, consider narrowing the criteria. As you identify the options, follow the tips on the next page.

TIPS FOR IDENTIFYING OPTIONS

- **Make sure that you identify all available options**. If you miss an option or simply leave it out, you may eliminate the option that provides the most feasible course of action. Your readers will assume that you carefully identified all the options available that meet the minimum criteria. Therefore, you have an ethical responsibility to identify every option.
- **Research thoroughly**. Use a variety of primary and secondary research techniques to ensure that you identify all available options. (See Chapter 5 for information on primary and secondary research techniques.)
- **Avoid the temptation to simplify the study by eliminating possible options**. You may feel tempted to eliminate options because you are on a tight schedule, you have a heavy workload, or you are unfamiliar with the subject. Remember, you may be eliminating the option that best meets the established criteria. For instance, you may be tempted to purchase a laptop when leasing one may actually cost the company less money in the long term because leasing keeps the hardware more current. You are ethically obligated to look at all the options that meet your criteria before you recommend the option that best meets (and possibly exceeds) your criteria.

Evaluating Options

When you have established the criteria and identified the options, you're ready to evaluate the options. You will need to gather information to determine how each option measures up to the criteria. You must make sure that you gather the same types of information for each option, so that you can evaluate the options equally. Let's again consider the laptops: you might do secondary research by looking at consumer magazines that compare laptops or by talking to end-users of each laptop. You might do primary research by going to stores and using the laptops or contacting vendors to see if they have additional features or packages that may exceed your minimum criteria—for example, features that may be incentives available only to corporate buyers.

Once you've gathered the information to evaluate your options, you will rank the criteria. How will you decide which option most closely meets or exceeds the criteria? You will need a method that eliminates as much subjectivity as possible. Some writers use a ranking system. A ranking system will not ensure objectivity, but it will ensure that you are looking at the same criteria for each option. For example, to evaluate the laptops, you might use a ranking system like that in Figure 18.22. You assign a value from 0 to 5, with 5 being the highest and 1 being the lowest.

Drawing Conclusions and Making Sound Recommendations

After you have evaluated your options based on criteria, you are ready to interpret the results of your evaluation. You may find that the conclusions, and thus your recommendations, are clear cut. However, often, the conclusions and the recommendation will not be clear cut.

Let's look at the laptops in Figure 18.22. With this ranking system, the Micro Express and the Apple Macbook Pro have the same total score. How do you decide which option to recommend? You could recommend both options. However, because you can only purchase one laptop, this approach does not work. You must recommend only one laptop. You could use cost as the overriding factor; but the difference in cost is insignificant. The Micro Express costs $2,799 and the Apple Macbook costs $2,899. When you established your criterion, you determined that if the cost of the options was within $150, you could look at the other criteria to make a recommendation. Figure 18.23 compares these two options based on criteria other than cost. In this compar-

FIGURE 18.22

A Ranking System for Comparing Criteria

Option	Cost	Processor Speed	Display Size (larger is preferred)	Weight (lighter is preferred)	Hard Drive Size	Total
Micro Express IFL9025	5	4	3	4	4	20
Lenovo ThinkPad T61p	2	4	3	5	1	15
HP Pavilion HDX	1	3	5	1	5	15
Dell Inspiron 1720	3	3	4	2	4	16
Apple Macbook Pro	4	5	4	4	3	20

FIGURE 18.23

Using Criteria Other than Cost to Evaluate Two Closely Ranked Options

Option	Processor Speed	Display Size (larger is preferred)	Weight (lighter is preferred)	Hard Drive Size	Total
Micro Express IFL9025	4	3	4	4	15
Apple Macbook Pro	5	4	4	3	16

ison, the Apple receives a higher total score. You can then recommend this option even though it has a slightly higher cost. In the workplace, cost is the overriding factor when evaluating options. If the difference in cost is not significant, you may be able to justify purchasing the higher-cost option using other criteria—especially if that option exceeds the minimum criteria. In this situation, you may also be able to use qualitative criteria that cannot be included in the ranking system. However, qualitative criteria generally cannot be the overriding criterion when cost is being considered.

You can present your recommendations in any of the following ways:
- **Recommend one option**. You might recommend that the company purchase the Apple MacBook Pro.
- **Rank order the options**. You might rank the laptops as follows: the Micro Express is the best option, the Apple Macbook Pro is the second-best option, the Dell Inspiron is the third-best option, and the Lenovo and HP Pavilion tie for the fourth-best option.
- **Categorize the options by their rankings based on certain criteria**. You might recommend the Apple MacBook as the best value or the Micro Express as the lowest cost.

Organizing the Results Section of a Feasibility Report

In feasibility reports, you can most effectively present the results by using a comparison/contrast organization. Using this basic organization, you have two patterns from which to choose:
- Comparing by criteria
- Comparing by options

Compare by criteria (see Figure 18.24) if you are evaluating only a few options. If you are evaluating more than three options, compare by options (see Figure 18.25). When you compare by criteria, you focus the comparison on the criteria, and the options are subsections of each criterion. When you organize by options, you focus on the options, and the criteria are the subsections. When you use either organization, you can help readers understand your results by creating a chart or table that compares each option criterion by criterion.

TWO SAMPLE REPORTS

Figure 18.26, pages 604-612, and Figure 18.27, pages 613-618, present reports that met the readers' needs and expectations as well as the writers' purpose. Figure 18.26 is an internal feasibility report prepared for an executive director at a telecommunications company. The writer analyzes three options for writing boilerplate text for Select Business Account proposals. *Boilerplate* refers to any text that will be repurposed with little or no change. Figure 18.27 includes the body of a recommendation report prepared for the U.S. Department of Energy by scientists at the Oak Ridge National Laboratory. The report details a study of fish and benthic macroinvertebrates in streams near a landfill.

FIGURE 18.24

Comparing by Criteria

Cost of the Laptop (*criterion*)
- Micro Express (*options*)
- Lenovo ThinkPad
- HP Pavilion
- Dell Inspiron
- Apple Macbook Pro

Processor Speed
- Micro Express
- Lenovo ThinkPad
- HP Pavilion
- Dell Inspiron
- Apple Macbook Pro

Display Size
- Micro Express
- Lenovo ThinkPad
- HP Pavilion
- Dell Inspiron
- Apple Macbook Pro

Weight
- Micro Express
- Lenovo ThinkPad
- HP Pavilion
- Dell Inspiron
- Apple Macbook Pro

Hard Drive Size
- Micro Express
- Lenovo ThinkPad
- HP Pavilion
- Dell Inspiron
- Apple Macbook Pro

FIGURE 18.25

Comparing by Options

Micro Express (*option*)
- Cost of the laptop (*criterion*)
- Processor speed
- Display size
- Weight
- Hard drive size

Lenovo ThinkPad
- Cost of the laptop
- Processor speed
- Display size
- Weight
- Hard drive size

HP Pavilion
- Cost of the laptop
- Processor speed
- Display size
- Weight
- Hard drive size

Dell Inspiron
- Cost of the laptop
- Processor speed
- Display size
- Weight
- Hard drive size

Apple Macbook Pro
- Cost of the laptop
- Processor speed
- Display size
- Weight
- Hard drive size

FIGURE 18.26

Feasibility Report with Front Matter

For internal for-
mat reports, you
may use the
memo form for
the "letter" of
transmittal.

Memo

November 15, 2008

To: Charlie Divine
                Executive Director

From: Neil Cobb
                Regional Manager

Subject: Feasibility report for hiring technical communication interns

I am please to submit the accompanying report "Hiring Technical Communication Interns: A
Feasibility Report" in response to your request. The report examines options for writing the
boilerplate text for our automated proposals.

For the report, I examined three options for writing the boilerplate: using the current proposal
center staff, hiring professional technical communicators, and hiring graduate students as
interns. I used the criteria of cost of producing the boilerplate, effect on the productivity of the
proposal center, the effect on the workplace, and the quality of the finished product. Based on
my examination, I recommend hiring two technical communication interns in January 2009 to
create boilerplate for the automated proposals.

If you have any questions, please call me at (214) 555-5555.

Source: Courtesy of Neil Cobb

FIGURE 18.26 CONTINUED

Hiring Technical Communication Interns: A Feasibility Report

Prepared for
Charlie Divine, Division Manager, ISM

Prepared by
Neil Cobb, Account Manager, ISM

November 2008

FIGURE 18.26 CONTINUED

Contents ii

Contents

FIGURE 18.26 CONTINUED

iii

Executive Summary

The writer
summarizes the
report and
states his
recommendation.

I have investigated the feasibility of using current staff, hiring freelance technical
communicators, and hiring student interns to generate low-cost boilerplate texts for
automated proposals. Because interns will work for low hourly wage with no benefits in exchange for experience, we can produce high-quality, low-cost boilerplate
text. By using this boilerplate text, the productivity of Proposal Specialists who serve
the Select Business Accounts (SBA) will ultimately increase.

Universities in the Dallas, Houston, and St. Louis areas have programs that can provide these interns. Ninety percent of the students in these programs are looking for
internship opportunities to complete their degree requirements. These programs will
provide a continuous resource for capable writers to serve as interns.

I recommend hiring two technical communication interns in January 2009 to create
boilerplate for SBA automated proposals.

FIGURE 18.26 CONTINUED

Introduction

Integrated Systems Marketing's proposal centers in Dallas, St. Louis, and Houston will publish over 700 proposals in 2008 while keeping expenses at 2007 levels. In 2008, we will challenge our Account Managers and Proposal Specialists to double 2007 production without increasing costs. To meet this challenge and improve productivity, we must change our procedures. We must implement these changes early in 2008 to increase our productivity.

The writer explains the background and the purpose of the report.

To begin meeting this challenge, we have begun automating the proposal process for Select Business Accounts. This preliminary work—writing client questionnaires, building graphics libraries, and developing price templates—is taking valuable resources from producing proposals. The most labor-intensive element of this work is still pending—writing the boilerplate for all Southwestern Bell products. We estimate that writing the boilerplate for each product will take 80 hours because the text will identify several introductory scenarios for various industries and all potential problems and benefits that the product will address.

To complete these boilerplate texts while maintaining our current level of production and quality that our clients and their customers now expect, we have three options:

The writer highlights the options in a list.

Option 1:   **Produce the boilerplate as an overlap with current staff.** We assume that we can assign one of our three Dallas Proposal Specialists to write the boilerplate in lieu of producing proposals.

Option 2:   **Hire professional technical writers to produce the boilerplate.** Freelance technical communicators who work on a contractual basis are readily available in the Dallas, Houston, and St. Louis areas.

Option 3:   **Hire graduate students as interns to produce the boilerplate.** Students seeking master's degrees are available from area universities.

The writer tells the reader what follows in the report.

In the report that follows, I examine these options and recommend the most cost-effective option that will allow us to increase our productivity without sacrificing quality and client satisfaction.

Methods for Evaluating the Options

To support my recommendations, I evaluate the feasibility of each option using four criteria.

The writer lists the criteria and then explains the methods.

- The cost of producing the boilerplate text
- The effect on the productivity of the proposal center
- The effect on the workplace
- The quality of the finished product

FIGURE 18.26 CONTINUED

To gather information for the evaluation, I studied the proposal development process currently used in the proposal center for Select Business Account clients and the proposals for these and major account clients to determine the quality of the proposals, especially the accuracy of the information and the quality of the writing. I interviewed proposal specialists in the proposal center, and an academic specialist in technical communication. The two proposal specialists provided information on the way work flows through our system in the proposal center. The five account representatives gave me their perspective on the proposal center's process and on the quality of the proposals. Finally, the academic specialist provided information on the availability, cost, and abilities of students from technical communication programs and of professional technical communicators who work on a freelance basis.

Results of the Evaluation

To determine the feasibility of hiring interns, I evaluated the three options according to the criteria of cost, productivity, quality of the finished proposal, and effect on the workplace.

Cost of Producing the Boilerplate Text

Option 1, using the current staff, is least expensive. Option 2, hiring freelance technical communicators, is the most expensive.

Option 1: Overlap with Current Staff
Proposal Specialists (SG-22) earn on the average $48,700 per year. Loaded for relief and pension benefits, their total compensation is equivalent to $72,000 per year, or $34.60 per hour. With Option 1, the initial boilerplate for a single product would cost approximately $2,800.[1] Since loaded salaries for the Proposal Specialist are an embedded cost, Option 1 would not affect expenses.

Option 2: Hire Technical Communicators on a Contractual Basis
Professional technical communicators charge between $35 and $50 per hour, depending on their experience. With this option, the initial boilerplate would cost between $2,800 and $4,000 and would raise our departmental expenses accordingly.

Option 3: Hire Student Interns
Graduate interns will work for $10-$15 per hour in exchange for on-the-job experience. These interns will be classified as part-time employees, so the company will not pay them benefits. With Option 3, the initial boilerplate text for a single product would cost between $800 and $1,200 and would raise our expenses accordingly.

Effect on the Productivity of the Proposal Center

Option 2, hiring freelance technical communicators, would be most productive. Option 1, using current staff would be the least productive.

Option 1: Overlap with Current Staff
When Account Representatives fill out forms and pricing tables, Proposal Specialists can produce the proposal in eight hours. If we take Proposal Specialists from their

[1] I am using an 80-hour estimate for development time for determining the cost of producing the boilerplate text.

The writer organizes the results by criteria.

The writer uses specific figures to clearly explain the cost of each option. He tells the reader how he arrived at the figures.

FIGURE 18.26 CONTINUED

The writer uses the same order for the options as in the introduction. The writer also uses the same order for the criteria throughout the results and clearly identifies each criterion and option in a heading.

regular proposal-writing tasks to write boilerplate text (80 hours of work), we will produce 10 fewer proposals every two weeks until the boilerplate is complete. This option will reduce potential revenues. With this option, the proposal center will not meet the immediate needs of the Account Representatives and thus discourage them from using the center.

This option does have an advantage. Because the proposal specialists are familiar with our products and the proposal-writing process, they may be able to create the boilerplate for a product in less than the estimated 80 hours.

Option 2: Hire Technical Communicators on a Contractual Basis
Professional technical communicators should be able to produce boilerplate for a single product with the 80-hour benchmark. Some of these writers may require less time, depending on the writer's expertise in telecommunications and how long they work for us. Option 2 may be the most productive scenario overall because it will free Proposal Specialists to continue writing proposals while spending minimal time working with the hired writers to develop and edit the boilerplate.

Option 3: Hire Student Interns
Like the professional technical communicators, the interns will lack experience with our products and proposal style. Because they are relatively less-experienced writers, we expect their productivity to be less than that of the professional communicators. However, student interns are professional writers in training. They will have considerable academic and practical experience from at least 20 hours of technical communication courses. Because much of the course work is "real-world" oriented, the difference in productivity may not be significant. Like Option 2, Option 3 will free Proposal Specialists to continue producing proposals although the specialists may spend more time guiding the process.

Effect on the Quality of the Finished Product
Options 2 (hiring freelance technical communicators) and 3 (hiring student interns) will provide the highest quality in the finished product.

Option 1: Overlap with Current Staff
Account representatives praise the effectiveness of SBA proposals, especially the "looks" of the documents. However, the quality of the text is presently inferior to that of proposals for major accounts. The content is technically correct—the Proposal Specialists know the products well—but they often write in passive voice and have difficulty writing clear prose. Unfortunately, because the Proposal Specialists know the material so well, they tend to let some difficult concepts and technical descriptions flow unedited into the final document.

Option 2: Hire Technical Communicators on a Contractual Basis
Professional technical communicators should greatly improve the quality of the written text of SBA proposals. Their writing skills and a fresh perspective on our documents would ensure that the proposals meet the needs of the non-technical readers.

FIGURE 18.26 CONTINUED

Conclusions   4

Option 3: Hire Student Interns
Interns would improve the quality of the written text for the same reasons discussed in Option 2. We would see work of higher quality because the interns' graduate advisers would evaluate the text. Their text would have to pass two layers of edits: edits from the advisers and from our Proposal Specialists.

Effect on the Workplace
Option 3, hiring student interns, could positively affect the workplace. Option 1, using current staff, and Option 2, hiring freelance technical communicators, could negatively affect the workplace.

Option 1: Overlap with Current Staff
Option 1 will affect the workplace negatively for two reasons. First, taking one Proposal Specialist from the group to write boilerplate will require the others to do additional work or to turn one of three prospective clients away. Neither result would be acceptable long term, and turning away clients could cripple the automation project before it starts. Second, choosing one Proposal Specialist to write the boilerplate might bruise the egos of the others. The choice could irreparably divide the group.

Option 2: Hire Technical Communicators on a Contractual Basis
Option 2 could help to avoid the negative effects of Option 1, but could also cause problems. The Proposal Specialists might resent management hiring another writer to do "their" work, especially if the freelance writers receive higher wages.

Option 3: Hire Student Interns
Like Option 2, Option 3 should prevent the negative effects of Option 1. Interns could also help us to avoid the problems of Option 2 because the Proposal Specialists would more readily accept the interns as subordinates and take on the responsibility of supervising their work. The interns themselves also will more readily accept the role as subordinates.

Conclusions

Each of the three options has advantages and disadvantages. Option 1 (produce the work as an overlap with current staff) maintains current expense levels and builds on current expertise to reduce the time for writing boilerplate text. However, Option 1 fails to appreciably improve the quality of the text and could disrupt the work group. Option 2 (hire professional technical communicators to do the work) offers the highest quality text at the highest level of productivity, yet could cause the most friction in the workplace. It also carries the highest price tag on a "per-unit" basis. Option 3 (hire graduate students as interns to do the work) delivers improved quality and productivity at minimal cost with the most positive potential effect on the workplace. The quality of text written by interns and the speed at which they produce it will rival that of the professional writers if we screen the applicants properly.

The writer presents and justifies the conclusions.

FIGURE 18.26 CONTINUED

Recommendations   5

Based on our evaluation of the options in light of the criteria, productivity and cost are secondary issues. The savings that we realize from automating the proposals will recover the up-front costs of creating the boilerplate text. Therefore, the center is primarily concerned with the quality of the boilerplate and the effect of the option on the workplace.

The writer states and justifies his recommendations.

Recommendations

Because Option 3 offers the highest quality for the dollar with potentially no impact on the morale of the work groups, we recommend hiring two technical communication interns in January 2009 to write boilerplate text for Select Business Account (SBA) automated proposals.

FIGURE 18.27

Recommendation Report

Because the readers are biologists, the writers use technical terminology the readers will understand and expect.

The writers list the specific sites studied.

The writers use lists and sub-heads to identify the methods for sampling benthic macroinverte-brates and fishes.

Biotic Characterization of Small Streams in the Vicinity of Oil Retention Ponds 1 and 2 Near the Y-12 Plant

Introduction

This report provides data on the aquatic biota in the streams near the oil retention ponds west of the Y-12 plant at Oak Ridge National Laboratory. Built in 1943, the Y-12 plant originally produced nuclear weapon components and subassemblies and supported the Department of Energy's weapon-design laboratories. In the production of the subassemblies, the plant used materials such as enriched uranium, lithium hydride, and deteride. The plant disposed of both solid and liquid wastes in burial facilities in Bear Creek Valley—approximately one mile west of the plant site. This report fulfills the Department of Energy's commitment to assess the aquatic biota near the man-made oil retention pond in the valley.

Methods

The Environmental Sciences Division conducted quantitative sampling of benthic macroinvertebrates and fish at the following sites:
- Bear Creek above and below the confluence with Stream 1A, a small tributary that drains Oil Retention Pond 1 (Stations 1 and 2, respectively)
- Stream 1A just above the confluence with Bear Creek (Station 3)
- Stream 2, a small uncontaminated tributary of Bear Creek that flows adjacent to Bear Creek Road (Station 4, the control station)

We also conducted qualitative sampling at the following sites in the watersheds of Oil Retention Ponds 1 and 2:
- Stream 1A, immediately below Pond 1 (Station 5)
- The diversion ditch that carries surface runoff from portions of Burial Grounds B, C, D, and the area north of Burial Ground A to Stream 1A just below the pond (Station 6)
- Stream 1A above the diversion ditch (Station 7)
- Oil Retention Pond 2 (Station 8)

We could not sample above or below Pond 2 because of insufficient flows. We did not sample Oil Retention Pond 1.

Methods for Sampling Benthic Macroinvertebrates

To sample the benthic macroinvertebrates at Stations 1B-4B (three samples per site), we followed these procedures:
1. Placed a 27 x 33 cm metal frame on the bottom of the stream in the riffle area.
2. Held a 363-m mesh drift net at the downstream end of the metal frame while agitating the stream bottom (within the frame) with a metal rod. The stream flow transported suspended materials into the net.
3. Washed the net three times with the stream water to concentrate the sample and remove fine sediments.
4. Transferred the sample to glass jars that contained approximately 10% formalin to preserve the sample.

In the laboratory, we followed these procedures to analyze the samples:
1. Washed each sample using a standard No. 35 mesh (500 m) sieve and placed the washed sample in a large white tray.

FIGURE 18.27 CONTINUED

2

The writers use specific, detailed language so that readers can duplicate their methods. This language also helps to justify their conclusions and recommendations.

2. Examined large pieces of debris (e.g., leaves and twigs) for organisms and then removed the debris not containing organisms.
3. Covered the contents of the tray with a saturated sucrose solution and agitated the tray to separate the organisms from the debris.
4. Identified all organisms that floated to the surface by taxonomic order or family.
5. Collectively weighed (to the nearest 0.1 g) the individuals in each taxonomic group.

Methods for Sampling Fishes

We used a Smith-Root Type XV backpack electroshocker to sample the fish community at Stations 1F-3F. This electroshocker can deliver up to 1200 V of pulsed direct current. We used a pulse frequency of 120 Hz at all times and adjusted the output voltage to the optimal value, based on the water conductivity at the site. We measured the conductivity with a Hydrolab Digital 4041. This instrument also concurrently measured the water temperature and pH.

At each of the sampling stations, we followed these methods to sample the fish community:
1. Made a single pass upstream and downstream using a representative reach. The length of the reach varied among sites (from 22 to 115 m).
2. Held captured fish in a 0.64-cm plastic-mesh cage until we completed the sampling.
3. Anesthetized the fish in the field with MS-222 (tricane methane sulfonate).
4. Counted the fish by species and collectively weighed the individuals of a given species to the nearest 0.5 g on a triple-beam balance.
5. Released the fish to the stream.

In a preliminary sampling we collected representative individuals of each species by seining and preserved the fish in 10% formalin. In the laboratory, we identified the fish using these methods:
1. Identified species using unpublished taxonomic keys of Etnier (1976).
2. Compared the mountain redbelly dace (*Phoxinus Oreas*) and the common shiner (*Notropis Cornutus*) with specimens collected from Ish Creek, a small stream on the south slope of West Chestnut Ridge, and identified as *Phoxinus Oreas* and *Notropis Cornutus*.

Results

This section presents results of the sampling and analyzing of the benthic macroinvertebrates and of the fishes in the study sites.

Benthic Macroinvertebrates

The writers use parenthetical notes to refer readers to tables.

Qualitative sampling at the sites near Oil Retention Pond 1 and in Pond 2 resulted in few samples (see Table I). We also found low densities in the quantitative samples taken from Stream 1A, which drains Oil Retention Pond 1, and in Bear Creek near the confluence with Stream 1A (see Tables II and III). Relatively high densities and biomass of benthic organisms appeared in Stream 2, a small uncontaminated tributary of Bear Creek (Station 4). Some of the differences in the composition of the benthic community between this site and others may have occurred because of sub-

FIGURE 18.27 CONTINUED

3

strate differences (e.g., the large amounts of detritus in Stream 2 when compared to the predominately small rubble and gravel at Stations 1 and 2). However, a depauperate benthic fauna existed in Bear Creek and Stream 1A.

Table I. Description of qualitative sampling conducted near oil Retention Ponds 1 and 2 west of the Y-12 Plant (previous study)

The writers summarize results in tables.

Station	Location	Method	Sampling Results
5	Stream 1A just below Oil Retention Pond 1	Kick-seining	No organisms found
6	Diversion ditch just west of Oil Retention Pond 1	Kick-seining	No organisms found
7	Stream 1A above the diversion ditch	Kick-seining	Few Isopoda; unidentified salamander
8	Oil Retention Pond 2	Dip-netting; removal of sediment/litter from margins of ponds	No organisms found

Table II. Total number and weight (g, in parentheses) of benthic macroinvertebrates in each of three 27 x 33 cm bottom samples collected from four sampling sites in the vicinity of Y-12 Oil Retention Pond 1

The writers use graphics to present the results.

Sample no.	Sampling Station			
	1B	2B	3B	4B
1	0	0	0	53(0.8)
2	0	2(0.1)	3(0.1)	46(2.2)
3	0	2(1.8)	0	25(1.0)
Mean no./m^2 (g/m^2)	0	14.9(7.1)	11.2(0.4)	501.3(15.3)
Substrate	Course gravel embedded in sand and silt; leaf packs uncommon	Same as Staion 1B	Sand, silt, mud, and detritus/leaves	Deep soft mud covered by leaves and woody debris

FIGURE 18.27 CONTINUED

4

Table III. Density (mean no./m^2) of various benthic macroinvertebrate taxa in bottom samples collected from four sampling sites in the vicinity of Y-12 Oil Retention Pond 1. Biomass (wet weight, g/m^2) in parentheses. NC=None Collected.

| Taxon | Sampling Station | | | |
	1B	2B	3B[a]	4B
Amphipoda	NC	NC	NC	67.3(0.6)
Annelida	NC	7.5(0.4)	NC	NC
Chironomidae	NC	NC	NC	273.1(0.7)
Decapoda	NC	NC	NC	22.4(7.1)
Isopoda	NC	NC	NC	86.1(1.9)
Oligochaeta	NC	NC	11.2(0.4)	NC
Sialidae	NC	3.7(1.5)	NC	NC
Tipulidae	NC	3.7(5.2)	NC	7.5(3.7)
Tricoptera	NC	NC	NC	44.9(1.3)

[a]Damselfly nymph collected by kick-seining.

FIGURE 18.27 CONTINUED

5

Fishes

The four fish species collected at the four sample sites commonly inhabit small streams on the Department of Energy Oak Ridge Reservation. For example, these four species were the most abundant fishes found by electrofishing in Ish Creek, a small, undisturbed tributary of the Clinch River that drains the south slope of Chestnut Ridge. The presence of fish in the lower reaches of Stream 1A is consistent with the results of a bioassay conducted on the water from Oil Retention Pond 1. This bioassay showed no mortality to bluegill sunfish after 96 hours (Giddings). The high density and biomass of fish at Station 3F may relate to the abundant perphyton growth observed in the winter and to the chemistry of the effluent from Oil Retention Pond 1.

We found no aquatic species listed as threatened or endangered by either the U.S. Fish and Wildlife Service or the State of Tennessee. However, the Tennessee Wildlife Resources Agency has identified the mountain redbelly dace (*Phoxinus Oreas*) as a species which, though not considered threatened within the state, may not currently exist at or near their optimum carrying capacity (see Table IV).

Table IV. Density (mean no./m^2) of various benthic macroinvertebrate taxa in bottom samples collected from four sampling sites in the vicinity of Y-12 Oil Retention Pond 1. Biomass (wet weight, g/m^2) in parentheses. NC=None Collected.

Species	Sampling Station		
	1F	2F	3F
Blacknose dace (*Rynichthys atratulus*)	8(21.0)	1(1.5)	2(2.5)
Common shiner (*Notropis cornutus*)	3(22.5)	0	1(6.5)
Creek chub (*Semotilus atromaculatus*)	42(204.5)	4(76.5)	10(64.0)
Mountain redbelly dace (*Phoxinus oreas*)	39(64.0)	1(3.5)	35(51.0)
Total (all species combined) Density (no./m^2) Biomass (g/m^2)	0.30 1.00	0.03 0.41	1.68 4.33
Physical characteristics of sampling site Length of stream sampled (m) Mean width (m) Mean depth (m) Conductivity (S/cm) Water temperature (°C) pH	115 2.7 19 260 8.5 7.1	91 2.2 13 1005 0.5 7.6	22 1.3 10 477 1.5 7.5

FIGURE 18.27 CONTINUED

6

The writers inter-
pret the results.

Conclusions

The Y-12 Plant operations have had an adverse impact on the benthic communities of Bear Creek and some of its tributaries. The low benthic densities at Station 2B just above the confluence with Stream 1A suggest that the source of impact is not limited to the effluent from Oil Retention Pond 1. The relatively low fish density at Station 2F also implies an upstream pertubation (such as the S-3 ponds).

Further evidence of upstream impact(s) is available from the results of a similar study conducted 10 years prior to this study. In this study, researchers sampled benthic macroinvertebrates and fish at two sites located 50 m above and 100 m below the Y-12 sanitary landfill site. The west end of the landfill is approximately 1.4 stream kilometers above the confluence of Stream 1A with Bear Creek. No benthic organisms or fish were collected at either of the two sites, and in situ fish bioassays conducted just above and 500 m below the landfill resulted in 100% mortality after 24 hours.

The results of the earlier study differ significantly from those of the present study. The current presence of fish at Station 2F, which is approximately 500 m below the site of the earlier bioassays, may indicate changes in water quality over the past 10 years. However, the occurrence of fish in this region of Bear Creek may only be a temporary phenomenon (such as from a storm), reflecting short-term changes in water quality. The limited sampling in the present study may not have detected such a phenomenon. In view of the information currently available, both explanations of the difference in results seem equally plausible.

The writers
present only the
recommendations
here.

Recommendations

We recommend additional studies to address several important issues:

- More extensive season sampling of the benthic macroinvertebrate and fish communities to obtain a complete inventory of the aquatic biota in Bear Creek watershed, to investigate the potential recovery of biotic communities down stream of the confluence of Stream 1A, and to identify specific sources of impact to aquatic biota in Bear Creek above the confluence with Stream 1A.
- In situ and acute bioassays to assist with identifying potential sources of impact.
- Chronic bioassays to determine the effects on biota of long-term exposure to effluent sources.
- Studies of storms and their role in transporting contaminants downstream and in establishing and/or recovering of the biotic communities in Bear Creek.

Source: Adapted from Loar, J. M. and D. K. Cox, *Biotic Characterization of Small Streams in the Vicinity of Oil Retention Ponds 1 and 2 Near the Y-12 Plant Bear Creek Valley Waste Disposal Area.* Jan. 31, 1984. Oak Ridge National Laboratory, Oak Ridge, TN.

CASE STUDY ANALYSIS
New Media and the Global Workplace

Background

"In the 1950s, the family care, TV, and supermarkets changed the way we thought about and shopped for daily provisions. In the 1960s and 1970s, the shopping mall replaced Main Street as the place where people gathered. In the 1980s toll-free phone numbers and database marketing created an explosion in direct-mail commerce and forced mall and department store managers to rethink their strategies. . . . As mass customization becomes a reality, the opportunity to add value to information-intensive products and services becomes limitless" (Cash). In the 1990s, the personal computer, technology, and global communication revolutionized how companies conducted daily business—how they distributed information to their employees and to those outside the company. As the use of personal computers and the Web grew, do did the "user's expectation for information availability on or through that machine" and the Web (Foy).

In this current decade, companies now have more than the personal computer and the Web for delivering information in the workplace. Information is now delivered through personal networking such as Facebook, through blogs, and through hand-held devices and cell phones. Because of these new media for delivering information, people are increasing demanding that organizations deliver information and even some services directly into their hands via these new media. While employees and consumers may expect almost universal access to information, the real issue for organization is the possibility of increased work and possibly mediocrity (Foy). For many organizations, the issue is twofold:
- Deciding how to best create, share, protect, and distribute information
- Making the right information easy to get.
- Selecting the appropriate media in a global context.

Organizations want to "put their best foot forward." Therefore, they want to put out information in the best format and deliver that information in the most appropriate media. To select that media, organizations must consider the reader, their organization, and the type of information. If the readers are from different countries and cultures, the selection of media becomes even more complicated. Many organizations are unsure how to "select the most appropriate communication medium in this global business context" (Lee and Lee).

Assignment

- Locate two or more articles that discuss how new media is changing how organizations make information available both internally and externally. At least one of these articles must address how organizations use new media to deliver information to readers in a global context.
- Write a brief summary of each article and e-mail the summaries to your instructor.

Cash, J.I., Jr. "A New Farmer's Market," *Information Week*, 26 Dec. 1994:60.

Foy, Patricia S. "The Reinvention of the Corporate Information Model." *IEEE Transactions on Professional Communication*, 39 (1996): 23-29.

Lee, Zoonky and Younghwa Lee. "Emailing the Boss: Cultural Implications of Media Choice." *IEEE Transactions on Professional Communication*, 52.1 (2009): 61-74.

EXERCISES

1. **Collaborative exercise**: Working with a team, identify a problem at your university, in your community, or at your workplace. Write a feasibility report that examines options for solving this problem. As your team works, follow these guidelines:
   - Make sure that all team members are familiar with the problem and can gather the needed information.
   - Establish reasonable and appropriate minimum criteria for evaluating the options.
   - Identify the options and decide what methods your team will use to gather the needed information for evaluating the options.
   - Assign each team member an information-gathering task. For example, if your team decides to develop and distribute a questionnaire, decide who will write the questionnaire, who will prepare and distribute the copies, and who will tabulate the returned questionnaires.
   - Use the online "Worksheet for Writing Reader-Focused Formal Reports" as you plan and evaluate your feasibility report, and follow the tips for establishing criteria and identifying options on pages 597 and 599, respectively.

Download a copy of the Worksheet for Writing Reader-Focused Formal Reports at www.kendallhunt.com/technicalcommunication.

2. **Collaborative exercise**: Working with a team of peers in your major field, identify a problem to research and then decide how to approach the research. For example, you might identify the methods you would use to research the problem, the means for verifying your results, and so on. Write a memo to your instructor describing the problem and your approach to researching it.

3. **Web exercise**: Many organizations, Web sites, and consumer groups compare products and make recommendations. For example, one of the better-known such publications is *Consumer Reports*. This publication has established criteria for evaluating consumer products such as cars, electronics, appliances, and computers. The editors recommend the best "option" based on their established criteria. Visit the Web site of *Consumer Reports*, at www.ConsumerReports.org, or find a similar group. After visiting a site,
   - select a product (it can be food, paper products, appliances, etc.)
   - establish criteria for evaluating options

- identify the options
- research the options
- evaluate the options
- write a feasibility report for consumers in your area recommending one of the options

4. Locate a formal report from a federal, state, or city government agency or organization; from a private organization; or from a department or group at your college or university. To find a report, ask officials at your university or visit the Web sites of government agencies or organizations. You can also ask reference librarians to help you locate reports. When you have located a report, write a memo to your instructor analyzing it. Be sure to identify the type of report (information, analytical, or recommendation) and the intended readers. Your memo should
   - evaluate the conventional sections of the body or the report
   - evaluate the front matter
   - evaluate the overall effectiveness of the report

   For more information on memos, see Chapter 12.

5. Write a letter of transmittal for a report that you are writing. Follow the paragraph-by-paragraph outline in "Tips for Writing the Letter of Transmittal" on page 574.

6. Prepare a table of contents for a report that you are writing. Include all first-level, second-level, and third-level headings.

7. Design a cover and title page for a report you are writing. As you design the cover, decide what materials and binding style you will use. When you have completed the cover and title page, turn in the following to your instructor:
   - a memo explaining the material you will use for the cover and justifying your design of the cover and title page: why you selected the graphical elements, what the graphical elements represent, whether the design is the organization's standard design (if applicable), and so on. (For information on memos, see Chapter 12.)
   - the cover printed on plain paper (indicate in the memo how you will present the final version of the cover: type of paper, binding, and so on)
   - the title page

8. Decide whether any material for a report that you are writing should appear in an appendix. Then write a memo to your instructor explaining your decision. (For information on memos, see Chapter 12.) If you decide to include material in an appendix, answer these questions in your memo:
   • Why is the material better placed in an appendix than in the body of the document?
   • What readers are likely to be interested in or to need the information presented in the appendix?
   • What is the purpose of the material that will appear in the appendix?

9. Decide whether the report you are writing should have an informative abstract or an executive summary. Then, write the abstract or the executive summary for your report.

REAL WORLD EXPERIENCE
Deciding Where to Live

Each spring, students at your college or university decide where they will live in the fall. Usually they have several options such as
• renting an apartment
• renting a house or duplex
• living with a family member, such as their parents
• living in a dorm on campus

These options may vary slightly, depending on the campus, the campus rules, and the students' personal situations.

Assignment

Your instructor will ask you to prepare either an individual or a team feasibility report recommending the best housing option for undergraduate students at your university.

Individual Reports

If your instructor asks you to prepare an individual feasibility report, follow these steps:

1. Revise the housing options listed in the "Background" section to match those available to students at your college or university.

2. Determine the criteria you will use to evaluate the options.
3. Gather the information necessary to evaluate each option, based on the criteria.
4. Analyze the information about each option according to the criteria.
5. Write a feasibility report for students at your college or university.

Team Reports

If your instructor asks your team to write a feasibility report, follow these instructions.

At the first team meeting
1. Determine the readers and the purpose of your report.
2. Establish the criteria you will use to evaluate the options.
3. Decide how you will identify all available options. You might assign each team member a research task.

At the second team meeting
1. Using the research that you've gathered, determine the options you will evaluate.
2. Decide how the team will gather the necessary information for evaluating the options.
3. Assign each team member an information-gathering task. (Each member should be ready to report on that information at the third team meeting.)

At the third team meeting
1. Analyze the information gathered by each member, and determine the best housing option for students.
2. Assign a section or part of a section of the report for each team member to write. (Each team member should complete his or her assignment before the next team meeting.)

At the fourth team meeting
1. Read all sections of each team member's report.
2. Compile one report.
3. Edit and proofread the report.

iStockphoto 2008.

C H A P T E R N I N E T E E N

19

Creating User-Focused Web Sites

You may be asked to create a Web site for your organization. In some organizations, you may be responsible for the entire process of creating the site, from planning to maintaining the site. In other organizations, you may work with a team to create the site; or you might work with outside consultants, directing them to create the site using the feedback and guidance of your team. Whether you are solely responsible or you work with a team, you will create a more effective site when you know the characteristics that make a site user-focused and you understand the overall process for creating that site.

CHARACTERISTICS OF A USER-FOCUSED WEB SITE

A user-focused Web site has these characteristics:

- **It is easy to scan.** Users "rarely read Web pages word by word; instead, they scan the page" (Nielsen, "How Users Read on the Web"). Users want to quickly scan a Web page, looking for keywords and materials without scrolling through pages of text-filled screens. A scannable Web page has highlighted keywords, sub-headings, and bulleted lists.
- **It uses concise language.** Users respond more positively to a Web site with fewer words. Because reading on a computer screen is about 25 percent slower than reading from paper, users don't want to read Web pages with lots of text, so use about 50 percent fewer words when writing for the Web (Nielsen, "Be Succinct").

- **It is easy to navigate.** Users want to know how to navigate between pages. They want to know where they are in the site and how to get back to the home page. They want the navigation bar and hot buttons to appear consistently on every page.
- **It is accessible.** Users may come from various cultures, may not be fluent in English, or may have disabilities. A successful Web site enables all users to access the information.
- **It is credible and trustworthy.** Because the Web is filled with unreliable information, make sure that readers see your site as credible. A credible and trustworthy site uses high-quality graphics, up-to-date links, worthwhile content, and effectively designed pages.

PLAN THE WEB SITE

As Figure 19.1 illustrates, the process of creating a Web site is a recursive process; you may revisit steps of the process as the needs of your users change or as you receive input from usability testing and users.

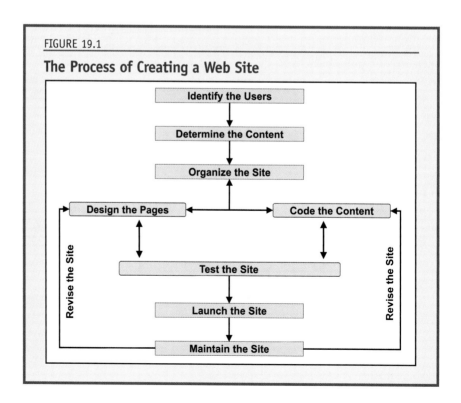

FIGURE 19.1

The Process of Creating a Web Site

In this section, we will focus on the following steps of the planning process:
- identifying the users and your purpose for the site
- determining the appropriate content

Identify the Users and the Purpose of the Site

To design a user-focused Web site, you need to first identify its purpose and its users. You might begin by thinking about the purpose of the site. For example, do you want users to buy products and services? Do you want users to learn how to use or install products? Do you want to provide answers to frequently asked questions or to provide information about your organization? The site may have more than one purpose; for example, you may want to provide information as well as answer frequently asked questions.

Successful Web sites also reflect an intimate knowledge of the users. These Web sites build relationships with the users through the Web, and the users often participate in site design. To identify your users, consider conducting user research. For example, if the site is for people who use your products and services, send them a questionnaire to find out what information they would use on the site. You might conduct a focus group with selected users. In these questionnaires and focus groups, you will learn more than the demographics of your users. You can also discover
- what your users want to learn or expect from a Web site
- how your users want information presented on the Web
- how your users search for information on the Web
- how your users might use the information that you plan to present on your site

As you analyze your users, ask these questions:
- **What do the users know about the subject of the site?** Do the users have similar backgrounds? For example, if you are creating a site for NASA scientists, you can assume that your users have similar backgrounds and some understanding of your subject. However, if you are creating a site for people interested in NASA, your users might be children, teachers, scientists, "amateur" scientists, and so on. Your users may or may not know much about your subject.
- **Why will users visit your site?** What do they want to do? Do they want to gather information, make a decision, or answer a question? Do

they want to perform a task, to link to other related sites, or perhaps to download information?

- **Are the users internal or external?** Will only people in your organization use your site? Or will people outside your organization use your site? If the users are external, what do they know about your organization? Are their attitudes positive or negative about your organization?
- **Are your users global?** Consider your users' culture and language. How will you meet the needs of your global users? For example, will your organization and your users benefit from putting the content of the site in multiple languages?
- **Have you considered whether your users have disabilities that will affect their ability to access the information on your site?** Design your site so that all potential users can access the information.

When you launch your site, you can use the site itself to gather feedback from your users. You can then refine your site based on that feedback.

Determine the Appropriate Content for the Site

You might think of creating a Web site like building a house. Before you begin drawing the plans, you need to know what you want in the house. For example, how many people will live in the house? Do you need one bedroom or two? Do you want the house to have two stories? If you want two stories, will all members of the household be able to use the stairs? Do you want the master bedroom next to the other bedrooms? Will you need a garage? How many cars should the garage accommodate? What is your budget? You need an idea of your budget, users, and needs before you can begin to design a house—or, in this case, a Web site.

As you consider the information and the elements that you want to include and that your users need, answer these questions:

- **Will you use existing information from paper documents?** If so, you will have to adapt this information for an online format. If the information is copyrighted, make sure that you have permission to use the information (see Chapter 2 and the Ethics Note in this chapter on page 640).
- **Will you have to create new content?** As you consider various types of content, determine how much time you will need to create the content, how much research you will have to do, and whether you will create the content alone or with a team.

- **Where will you get graphics for the site?** Do you have graphics that you can use? If not, will you or someone else create those graphics? If you will use existing graphics, do you have permission to use those graphics? Are they copyrighted?
- **Will you include audio on your site?** If so, will your budget allow for creating the audio? Do you have the tools to create professional audio?
- **What is the budget for creating the site?** How will that budget affect the graphics, audio, design, and so on?

> *A successful Web site reflects an intimate knowledge of its users, building on that relationship and incorporating feedback received from them.*

You may think of other questions that you need to consider as you think about the content for the site. You can also use the tips below to help you determine the appropriate content for your purpose and your users.

✔ TIPS FOR DETERMINING THE CONTENT FOR THE SITE

- **Select content that is interesting and relevant to the users.** Many Web authors fail to filter out irrelevant or uninteresting content because they can easily put the contents of existing paper documents into hypertext markup language (HTML) and place them on the site. These documents may be appropriate for paper, but they might not serve Web users well. Irrelevant or uninteresting content can include graphics or audio and video, as well as text. *Just because the information is available does not mean that it belongs on the site.*
- **Adapt existing content created for other media.** You may find that you already have some content for your Web site. For example, you can convert content, such as text from online presentations, for Web pages. Some Web designers simply scan in paper documents and load them onto their sites. These documents may work effectively on paper, but not on the Web. Instead, decide what information from the paper document is appropriate for your purpose, your users, and the Web format. Then adapt that information for your site.
- **Decide if the content is best presented on paper or online.** You may want to leave some documents in print. Print is still a viable medium for some types of information and documents. Your content may best meet readers' needs in print rather than in an online medium.
- **Make sure that all content helps the site achieve your purpose and meet your users' needs.** The text, graphics, and links should relate directly to the purpose of the site. For example, the former Web site of the IRS (Internal Revenue Service) began with the title "The Digital Daily" in a newspaper-style layout. This layout and the title didn't help the site achieve its purpose because many users did not recognize the site as the IRS's Web site.

Evaluate a Web site in the Interactive Student Analysis online at www.kendallhunt.com/technicalcommunication.

DESIGN THE SITE

To design a user-focused Web page, you will use many of the principles that you learned for designing visual information and graphics in Chapters 10 and 11. This section focuses on how you can adapt those principles for the Web. To design an effective Web page, focus on these areas:

- Create an effective online layout.
- Write concise, scannable content.
- Make the site easy to navigate.
- Make the site accessible.
- Create a credible, trustworthy site.

Create an Effective Online Layout

To create an effective online layout, use the principles of contrast, repetition, alignment, and proximity (see Chapter 10). These principles work just as effectively in online pages as in paper documents. The following guidelines will help you adapt these principles for Web pages:

- **Use ample white space.** You can create contrast on the page by including ample white space (blank space). If a Web page is filled with text and/or graphics, readers may not want to read the page or will not know what information is important. User-focused Web pages use white space to highlight information and to organize information. Don't be tempted to fill the page just because you can. The sample Web pages throughout this chapter demonstrate effective use of white space.
- **Group related information together.** Use the principle of proximity to group related information (see Chapter 10). The Web page in Figure 19.2, effectively groups related information, so that users can easily scan the page and locate the information that relates to their needs. For example, if you are an eyecare provider, you can quickly find the information related to you.
- **Use a consistent design for the pages.** The pages should look like they belong to the same Web site, so select an appropriate color scheme and use it consistently throughout the site. Use the same color for the same elements and the same background color for every page. For example, the two Web pages in Figure 19.3, page 632, from the Balfour Beatty Construction have a consistent page design: the navigation bar, and the

FIGURE 19.2

Web Page with Related Information Grouped Together

company logo are in the same place on both pages, and the pages use the same colors.

- **Include an informative title at the top of every page** (Farkas and Farkas; Spyridakis). An informative, specific title "helps orient users" (Spyridakis). When users follow a link to a page, they are looking for information that will confirm that they have arrived at the content they expected. Without a title to guide them, readers will be confused. For example, Spyridakis points to the former U.S. Internal Revenue Service (IRS) Web site. The title of the page appeared to be "The Digital Daily." The page did not clearly specify that the user had arrived at the IRS home page. The Web pages in Figure 19.3, page 632, have an informative title on each page: "Careers" and "Hospitality."
- **Include the site name or logo on every page.** The title or logo should appear in the same place on every page to maintain site identity (Farkas and Farkas). The Web pages in Figure 19.3, page 632, show the con-

struction company's logo consistently in the top left-hand corner and the titles in the rectangle just below the logo.

- **Include an informative header and footer on every page.** Headers and footers add consistency and credibility to your site. The footer should include the copyright information and the title of the organization that is publishing the site. It may also include a link to the home page and the name of the Web master or page editor. Figure 19.4, page 633, is the footer from the NASA site. This footer includes the date that the page was last updated, the page editor, the logo, and the links to a site map and other relevant information and government sites.

- **Use graphics and color appropriate for the users and the topic.** Select graphics and colors that relate to the topic. For example, the graphics in the Balfour Beatty Construction site relate to the topic (see Figure 19.3). The bright colors are appropriate for the topic and the users. These same colors and graphics would be less appropriate for the Eye Care Web site in Figure 19.2.

FIGURE 19.3

Consistent Page Layout

Source: Downloaded from the World Wide Web:
http://www.balfourbeattyus.com/portfolio/
Hospitality.htm and
http://www.balfourbeattyus.com/Careers/careers.htm

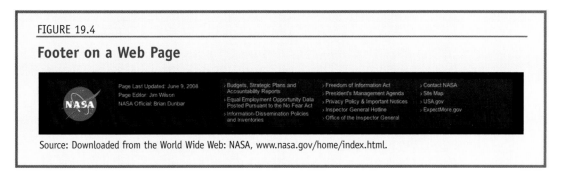

FIGURE 19.4

Footer on a Web Page

Source: Downloaded from the World Wide Web: NASA, www.nasa.gov/home/index.html.

- **Use graphics that download quickly.** Try creating an individual file for each graphic. Use a thumbnail sketch of the graphic with links to the image so users can click on it to link to the larger version of the graphic (Wilkinson). You can find examples of thumbnail graphics on map and weather sites.

Write Concise, Scannable Content

Web users scan for information rather than reading an entire paragraph; write content that allows them to quickly scan the page. As you develop the content, follow these guidelines:

- **Write in small "chunks" and short pages** (Bradbury; Spyridakis). Small chunks help users scan pages. Break long paragraphs and pages into shorter ones. Put less information on each page and one idea in each paragraph (Bradbury and Nielsen, "How Users Read on the Web"). In this manner, users can "more easily find the information they need and read and retain it" (Spyridakis).
- **Use bulleted items and lists** (Bradbury and Nielsen, "How Users Read on the Web"). Bulleted items and lists visually break up text, make pages easier to read, and guide users to information.
- **Include highlighted keywords.** These words help users scan for information.
- **Use subheadings.** Subheadings categorize information and help users scan the page. The "visitor information" page from the Denver Zoo Web site includes subheadings for admission and parking information (see Figure 19.6, page 636).
- **Use simple words and short, concise sentences.** Use words that users can easily read and understand (see Chapter 8). Eliminate any unnec-

essary detail and use only a few examples of a concept; don't provide "exhaustive coverage" (Spyridakis).

- **Use active-voice verbs whenever possible and appropriate.** Active voice helps users move more quickly through Web content (see Chapter 7).
- **Include an introduction or introductory sentence that identifies the purpose of the site and specifies the intended users** (Spyridakis). On the home page, introduce the purpose of the site. If you're writing for an organization, introduce the organization and state what it does or include a link to a page that gives information about the organization. This page might be called "About Us." Figure 19.7 on page 637 is an example of a Web page that illustrates an effective introduction to an organization.

Make the Site Easy to Navigate

With a book or magazine, you can see the entire document at once: you can flip from the beginning to the end. However, with a Web document, you can't see the "end" or even the "middle." You can't see the entire site at one time as you can with a book or a magazine. Unlike a print document, a Web site does not have a linear organization with pages numbered consecutively from the first page to the last page. However, a site does have an organization, and you have to tell and show your users how to navigate between pages and how to get back to the beginning, or the home page. Follow these guidelines to make your site easy to navigate:

- **Include a site map.** A site map shows the "global structure" of your site. It lists the pages in the site, classified into categories. Your site map will be more effective if you include a "You are here" or "Last page visited" marker (Farkas and Farkas). Figure 19.5, page 635, from the City Zoo site illustrates a partial site map.
- **Include an index.** An index is an alphabetical listing of the pages and topics on your Web site. Often users can simply select the beginning letter of their search item rather than having to scroll through the site to find what they want.
- **Provide a link to the home page on every page.** Put the link in the same place on every page. You can also use your organization's logo as the link as on the Denver Zoo web pages, Figure 19.6, page 636. You can also simply use a link to "Home."

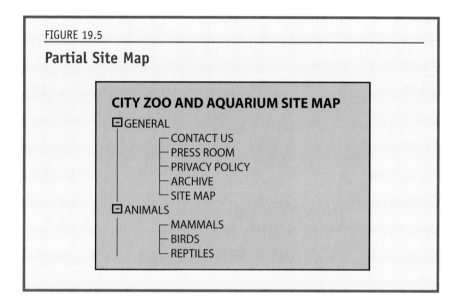

FIGURE 19.5

Partial Site Map

CITY ZOO AND AQUARIUM SITE MAP
GENERAL
⊟ CONTACT US
PRESS ROOM
PRIVACY POLICY
ARCHIVE
SITE MAP
ANIMALS
MAMMALS
BIRDS
REPTILES

- **Include a navigation bar on every page.** Locate this bar in the same place on each page. For example, the navigation bar on the Denver Zoo site, page 636, appears at the top of every page. On many sites, the navigation bar appears on the left-hand side of the page and has a vertical, rather than horizontal, orientation as on the DFW Airport site in Figure 19.8, page 638.
- **Include a search function on every page.** You can help users navigate the site by including a search box on every page. Be sure to locate the search function in the same place on every page.

Make the Site Accessible

Your site should be accessible on several levels: accessible for users with varied types of technology, accessible for users of various cultures, and accessible for users with disabilities. Figure 19.9, page 639, is an excellent example of a site that caters to a wide scope of users. To create an accessible site, follow these guidelines:

- **Include a text-only version of all visual information.** Some users may use a text-only version of your site because they have a slow connection to the Internet or their browsers are set to view text only. Other users may have limited ability to view the site because they are using a hand-

FIGURE 19.6

Page from the Denver Zoo Web Site

The page is easy to navigate. It includes two navigation bars: the horizontal bar that appears on every page and the vertical bar that provides a table of contents for the information in the education section.

Even though the page is information-rich, users can easily scan the page.

The page includes a "quick links" feature to help users easily navigate frequently accessed areas of the Web site.

The information is chunked using subheadings.

The page is accessible because it uses text to accompany the photos.

The page is credible because of the high-quality photos.

The photos and the colors are relevant to the topic and appropriate for the users.

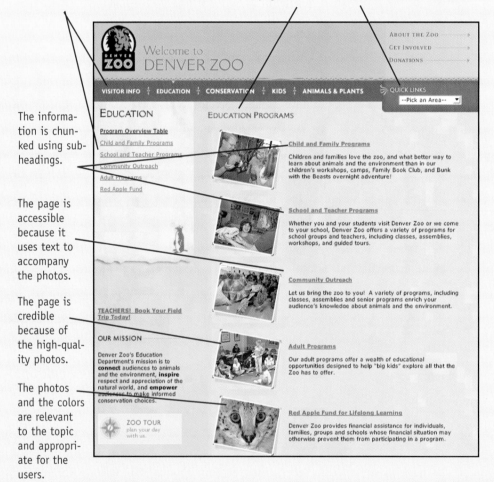

FIGURE 19.7

Home Page for Dow Chemical Company

The logo is easy to see and appears in the header of every page.

The page is easy to navigate because of the navigation bar on the left-hand side and the search box in the top right-hand corner.

The page uses the design principle of repetition by repeating the red from the logo in the heading and bullets.

The page orients the reader with a "you are here."

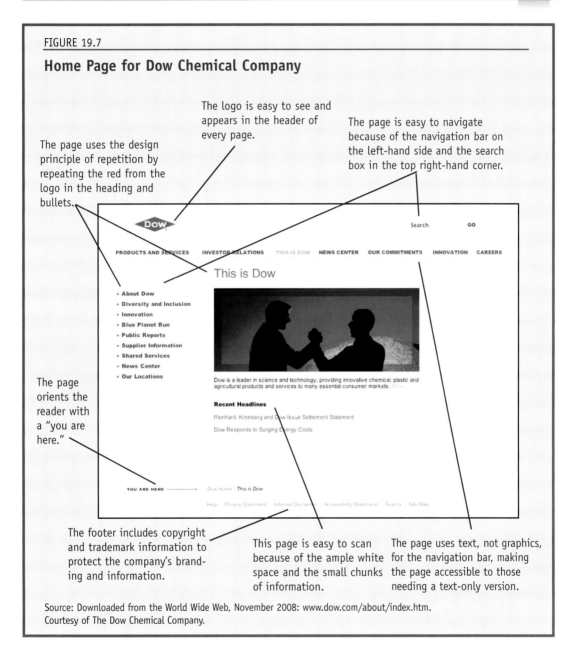

The footer includes copyright and trademark information to protect the company's branding and information.

This page is easy to scan because of the ample white space and the small chunks of information.

The page uses text, not graphics, for the navigation bar, making the page accessible to those needing a text-only version.

Source: Downloaded from the World Wide Web, November 2008: www.dow.com/about/index.htm.
Courtesy of The Dow Chemical Company.

FIGURE 19.8

Home Page for the Dallas/Fort Worth International Airport
(Design replaced in June 2008)

The logo is clear and appears on every page to create consistency.

The site creates a sense of credibility and trustworthiness through the date in the top right-hand corner, the link to Web support, and the high-quality graphics.

The page has a clear navigation bar.

The footer includes copyright information and identifies the organization publishing the site.

The page is easy to scan because the information appears in small chunks with subheadings.

The language is concise.

Source: Downloaded from the World Wide Web: Dallas/Fort Worth International Airport, www.dfwairport.com/.
Design replaced in June 2008

held device or they are visually impaired. For example, if you rely solely on a graphic or on color to indicate a link, colorblind users may not find the link.

- **Don't rely on color or graphics to convey information.** For example, if you use blue to indicate a link, also underline the link so colorblind users can see the link.

- **Allow users to adjust the size of the text.** You can allow users to make the type larger or smaller. For example, the Balfour Beatty Construction Web site includes two buttons that allow users to make the text larger or smaller (see Figure 19.3, page 632). Notice that the Web site authors use both a button and the words "text size" to indicate that users can adjust the text.
- **Be sensitive to readers of other cultures**. Use Simplified Technical English. Avoid idioms and brand names. (See Chapter 8 for more information on writing for readers of other cultures and for readers who are not fluent in English.)

FIGURE 19.9

Welcome Page for IKEA

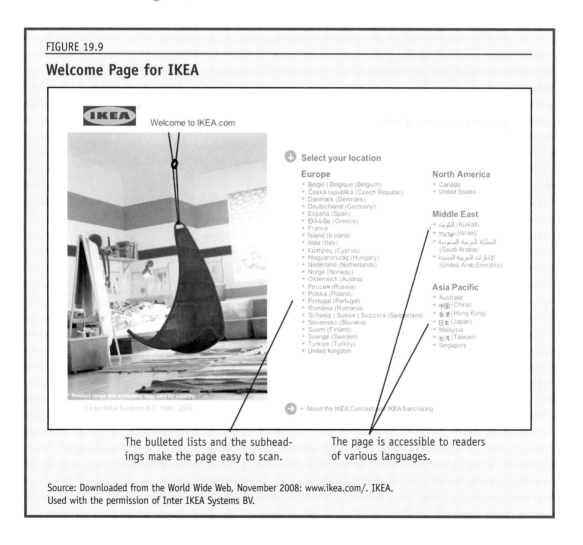

The bulleted lists and the subheadings make the page easy to scan.

The page is accessible to readers of various languages.

Source: Downloaded from the World Wide Web, November 2008: www.ikea.com/. IKEA. Used with the permission of Inter IKEA Systems BV.

ETHICS NOTE

Copyright, Intellectual Property, and the Web

Have you ever found a graphic that you liked on a Web site and used it on your personal Web site? Have you copied code from a Web site and used it for your organization's Web site? Have you ever used a company's logo or trademark from a Web site? If so, you could be guilty of copyright, patent, or trademark infringement. U.S. and international laws protect every element of a Web site: the text, graphics, and the code (Le Vie). Even if the authors of a site haven't filed a formal copyright application, the law protects their work. Along with copyrights, the law protects an organization's patents, trademarks, and trade secrets. You or your organization can "bruise its reputation by infringing on someone else's copyright" or intellectual property, and you could face legal penalties for it (Le Vie).

You must protect your Web site, your intellectual property, and your organization's products and services. For example, aspirin originally started as a trademarked product, but the manufacturer failed to protect its trademark, and aspirin became a name used by many manufacturers. To protect your Web site and your intellectual property—or that of your organization—follow these guidelines to maintain the uniqueness of your product or service:

- Place a copyright notice in the footer of every Web page, not just on the home page.
- Link from the word "copyright" to another Web page that defines what you own on the site (Le Vie).

You must also protect yourself and your company from the legal penalties and bruised reputations that occur when you don't respect the copyrights and intellectual property of others:

- Obtain permission for any information, graphic, or code that you use from another site or printed document.
- Place the permission, copyright, and trademark information in a conspicuous place.
- ***Do not use any information, graphic, or code without the written permission of its authors.***

Create a Credible and Trustworthy Site

To encourage users to read your site and visit it again, you must establish credibility and gain their trust. According to Spyridakis, "the credibility of a Web page is easily affected by the presence or absence of certain content.

For some, the credibility of a company or institution may begin with its Web site. People may choose a graduate school or a future employer or purchase a specific brand of product in part because of how credible the company or agency appears to be in its Web site." To make your site credible and trustworthy, follow these guidelines:

TIPS FOR CREATING A CREDIBLE AND TRUSTWORTHY SITE

- **Include your name and/or the name of the organization that is publishing the site, as well as the contact information.** If you're writing for an organization, include the name and purpose of the organization and the contact information for the organization. If you are publishing the site, include your credentials—some kind of affiliation or title that the user can check (Spyridakis).
- **Include the date that you launched the site or the date you last updated the site.** Users want to know if the information is current. If you launch a site and don't update it frequently, the site will have little, if any, credibility. Users will not be able to trust that the information is accurate or up to date.
- **Check all links regularly to make sure that they still work.** If you have linked to related sites, make sure those links still work and that the linked sites are relevant and current.
- **Make sure the information on your site is accurate and free of typographical errors, spelling errors, and other mistakes.** These errors may cause users to assume that the site is not credible. These errors can also "draw the user's attention away from the important information on the page and raise doubts about the author's credibility" (Spyridakis).
- **Use high-quality graphics.** High-quality graphics add credibility. Web sites without high-quality graphics look amateurish and unprofessional, making the content and the site look untrustworthy.
- **Use a professional tone that is appropriate for your purpose and users.** Avoid hyped-up, unsubstantiated language, such as "scientific breakthrough" or "best weather site on the Web."
- **Cite sources.** If you use information from other sources, tell users where you found the information.
- **Respond to reader mail.** If you offer users the opportunity to contact the organization or to provide feedback about the site, you must respond to user inquiries in a timely manner. You will anger, frustrate, and alienate users if you don't follow up on their inquiries. Offer users the option to send you e-mail only if you can respond to them in a timely manner.

TAKING IT INTO THE WORKPLACE
Creating an ePortfolio

In Chapter 13 you learned about electronic career portfolios, often called ePortfolios. These portfolios are a Web-based collection of materials related to your job search and your career. As you begin your job search, employers may ask you for an ePortfolio, or you may want to provide employers with additional information about you and your work.

iStockphoto 2008.

Assignment

1. Search the Web for good examples of ePortfolios. The examples should have the characteristics of a user-focused Web site.
2. In an e-mail to your instructor, send links to three excellent ePortfolios.

An ePortfolio makes your work samples, resume, and other career information readily available for review on a Web site.

3. Create your own ePortfolio. At a minimum, your portfolio should include
   a. an introductory (home) page
   b. a resumé with highlighted keywords
   c. several samples of your work
   You may include other relevant information. If you are unfamiliar with coding a Web site, you can locate tutorials on the Web or use Web-authoring software available at your college or university.
4. Test your ePortfolio.
5. Determine what users will have access to your portfolio. Because you will have personal information on your portfolio, you should restrict access only to those whom you trust: such as friends, family, professors, and potential employers.
6. Launch your ePortfolio. Send an e-mail to your instructor with a link to your ePortfolio.

CASE STUDY ANALYSIS
Web Sites, Sustainability, and Environmental Stewardship

Background

The United Nations defines sustainability as "development that meets the needs of the present without compromising future generations to meet their own needs."[1] The Environmental Protection Agency describes sustainability with the term "environmental stewardship." The Environmental Protection Agency describes how companies and non-profit organizations can be stewards of the environment:

> From the way they manage their operations to the products and services they offer customers to the projects and activities they support in their communities, businesses and other institutions can play an important role in protecting the environment and preserving natural resources.

Indeed, major U.S. companies are using their Web sites to showcase their environmental stewardship and their commitment to sustaining the environment.

If you visit the Web sites of most U.S. global companies, you will find a section on environmental responsibility. The companies include sections such as "Green Initiatives" (Google), "Environment & Society" (Shell Oil), "Environmental Sustainability" (Microsoft), and "Sustainability—Protecting our Resources" (Kraft Foods). These environmentally-focused pages are not limited to one industry. These pages provide an opportunity for consumers to see how an organization views environmental stewardship.

[1] Kraft Foods. "Sustainability—Protecting our Resources" http://www.kraftfoodscompany.com/About/sustainability/ Accessed 18 April 2009.

Assignment

1. Locate the Web site for three global U.S. companies. These Web sites must include a section on sustainability and/or environmental stewardship. These Web sites must be from three different industries. For example, you cannot use only Web sites from oil companies or food companies. Instead, you might use a Web site from an oil company, a food company, and a software company.

2. Compare the three Web sites' approach to presenting the company's commitment to the environment (See Chapter 6). Your comparison should answer the following questions:
   a. What are the companies' commitment to the environment as presented on their Web sites?
   b. Who are the intended readers?
   c. Is the information credible? Explain your answer.
   d. What persuasive appeals do the Web sites use? (See Chapter 9.)
   e. Which of the Web sites best presents the information for the intended readers?

EXERCISES

1. **Web exercise**: Locate two Web sites on the same topic: one should be credible and trustworthy, and one should be less credible and not trustworthy.

   Download a copy of the Worksheet for Creating User-Focused Web Sites at www.kendallhunt.com/technicalcommunication.

   a. Write a memo to your instructor comparing and contrasting the two sites. Be sure to include specific examples from the sites. If you need help with the comparison/contrast pattern of organization, see Chapter 6.
   b. In your memo, include the URL for each Web page and attach a copy of the home pages.

2. **Web exercise**: Locate the Web site for a national or local nonprofit organization such as the Red Cross, Habitat for Humanity, or the American Heart Association. You can find nonprofit organizations in your local community or on a national level. Analyze the following aspects of the site:
   a. consistency of the page design
   b. color (you might want to review the information on color in Chapter 11)
   c. ease of navigation
   d. style of writing
   e. appropriateness of content

   Write an informal report to your instructor citing problems with the site and recommending solutions. Print a copy of the organization's home page to attach to your report. (You might review the problem/solution pattern of organization discussed in Chapter 6 and the information on informal reports in Chapter 17.)

3. **Web exercise**: Evaluate the Web site of your hometown and compare it to the sites of two cities of similar size. As part of your evaluation, consider whether the site has the characteristics of a user-focused Web site:
   a. Is it easy to scan?
   b. Does it use concise language?
   c. Is it easy to navigate?
   d. Is it accessible?
   e. Is it credible and trustworthy?

Write an informative informal report to the Webmaster, mayor, or city manager summarizing your evaluation and comparisons. Turn in a copy of your report to your instructor. (See Chapter 17 for information on informal reports.)

4. **Web exercise**: Expand your evaluation of your hometown's Web site. Determine
   a. if the site has appropriately protected its copyright and other intellectual property, such as its logo (its branding)
   b. if the site may have infringed on the copyrighted information or trademarks of others

Based on your evaluation, write a memo to the Webmaster explaining what you were looking for and whether the site adequately protects the city's intellectual property and branding and properly uses copyrighted information. Turn in a copy of your report to your instructor.

REAL WORLD EXPERIENCE
Helping a Nonprofit Organization in Your Community

Nonprofit organizations are groups that provide a service without making a profit. Nonprofit organizations include groups such as Little League, Habitat for Humanity, Boys and Girls Clubs, and Big Brothers and Big Sisters. They could also be groups that rescue animals or feed the homeless. Some nonprofit organizations don't have the budget or expertise to create a Web site. Even local and regional offices of some larger nonprofit organizations may not have the time, money, or staff to create and maintain a site. Sometimes these organizations have Web sites, but they are poorly designed and, therefore, are not credible and trustworthy. In your community, you can most likely locate nonprofit organizations that would benefit from a Web site or from a redesigned Web site.

Assignment

Locate a nonprofit organization in your community.[1] You might go to the Web site for your city; this site may have links to contacts for local nonprofit organizations or for groups needing volunteers. You can also visit with volunteer groups in your community to find names of nonprofit groups. Some colleges and universities have volunteer groups that help local nonprofit organizations. When you have located a nonprofit organization, do the following:

- Contact the organization and make an appointment to talk to a representative about designing a Web site.
- Prepare interview questions before you arrive for the appointment (see Chapter 5 for information on interviewing). Be sure to develop questions about what types of information the organization would like to have on the Web site. You can also ask the representative how the organization and its users would use the site.
- Identify the users and the purpose of the site.
- Select and create the content for the site. If you receive printed documents for the site, remember to adapt them for the Web.
- Create a site map.
- Design the home page and the additional pages.
- Test your Web site and revise as necessary. You might ask the staff at the organization to test the site.
- Present your Web site to the organization and to your instructor. (The organization may launch the site on the Web; you will only create it and test it.)

[1] Your instructor may have you work with a team on this assignment.

CHAPTER TWENTY

20

Preparing and Delivering Oral Presentations

Oral presentations are an important part of technical communication; these presentations allow professionals to interact with their audience and to sell their ideas. You might make an oral presentation to any of these audiences:

- **Coworkers and decision-makers in your organization.** For example, you might make a formal presentation to upper-level management on the results of a feasibility study. You might give an informal presentation on new software that the organization has adopted for tracking projects; or in a staff meeting, you might give an extemporaneous status report about a project.

- **Customers and clients.** You might make a presentation about a project or product to customers and clients.

- **Peers at a professional meeting.** You might present the findings of a research project at a professional meeting. Your audience would be your *peers*—people who do the same type of work that you do or who have a similar educational background.

- **General public.** You might make a presentation to a non-profit, civic, or government group.

Whether you are speaking to coworkers, customers, clients, peers, or the public, the guidelines in this chapter will help you create and deliver memorable presentations.

UNDERSTANDING THE TYPES OF ORAL PRESENTATIONS

Professionals might make any of the following types of oral presentations:

- **Impromptu.** You do not plan an impromptu presentation. You decide what you will say while you are speaking. You might give an impromptu presentation in a staff meeting when someone asks you about your research or wants an update on a project. You would talk briefly and then answer questions.
- **Extemporaneous.** You plan an extemporaneous presentation and deliver it (perhaps using notes or an outline) in a conversational style. Most audiences prefer this type of presentation to the scripted and memorized types.
- **Scripted.** You write a script for the presentation and read the script to the audience.
- **Memorized.** You write a script; but instead of reading it, you memorize it. You don't use notes as you make the presentation. Most professionals avoid memorized speeches because they make speakers look too stiff and formal (Pfeiffer). You also risk losing your place or forgetting an important point.

Figure 20.1 presents advantages, disadvantages, and guidelines for each type of presentation. This chapter will focus on the two types of presentations that are most preferred and delivered in the workplace: extemporaneous and scripted.

PREPARE WITH THE AUDIENCE AND THE OCCASION IN MIND

The process of preparing an oral presentation is similar to preparing most technical documents. You analyze your audience and occasion (situation), gather and organize information, and prepare visual aids. To prepare the presentation, follow these guidelines:

- Find out about the audience.
- Determine the purpose of the presentation.
- Plan for only the information that the audience can absorb.
- Organize the presentation.
- Anticipate the audience's needs and questions.
- Rehearse the presentation.
- Prepare for emergencies.

FIGURE 20.1

Advantages, Disadvantages, and Guidelines for Oral Presentations

Type of Presentation	Advantages	Disadvantages	Guidelines
Impromtu	• Delivered in a relaxed, conversational manner	• May be disorganized because the speaker can't prepare in advance • May be rambling and unfocused	• Stop and think before speaking • Ask the audience questions to determine what they want you to speak about
Extemporaneous	• Prepared ahead of time • Delivered in a relaxed, conversational manner • Allows the speaker to adjust the presentation in response to the audience's reactions • Takes less time to create than a scripted presentation	• Can easily run over the time limit • May cause the speaker to leave out information	• Rehearse the presentation • Use visual aids to guide you as you give the presentation (these aids will also help the audience) • Define any new information or terminology • Use simple, not fancy, language • Prepare notes or an outline
Scripted	• Prepared ahead of time • Allows the speaker to deliver complete, accurate information • Helps the speaker to stay within the time limit	• Often delivered in an unnatural, boring manner • Doesn't allow the speaker to adjust to the audience's reactions • Takes a long time to prepare	• Use for situations where the audience expects precision • Use visual aids and examples • Define any new information or terminology • Use simple, not fancy, language
Memorized	• Prepared ahead of time • Can allow the speaker to deliver complete, accurate information • Helps the speaker to stay within the time limit	• Makes the speaker seem stiff and formal • Without notes, the speaker may lose his or her place or forget part of the presentation • Takes a long time to prepare	• When possible, select either the extemporaneous or scripted instead of the memorized presentation

Find Out about the Audience

To deliver an effective presentation, you have to know who will be listening. To find out as much as possible about the audience, consider these questions:

- **What do you know about the demographics of your audience?** *Demographics* are the characteristics of a group such as the size of the group, the average age, the gender, the educational backgrounds, where they work, etc. The more you know about the demographics of your audience, the better you can deliver an audience-focused presentation.

- **What does the audience know about your topic?** Find out what, if anything, your audience knows about your topic. With this information, you can determine what technical terminology you can use, what the audience will understand, and how much background, if any, your audience will need to understand your presentation.

- **What is your audience's attitude about your topic?** Do they have a positive, negative, or neutral attitude toward your topic? Will the members of the audience have different attitudes about the topic? For example, if you are delivering a presentation on cost-cutting initiatives for your department, will the audience favorably receive your message, or will they feel threatened? The answers to these questions will help you know what types of information to include, what information to emphasize, and so on.

- **What is the audience's impression of you or your organization?** The audience may already have developed an impression of your organization. That impression may be positive or negative. Will your audience be hostile? Will they be enthusiastic? Have they not yet formed an impression of you or your organization?

- **Why is the audience listening to the presentation?** The audience may attend the presentation simply because they have to; or they may attend because they want to gather information, to make a decision, or to learn how to do something.

- **What does the audience expect from your presentation?** Does the audience expect a formal presentation or an informal one? Does your audience expect certain types of visual aids? Be sure that your presentation is appropriate to the occasion and meets the expectations of your audience.

Determine the Purpose of the Presentation

Once you have identified your audience and occasion, spend some time considering your purpose:

- **Why are you giving the presentation?**
- **What do you want to accomplish?** What are your goals for the presentation? Can you accomplish those goals in the time allotted for the presentation? If not, adjust the goals so you can achieve them in the time allotted.
- **Is your goal to inform, to persuade, or both?** For example, if you are delivering a presentation on how to use a new product that the audience has just bought from your organization, your purpose is to inform. However, if you are delivering a presentation on the benefits of this new product to an audience that might purchase it, your purpose is to persuade.
- **What do you want your audience to do with the information?** (Reiffenstein, "Five Things") For example, do you want your audience to purchase a product, to make a decision, or to simply gather information?

> *To deliver an effective presentation, you must know your audience; determine your purpose; and plan, organize, and rehearse your material.*

Plan for Only the Information that the Audience Can Absorb

Listening to information takes twice as long as reading that same information. If your audience can read 10 pages in 8 minutes, they can comprehend the same information as listeners in about 16 minutes. Similarly, the average adult's attention span is about 18 minutes, so you should include only the amount of information that the audience can absorb in that time (Reiffenstein, "Five Things"). If your talk is more than 20 minutes long, break up the presentation with interactive exercises, lengthen the question-and-answer time, or perhaps supplement the presentation with handouts (Reiffenstein, "Five Things").

Follow these guidelines to prepare a presentation that centers on what the audience can absorb:

- **Condense your presentation into its key points.** Don't try to give the audience every bit of information you have about a topic or all the exhaustive details. Instead, select the key points and present those. If necessary, you can refer the audience to your written research, to published articles, or to handouts that you have prepared.
- **Stay within your allotted time and speak to the requirements of the occasion.** For example, if you are speaking at a professional conference, the conference organizers will tell you how long to speak. They expect you to speak for no more than the allotted time. If they give you 30

minutes, they expect you to make your presentation and answer questions during those 30 minutes. If you take more than your allotted time, other speakers may not have their full 30 minutes. When you speak longer than your allowed time, you are being inconsiderate and most audiences will begin to "tune you out."

- **Plan the presentation to take slightly less than the allowed time.** Look for ways to tighten your presentation, so you have time to answer the audience's questions. Most audiences prefer a presentation that is a few minutes short rather than a few minutes long.

Organize the Presentation

Because most presentations have a specified time period, you should prioritize the information according to the needs and expectations of the audience and your purpose. Imagine that you are studying the effects of mercury emissions on health. At the end of the study, you will present your results to a group of public health experts. In a written document, you provide background information, methods, results, conclusions, and recommendations. However, the audience of public health experts will be interested primarily in your conclusions and recommendations. So if you have only 20 minutes for your presentation, you should spend the majority of your time on the conclusions and recommendations. You would briefly state the purpose of your research and how you conducted the research. You would only summarize the methods, if you mention them at all. Instead, you might refer the audience to a handout that describes, in detail, your methods.

Once you have prioritized the information that you will present, prepare an outline of your presentation—as you would for a written document. Some speakers use a storyboard approach for their outline. A *storyboard* is a "sketch" of the document or presentation; it maps out each section or module of your document or presentation, along with the accompanying visual aids. For an oral presentation, the storyboard includes an outline on the left side of the page and a list of the corresponding visual aids on the right side. You could include a description of the visual aids, or you could include small printouts or thumbnail sketches of the visual aids. Figure 20.2 illustrates a storyboard for part of a presentation on writing effective e-mail. On the left side of the storyboard, the writer includes the outline,

---

FIGURE 20.2

Excerpt from a Storyboard for an Oral Presentation

Introduction • What we will talk about • Consider how your e-mail represents you and your organization • Organize your e-mail to guide your readers • Use a tactful tone	**Slide**: title of presentation **Slide**: list of these four topics that I will cover in the presentation. Use a bulleted list for the topics
Consider how your e-mail represents you and your organization • Determine the type of readers: coworkers, managers, and customers • Put yourself in the reader's shoes so you can more easily solve problems and get the results you want • Consider these examples	**Slide**: introduce this section **Slide**: details of putting yourself in the reader's shoes **Slides**: sample e-mails (two or three examples—one per slide)
Organize your e-mail to guide your readers • In the first paragraph, present the main message • In the middle paragraph(s), explain the main message • In the final paragraph, close the e-mail	**Slide**: transition slide **Slide**: details of the first paragraph. Refer readers to sample e-mail in handouts **Slide**: details of the middle paragraph. Refer readers to sample e-mail in handouts **Slide**: details of the final paragraph. Refer readers to sample e-mail in handouts

and on the right side, the slides and handouts that will accompany the information in the outline. If you plan to use Microsoft *PowerPoint*® slides, you can use the outline function. Figure 20.3, page 656, illustrates a portion of a *PowerPoint*® outline for an e-mail presentation.

When you have prepared your outline and/or storyboard, you can prepare your notes for the presentation. To prepare the notes, you might
- prepare note cards
- prepare speaker notes, if you're using presentation-graphics software such as Microsoft *PowerPoint*®. Figure 20.4, page 657, shows speaker notes for the presentation on e-mail.

Provide Previews, Transitions, and Summaries

Your presentation needs a beginning, a middle, and an end.
- **Begin your presentation with a preview.** In the preview, tell (and possibly show) the audience what you will be talking about in the presentation.

FIGURE 20.3

Excerpt from an Outline Created in Microsoft *PowerPoint*®

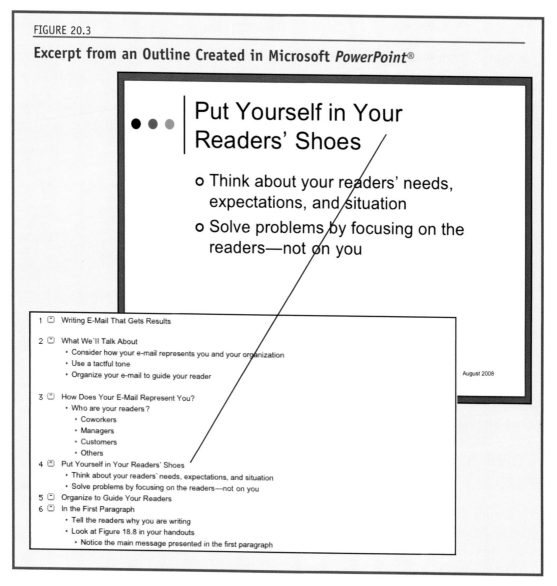

- **Introduce the key points you will discuss (in the order in which you will discuss them).** For example, you might say, "This morning, I will talk about X, Y, and Z." Use a slide that lists X, Y, and Z. Then discuss X, Y, and Z in the order in which you listed them on the slide.
- **Use clear transitions between topics.** *Transitions* tell the audience that you are changing the topic. Without transitions, the audience may mis-

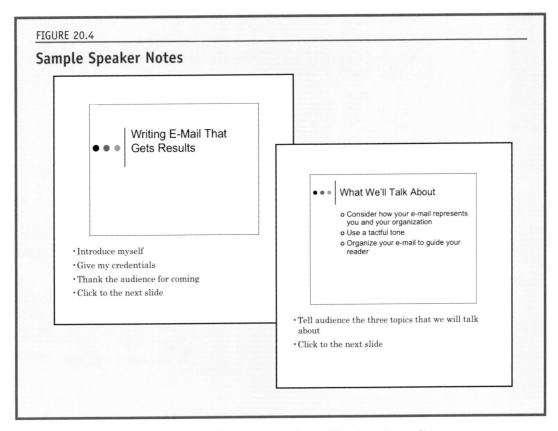

FIGURE 20.4

Sample Speaker Notes

understand your words, get confused, or lose focus. To alert the audience to a change, use transitions such as "My second point is" and "Next, I will discuss…." You can also use visual aids, such as transition slides, to signal a change in topics.

- **Summarize your presentation at the end.** Tell the audience what you've told them by summarizing your key points. You can use a visual aid to emphasize this summary.

Anticipate the Audience's Needs and Questions

To prepare an audience-focused presentation, put yourself in your audience's shoes and consider their needs, concerns, and perspectives. As presenters, we tend to focus on ourselves rather than on the audience. Kathy Reiffenstein, author of "Five Things Not to Do in Front of an Audience," explains: "we worry that we won't look knowledgeable, we won't be able to answer some questions, we'll appear nervous, or we'll forget what we wanted to say." While these worries are valid, the focus is wrong. Instead, we should worry about anticipating and meeting the needs of the audience. "Your presentation is not about you, it's **all** about your audience" (Reiffenstein, "Five Things"). To ensure that you focus your presentation on the audience, follow these guidelines:

- Define any terminology the audience may not know. If audience members aren't experts in your field, avoid technical terminology when possible.
- Use specific, unambiguous examples to help audience members understand complicated or abstract concepts.
- Explain any information that may be new to the audience.
- Clarify and support unfamiliar information, especially if the audience may disagree with or try to reject it.

Rehearse the Presentation

Spend time rehearsing and timing your presentation and checking out the room and its equipment. Follow the tips on the next page as you rehearse and prepare for your presentation.

Prepare for Emergencies

Your presentation may not go as planned. Be prepared for emergencies and be flexible. You can lessen the likelihood of some emergencies by doing the following:

- Take an extra light bulb for the overhead projector or slide projector.
- Bring a backup of your slides and handouts.
- Have a backup plan in case the computer, projector, or DVD player doesn't work.
- Prepare notes. If you use note cards, number the cards. If you use speaker notes from presentation-graphics software, the pages should automatically be numbered. You may not need to refer to your notes often, but they will help if you lose your place while speaking.

PREPARE AUDIENCE-FOCUSED VISUAL AIDS

Visual aids give the audience something to focus on while you're speaking. Audiences often have difficulty staying focused on an oral presentation even when the topic is interesting. However, guard against using too many

TIPS FOR REHEARSING AND PREPARING THE EQUIPMENT

- **Make your rehearsal realistic.** If possible, rehearse your presentation with the actual equipment you will use and in the room where you will speak. As you rehearse, practice with any visual aids you will use.
- **Time your presentation.** Most presentations have an allotted time; so as you rehearse, time your presentation. If the presentation is too long, cut some of the text or some of the visual aids. Allow time for questions from your audience, and stay within your allotted time.
- **Determine where you will use the visual aids.** Mark your note cards or speaker notes for where to present each visual aid.
- **Check out the room and its equipment.** If you are speaking in an unfamiliar setting, check it out before you speak. Find the electrical outlets (if needed), look at the lighting, and determine whether you will need a microphone or a pointer.
- **Practice setting up the computer and projection equipment.** The setup can take more time than you think, so leave yourself plenty of time.
- **Review how to operate the presentation software.** If you plan to use Microsoft *PowerPoint®*, practice these techniques to ensure that you don't distract the audience:
  - **Press Ctrl-H after the slideshow has started to keep the arrow or pointer from appearing on screen**. When you move the mouse, the arrow or pointer appears on screen—distracting the audience. To cancel, press Ctrl-A. If the arrow appears while you are presenting, don't press the escape button. The escape button stops the slide show.
  - **Pause the presentation by pressing "w" for a white screen or "b" for a black screen**. To resume the presentation, press the "w" or "b" key again.
- **Practice using the remote control for advancing the slides.** Each remote control works differently, so spend some time getting familiar with it before giving your presentation.

visual aids. If you use too many, the audience may focus on the visual aids instead of on your message. Visual aids enhance your presentation by

- keeping the audience focused on what you are saying
- helping the audience to remember key points and to follow the organization of your presentation
- helping you to stay with your planned organization, to remember what

you want to talk about, and to use only the allotted time (especially with an extemporaneous presentation)

- helping you to concisely explain ideas, concepts, products, or technical information

Many types of media are available to help you visually enhance your presentation. Figure 20.5, pages 662-663, describes the most commonly used and appropriate media and presents the advantages, disadvantages, and guidelines for using each type. Tips for creating effective visual aids appear on pages 664-665.

DELIVER A MEMORABLE PRESENTATION

To deliver a memorable presentation and to put yourself and your audience at ease, follow these strategies.

Help the Audience Enjoy Your Presentation

Use the tips below to help the audience enjoy and focus on your presentation. If you are well prepared and have rehearsed your presentation, these guidelines will be easy to follow because you will feel comfortable and confident.

TIPS FOR HELPING THE AUDIENCE ENJOY YOUR PRESENTATION

- **Talk slowly and distinctly.** Make sure the audience can understand your words.
- **Look the audience in the eye.** Audiences tend to be suspicious of speakers who don't maintain eye contact.
- **Speak with enthusiasm and confidence.** The audience doesn't want to listen to someone who seems uninterested or who lacks confidence.
- **Avoid verbal pauses ("um," "ah," "uh," "you know").** Rehearsing will help you eliminate these.
- **Don't read the slides.** When you read the slides, you are not maintaining eye contact and you are not interacting with the audience.
- **Introduce yourself before the presentation begins.** You might shake hands or make casual conversation.

Field Questions Effectively

If you have prepared well and anticipated the audience's questions, you will be ready to answer their questions. At the end of your presentation, ask the audience if they have any questions. If appropriate for the occasion, you can also encourage the audience to ask questions during the presentation. For example, you can give the audience the opportunity to ask questions if they don't understand something.

As you field questions, follow these guidelines:
- Repeat the question, so you can make sure that you heard it correctly and that everyone in the audience knows the question that you will be answering.
- Take a few seconds to think before you answer, so you can respond in an organized, clear manner.
- If you don't know the answer to a question, say, "I don't know, but I'll find out" or "I don't know, but I will get you an answer." Your audience will sense if you are "making up" an answer. You will gain the respect of your audience if you are honest and ethical.
- If someone disagrees with you or criticizes your presentation, be respectful.

FIGURE 20.5

Types of Media for Visual Aids

Whiteboard

☐ Advantages
  ▪ Flexible
  ▪ Excellent for informal presentations
  ▪ Excellent for incorporating the audience's ideas
☐ Disadvantages
  ▪ Appropriate only for small rooms and audiences
  ▪ Requires a room with good lighting
☐ Guidelines
  ▪ Write legibly so everyone can read it
  ▪ Look at the audience, not at the board
  ▪ Don't forget to bring dry-erase markers

Poster

☐ Advantages
  ▪ Flexible
  ▪ No equipment required
  ▪ Inexpensive
☐ Disadvantages
  ▪ Appropriate only for small rooms and audiences
  ▪ Unprofessional appearance
  ▪ Hard to transport
☐ Guidelines
  ▪ Use at least 20" x 30" sturdy poster board
  ▪ Use a simple, uncluttered design
  ▪ Make sure every person in the room can read it
  ▪ Use only in rooms with good light
  ▪ Use bright colors for greater contrast and readability

Flip Chart

☐ Advantages
  ▪ Flexible
  ▪ Excellent for incorporating audience's ideas
  ▪ Inexpensive
☐ Disadvantages
  ▪ Appropriate only for small rooms and audiences
  ▪ Unprofessional appearance
☐ Guidelines
  ▪ Use bright colors
  ▪ Keep the flip chart simple
  ▪ Don't put too much on one page
  ▪ Use only in rooms with good light
  ▪ Make sure every person in the room can read it
  ▪ Don't forget to bring markers

Handouts

☐ Advantages
  ▪ Excellent for presenting complex information
  ▪ Helps the audience to remember the information
☐ Disadvantages
  ▪ Distracts the audience from listening
☐ Guidelines
  ▪ Give the audience the handouts at the end of the presentation
  ▪ Make sure the handouts are error free and have a professional appearance
  ▪ Number the pages

FIGURE 20.5 CONTINUED

Opaque Projection

- Advantages
  - Excellent for displaying paper documents
  - Requires little if any expense
  - Flexible—allows you to incorporate documents from the audience
- Disadvantages
  - Poor image quality with many projectors
  - Appropriate only for small rooms and audiences
  - Room must be dark
- Guidelines
  - Leave the projector on until you are finished
  - Make sure the audience is close enough to see the images
  - Plan for a light source so you can see your notes

Slide Projection

- Advantages
  - Professional appearance
  - Excellent for incorporating charts, tables, photographs, and diagrams
  - Easy to operate
- Disadvantages
  - Expensive to produce slides
  - Equipment may fail—have a backup
  - Room must be dark
- Guidelines
  - Put the slides in sequence before you begin speaking
  - Plan for a light source you can use to see your notes
  - Use a laser pointer

Computer Projection

- Advantages
  - Professional appearance
  - Versatile—multi-media capability
  - Legible in small and large rooms
  - Easy to transport
- Disadvantages
  - Room must be somewhat dark
  - Equipment may fail—have a backup
- Guidelines
  - Use a consistent design
  - View the presentation before speaking
  - Rehearse using the equipment
  - Bring more than one copy of the presentation file
  - "Pack" the presentation to ensure that you can show it

Film and Video

- Advantages
  - Excellent for presentations needing moving images and sound
  - Versatile
- Disadvantages
  - Expensive and time consuming to produce
  - Requires high-quality equipment to produce professional quality
  - Requires expertise to produce
- Guidelines
  - Introduce the film or video—tell audience what they will be viewing
  - Practice setting up and operating the equipment
  - Have a backup in case the equipment fails

TIPS FOR CREATING EFFECTIVE VISUAL AIDS

- **Use only the visual aids your audience will need to understand your key points.** Don't complicate your presentation with unnecessary or overly complex visual aids.
- **Select simple visual aids.** Each visual aid should present only one main idea. The audience should be able to easily and quickly comprehend the information that the visual aid is presenting.
- **Make sure each visual aid has a specific purpose.** Each visual aid should relate directly to the key points of your presentation. Avoid using visual aids simply to entertain.
- **Use brief phrases rather than sentences.** Sentences can clutter a visual aid and make it illegible. Instead, use phrases. The phrases should be clear enough that the audience can understand the visual aid after the presentation:

  Sentences   Hydroelectric power has these disadvantages:
                       It causes the loss of wildlife habitats.
                       It takes significant amounts of land for constructing the
                       needed reservoir.

  Phrases   Disadvantages of hydroelectric power:
                       Loss of wildlife habitats
                       Loss of land

- **Make sure everyone in the audience can read the visual aids, not just those sitting up front.** Remember that less is more, especially with slides created with presentation software such as Microsoft *PowerPoint*®. For example, instead of trying to fit all the information about a specific point on one slide by using a smaller type size or narrow margins, use two or more slides. Let's look at some examples: Figure 20.6 has too many words and the type is too small. Figure 20.7, page 666, has ample margins and uses an appropriate type size.
- **Use legible type.** Use legible typefaces and type size. If you are using presentation-graphics software, such as Microsoft *PowerPoint*®, follow these guidelines:
  - Use 28-32 point type for text.
  - Use 36-44 point type for headings.
  - Use sans serif type, such as Tahoma, Arial, and Verdana, as they take less time to read on-screen.
  - Use upper- and lowercase letters, not all capital letters.
  - Avoid shadowed, underlined, or outlined type.
  - Use heavier, filled-in bullets, as they are easier to read (Reiffenstein, "Harness").
  - Use contrasting colors for the text and the background. For example, if you use a white or light background, use a dark type, preferably black or navy.

- **Use slide designs that are appropriate for the occasion and don't distract or make reading difficult.** Some of the designs available in presentation-graphics software are distracting and make text difficult to read. Select backgrounds without images behind the words and with appropriate colors for your presentations. The background in Figure 20.8, page 666, makes reading difficult, whereas the background in Figure 20.7, page 666, enhances reading.
- **Make sure that the visual aids are free of errors and contain correct, accurate information.** Your visual aids represent you and your organization. If they contain errors in grammar and style, or if they contain incorrect information, you will lose credibility. Your audience may not trust you or the information that you are presenting because they may believe that if the visual aids are sloppy or incorrect, your information, research, service, or product is also sloppy or incorrect.

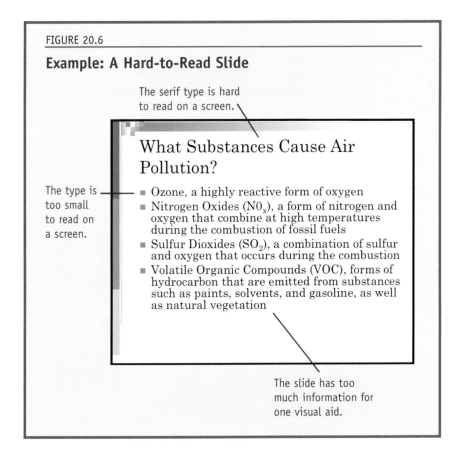

FIGURE 20.6

Example: A Hard-to-Read Slide

The serif type is hard to read on a screen.

What Substances Cause Air Pollution?

The type is too small to read on a screen.

- Ozone, a highly reactive form of oxygen
- Nitrogen Oxides (NO_x), a form of nitrogen and oxygen that combine at high temperatures during the combustion of fossil fuels
- Sulfur Dioxides (SO_2), a combination of sulfur and oxygen that occurs during the combustion
- Volatile Organic Compounds (VOC), forms of hydrocarbon that are emitted from substances such as paints, solvents, and gasoline, as well as natural vegetation

The slide has too much information for one visual aid.

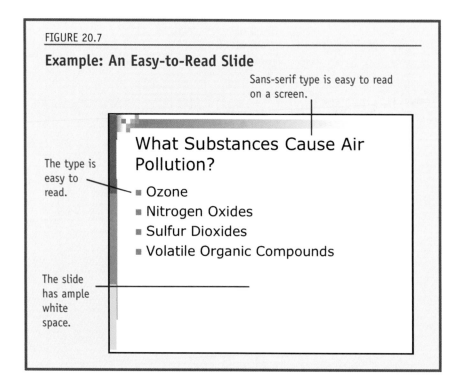

FIGURE 20.7

Example: An Easy-to-Read Slide

Sans-serif type is easy to read on a screen.

What Substances Cause Air Pollution?

The type is easy to read.

■ Ozone
■ Nitrogen Oxides
■ Sulfur Dioxides
■ Volatile Organic Compounds

The slide has ample white space.

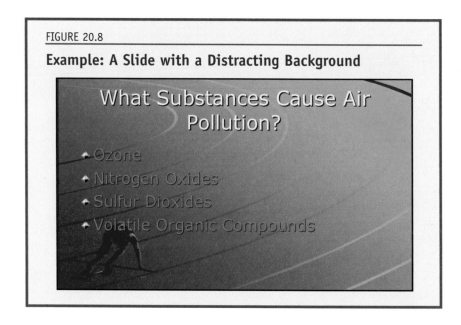

FIGURE 20.8

Example: A Slide with a Distracting Background

What Substances Cause Air Pollution?

◆ Ozone
◆ Nitrogen Oxides
◆ Sulfur Dioxides
◆ Volatile Organic Compounds

TAKING IT INTO THE WORKPLACE
Taking Cues from the Audience

Your audience can give you cues about whether you are meeting their expectations. Karl Walinskas, author of *Reading Your Audience,* suggests that you take those cues and adjust how you are delivering your presentation. He identifies three cues the audience will give you:

Audience Cue	What it Means	How to Adjust
Shut eyelids	Boredom, fatigue	Change pace, volume, and subject matter; use humor to get them laughing.
Wandering eyes	Distraction	Use dramatic action; call attention to an important point and ask for audience focus; use humor.
Mass exodus	Boredom; they've heard it before	Change tactics; use pointed humor; do something dramatic to reconnect; add controversy; move on to the next point; work on content for next time.
Leaning back in seats	Apathy; waiting for something better	Use dramatic action; insert an exercise to involve them; use humor.
Shaking heads	Disagreement	Confront a select head-shaker ("You disagree? Tell us why."); offer an alternative viewpoint that others embrace (even though you may not).
No questions during Q & A	Disinterest, confusion, hesitation	Plant seed questions with several people in the audience ahead of time to get the ball rolling; call on people who you read as being most engaged during the presentation.

- **"The eyes have it."** Your first clue to audience interest is the eyes of each person. (If the room and the audience are large, you can use the people in the first few rows.) Walinskas suggests making sure "their eyes are *open!*...Shut eyelids mean a bored crowd." Is the audience looking around the room or at their laps rather than at you and your visual aids? If so, change the pace or volume of your speech.
- **"Actions speak louder than words."** The audience's body language can also give you cues. For example, if people in the audience are leaning back in their chairs getting comfortable (perhaps for a nap), you might want to change the pace of your presentation or get the audience involved. You might ask them to look at something on a screen or in a handout.
- **"The engagement factor."** Walinskas explains that "the level to which your audience participates is a critical factor in determining how well they are receiving" your presentation. For example, even if you have asked them to hold questions until the end, someone in the audience may be so engaged that he or she can't wait. This signal tells you and the rest of the audience that your presentation is engaging.

Assignment

Attend an oral presentation and evaluate how well the speaker delivers the presentation, based on audience cues. Write a memo to your instructor including the name of the speaker, the title and date of the presentation, and your evaluation.

Compiled from information downloaded from the World Wide Web: www.speakingconnection.com. Walinkskas, Karl, The Speaking Connection.

CASE STUDY ANALYSIS
An Award-Winning *PowerPoint*® Hits Theaters

Background

Few people would have bet that a *PowerPoint*® presentation could win an Academy Award. Yet, in 2007, that's exactly what happened.

After bowing out of the political scene, former Vice President, Al Gore, wanted to spread the word about global warming. He created a *PowerPoint*® presentation that became a book and an Academy-Award-winning movie.

Even though the medium changed (from oral presentation given in person) to video (watched *en masse* in theaters), the format of Gore's presentation remained virtually the same; the video was a film of Gore giving his oral presentation to a group.

Regardless of your beliefs about Al Gore or global warming, Gore's film, *An Inconvenient Truth*, is a classic study in oral presentations. Gore, who has been criticized for his droll vocal delivery style, uses many of the elements discussed in this chapter—including humor, varying intensity and pace and visual aids—to create an engaging presentation.

Assignment

Meet with a group of students (or as a class) and watch *An Inconvenient Truth*.
1. Take notes on the elements used by Gore in his presentation.
2. Write an evaluation of Gore's presentation—*do not write about your thoughts on global warming*—and turn it in to your instructor.
    a. In your evaluation, describe the audience to which Gore directs his presentation; the level of knowledge his audience has; the depth in which Gore covers the topic; the quality of the video used in the presentation; the type used in the slides; the delivery style and pace, etc.
    b. Discuss what works and what doesn't in terms of oral presentations and style.

EXERCISES

1. Write instructions for creating a slide presentation using presentation-graphics software, such as Microsoft *PowerPoint®*, available at your college or university. The users of your instructions are your classmates. If you need help writing instructions, refer to Chapter 15.

Download a copy of the Worksheet for Preparing and Delivering Oral Presentations at www.kendallhunt.com/ technicalcommunication.

2. Using presentation-graphics software, such as Microsoft *PowerPoint®*, create a custom slide design that you can use as a master slide for an oral presentation. Make sure that your slide design follows the tips on pages 664-665.

3. Create an oral presentation on one of the following topics. You will give your presentation to your classmates. Your presentation should include visual aids. Your instructor will tell you the time allotted for the presentation.
   a. Define a concept in your major.
   b. Explain a procedure you commonly follow at your workplace or will use as a professional in your field.
   c. Present the conclusions and/or results of a report that you prepared for your technical communication class.
   d. Present a visitors' guide to your hometown.
   e. Explain a procedure that students follow at your college or university.

4. Using the online evaluation sheet, evaluate the presentations of your classmates. Your instructor may ask you to complete this evaluation online and attach it to an e-mail. The evaluation sheet is available on the Web site for this book.

Download an evaluation sheet online at www.kendallhunt.com/ technicalcommunication.

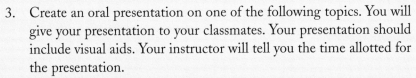

REAL WORLD EXPERIENCE
Working with a Team to Prepare an Oral Presentation

You and your team have developed a product or service that community groups might use for fundraising. You want to get people to buy your product or service, so you have decided to prepare an oral presentation introducing your product or service to groups in your community. You and your team will work together to prepare and deliver the presentation. The purpose of your presentation is to

- introduce your product or service
- explain the advantages of your product or service
- persuade the group to use your product or service for its next fundraising project

Assignment

Your team will complete the following steps:

Download an evaluation sheet online at www.kendallhunt.com/technicalcommunication.

1. Create a product or service that you can use for your presentation. (You don't have to actually create a product; you can simply describe the product.)
2. Decide what community group would likely use your product or service for fundraising.
3. Analyze the audience of your presentation. (This audience would be the group that you identified in Step 2.)
4. Plan your presentation.
5. Create appropriate visual aids.
6. Rehearse the presentation. Be sure to assign roles to each member of the team. Every member should have a speaking part in the presentation.
7. Deliver the presentation to your instructor and your classmates. They will use the evaluation sheet (available on the Web site for this book) as they listen to your presentation.

APPENDIX A: DOCUMENTING YOUR SOURCES

Your organization or your instructor may have a preferred style for documenting sources. Find out what style they expect and use that style. Many instructors will expect you to use one of the following styles, which appear in this appendix:

- *Publication Manual of the American Psychological Association*, 5th ed. (Washington: APA, 2001). This style is used in the social sciences and other fields.
- *MLA Handbook for Writers of Research Papers*, 5th ed. (New York: MLA, 1999). This style is used in the humanities.
- *Scientific Style and Format: The CSE Manual for Authors, Editors, and Publishers*, 7th ed (Council of Science Editors 2006). CSE style is one of many numerical documentation styles.

APA STYLE

When you document information using APA style, you should consider two areas: citing the information in the text and preparing the references at the end of the document.

Citing Information Using APA Style

When using APA style to cite information in a document, you typically will include the following information:
- author's last name
- year the source was published
- page number(s) if you are giving a specific fact, idea, or quotation

For example, a citation might read as follows:

> Thomas (2007) identified the reaction times while working in Alaska.
> Documents with an illogical structure make readers' tasks more difficult (Cobb 2008).

Boswood (1999) states that "the professional applies a body of knowledge by exercising a range of skills in an ethical manner" (116).

The textual citations may vary depending on the type of information and context. If the models above don't fit your information, consult the *Publication Manual of the American Psychological Association*.

If you are citing a personal communication, use this format:

N. Cobb (personal communication, June 29, 2008) suggested that . . .

Preparing the Reference List Using APA Style

A reference list gives readers the information they need to find each source that you have cited in your document. Each entry in the reference list normally includes the following:
- author's name
- year of publication
- title of publication
- publishing information

In the reference list, include only those sources that you have actually used and cited in your document. Leave out sources that you used for background reading. When you prepare your reference list, follow these guidelines:
- **Put the sources in alphabetical order by the author's last name**. If you have more than one source from the same author, arrange the sources by date—beginning with the earliest source and moving to the latest (or most recent) source. If the sources are from the same year, use lowercase letters to distinguish the articles (for example, Johnson 2008a, Johnson 2008b, and so on).
- **Use only initials for the author's first and/or middle name** (for example, Smith, E., not Smith, Ellen).
- **Capitalize only the first word of each title (and the first word of the subtitle, if necessary)**. This guideline applies to all titles.
- **Italicize or underline the names of journals, magazines, newspapers, and books**. Also italicize or underline the volume and issue numbers. Use the underlining or italics consistently.
- **End each reference with a period**.

- **Use a hanging indent of five to seven spaces to indicate the second (and subsequent) lines of a reference**. You can see this indent in the examples throughout this section.
- **Give the complete page numbers when citing a range of page numbers**. For example, write 345-352, not 345-52.

Books and Brochures

For books, include the following information in this order:
- last name(s) and initials of the author(s)
- publication year
- title (in italics)
- publication information (city and publisher)

Books by One Author
Beaufort, A. (2000). *Writing in the real world: Making the transition from school to work*. New York: Teachers College Press.

Book by Multiple Authors
Separate the authors' names with commas and use an ampersand (&) between the final two authors' names.

Miller, J. D., & Kimmel, L. G. (2001). *Biomedical communication: Purposes, audiences, and strategies*. San Diego: Academic.

Book in Edition Other Than First
When a book is in a later edition, include the edition of the book in parentheses.

Williams, J. (2006) *Style: Ten lessons in clarity and grace*. (9th ed.) Boston: Longman.

Book Issued by an Organization
American Psychological Association. (2001). *Publication manual of the American Psychological Association* (5th ed.). Washington, DC: American Psychological Association.

An Edited Book

When referencing an edited book, include the following information in this order:

- name(s) of the editor(s), following the same guidelines for multiple editors as for multiple authors
- publication year
- title of the book (in italics)
- publication information (city and publisher)

Kynell, T. C., & Moran, M. G. (Eds.). (1998). *Three keys to the past: The history of technical communication.* Stamford, CT: Ablex.

Book, No Author or Editor

Merriam-Webster's Medical Desk Dictionary. (2002). Springfield, MA: Merriam-Webster, Inc.

Journals, Magazine Articles, and Newspaper Articles

Include the following information in this order:

- author's last name and initials. If the article does not have an author, alphabetize the reference by the title of the article, ignoring all articles (*a, an,* and *the*).
- article's publication year. For magazines and newspapers, include as much detail as possible.
- title of the article
- title and volume number of the publication
- page number(s) of the article

Journal Article, One Author

Smith, E. O. (2000). Points of reference in technical communication scholarship. *Technical Communication Quarterly, 9,* 427-453.

Journal Article, Multiple Authors

Separate the authors' names with a comma and an ampersand (&) and use the same format for each author's name: *last name first.*

Constantinides, H., St. Amant, K., & Kampf, C. (2001). Organizational and intercultural communication: An annotated bibliography. *Technical Communication Quarterly, 10,* 31-58.

When a journal article has six or more authors, list the first three authors' names and finish the entry with *et al.*

Smith, E. O., Audrain, S., Bowie, J., et al. (2001). 2000 ATTW bibliography. *Technical Communication Quarterly*, 10, 447-479.

Magazine Article
Hindo, B. The empire strikes at silos. (2007, August 20 & 27). *BusinessWeek*, 63-65.

Newspaper Article
For the pages of a newspaper article, use *p.* for a single-page article and *pp.* for a multipage article. If the article runs on continuous pages, separate the page numbers with an *en dash* (pp. 1-2). If the article runs on discontinuous pages, separate the page numbers with a comma (pp. 1, 3).

Young, J. R. (2000, November 9). Going to class in a 3-D lecture hall. *New York Times* (Late ed.). p. G8.

Article or Chapter in an Edited Book
When referencing an article or a chapter within an edited book, include the following information in this order:
- last name and initials of the author
- publication year of the book
- title of the article or chapter
- the word *In* followed by the name of the editor (beginning with the editor's first initials) and the abbreviation *Ed.* in parentheses
- title of the book
- inclusive page numbers of the article or chapter
- book's publication information (city and publisher)

Ornatowski, C. M. (2000). Ethics in technical/professional communication: From telling the truth to making better decisions in a complex world. In M. A. Pemberton (Ed.). *The ethics of writing instruction: Issues in theory and practice* (pp. 139-166). Stamford, CT: Ablex.

Journal, Magazine, or Newspaper Article with No Author Cited
Vista's Edge, Inc., wins technical writing award (2001, September 3). *San Diego Business Journal*, 36, 10.

Electronic Documents

Document from a Web Site

When referencing a non-periodical document from a Web site, include as much of the following information as possible:

- author of the document (if you cannot find the name of the author, begin with the title of the document)
- publication date or most recent update (or *n.d.* if the document doesn't have a publication date)
- title of the document (in italics)
- date you retrieved the document
- URL. If necessary, you may break a URL to a new line after a slash or before a period; do not insert a hyphen at the break. Omit the period following the URL.

Hansen, K. (n.d.). *Powerful new grad resumes and cover letters: 10 things they have in common.* Retrieved July 30, 2008, from http://www.col-legerecruiter.com/pages/articles/article597.php

Article in an Online Periodical

If you referenced an article online, provide the same information as you would for articles appearing in print (see page 674). If the article appears in print,

- do not include the URL
- include the words *Electronic version* in brackets after the title of the article

Roy, D. (2008). Designing procedural graphics for surgical patient-educa-tion modules: An experimental study [Electronic version]. *Technical Communication Quarterly*, 17(2), 173-201.

If the article is not available in print or if you are citing a different version from the printed version, include the

- date you retrieved the article
- URL. If the article is from a searchable Web site, include the URL for the site. If the article is not from a searchable site, include the URL for the article.

Hesseldahl, A. (2008, June 24). Technology: It's where the jobs are. *BusinessWeek.* Retrieved July 28, 2008, from http://businessweek.com

Article Retrieved from a Database

Provide the same information as you would for articles appearing in print (see page 674) along with the following information:

- date you retrieved the article
- publication information
- name of the database
- item number if available

Morrice, J. (2008, March 1). Human influences on water quality in great lakes coastal wetlands. *Environmental Management,* 41(3), 347-357. Retrieved July 30, 2008, from Agricola database (IND44015382).

E-Mail Message

When referencing an e-mail message in the text, cite the author of the message in the same way you would cite the source of a personal communication. Do not include the e-mail in the reference list.

Message Posted to Online Forums, Discussion Groups, or Electronic Mailing Lists

If the posting is not archived (you cannot retrieve it), cite the posting as a personal communication and do not include it in the reference list. If you can retrieve the posting from an archive, provide the following information:

- author's name
- exact date of the posting
- title or subject line of the posting
- any identifiers such as a message number in brackets
- the words *Message posted to* followed by the URL. Omit the period following the URL.

DuBay, W. (2008, July 30). Journal readability. Message posted to http://lyris.ttu.edu/read/?forum=attw-1

Other References

Article from a Volume of Proceedings

When referencing an article from a volume of conference proceedings, include the following information in this order:

- author's last name and initials
- publication date of the proceedings

- title of the article
- title of the published proceedings
- publication information (city and publisher)

DeLoach, S. (2001). An overview of HTML-based help. *Proceedings of the 48th International Technical Communication Conference.* Fairfax, VA: Society for Technical Communication.

Brochure, Corporate Author

When referencing a brochure written by a company or organization, include the following information in this order:
- corporation (company or organization)
- publication year
- title of the brochure (in italics) with the word *Brochure* in brackets following the title
- publication information (city and corporate publisher).

IBM. (2008). *IBM Annual Report 2008.* [Brochure]. White Plains, NY: IBM.

Government Document

When referencing a document published by the Government Printing Office (GPO), include the following information in this order:
- government agency that released the document
- publication year
- title of the document (in italics)
- publication information (city and publisher).

U.S. Department of Justice. (2000). *Bellows report: Final report of the attorney general's review team on the handling of the Los Alamos National Laboratory investigation.* Washington, DC: U.S. Government Printing Office.

Technical Report

When referencing a technical report, include the following information in this order:
- author's last name and initials
- publication year

- title of the report (in italics)
- series name (if any) and number of the report
- publication information (city and publisher)

Scott, D. D. (2000). *Archeological overview and assessment for Wilson's Creek National Battlefield, Greene and Christian Counties, Missouri.* (Midwest Archeological Center technical report No. 66). Lincoln, NE: U.S. Department of the Interior, National Park Service, Midwest Archeological Center.

Personal Interviews and Personal Correspondence

Treat personal interviews and correspondence like personal communications and do not include them in the reference list. Instead, cite them in the text as personal communications.

Published Interviews

Include the following information in this order for published interviews:
- person(s) interviewing
- date
- title. If the interview doesn't have a title, include the word *Interview* with the subject's name in brackets.
- publication information

Zachary, M., & Thralls, C. (2004). An interview with Edward R. Tufte. *Technical Communication Quarterly, 13*(4), 447-462.

MLA STYLE

When you document information using MLA style, you need to understand how to cite information in the text and how to prepare the "Works Cited" page.

Citing Information in the Text Using MLA Style

When citing information in a document, you typically will include the following information:
- author's last name
- page number being referenced

For example, a citation might read as follows:

> Thomas identified the reaction times while working in Alaska during the 1950s (14-28).
>
> Documents with an illogical structure make readers' tasks more difficult (Cobb 18).

If you're citing an entire source, you would include only the author's last name. The textual citations may vary depending on the type of information and context. If the models above don't fit your information, consult the *MLA Handbook for Writers of Research Papers*.

Preparing the List of Works Cited Using MLA Style

A list of works cited gives readers the information they need to find each source that you have cited in your document. Each entry in the list normally includes the following:
- author's name
- year of publication
- title of publication
- publishing information

In your works cited, include only those references that you have cited in your document. You should not include references that you used for background reading. When you prepare your list of works cited in MLA style, follow these guidelines:
- **Put the references in alphabetical order by the author's last name**. If you are citing two or more references by the same author, arrange them by title. If you are citing references by an organization, alphabetize the work by the first important word in the name of the organization. For example, if the American Heart Association wrote the reference, alphabetize by the word *American*.
- **Use title-case capitalization**. Capitalize each important word in the titles and subtitles.
- **Italicize or underline the names of journals, magazines, newspapers, and books**. Use the italics or underlining consistently.
- **Put the titles of articles and other short works in quotation marks**.
- **End each reference with a period**.
- **Use a hanging indent of one-half inch to indicate the second (and**

subsequent) lines of a reference. You can see this indent in the examples throughout this section.

- **Give only the last two digits of the page number when citing a range of page numbers**. For example, write 345-52, not 345-352. Don't include *p.* or *pp.* to indicate page(s).
- **Follow this format for dates: day, month, year, without commas** (29 June 2002).

Books

When referencing a book, include the following information in this order:
- author's name
- title of the book (in italics)
- publication information including the publisher's location and name and the date.

Book by One Author

Beaufort, Anne. *Writing in the Real World: Making the Transition from School to Work*. New York: Teachers College Press, 2000.

Book by Multiple Authors

- List the authors' names in the order they appear on the title page (which is not necessarily in alphabetical order).
- List the first author's name with the last name first followed by a comma, and then the next author's name beginning with the first name.
- Separate the last two authors with the word *and*. If a book has more than three authors, list the first author's name only followed by *et al.*

Miller, Jon D., and Linda G. Kimmel. *Biomedical Communication: Purposes, Audiences, and Strategies*. San Diego: Academic, 2001.

Book in Edition Other Than First

Williams, Joseph. *Style: Ten Lessons in Clarity and Grace*. 8th ed. Boston: Longman, 2006.

Book Issued by an Organization

Include the following information in this order:
- name of the organization

- title of the book (in italics)
- publication information including the publisher's location and name and the date. The organization and the publisher may be the same.

American Psychological Association. *APA Style Guide to Electronic References*. Washington, D.C.: American Psychological Association, 2007.

Book Compiled by an Editor or Issued Under an Editor's Name
Include the following information in this order:
- editor(s) names followed by the abbreviation *ed.* or *eds.*
- title of the book (in italics)
- publication information including publisher's location and name and date

Kynell, Teresa C., and Michael G. Moran, eds. *Three Keys to the Past: The History of Technical Communication.* Stamford, CT: Ablex, 1998.

Book, No Author or Editor
Merriam-Webster's Medical Desk Dictionary. Springfield, MA: Merriam-Webster, 2002.

Multiple Books by the Same Author
For the second and subsequent books by the same author, use three hyphens instead of the author's name.

Horton, William. *E-Learning by Design.* San Francisco, CA: John Wiley & Sons, 2006.
—. *The Icon Book: Visual Symbols for Computer Systems and Documentation.* San Francisco, CA: John Wiley & Sons, 1994.

Journals, Magazine Articles, and Newspaper Articles
When referencing articles, include the following information in this order:
- full name(s) of the author(s) (last name first)
- title of the article (inside quotation marks)
- title of the journal, magazine, or newspaper (in italics)
- publication information, publication date, and page numbers

Journal Article, One Author

Roy, Debopriyo. "Designing Procedural Graphics for Surgical Patient-Education Modules: An Experimental Study." *Technical Communication Quarterly* 17.2 (2008): 173-201.

Journal Article, Multiple Authors

- List the authors' names in the order they appear in the publication (which is not necessarily in alphabetical order).
- Put the first author's name with last name first followed by a comma, and the other authors' names beginning with the first name.
- Separate the last two authors with the word *and*.

Popham, Susan L. and Sage Lambert Graham. "A Structural Analysis of Coherence in Electronic Charts in Juvenile Mental Health." *Technical Communication Quarterly* 17.2 (2008): 149-72.

When referencing a journal article with more than three authors, cite only the first author's name in full and add *et al.*

Smith, Elizabeth Overman, et al. "2000 ATTW Bibliography." *Technical Communication Quarterly* 10.3 (2001): 447-79.

Journal, Magazine, or Newspaper Article with No Author Cited

"Vista's Edge, Inc., Wins Technical Writing Award." *San Diego Business Journal* 3 Sep. 2001: 10.

Magazine Article

Do not list the magazine's volume and issue numbers. If the article is on more than one page, list the first page followed by a plus sign (+) with no intervening space.

Hindo, Brian. "The Empire Strikes at Silos." *BusinessWeek* 20 & 27 Aug. 2007: 63+.

Newspaper Article

If the city of publication is not included in the name of a locally published newspaper, add the city in square brackets following the newspaper title.

If the article is on more than one page, list the first page followed by a plus sign (+) with no intervening space.

Young, Jeffrey R. "Going to Class in a 3-D Lecture Hall." *New York Times* 9 Nov. 2000, late ed.: G8.

Article or Chapter in an Edited Book

When referencing an article or chapter in an edited book, include the following information in this order:

- author's full name
- title of the article or chapter (in quotation marks)
- title of the edited book (in italics)
- *ed.* for editor followed by the name(s) of the editor(s)
- publication information including the publisher's location and name, the date, and the page numbers of the article or chapter

Ornatowski, Cezar J. "Ethics in Technical/Professional Communication: From Telling the Truth to Making Better Decisions in a Complex World." Ed. Michael A. Pemberton. *The Ethics of Writing Instruction: Issues in Theory and Practice.* Stamford, CT: Ablex, 2000. 139-66.

Electronic Documents

Document from a Web site

When referencing an electronic document from a Web site, include the following information in this order:

- author's full name (if one is given)
- title of the document (in quotation marks)
- title of the Web site (in italics)
- publication date (if one is given)
- name of the publishing organization
- date you retrieved the document
- URL (inside angled brackets.) Close the reference with a period.

Hansen, Randall S. "Scannable Resume Fundamentals: How to Write Text Resumes." *Quintessential Careers.* 2008. Quintcareers.com. 30 July 2008 < http://www.quintcareers.com/scannable_resumes.html>.

Article in an Online Periodical

If you referenced an article online, provide the same information as you would for articles appearing in print (see page 674) along with the following additional information
- date you retrieved the article
- URL (between angled brackets)

Hesseldahl, Arik. "Technology: It's Where the Jobs Are." *BusinessWeek* 24 June 2008. 28 July 2008 <http://www.businessweek.com/technology/content/jun2008/tc20080623_533491.htm?campaign_id=rss_tech>

Article Retrieved from a Database

When referencing an article that you retrieved from an electronic database, provide the same information as you would for articles appearing in print (see page 674) along with the following information:
- name of the database
- library where you retrieved the article
- URL for the database or service

Morrice, John A. "Human Influences on Water Quality in Great Lakes Coastal Wetlands." *Environmental Management*, 41.3 (2008): 347-57. *Agricola*. EBSCOhost. University of North Texas, Denton. 30 July 2008 <https://libproxy.library.unt.edu:9443/login?url=http://search.ebscohost.com/login.aspx?direct=true&db=agr&AN=IND44015382&site=ehost-live&scope=site>.

E-Mail Message

When referencing an e-mail message, include the following information in this order:
- name of the writer
- title of the message taken from the subject line
- name of the recipient
- date of the message

Sims, Patrick. "Re: Graduate Program Questions." E-mail to Susan Audrain. 14 Nov. 2008.

Message Posted to Online Forums, Discussion Groups, or Electronic Mailing Lists

When referencing a message posted to an online forum, include the following information in this order:

- author's full name
- subject of the message followed by the description *Online posting*
- date of the message posting
- name of the online forum
- URL in angled brackets. Close the reference with a period.

DuBay, William. "Journal Readability." Online posting. 30 July 2008. ATTW-L. 2 Aug. 2008 <http://lyris.ttu.edu/read/?forum=attw-l>.

Other References

Article from a Volume of Proceedings

When referencing an article from a volume of conference proceedings, include the following information in this order:

- author's full name
- title of the article (in quotation marks)
- title of the proceedings (in italics)
- publication information including the publisher's location and name and the date

DeLoach, Scott S. "An Overview of HTML-Based Help." *Proceedings of the 48th International Technical Communication Conference.* Fairfax, VA: Society for Technical Communication, 2001.

Brochure, Corporate Author

Treat a brochure or pamphlet the same way you would treat a book (see page 681).

IBM. *IBM Annual Report 2008.* White Plains, NY: IBM, 2008.

Government Documents

When referencing a document published by a government agency, include the following information in this order:

- name of the government agency that released the document
- title of the document (in italics)

- publication information including the publisher's location and title and date

U.S. Department of Justice. *Bellows Report: Final Report of the Attorney General's Review Team on the Handling of the Los Alamos National Laboratory Investigation.* Washington, D.C.: U.S. Government Printing Office, 2000.

Interviews

Include the following information in this order for a published interview:
- person(s) interviewed
- title (in quotation marks). If the interview doesn't have a title, include the word *Interview*, followed by a period.
- publication information

Tufte, Edward R. "An interview with Edward R. Tufte." *Technical Communication Quarterly* 13.4(2004): 447-62.

If you are citing a personal interview, include the following information in this order:
- person(s) interviewed
- the words *Personal interview* followed by a period
- date

Cobb, Neil. Personal interview. 29 June 2008.

NUMBERED DOCUMENTATION STYLES

In a numbered documentation style, each reference is assigned a number the first time it is cited. The writer uses this number when referencing the article in the text. Numbered documentation styles are common in the physical sciences such as chemistry, physics, and geology and in the applied sciences such as computer science, engineering, and medicine. Each discipline has its own preferred numbered style described in manuals such as
- American Mathematical Society, *A Manual for Authors of Mathematical Papers*
- American Chemical Society, *The ACS Style Guide for Authors and Editors*

- American Medical Association, *Manual of Style*

Many disciplines use the style guide published by the Council of Science Editors, *The CSE Manual for Authors, Editors, and Publishers*. This manual outlines the guidelines for a numbered-citation style. It also includes a name-year citation style that basically duplicates the APA documentation style.

Citing Information in the Text Using CSE Numbered Style

In the CSE numbered citation style, you cite the reference using a super-script number immediately following the reference.

> This knowledge is based on extensive epidemiological studies of thousands of underground miners exposed to radon[1-3], carried out over more than fifty (50) years world-wide, including miners in the United States and Canada. One particular study by Wheeler[4] observed miners at relatively low exposure to radon.

When you are referring to two or more sources in a single citation (as in the above example), separate the numbers by a hyphen if they are in sequence. If the references are not in sequence, separate them with a comma ([2,4,10]).

Preparing the List of References in CSE Style

When preparing the list of references in CSE style, follow these guide-lines:

- **List the references in the order in which they appear in the text**. Unlike APA and MLA styles, the references do not appear in alpha-betical order.
- **List the author's name with the last name first and use initials for the first and middle names**. Do not use a comma between the last name and the initials.
- **For the titles of books and articles, capitalize only the first word of the title and all proper nouns**. Do not underline, italicize, or use quo-tation marks for titles of books or articles.
- **For the titles of journals, abbreviate the titles of journals that consist of more than one word**. Capitalize all the words and abbreviations. Do

not underline or italicize the titles.
- **For the publication information, include the publisher's location and name and the date published.**
- **Include the complete page ranges for articles and chapters**: 148-170 not 148-70. For chapters in edited volumes, use *p.* before the page numbers.

Books

For all books, include the following information in this order:
- name of the author(s)
- title of the book (no italics)
- publication information including the publisher's location and name and date published
- number of pages

Books by One Author
1. Beaufort A. Writing in the real world: Making the transition from school to work. New York: Teachers College Pr; 2000. 239p.

Books by Multiple Authors
2. Patterson K, Grenny J, McMillan R, Switzler A. Crucial conversations: tools for talking when stakes are high. New York: McGraw-Hill; 2002. 240p.

Books in Editions Other than the First
Add the edition number along with *ed.* after the title.

3. Williams J. Style: ten lessons in clarity and grace. 9th ed. Boston: Longman; 2006. 304p.

Book Issued by an Organization
4. American Psychological Association. APA style guide to electronic references. Washington, D.C.: American Psych Assoc; 2007. 24p.

Book Compiled by an Editor or Issued Under an Editor's Name
5. Mirel B, Spilka R, editors. Reshaping technical communication. Mahwah, NJ: Lawrence Erlbaum; 2002. 216p.

Book, No Author or Editor

Begin the reference with the title of the book

6. Merriam-Webster's Medical Desk Dictionary. Springfield, MA: Merriam-Webster; 2002. 928p.

Journals, Magazine Articles, and Newspaper Articles

When referencing articles, include the following information in this order:

- name of the author(s). For an article with up to 10 authors, list the names of all the authors. For an article with 11 or more authors, list the first 10 authors followed by a comma and *et al.*
- title of the article (no quotation marks)
- title of the journal (no italics)
- publication date, publication information, and page numbers

Journal Article, One Author

8. Roy D. Designing procedural graphics for surgical patient-education modules: an experimental study. Technical Commun Q 2008; 17(2):173-201.

Journal Article, Two or Three Authors

List the authors' names in the order they appear in the publication (which is not necessarily in alphabetical order).

9. Popham SL, Sage LG. A structural analysis of coherence in electronic charts in juvenile mental health. Technical Commun Q 2008; 17(2):149-172.

Magazine Article

10. Hesseldahl A. Technology: it's where the jobs are. BusinessWeek. 2008 Jun 24:24.

Newspaper Article

11. Young JR. Going to class in a 3-d lecture hall. New York Times (late ed) 2000 Nov 9; Sect. G:8 (col. 2).

Article or Chapter in an Edited Book
Include the following information in this order:
- name(s) of the author of the article or chapter
- title of the article or chapter
- *In* followed by a colon and the book editor(s) and the book title
- publication information including the publisher's location and name, the date, and the page numbers of the article or chapter

12. Ornatowski CJ. Ethics in technical/professional communication: from telling the truth to making better decisions in a complex world. In: Pemberton MA, editor. Ethics of writing instruction: issues in theory and practice. Stamford (CT): Ablex; 2000. p. 139-166.

Electronic Documents
Document from a Web Site
13. Hansen K. Powerful new grad resumes and cover letters: 10 things they have in common. [Internet] DeLand (FL): Quintessential Careers; c2008 [cited 2008 Jul 30]; [about 5 screens]. Available from: http://www.collegerecruiter.com/pages/articles/article597.php

Article in an Online Journal or Magazine
14. Isaacs FJ, Blake WJ, Collins JJ. Signal processing in single cells. Science [Internet]. 2005 Mar 25 [cited 2008 Aug 1];307(5717):1886-1888. Available from: http://www.sciencemag.org/cgi/content/full/307/5717/1886

Article Retrieved from a Database
15. Morrice J. Human influences on water quality in great lakes coastal wetlands. Environmental Management 2008 March 1;41(3):347-57. In: Agricola [database on the Internet]. New York: Springer-Verlag c2008- [cited 2008 Jul 30]: Available from: http://dx.doi.org/10.1007/s00267-007-9055-5; Article IND44015382.

E-Mail Message
"CSE recommends not including personal communications such as e-mail in the reference list" (Hacker). Instead, include a parenthetical note in the text as follows: (2008 e-mail to author; unreferenced).

Message Posted to Online Forums, Discussion Groups, or Electronic Mailing Lists

16. DuBay W. Journal readability. In: ATTW-L [discussion list on the Internet]. [Lubbock (TX): Association of Teachers of Technical Writing]; 2008 Jul 30, 11:20 am [cited 2008 Aug 1]. [about 6 paragraphs]. Available from: http://lyris.ttu.edu/read/?forum=attw-l

Other References

Article from a Volume of Proceedings

17. DeLoach SS. An overview of HTML-based help. Proceedings of the 48th International Technical Communication Conference. Fairfax (VA): Society for Technical Communication, 2001. p. 26-30.

Brochure, Corporate Author

Treat a brochure or pamphlet the same way you would treat a book (see page 681).

18. American Funds. The income fund of America: annual report for the year ended July 31, 2007. White Plains, NY: American Funds; 2008. 36p.

Government Documents

When referencing a government document, include the following information in this order:

- name of the government agency that released the document
- title of the document
- description of the report (if any)
- publication information including the publisher's location and title and date
- information that identifies the document, such as a document number
- the phrase *available from* followed by the name and location of the publishing organization

19. EPA's report on the environment 2008. Washington: U.S. Environmental Protection Agency; 2008. 320p. Available from: EPA, Washington; EPA/600/R-07/045F.

APPENDIX B: COMMON SENTENCE ERRORS, PUNCUTATION, AND MECHANICS

This appendix presents information about common sentence errors, punctuation, and mechanics. The topics within each section appear in alphabetical order, and an abbreviation (such as cs for comma splice) accompanies each topic. You can use these abbreviations as you edit your own documents or those of your team members. This appendix only briefly reviews grammar, usage, and mechanics; if you want complete information on these topics, consult a handbook for grammar and style.

COMMON SENTENCE ERRORS

This section presents common sentence errors and suggests ways to eliminate these errors.

Agreement Errors: Pronoun and Referent (agr P)

A pronoun should refer clearly to a specific noun or pronoun—its referent (also called its *antecedent*)—and should agree in number and in gender with that referent.

Correct	**Gwen** paid cash for **her** new **car** although **it** cost more than **she** was hoping to pay.
Correct	The **students** received a trophy for **their** class project.

When you use a pronoun, make sure that its referent is clear.

Vague	Douglas told Bill that he should move his car.
Clear	Douglas told Bill, "I should move my car."
	Douglas told Bill, "You should move your car."
	Douglas told Bill, "I should move your car."

Pronoun-referent agreement becomes especially tricky with indefinite pronouns (such as *each, everyone, anybody, someone,* and *none*) and collective nouns. When an indefinite pronoun is the referent, the pronoun is singular, as in this example:

Incorrect	**Each** student will receive **their** diploma through the mail.
Correct	**Each** student will receive **his or her** diploma through the mail.

When a collective noun is the referent, determine whether the noun is singular or plural in its context. A collective noun may take a singular or a plural pronoun as its referent.

Incorrect	The **university** will begin a new e-mail service for **their** students.
Correct	The **university** will begin a new e-mail service for **its** students.

In this sentence, *university* refers to a single unit, not to individual members of the university community. In this context, it is singular, and the pronoun referring to it must also be singular.

A collective noun can also be plural, as in this example:

Incorrect	The **faculty** can pick up **its** paychecks on Friday.
Correct	The **faculty** can pick up **their** paychecks on Friday.

Here, *faculty* is a collective noun referring to the faculty in a context that emphasizes the individuals who comprise the group. The faculty wouldn't pick up their checks as a group.

Agreement Errors: Subject and Verb (agr sv)

The subject and verb should agree in number. Often writers create subject-verb agreement errors when the verb follows a prepositional phrase:

Incorrect	The **consequence** of the accidents trouble several board members.
Correct	The **consequence** of the accidents **troubles** several board members.

The noun *accidents* in the prepositional phrase *of the accidents* does not affect the number of the verb; only the subject of the sentence affects the number of the verb.

Comma Splice (cs)

A *comma splice* occurs when writers incorrectly use a comma to link two independent clauses, as in this example:

Comma splice   We baked 10,000 pretzels, we dipped them in dark chocolate.

To correct a comma splice:
- Change the comma to a period followed by a capital letter:
  Correct   We baked 10,000 pretzels. We dipped them in dark chocolate.
- Change the comma to a semicolon:
  Correct   We baked 10,000 pretzels; we dipped them in dark chocolate.
- Leave the comma and after it add an appropriate coordinating conjunction (*and, or, nor, so, for, yet, but*):
  Correct   We baked 10,000 pretzels, and we dipped them in dark chocolate.
- Add a subordinating conjunction to create a sentence consisting of one dependent and one independent clause. A *dependent clause* has a subject and a verb but can't stand alone; an *independent clause* has a subject and a verb and can stand alone. In the following example, the dependent clause begins with *after*:
  Correct   After we baked 10,000 pretzels, we dipped them in dark chocolate.

Modification Errors: Dangling Modifiers (dgl)

See Chapter 8.

Modification Errors: Misplaced Modifiers (mm)

See Chapter 8.

Lack of Parallelism (//)

Use parallel structure when you put items in a series or in a list. All the items in a series or list must be *parallel*—that is, they must have the same grammatical structure. If the first item is a verb, the remaining items must be verbs. If the first item is a noun, the remaining items must be nouns.

Not parallel   To complete the course, you will **write a research paper, take four exams, two collaborative projects**, and **participation**.

The first two items in the series are verb phrases, the third item is a noun phrase, and the fourth item is an unmodified noun. To make items in this series parallel in structure, the third and fourth items must be verb phrases:

Parallel   To complete the course, you will **write** a research paper, **take** four exams, **complete** two collaborative projects, and **participate** in class discussions.

For more information on parallelism, see the discussion of parallelism and headings in Chapter 7.

Run-On Sentences

A *run-on sentence* occurs when two or more independent clauses appear together without any punctuation (*independent clauses* have a subject and a verb and can stand alone). To correct a run-on sentence, you can use the same techniques you would use to correct a comma splice.

Run-on   We baked 10,000 pretzels we dipped them in dark chocolate.

Correct   We baked 10,000 pretzels. We dipped them in dark chocolate.

Correct   We baked 10,000 pretzels; we dipped them in dark chocolate.

Correct   We baked 10,000 pretzels, and we dipped them in dark chocolate.

Correct   After we baked 10,000 pretzels, we dipped them in dark chocolate.

Sentence Fragment

A *sentence fragment* is an incomplete sentence. Sentence fragments usually appear because the writer has left out the subject or the verb or failed to write an independent clause, which can stand alone.

Fragments Resulting from Missing Subjects

Fragment	Detached the coupon from the statement.
Correct	Norma detached the coupon from the statement.
Correct	Detach the coupon from the statement.

The fragment lacks a subject; no actor is doing the detaching. The first complete sentence has a subject—Norma. The second complete sentence has an understood *you* as its subject.

Fragments Resulting from Missing Verbs

Fragment	Norma detaching the coupon from the statement.
Correct	Norma is detaching the coupon from the statement.

The fragment lacks a verb; the *-ing* form requires *is, was,* or *will be* to function as a verb in a complete sentence.

Fragment	The power surge caused by the thunderstorm.
Correct	The power surge caused by the thunderstorm damaged my computer.

In the fragment, *caused* functions as an adjective, not as a verb.

Fragments Resulting from Dependent Clauses

Fragment	You can use the cellular telephone. **If you charge the battery.**
Correct	You can use the cellular telephone if you charge the battery.

If you charge the battery (a dependent clause) cannot stand alone as a sentence because it begins with the subordinating word *if.* An **independent clause** has a subject and a verb, can stand alone, and does not begin with a subordinating word.

Verb Tense Errors

Writers of technical material often misuse the present and the past perfect tenses and shift tense unnecessarily.

Present Tense

Use the present tense to describe timeless principles and recurring events.

Incorrect	In 1997, the Mars *Pathfinder* scientists discovered that the climate of Mars was extremely cold.
Correct	In 1997, the Mars *Pathfinder* scientists discovered that the climate of Mars is extremely cold.

The scientists made their discovery in the past (in 1997) but the climate of Mars continues to be cold.

Past Perfect Tense

Use the past perfect tense (indicated by *had*) to indicate which of two past events occurred first.

Correct	The presentation **had started** when we found the overhead projector.

The writer uses the past perfect tense to clarify that when the presentation started, they had not found the overhead projector.

Unnecessary Shifts in Tense

Within a sentence, do not change tense unnecessarily.

Incorrect	He tested the new hardware, loaded the software, **adjusts** the computer settings, and waited for the network to respond.

Needless shifts in tense distract readers. In this example, the tense of the four verbs should be the same:

Correct	He **tested** the new hardware, **loaded** the software, **adjusted** the computer settings, and **waited** for the network to respond.
Correct	He **tests** the new hardware, **loads** the software, **adjusts** the computer settings, and **waits** for the network to respond.

PUNCTUATION

This section presents information on how to use the apostrophe, colon, comma, dash, exclamation point, hyphen, parentheses, period, question mark, quotation marks, and semicolon.

Apostrophe (ap)

Use the apostrophe to indicate possession, to create some plural forms, and to form contractions.

Apostrophes to Indicate Possession

You can use apostrophes to indicate possession in the following situations:

- To create the possessive form of *most* singular nouns, including proper nouns, use an apostrophe and s, as in these examples:

  gas's odor

  Charles's calculator

  student's book

  If adding an apostrophe and *s* would create an *s* or *z* sound that is hard to pronounce, add only an apostrophe as in *Moses'*. (Try pronouncing *Moses's* and then *Charles's* to see the difference.) When a plural noun does not end in *s*, add an apostrophe and *s*. When a plural noun does end in *s*, add only an apostrophe:

  men's   students'

  children's   members'

- To indicate joint possession, add an apostrophe and *s* to the last noun. To indicate separate possession, add an apostrophe and *s* to each noun.

  Joint possession   John and Stephanie's multimedia
  presentation

  Separate possession John's and Stephanie's multimedia presentations

- To create the possessive form of pronouns, add an apostrophe and *s* only to indefinite pronouns. Personal pronouns and the relative pronoun *who* have special forms that indicate possession.

Possessives of Indefinite Pronouns   **Possessives of Other Pronouns**

anyone's   mine (my)

everybody's   his, hers (her)

everyone's   yours (your)

nobody's   ours (our)

                                         its

no one's   theirs (their)
other's (also others')   whose

Notice that the possessive form of *it* does not have an apostrophe. When you add an apostrophe and *s* to *it*, you create *it's*, the contraction for *it is*.

Incorrect   The city does not believe the pollution is **it's** problem.
Correct   The city does not believe the pollution is **its** problem.

Apostrophes to Create Plural Forms

Use an apostrophe to create the plural form of letters and numbers:
 a's and b's
 8's and 5's (or 8s and 5s)

Some organizations prefer omitting the apostrophe in plural numbers. Check your organization's style guidelines to determine what your organization prefers.

Apostrophes to Form Contractions

Use an apostrophe to indicate the omission of a letter or letters in a contraction.

cannot = can't   who is = who's
you are = you're   let us = let's
it is = it's   they are = they're
does not = doesn't   she will = she'll

Brackets

Use brackets in the following situations:

To indicate that you've added words to a quotation:

Correct   The press release said, "They [Thompson and Congrove] voted against the amendment."

To identify parenthetical information within parentheses:

Correct   (For more information, see *The Publication Manual of the American Psychological Association* [Washington, D.C.: APA, 2001].)

Colon

You can use colons to introduce quotations and lists; to introduce words, phrases, and clauses; and to observe other stylistic conventions.

Colons to Introduce Quotations

Use a colon to introduce a long or formal quotation:

Correct   In the Gettysburg Address, Lincoln began: "Four score and seven years ago our fathers brought forth on this continent, a new nation, conceived in Liberty, and dedicated to the proposition that all men are created equal."

Colons to Introduce Lists

Use a colon to introduce a list when the introductory text would be incomplete without the list. A complete sentence must come before the colon:

Incorrect   For the user testing, you will need: the beta version of the software, a flash drive, and a notepad.

Correct   For the user testing, you will need the following items: the beta version of the software, a flash drive, and a notepad.

Colons to Introduce Words, Phrases, and Clauses

Use a colon to introduce a word, phrase, or clause that illustrates or explains a statement:

Correct   Our manager asked the following people to attend the meeting: production editor, art editor, and copy editor.

Correct   He suggested this solution: balancing the turbine to eliminate the vibration.

The text before a colon must have a subject and verb and must be able to stand alone.

Incorrect   We discovered problems in: the piping system and the turbine.

| Correct | We discovered problems in the piping system and the turbine. |

In the incorrect example, *We discovered problems in* cannot stand alone; therefore, the colon is incorrect.

Other Conventional Uses of Colons

- **Salutations.** Use a colon after the salutation (with a title such as Dr., Mr., or Ms.) in a letter:
  Dear Mr. Johnson:
- **Time**. Use a colon to separate hours and minutes:
  8:30 A.M.
- **Subtitles**. Use a colon to separate the main title from a subtitle:
  Creating Web Pages: A Handbook for Beginners

Comma

The following guidelines will help you to use commas correctly.

Commas to Separate the Clauses of a Compound Sentence

A *compound sentence* has two or more independent clauses (*independent clauses* have a subject and verb and can stand alone). Use a comma to separate the clauses of a compound sentence when a coordinating conjunction (*and, or, for, nor, but, so, yet*) links those clauses.

| Correct | We distributed 500 surveys to the shoppers, but we expect only 20 percent to return the surveys. |

Often, the comma between the clauses of a compound sentence prevents readers from at first thinking that the subject of the second clause is an object of the verb of the first clause:

| Incorrect | Bob will use the test results and the survey results will help him to prepare a prototype of the software. |
| Correct | Bob will use the test results, and the survey results will help him to prepare a prototype of the software. |

Without the comma before *and*, readers at first may think that Bob will use both the test results and the survey results. The comma signals that an independent clause, not the object of the verb *use*, follows *and*.

Commas to Separate Items in a Series

Use commas to separate items in a series composed of three or more items:

Correct   The assistant will deliver, collect, and tally the questionnaires.

The comma before the coordinating conjunction *and* is optional; however, many style manuals encourage writers to use the comma to distinguish items, to prevent ambiguity, and to prevent misreading.

Commas to Set Off Introductory Words, Phrases, or Dependent Clauses

Generally, use a comma to set off an introductory word, phrase, or dependent clause from the main clause:

Correct   Therefore, NASA launched the shuttle two hours later. (introductory word)

Correct   To localize documents, some companies hire translation agencies. (introductory phrase)

Correct   Because the team lost the debate, the school will not receive the prize money. (introductory dependent clause)

A comma after an introductory clause can prevent misreading:

Incorrect   After we completed bathing the cat jumped into the tub.

Correct   After we completed bathing, the cat jumped into the tub.

Without the comma, readers at first might think that the cat was being bathed. If the introductory text is short and readers can't misunderstand it, you can omit the comma.

Commas to Set Off Nonrestrictive Modifiers

A ***nonrestrictive modifier*** is not essential to the meaning of a sentence. Writers can omit a nonrestrictive modifier and readers will still understand the sentence. When writers omit a restrictive modifier, they change the meaning of the sentence.

Restrictive	Homeowners **who don't pay their property taxes** risk severe penalties.
Nonrestrictive	Homeowners, **whether novice or experienced**, can benefit from the seminar on home equity.

The restrictive modifier makes clear that homeowners who don't pay their property taxes risk severe penalties; homeowners who do pay their taxes will not risk penalties. The writer restricts, or limits, the homeowners to those who don't pay their property taxes. The restrictive modifier is essential, and commas should not be used. The nonrestrictive modifier is not essential, and commas should be used.

Commas to Separate Coordinate Adjectives

Use a comma to separate *coordinate adjectives*—adjectives that modify the same noun equally.

Correct	The company will test this fast, powerful computer next week.
Correct	The new design incorporates a bright, rectangular screen.

When adjectives are coordinate, the sentence would still make sense if you replace the comma with the coordinating conjunction *and*. When adjectives are not coordinate, do not separate the adjectives with a comma. Adjectives are not coordinate when the noun and the adjective closest to the noun are closely associated in meaning, as in the following examples:

Incorrect	We will begin the test after the second, special session.
Correct	We will begin the test after the second special session.

In this example, the adjective *second* modifies the combination of the adjective *special* and the noun *session*.

Other Conventional Uses of Commas

* **Dates**. Use commas to separate the parts of a date.

Correct	After Friday, January 1, 2008, you may use your corporate card to charge your tickets and meals.

Notice the comma after 2008. If you do not include the day (January 2008), omit the comma between the month and year. If you include the day before the month (1 January 2008), then don't use a comma.

- **Towns, states, and countries**. Use commas to separate the parts of an address. In the following example, notice the comma after Wisconsin.

Correct	The senator from Madison, Wisconsin, asked the first question.

- **Titles of persons**. Use commas before and after a title that follows a person's name.

Correct	Joseph Gerault, Ph.D., will address the board of directors on Tuesday.

- **Direct address**. Use a comma or commas to set off nouns used in direct address.

Correct	My friends, I am happy to report the results of the second test.
Correct	If you are willing to talk, Thomas, we will select a time convenient for you.

- **Quotations**. Use a comma to introduce most quotations.

Correct	According to John Keyes, "Color grabs a reader's attention before the reader understands the surrounding informational context."
Correct	They asked, "How long will the network be down?"

- **Interjections and transitional adverbs**. Use a comma or commas to separate interjections and transitional adverbs from the other words in a sentence:

Correct	Well, we did not budget any money for the new generator.
Correct	Therefore, we must wait until the next budget period to purchase the generator.
Correct	The old generator, however, is still fairly reliable.

Dash

Use a dash or dashes to emphasize a parenthetical statement or to indicate a sharp change in thought or tone.

| Correct | The United States is a locale, China is a locale, and India is a local—each has its own set of rules and cultural experiences. |
| Correct | The judge found the company guilty of deceptive advertising—as I remember. |

Exclamation Point

Place an exclamation point at the end of an exclamatory sentence—a sentence that expresses strong emotion.

| Correct | The new physics building, originally budgeted for $1.5 million, cost more than $5.5 million! |

Because technical communication strives for objectivity, you will rarely use exclamation points in technical documents.

Hyphen

Use hyphens to form compound words, adjectives, fractions, and numbers and to divide words at the end of a line.

Hyphens in Compound Words

A *compound word* consists of two or more words. Some, but not all, compound words are hyphenated. If you are unsure about whether to hyphenate a compound word, check your dictionary.

Hyphenated	Not Hyphenated
up-to-date	workplace
editor-in-chief	proofread
self-image	bulletin board

Hyphens to Form Compound Adjectives

A *compound adjective* is two or more words that serve as a single adjective before a noun.

twenty-one-inch monitor   **up-to-the-minute** news
self-induced attack   **reader-focused** sentences
black-spotted kitten   **general-to-specific** pattern

Hyphens in Fractions and Compound Numbers

Use hyphens to connect the numerator and denominator of fractions and to hyphenate compound numbers from twenty-one to ninety-nine when spelling out numbers.

three-fourths   twenty-three
one-third   seventy-seven

Hyphens for End-of-Line Word Breaks

Use a hyphen to divide a word at the end of one line and continue it on the next line. Divide words only between syllables. Consult a dictionary to identify correct syllable breaks. Your word-processing software should automatically divide the words between syllables.

Correct   Documents can change across cultures just as body lan-
                   guage, everyday expressions, and greetings change.

Whenever possible, avoid breaking a word at the end of a sentence. You can avoid end-of-line hyphens by using an unjustified right margin. You also can set your word-processing software not to hyphenate any words at the ends of lines.

Parentheses ()/

Use parentheses—always in pairs—in the following situations:
- To enclose supplementary or incidental information:
  Correct   Please e-mail me (jsmith@luminant.com) when you
                     complete your section of the report.
  Correct   To readers in the United States, *EPA* (for Environmental
                     Protection Agency) and *IRS* (for Internal Revenue
                     Service) are common acronyms.
- To enclose numbers and letters used to identify items listed within a sentence:
  Correct   To log on to the network, (1) type your login name,
                     (2) press the tab key, and (3) type your password.

Parentheses are unnecessary when you display a list vertically:

Correct   To log on to the network, complete these steps:
1. Type your login name.
2. Press the tab key.
3. Type your password.

Period

Use a period at the end of most sentences, after most abbreviations, and as a decimal point.

Periods to Create an End Stop

Put a period at the end of any sentence that does not ask a direct question or express strong emotion (an exclamation):

Correct   The Web has changed the way companies communicate with their employees.

Periods After Abbreviations

- Use a period after most abbreviations:
  Ph.D.   etc.
  J.D.   U.S.
- Omit periods from abbreviations for the names of organizations such as corporations and government and international agencies:
  GM (for General Motors)
  NCAA (for National Collegiate Athletic Association)
  FBI (for the Federal Bureau of Investigation)
  UN (for the United Nations)
- Omit periods from acronyms—words (that you can pronounce) formed from the initial letters of the words in a name:
  NASA
  WHO
  DARE

Periods as Decimal Points

- Use a period in decimal fractions and as a decimal point between dollars and cents:
  6.079   69.8%
  .05 $789.40

Question Marks

- Put a question mark at the end of a sentence that asks a direct question:
  Correct   How many volunteers participated in the survey?
- Don't put a question mark at the end of an indirect question:
  Incorrect   The director asked how many volunteers participated in the survey?
  Correct   The director asked how many volunteers participated in the survey.
- When a question mark appears within quotation marks, don't include any other end punctuation:
  Correct   The director asked, "How many volunteers participated in the survey?"

Quotation Marks (""/)

Enclose short quotations and the titles of some published works in quotation marks. Many of us know when to use quotation marks but have trouble knowing how to use other marks of punctuation with them; therefore, this section also presents conventions for punctuation that accompanies quotation marks.

Quotation Marks to Enclose Short Quotations

Enclose a quotation within quotation marks when it is short enough to fit within a sentence and takes up no more than three lines of text:

Correct   According to Thompson, "Monarch butterflies have reddish-brown, black-edged wings."

When a quotation is longer than three lines, follow these guidelines:
- Indent the quotation ten spaces from the left-hand margin.
- Omit the quotation marks. The indentation serves the same purpose as the quotation marks enclosing a short quotation.
- Introduce the quotation with a complete sentence followed by a colon.
  Correct   Thompson (2000) writes the following about monarch butterflies:

  Monarch butterflies have reddish-brown, black-edged wings. The larvae of these butterflies feed on milkweed. These butterflies migrate hundreds of miles through

North America. They have been sighted as far south as Mexico and as far north as Canada. (261)

Quotation Marks Around the Titles of Some Works

Place quotation marks around titles of articles from journals, newspapers, and other periodicals:

Correct	Tumminello and Carlshamre's article "An International Internet Collaboration". . .

Conventional Punctuation with Quotation Marks

Follow the conventions presented below when using quotation marks with other punctuation.

- **Commas and periods**. Put commas and periods *inside* the quotation marks.

Correct	Joanna Tumminello and Par Carlshamre wrote "An International Internet Collaboration."
Correct	He cited "An International Internet Collaboration," an article by Joanna Tumminello and Par Carlshamre.

- **Semicolons and colons**. Put semicolons and colons *outside* the quotation marks.

Correct	Joanna Tumminello and Par Carlshamre wrote "An International Internet Collaboration"; this article includes valuable information about collaborating to complete a research project.

- **Question marks, dashes, and exclamation points**. Put question marks, dashes, and exclamation points inside the quotation marks when they apply to the quoted material only and outside the quotation marks when they apply to the entire sentence.

Correct	She asked, "Have you completed the audit?" (inside)
Correct	Did she ask, "Have you completed the audit"? (outside)

Semicolon ⟨ ;/ ⟩

Use semicolons in the following situations.

Semicolons to Link Independent Clauses

Place a semicolon between two independent clauses not linked by a coordinating conjunction (*and, or, nor, so, for, but, yet*):

| Incorrect | The newest version of the software has more options; but it requires more memory and a faster processor. |
| Correct | The newest version of the software has more options; however, it requires more memory and a faster processor. |

Semicolons to Separate Items in a Series

Use a semicolon to separate the items in a series when any of the items already has internal punctuation:

| Correct | The production team consists of the following people: Patrick Sims, managing editor; Norma Rowland, production editor; Gwen Chavez, copy editor; and Thomas Thompson, art editor. |

MECHANICS

This section includes information on how to use abbreviations, capitalization, italics, and numbers.

Abbreviations

Generally, use abbreviations only when your readers are familiar with them. If you must use abbreviations, attach a list explaining what each one means. If you are uncertain about how to use an abbreviation, spell out the term. When using abbreviations, follow these guidelines:

- Use the singular form for most units of measure even when the word would be plural if spelled out:

| psi | means either "pound per square inch" or "pounds per square inch" |
| oz | means either "ounce" or "ounces" |

  Use a period after the abbreviation for clarity if readers might confuse an abbreviation for another word. Otherwise, omit the period from technical abbreviations.
- Spell out short or common terms such as *ton* and *acre*.
- Abbreviate units of measurement only when a number precedes them:

| Incorrect | How many sq ft? |
| Correct | How many square feet? |

Incorrect	10 square feet
Correct	10 sq ft

Capitalization

Follow standard capitalization conventions. The conventions listed here are the most important ones; for a more complete list, consult your dictionary. Most dictionaries put the list in the end matter. You can also consult the style guides mentioned in Appendix A.

- Capitalize proper nouns, such as personal names, formal titles, place names, languages, religions, organizations, days of the week, and months:
  Kathryn Sullivan (personal name)
  American Association of Mechanical Engineers (organization)
  Chief Counsel (formal title)
  Europe (place name)
  Monday, Tuesday (days of the week)
  Chinese (language)
  Catholicism (religion)
  January, February (months)
- Don't capitalize seasons, compass directions (unless the reference is to a geographic region), and areas of study (unless the area already is a proper noun):
  winter, spring, summer, fall (seasons)
  We traveled north through Wyoming. (direction)
  The storm hit the Pacific Northwest. (geographic region)
  the study of language (area of study)
  the study of the French language (area of study)
- Capitalize the first word, the last word, and every important word in titles and headings:
  Technical Communication in the Information Age (title)
  Research on Electronic Mail and Other Media (heading)
  If you are using the titles in a list of references, follow the guidelines for that style-such as APA, CSE, or MLA.

Italics (*ital*)

Use italics in the following instances:

- For Latin scientific names:
  Lagerstroemia indica (crepe myrtle)
  Tryngites subruficollis (buff-breasted sandpiper)
- For the titles of books, plays, pamphlets, periodicals, manuals, radio and television programs, movies, newspapers, lengthy musical works, trains, airplanes, ships, and spacecraft:

  War and Peace (book)   *Madame Butterfly* (musical work)
  Hamlet (play)   *Titanic* (ship)
  American Idol (television program) *Apollo V* (spacecraft)
  New York Times (newspaper)
- For foreign words that are not widely considered to be part of the English language. If you are not sure, consult a dictionary.
  The county levied an *ad valorem* tax.
- For words, letters, and numbers that you are referring to:
  Use the coordinating conjunctions *and, so, nor, but, yet, for, or.*
  The students should work on writing lowercase *a* and *d* and the number *8*.

Numbers (*num*)

Guidelines for using numbers vary widely and in many instances differ from one field to another. Follow the standard practices of your field or organization, and use numbers consistently throughout each document. These guidelines will apply to most technical documents:

- When a number is the first word of a sentence, do not use numerals. Either spell out the number or, if you can't express it in two words, rewrite the sentence.

  Incorrect   25 years ago, we began offering this degree.
  Correct   Twenty-five years ago, we began offering this degree.
  Incorrect   One thousand seventy-five of the 6,500 people we
                  contacted returned the questionnaire.
  Correct   Of the 6,500 people we contacted, 1,075 returned the
                  questionnaire.

- Use numerals for days and years in dates, exact sums of money, exact times, addresses, percentages, statistics, scores, and units of measurement:

  March 31, 1999 or 31 March 1999 (dates)

  $6,432.58 (money)

  6:34 P.M. (time)

  2600 St. Edwards Court (address)

  67 percent or 67% (percentage)

  a mean of 13 (statistic)

  a total score of 98.7 (score)

  37°P (unit of measurement)

- When mentioning rounded-off figures, use words:

  about five-million dollars

  approximately nine o'clock

- When using two different numbers back to back, use numerals for one and spell out the other:

  nine 2-inch screws

REFERENCES

"AECMA Simplified English Description." AECMA, 2007.
http://www.aecma.org/publications/senglish/sengbrc.htm

American National Standards Institute. "American National Standards for Product Safety Signs and Labels." American National Standards Institute, 1989.

American Psychological Association. *APA Style Guide to Electronic References*. Washington, D.C.: American Psychological Association, 2007.

Anderson, Paul. "What Survey Research Tells Us About Writing at Work." *Writing in Nonacademic Settings*. Ed. Lee Odell and Dixie Goswami. New York: Guilford, 1985, 3-84.

Baker, William H. "How to Produce and Communicate Structured Text." *Technical Communication*, 41 (1994): 456-66.

Barabas, Christine. *Technical Writing in a Corporate Culture: A Study of the Nature of Information*. Norwood: Ablex, 1990.

Barthon, Greg. "Eat the Way Your Mama Taught You."
http://www.mcneilml.com/html/referencelibrary_files/a_mama.htm
11/28/2007

Beal, Vangie 'Aurora'. "All About Phishing." Retrieved 14 July 2008, from http://www.webopedia.com/DidYouKnow/Internet/2005/phishing.asp

Beauchamp, Tom L., and Norman E. Bowie. *Ethical Theory and Business*. 2nd ed. Englewood Cliffs: Prentice Hall, 1983.

Behn, Bill. Quoted in "15 Biggest Job Seeker Mistakes." Zupek, Rachel. 17 Dec. 2007. www.careerbuilder.com. http://www.careerbuilder.com/Article/CB-769-Job-Search-15-Biggest-Job-Seeker-Mistakes/?ArticleID=769&cbRecursionCnt=1&cbsid=0c7e0bfa4fe8443dac79 59312a9aa828-280460727-KH-5&ns_siteid=ns_us_g_15_Job_Seeker_Mistakes

Beil, Laura. "Change of Heart: New Insights Gained as Cardiovascular Research Shifts More to Women." *Dallas Morning News* (6 Feb. 1995): 6D.

Benson, Philippa J. "Writing Visually: Design Considerations in Technical Publications." *Technical Communication*, 32.4 (1985): 35-39.

Berger, Joseph. "Consultant to First Lady Admits Error on Resume." *New York Times* (26 June 1996): B9.

Blain, Jennifer, and Taylor Lincoln. "Make Yourself Essential." *Intercom* (April 1999): 9-11.

Boeing, "What is Simplified English?" 2008.
http://www.boeing.com/phonton/sechecker/se/html

Bolles, Richard Nelson. *What Color is Your Parachute? 2009*. Berkeley, CA: Ten Speed Press, 2009, 33.

Bosley, Deborah S. "International Graphics: A Search for Neutral Territory." *Intercom* (Aug.-Sept. 1996): 4-7.

Bradbury, S. Gayle. "Writing for the Web." *Intercom* (Nov. 2000): 24.

Braffman-Miller, Judith. "When Medicine Went Wrong: How Americans Were Used Illegally as Guinea Pigs," *USA Today,* March 1995: 84.

Brockmann, R. John. *Writing Better Computer User Documentation: From Paper to Hypertext Version 2.0.* New York: Wiley, 1990.

Cadillac, "CTS Coupe Concept Celebrates Cadillac's Design Renaissance." Retrieved on 18 July 2008, from http://www.cadillac.com/cadillacjsp/experience/news_ctscoupe.jsp?evar10 =CTS_MODEL_HOMEPAGE_MASTHEAD_CTS

Caher, John M. "Technical Documentation and Legal Liability." *The Journal of Technical Writing and Communication,* 25 (1995): 5-10.

Cash, J. I., Jr. "A New Farmers' Market." *InformationWeek,* 26 Dec. 1994: 60.

Charney, Davida, Lynee Reder, and Gail Wells. "Studies of Elaboration in Instructional Texts." *Effective Documentation: What We Have Learned from Research.* Ed. Stephen Doheny-Farina. Cambridge: MIT, 1988, 47-72.

Cobb, Neil. E-mail interview. 8 Dec. 2001.

——. Telephone interview. 20 May 2008.

Coleman. *Gas Barbecue Use, Care and Installation Manual.* Neosho: Coleman, 1999.

Couture, Barbara, and Jone Rymer. "Situation Exigence: Composing Processes on the Job by Writer's Role and Task Value." In *Writing in the Workplace: New Research Perspectives.* Ed. Rachel Spilka. Carbondale, IL: Southern Illinois University Press, 1993, 4-20.

Crenshaw, W. Elmo, and Bruce Lawhorn. "Tick-borne Diseases of the Dog." L-22667, rpt. 10M-7-88. College Station: Texas Agricultural Extension Service.

Daugherty, Shannon. "The Usability Evaluation: A Discount Approach to Usability Testing." *Intercom* (Dec. 1997): 16-20.

Davidson, Eli. Quoted in Zupeck, Rachel. "15 Biggest Job Seeker Mistakes" http://www.careerbuilder.com/JobSeeker/Careerbytes/CBAarticle 4/22/2008

Dombrowski, Paul. *Ethics in Technical Communication.* Needham Heights: Allyn & Bacon, 2000.

Dragga, Sam. "Classifications of Correspondence: Complexity Versus Simplicity." *The Technical Writing Teacher,* 18.1 (1991): 1-14.

Duin, Ann Hill. "How People Read: Implications for Writers." *The Technical Writing Teacher,* 15 (1988): 185-93.

——. "Reading to Learn and Do." Proceedings of the 35th International Technical Communication Conference, May 10-13, 1988, Philadelphia. Washington: Society for Technical Communication, 1988.

Ede, Lisa, and Andrea Lunsford. *Singular Texts/Plural Authors: Perspectives on Collaborative Writing.* Carbondale: Southern Illinois UP, 1990.

Eiseman, Leatrice. *Color: Messages and Meanings: A Pantone Color Resource.* Glouchester, MA: Hand Books Press, 2006.

Farkas, David K., and Jean B. Farkas. "Guidelines for Designing Web Navigation." *Technical Communication,* 47.3 (2000): 341-358.

Farr, Michael. "Making Your Resume E-Friendly: 10 Steps." http://careerbuilder.com/JobSeeker/Careerbytes/CBArticle.aspx?articleID=40 3&pf=true 5/3/2008

Felker, Daniel B., et al. *Guidelines for Document Designers*. Washington: American Institutes for Research, 1981.

Forrester, S. Dru, and Bruce Lawhorn. *Canine Epilepsy*. (College Station: Texas Agricultural Extension Service, n.d.)

Foy, Patricia S. "The Reinvention of the Corporate Information Model." *IEEE Transactions on Professional Communication*, 39 (1996): 23-29.

Fugate, Alice E. "Writing for Your Web Site: What Works, and What Doesn't." *Intercom* (May 2001): 39-40.

Golen, Steven, Celeste Powers, and M. Agnes Titkemeyer. "How to Teach Ethics in a Basic Business Communication Class-Committee Report of the 1983 Teaching Methodology and Concepts Committee, Subcommittee 1." *Journal of Business Communication*, 22.1 (1985): 75-83.

Gomes, Lee. "Advanced Computer Screens Have Age-Old Rival." *San Jose Mercury News* (21 Feb. 1994).

GovLeaders.org. "The 6 Golden Rules of Meeting Management." http://govleaders.org/meetings.htm 5/20/2008

Green, Marianne. "Design Your Resume to Land an Internship." 20 Nov. 2008. http://www.jobweb.com/resumesample.aspx?id=250

Greenly, Robert. "How to Write a Resume." *Technical Communication* (1993): 42-48.

Hacker, Diana. "Documenting Sources: CSE Style: Biology and Other Sciences." Retrieved on 30 July 2008 from http://www.dianahacker.com/resdoc/p04_c11_o.html

Hahn, Harley. "What Is Usenet?" 9 Nov. 2001. http://www.harley.com/usenet/whatis-usenet.html

Halpern, J. W. "An Electronic Odyssey." *Writing in Nonacademic Settings*. Ed. Lee Odell and Dixie Goswami. New York: Guilford, 1985, 157-201.

Hansen, James B. "Editing Your Own Writing." *Intercom* (Feb. 1997): 14-16.

Hansen, Katharine. "*Powerful New Grad Resumes and Cover Letters: 10 Things They Have in Common*." http://www.quintcareers.com/new_grad_resumes.html 5/3/2008

Hansen, Randall S. "Scannable Resume Fundamentals: How to Write Text Resumes." http://www.quintcareers.com/scannable_resumes.html 5/3/2008

Hansen, Randall S., and Katherine Hansen. "The Importance of Good Writing Skills." 31 Aug. 2001. http://www.quitcareers.com/writing/skills.html.

Haramundanis, Katherine. *The Art of Technical Documentation*. Maynard: Digital Press, 1992.

Hayes, John R., and Linda S. Flower. "On the Structure of the Writing Process." *Topics in Language Disorders*, 7 (1987): 19-30.

Helyar, Pamela S. "Product Liability: Meeting Legal Standards for Adequate Instructions." *Journal of Technical Writing and Communication*, 22.2 (1992): 125-47.

Holland, V. Melissa, Veda R. Charrow, and William W. Wright. "How Can Technical Writers Write Effectively for Several Audiences at Once?" *Solving Problems in Technical Writing*. Ed. Lynn Beene and Peter White. New York: Oxford UP, 1988, 27-54.

Horton, William. "The Almost Universal Language: Graphics for International Documents." *Technical Communication,* 40 (1993): 682-93.

——. *Illustrating Computer Documentation.* New York: Wiley, 1991.

——. "Overcoming Chromophobia: A Guide to the Confident and Appropriate Use of Color." *IEEE Transactions on Professional Communication,* 34 (1991): 160-71.

Ireland, Susan. "Electronic Resume Guide." 20 Nov. 2008. http://susanireland.com/eresumeguide/index.html

Kellogg, Ronald T. "Attentional Overload and Writing Performance: Effects of Rough Draft and Outline Strategies." *Journal of Experimental Psychology: Learning, Memory, and Cognition,* 14 (1988): 355-65.

Keyes, Elizabeth. "Typography, Color, and Information Structure." *Technical Communication,* 40 (1993): 638-54.

Kintsch, Eileen. "Macroprocesses and Microprocesses in the Development of Summarization Skill." *ERIC* Document ED305613. Washington: Educational Research Information Center, 1989.

Klein, Fred. "Beyond Technical Translation: Localization." *Intercom* (May 1997): 32-33.

Koop, W. E., and R. L. DubIe. "Thatch Control in Home Lawns." College Station: Texas Agricultural Extension Service, 1982.

Kraft Foods. "Kraft Foods Code of Conduct for Compliance and Integrity" http://www.kraft.com/assets/pdf/KraftFoods_CodeofConduct.pdf.

Krause, Jim. *Color Index.* How Design Books, Cincinnati: F & W Publications, 2002.

Krull, Robert, and Jeanne M. Hurford. "Can Computers Increase Writing Productivity?" *Technical Communication,* 34 (1987): 243-49.

Krull, Robert, and Philip Rubens. "Effects of Color Highlighting on User Performance with Online Information," *Technical Communication,* 33 (1986): 268-69.

Lakoff, R. T. "Some of My Favorite Writers Are Literate: The Mingling of Oral and Literate Strategies in Written Communication." Spoken and Written Language. Ed. D. Tannen. *Advances in Discourse Processes,* Series 9. Norwood: Ablex, 1982: 239-60.

Langewiesche, William. "The Lessons of ValuJet 592." *The Atlantic Monthly.* 281.3 (1998): 81-98.

Lawrence, Steve, and Lee Giles. "Accessibility of Information on the Web." *Nature,* 400 (1999), 107-09.

Lay, Mary M. "Nonrhetorical Elements of Layout and Design." *Technical Writing: Theory and Practice.* Ed. Bertie E. Fearing and W. Keats Sparrow. (New York: MLA, 1989): 72-89.

Lee, Zonky and Younghwa Lee. "Emailing the Boss: Cultural Implications of Media Choice." *IEEE Transactions on Professional Communication,* 52.1 (2009): 61-74.

Le Vie, Donald S., Jr. "Internet Technology and Intellectual Property." *Intercom* (Jan. 2000): 20-23.

Li-Ron, Yael. "Office Assistant: Dog or Genius?" *PC World Online Edition* (April 1998).

Locker, Kitty O. *Business and Administrative Communication*. Homewood: Irwin, 1989.

Lorch, Robert E, and Elizabeth Pugzles Lorch. "Online Processing of Text Organization." *ERIC* Document ED245210. Washington: Educational Research Information Center, 1984.

Martin, Cynthia J. "Individually and as Executrix of Eugene J. Martin, Deceased, v. Arthur Hacker, et al., and Chelsea Laboratories, Inc., et al." *83 NY2nd I 23*, Nov. 1993.

Martinez, Benjamin, and Jacqueline Block. *Visual Forces*. Englewood Cliffs: Prentice Hall, 1988.

Men's Wearhouse, Inc. ("MW"). "The Half Windsor." *How to Tie a Tie* brochure. 6/5/09. Vancouver, B.C.

Mirshafiei, Mohsen. "Culture as an Element in Teaching Technical Writing." *Technical Communication*, 41.2 (1994): 276-82.

Nielsen, Jakob. "How Users Read on the Web." Alertbox for October 1, 1997. http://www.useit.com/alertbox/9710a.html 5/27/2008

——. "The Web Backlash of 1996." Alertbox for September 18, 1997. http://www.useit.com/alertbox19604.html9/18/1997

——. "Be Succinct! (Writing for the Web)." Alertbox for March 15, 1997. http://www.useit.com/alertbox/9703b.html 5/27/2008

Novello, Don. *The Lazlo Letters*. New York, NY: Workman, 1992.

O'Conner, Patricia T. *Woe is I*. New York, NY: Riverbend Books, 2003.

Ong, W. J. "Literacy and Orality in Our Times." *The Writing Teacher's Sourcebook*. Ed. G. Tate and Edward P. J. Corbett. New York: Oxford UP, 1981, 36-48.

Oracle, *OptQuest for Crystal Ball 2000 User Manual*, pp 7-10. Retrieved November 2008 from www.decisioneering.com

Parker, Roger, and Patrick Berry. *Looking Good in Print*. 4th ed. Scottsdale: The Coriolis Group, 1998.

Parson, Gerald M. "A Cautionary Legal Tale: The Bose v. Consumers Union Case." *The Journal of Technical Writing and Communication*, 22 (1992): 377-86.

Perl, Sondra. "The Composing Processes of Unskilled College Writers." *Research in the Teaching of English*, 13 (1979): 317-36.

Pfeiffer, William S. *Pocket Guide to Public Speaking*. Upper Saddle River: Prentice Hall, 2002.

Plumb, Carolyn, and Jan H. Spyridakis. "Conducting Survey Research in Technical Communication." Technical Communication 39.4 (1992): 625-637.

Quain, John R. "Time for Face Time." *Fast Company* Oct./Nov. 1997: 232.

Raign, Kathryn, and Brenda Sims. "Gender, Persuasion Techniques, and Collaboration." *Technical Communication Quarterly*, 2.1 (1993): 89-104.

Raytheon. "Ethics Quick Test." 1 Jan. 2008 http://www.raytheon.com/ Stewardship/ethics/ethics_answers/test/index.html

Redish, Janice (Ginny). "Adding Value as a Professional Technical Communicator." *Technical Communication*, 42.1 (1995): 26-39.

Redish, Janice C., and David A. Schell. "Writing and Testing Instructions for Usability." In *Technical Writing Theory and Practice*. Eds. Bertie E. Fearing and W. Keats Sparrow. New York: Modern Language Association, 1987, 61-71.

Reiffenstein, Kathy. "Five Things Not to Do in Front of an Audience."
http://www.presentations.com/msg/content_display/presentations/e3ieeeb821
466c455148358e3fa766ed4d3 5/22/2008

——. "Harness the Power of PowerPoint Presentations." *WorldWIT Newsletter*, 6
November 2006. http://www.andnowpresenting.us/id81.html

Rottenberg, Annette T. *Elements of Argument.* 3rd ed. New York: St. Martin's, 1991.

Rubens, Philip M. "Reinventing the Wheel? Ethics for Technical Communicators."
Journal of Technical Writing and Communication 11 (1981): 329-39.

Ruggero, Ed, and Haley, Dennis F. *The Leader's Compass.* 2nd ed. King of Prussia,
PA: Academy Leadership Books, 2005.

Samuels, Marilyn Schauer. "Scientific Logic: A Reader-Oriented Approach to
Technical Writing." *Journal of Technical Writing and Communication,* 12.4
(1982): 307-28.

Schrage, Michael. *No More Teams! Mastering the Dynamics of Creative Collaboration.*
New York: Doubleday Business, 1995.

Scudder, Joseph N., and Patricia J. Guinan. "Communication Competencies as
Discriminators of Superiors' Ratings of Employee Performance." *Journal of
Business Communication,* 26.3 (1989):217-29.

Selzer, Jack. "Arranging Business Prose." *Writing in the Business Professions.* Ed. Myra
Kogen. Urbana: NCTE, 1989.

Shimberg, H. Lee. "Technical Communicators and Moral Ethics." *Technical
Communication,* 27 (1980): 10-12.

Shroyer, Roberta. "Actual Readers Versus Implied Readers: Role Conflicts in Office
97." *Technical Communication,* 47.2 (2000): 238-40.

Simon, Jerold. "How to Write a Resume." N.p.: International Paper Company, 1981.

Sims, Brenda R. "Electronic Mail in Two Corporate Workplaces." *Electronic Literacies
in the Workplace: Technologies of Writing.* Eds. Patricia Sullivan and Jennie
Dautermann. Urbana: NCTE, 1996, 41-64.

——. "Linking Ethics and Language in the Technical Communication Classroom."
Technical Communication Quarterly, 2.3 (1993): 285-99.

Sims, Brenda R., and Stephen Guice. "Differences between Business Letters from
Native and Non-Native Speakers of English.*" The Journal of Business
Communication,* 29.1 (1992): 23-39.

Sims, William W. Personal interview. 12 March 2008.

——. Personal interview. 12 July 2008.

Snow, Kathie. "To Ensure Inclusion, Freedom, and Respect for All, It's Time to
Embrace People First Language." www.disabilityisnatural.com 2/22/2008

Solar Turbines. "Gas Turbine Overview." Solar Turbines. Retrieved on 28 March
2009 from http://mysolar.cat.com/cda/layout?m=35442&x=7.

Spivey, Nancy Nelson, and James R. King. "Readers as Writers Composing from
Sources." *Reading Research Quarterly,* 24.1 (1989): 7-26.

Spyridakis, Jan H. "Guidelines for Authoring Comprehensible Web Pages and
Evaluating Their Success." *Technical Communication,* 47.3 (2000): 359-382.

Stimpson, Brian. "Operating Highly Complex and Hazardous Technological
Systems Without Mistakes: The Wrong Lessons from ValuJet 592."
Manitoba Professional Engineer. October 1998. <http://www.cns-
snc.ca/branches/manitoba/valujet.html>

Strunk, William and White, E.B. *The Elements of Style*, 4th ed. Longman, 2000.

Sutton, Robert I. *The No Asshole Rule: Building a Civilized Workplace and Surviving One That Isn't.* New York: Warner Business Books, 2007.

Taylor, Barbara M., and Richard W. Beach. "The Effects of Text Structure Instruction on Middle-Grade Students' Comprehension and Production of Expository Text." *Reading Research Quarterly,* 19.2 (1984): 134-46.

Tebeaux, Elizabeth and Driskill, Linda. "Culture and the Shape of Rhetoric: Protocols of International Document Design" in *Exploring the Rhetoric of International Professional Communication: An Agenda for Teachers and Researchers.* Ed. Carl R. Lovitt with Dixie Goswami. New York: Baywood, 1999, 211-252.

Tumminello, Joanna, and Par Carlshamre. "An International Internet Collaboration." *Technical Communication,* 43.4 (1996): 413-18.

Unite for Sight. Retrieved November 2008. http://www.uniteforsight.org/course/cornealdisease.php

University of Maryland Baltimore County. "A Survey of the Frequency, Types, and Importance of Writing Tasks in Four Career Areas." 29 Aug. 2001. http://userpages.umbc.edu/-rachdl/oral.html

Walinskas, Karl. "Reading Your Audience." *Intercom* (Dec. 2001): 23-24.

Webopedia. http://www.webopedia.com/TERM/m/mouse.html

White, Jan. *Visual Design for the Electronic Age.* New York: Watson-Guptill, 1988.

Wicclair, Mark R., and David K. Farkas. "Ethical Reasoning in Technical Communication: A Practical Framework." *Technical Communication,* 31 (1984): 15-19.

Wilkinson, Theresa A. "How to Increase Performance on a Web Site." *Intercom,* 47.1 (2000): 38, 40.

Williams, Joseph M. *Style: Ten Lessons in Clarity and Grace.* 9th ed. New York: Longman, 2006.

Williams, Robin. *The Non-Designer's Design Book.* 3rd ed. Berkeley, California: Peachpit Press, 2008.

Wilson, Gina. E-mail interview. 20 Nov. 2001.

Winsor, Dorothy A. "The Construction of Knowledge in Organizations: Asking the Right Questions about the Challenger." *Journal of Business and Technical Communication,* 4.2 (1990): 7-20.

Yeo, Sarah C. "Designing Web Pages That Bring Them Back." *Intercom,* 43.3 (1996): 12-14.

Zachary, Lois. "Rekindling the Art of Persuasion." Retrieved on 18 July 2008, from http://www.leadservs.com/artofpersuasion.html

Zimmerman, Donald E., Michel Lynn Muraski, and Michael D. Slater. "Taking Usability Testing to the Field." *Technical Communication,* 46.4 (1999): 495-500.

Zupek, Rachel. "15 Biggest Job Seeker Mistakes." http://www.careerbuilder.com/JobSeeker/Careerbytes/CBArticle.aspx?articleID=769&pf=true 5/3/2008

INDEX